International Conference on Sustainable Innovations in Civil & Mechanical Engineering (ICSICME-2020)

IOP Conference Series: Materials Science and Engineering
Volume 814

Pandharpur, India
2 - 4 January 2020

Editors:

Kailash J. Karande
Sampat G. Deshmukh
Swanand G. Kulkarni

ISBN: 978-1-7138-3606-3
ISSN: 1757-8981

Printed from e-media with permission by:

Curran Associates, Inc.
57 Morehouse Lane
Red Hook, NY 12571

Some format issues inherent in the e-media version may also appear in this print version.

Institute of Physics
Dirac House, Temple Back
Bristol BS1 6BE UK

Phone: 44 1 17 929 7481
Fax: 44 1 17 920 0979

techtracking@iop.org

Additional copies of this publication are available from:

Curran Associates, Inc.
57 Morehouse Lane
Red Hook, NY 12571 USA
Phone: 845-758-0400
Fax: 845-758-2633
Email: curran@proceedings.com
Web: www.proceedings.com

TABLE OF CONTENTS

IOP Conf. Series: Materials Science and Engineering **814** (2020) 011001 doi:10.1088/1757-899X/814/1/011001

Preface

The development of any contrived device, structure or product is conceivable due to material science; hence, it is thrust area of research. Engineers play a vital role in sustainable development by scheduling and execution that preserve natural resources, cost-effectiveness and support human and natural environment. Materials and Energy segments are continuously demanded by Science and Technology. Necessity of state-of-the-art materials for aerospace, automobile and various industries worldwide made many opportunities. Therefore, novel materials which are environmental friendly, inexpensive and energy efficient should get developed worldwide. Innovations and developments without sustainability have no significance in present eras. Therefore, thought process should be directed towards sustainability.

We are honoured to state that International Conference on Sustainable Innovations in Civil and Mechanical Engineering, ICSICME-2020 will benefit to expand realistic and achievable Innovations in the arena of materials of Civil and Mechanical Engineering which in the long run aimed to bring together scientists, researchers, technologists and industrialists across the world to exchange their views and ideas for better possible scientific collaborations in the domain of interest. This will also help to expand the network and exchange knowledge for future collaborations. The scientific programme of the conference includes keynote speech, plenary and invited talks along with papers in the form of oral and poster presentations. We hope that they would have a very successful conference and vastly benefited by its deliberations.

It gives us immense pleasure to thank the International, National and Technical Advisory committees as well as reviewer for their generous cooperation and advice. We wish to express our sincere thanks for participating in ICSICME-2020 and sharing your valuable ideas in the field. We are grateful to media partners Conal Conference Alerts and EasyChair for wide promotional of ICSICME-2020. We acknowledge the Publisher IOP Publishing Ltd., UK for journal publication support extended to ICSICME-2020. We also express sincere thanks to all sponsors and well-wishers for their valued support and co-operation. Also, we are very much thankful to Ministry of External Affairs, Govt. of India, New Delhi, for granting political clearance to organize conference.

We would also like to acknowledge the dedication of all the teaching, nonteaching staff of SKN Sinhgad College of Engineering, Pandharpur, India. We once again thank everyone for their co-operation, support directly and indirectly in making ICSICME-2020 success.

Thank you.

Dr. Kailash J. Karande
Dr. Sampat G. Deshmukh
Dr. Swanand G. Kulkarni

Editors

28th February, 2020
Pandharpur, India

IOP Conf. Series: Materials Science and Engineering **814** (2020) 011002 doi:10.1088/1757-899X/814/1/011002

International Conference on Sustainable Innovations in Civil and Mechanical Engineering (ICSICME-2020)

02-04 January, 2020

Organised by

DEPARTMENT OF CIVIL AND MECHANICAL ENGINEERING,

SKN SINHGAD COLLEGE OF ENGINEERING

Korti, Pandharpur- 413304, Maharashtra, India.

www.sknscoe.ac.in

IOP Conf. Series: Materials Science and Engineering **814** (2020) 011002 doi:10.1088/1757-899X/814/1/011002

Chief Patrons

Hon. Prof. M. N. Navale	Chairman, Savitribai Phule Shikshan Prasarak Mandal, Kamalapur, Maharashtra, India.
Hon. Dr. Sunanda M. Navale	Vice Chairman, Savitribai Phule Shikshan Prasarak Mandal, Kamalapur, Maharashtra, India.
Hon. Mrs. Rachana Navale-Ashtekar	Member, Savitribai Phule Shikshan Prasarak Mandal, Kamalapur, Maharashtra, India.
Hon. Mr. Rohit M. Navale	General Secretary, Savitribai Phule Shikshan Prasarak Mandal, Kamalapur, Maharashtra, India.
Hon. Mr. Sanjay S. Navale	Jt. Secretary, Savitribai Phule Shikshan Prasarak Mandal, Kamalapur, Maharashtra, India.

Chairperson

Hon. Dr. Kailash J. Karande	Principal, SKN Sinhgad College of Engineering, Pandharpur, Maharashtra, India.

Convener

Dr. Sampat G. Deshmukh	Dean (Publication), SKN Sinhgad College of Engineering, Pandharpur, Maharashtra, India.

Co-convener

Prof. Shriganesh S. Kadam	Department of Civil Engineering, SKN Sinhgad College of Engineering, Pandharpur, Maharashtra, India.
Prof. Mahendra M. Katkar	Department of Mechanical Engineering, SKN Sinhgad College of Engineering, Pandharpur, Maharashtra, India.

IOP Conf. Series: Materials Science and Engineering **814** (2020) 011002 doi:10.1088/1757-899X/814/1/011002

International Advisory Committee

Dr. Kia Parsa	Azad University, Tehran.
Prof. (Dr.) Mohan L. Kolhe	University of Agder, Kristiansand, Norway.
Prof. (Dr.) M. A. Tawfik	Zagazig University, Zagazig, Egypt.
Dr. Chetan S. Jarali	National Aerospace Laboratory, Bangalore, India.
Dr. Vipul Kheraj	S. V. National Institute of Technology, Surat, India.
Dr. Shrikant P. Kallurkar	Atharva College of Engineering, Mumbai, India.
Dr. Achchhe Lal	S. V. National Institute of Technology, Surat, India.
Dr. Jyoti V. Meghani	S. V. National Institute of Technology, Surat, India.
Dr. Arvind V. Deshpande	Sinhgad Technical Education Society, Pune, India.
Dr. Sandip S. Inamdar	Vishwaniketan's Institute of Management Entrepreneurship & Engg. Technology, Mumbai, India.
Dr. Shashikant D. Lokhande	Sinhgad College of Engineering, Vadgaon, Pune, India.
Dr. Aanand K. Bewoor	Cummins College of Engineering, Pune, India.

National Advisory Committee

Dr. Pradeepkumar K. Agarwal	Maulana Azad National Institute of Technology, Bhopal, India.
Dr. Mohammad Talha	Indian Institute of Technology, Mandi, India.
Dr. Subhash D. Awale	Babasaheb Ambedkar Technological University, Lonere, India.
Dr. Sanjay N. Talbar	S.G.G.S. Institute of Engg. & Technology, Nanded, India.
Dr. Shasikant A. Halkude	Walchand Institute of Technology, Solapur, India.
Dr. Vilas V. Karjinni	KIT College of Engineering (Autonomous), Kolhapur, India.
Dr. V. G. Kalpana	Coimbatore Institute of technology, Coimbatore India.
Dr. Kirankumar R. Jagtap	Sinhgad Institute of Technology & Science, Narhe, India.
Dr. Appaso M. Gadade	Walchand College of Engineering, Sangli, India.
Dr. Nisha Radhakrishnan	National Institute of Technology, Tiruchirappalli, India.
Dr. Bharat C. Routara	Kalinga Institute of Industrial Tech. Uni., Bhubneshwar, India.
Dr. Arabinda Sharma	BRCM College of Engg. and Tech., Bahal, Harayana, India.
Dr. Manik K. Rodage	SSGS Institute of Engineering and Technology, Nanded, India.
Dr. Neeraj Grover	Thapar Institute of Engineering and Technology (Autonomous), Patiala, India.

IOP Conf. Series: Materials Science and Engineering **814** (2020) 011002 doi:10.1088/1757-899X/814/1/011002

Technical Advisory Committee

Dr. G. Suganya Priyadharshini	Coimbatore Institute of Technology, Coimbatore India.
Dr. Atul A. Sagade	Solar Energy Research Laboratory, Pandharpur, India.
Dr. Ravindra T. Vyavahare	SKN Sinhgad College of Engineering, Pandharpur, Maharashtra, India.
Dr. Shrirang N. Tande	Walchand College of Engineering, Sangli, India
Dr. Chetan P. Pise	SKN Sinhgad College of Engineering, Pandharpur, Maharashtra, India.
Dr. Sham S. Kulkarni	SKN Sinhgad College of Engineering, Pandharpur, Maharashtra, India.
Dr. Bhagyesh B. Deshmukh	Walchand Institute of Technology, Solapur, India
Dr. Sachin D. Khandekar	Sinhgad College of Engineering, Vadgaon (Bk), India
Dr. Pradip V. Jadhav	Bharati Vidyapeeth Deemed University, Pune , India
Dr. Sandip A. Kale	Trinity college of Engineering, Pune, India
Dr. Pradeep B. Kodag	Sinhgad College of Engineering Vadgaon (Bk), India
Dr. Swati A. Kulkarni	Army Institute of Technology,Pune
Dr. Sumant K. Kulkarni	D. Y. Patil Institute of Engineering & Technology, India.
Dr. Balasaheb S. Gandhare	SKN Sinhgad College of Engineering, Pandharpur, Maharashtra, India.
Dr. Amarsinh B. Kanase-Patil	Sinhgad College of Engineering, Pune, India.
Dr. Sunildatta N. Kulkarni	Walchand Institute of Technology, Solapur, India
Dr. Basavarajappa D. N.	GM Institute of Technology, Davangere, India.
Dr. S. C. Kamate	Vishvesvaraya University, Belgaum, Karnataka, India.

Dr. Kailash J. Karande
Dr. Sampat G. Deshmukh
Dr. Swanand G. Kulkarni

Editors

IOP Conf. Series: Materials Science and Engineering **814** (2020) 011003 doi:10.1088/1757-899X/814/1/011003

Dr. Kailash J. Karande

Dr. Kailash J. Karande is the Director and Principal of SKN Sinhgad College of Engineering, Korti, Pandharpur, India since 2011. He completed Ph.D. from SRTM University, Nanded, India in the field of Biometric Face Recognition using Independent Component Analysis. His research interest encompasses a wide range of activities with a focus in face recognitions and developed a comparative approach in ICA analysis. A strong electronics background combined with FRT has enabled him and his team to produce PCA-ICA modules for specific applications. He contributed by examining the underlying feasibility of devices for specific applications, to specifically designed frequency receptors and harvesting for technology solutions, the group looks to solve challenging problems with wide contributions to society. In 2011 Dr. Karande is passionate about enabling technology to provide RF microstrip patch design with a view of alleviating poverty in the field of energy science and better quality of life. Areas in which he and his group contribute include: ICA-PCA Face recognition models, RF and microwave energy materials, nano-biotechnology and nanotechnology. He is working as Principal Investigator and Co-PI on various projects funded by PAH Solapur University, Solapur. He has published 4 books and 80+ research papers in various National and International journals. His research collaborations are in progress with National and International partners-both from academia and industry. In 2010, he received European fellowship of Erasmus Mundus Program at Aalborg University, Denmark. He has acted as advisor/session chair to many National and International events held at Singapore, Denmark and Europe. He was honoured by Albert Nelson Lifetime Achievement Award. Also he was awarded as Best Principal from Punyashlok Ahilyadevi Holkar Solapur University, Solapur. He is credited by various positions like editor for many research journals like IJSMER, IJRSIT, IJRMBSS, IJAIEE, IJAIWCT, IJAIIPT, and IJAIEEE.

Dr. Sampat G. Deshmukh

Dr. Sampat G. Deshmukh is holding Dean Publication position at SKN Sinhgad College of Engineering, Pandharpur, India. He has completed his Ph.D. from Sardar Vallabhbhai National Institute of Technology (SVNIT), Surat, India, giving his societal contribution in the field of Nanoscience and Nanotechnology with design of Cu_3BiS_3 thin films for solar cell applications. He has twenty three years' experience of research, teaching and industrial. He gives his plentiful contribution in the development of research lab in the area of Nanoscience and Nanotechnology at SKN Sinhgad College of Engineering, Pandharpur. Dr. Sampat and his team have continuously working for their great contribution in the field of material science. He has published 13 research articles in journals and 15 papers in various National and International conferences. He is Editor Board Member of journals-Electronics Science Technology and Application (ESTA), Journals of Frontiers of Mechatronical Engineering. Dr. Deshmukh credited by Best Paper award in ICEM-2014 at Chennai, India and consolation award in ICC-2015 at Bikaner, India. Presently, he is working on project 'Investigation on Optical and Electrical Properties of Nano-crystalline Cu_3BiS_3 Thin Films for Environment friendly Photovoltaic Solar cell' funded by Punyashlok Ahilyadevi Holkar Solapur University, Solapur, India under Seed Money Funding Scheme. He developed many research collaborations, some of them are in progress with National and International partners.

IOP Conf. Series: Materials Science and Engineering **814** (2020) 011003 doi:10.1088/1757-899X/814/1/011003

Dr. Swanand G. Kulkarni

Dr. Swanand G. Kulkarni is holding a position of Dean (Research and Development) at SKN Sinhgad College of Engineering with the overall responsibility of various project proposals, funded projects, research proposals and innovative engineering practices in the Institute. He has completed Ph.D. from Sardar Vallabhbhai National Institute of Technology (SVNIT), Surat, India in the area of composite material. He has total 18 years' experience in industry and academics, out of it about 10 years' research experience in processing, characterisation and modelling of composite materials, mainly in sectors of particulate metal matric composites. He has worked as Principal Investigator on funded Seed money research project of Solapur University, Solapur on 'Utilization of sugar factory industry waste in enhancing wear resistance material properties for automotive application' as well as Principal Investigator on funded research project of Department of Biotechnology on 'Use of Foldscope for taking microscopic images of composite material to identify microstructural details like matrix and reinforcement phases, porosity, cluster formations grain formations etc'. He is member of Indian Society for Technical Education, Institute of Engineers India, Kolkatta and Indian Association for Computational Mechanics. He has 23 journal publication and 10 conference publications. He has participated in the University level project competitions by providing guidance and credited by a winning design model. He has made contributions in the setup of manufacturing and material testing lab with modified design of Pinon on disc wear test apparatus as well as hot press machine in the areas of his expertise.

IOP Conf. Series: Materials Science and Engineering **814** (2020) 011004 doi:10.1088/1757-899X/814/1/011004

Inauguration of ICSICME-2020 Souvenir

Keynote Address by Dr. M. A. Tawfik, Zagazig University, Egypt

IOP Conf. Series: Materials Science and Engineering **814** (2020) 011004 doi:10.1088/1757-899X/814/1/011004

Delegates and Participants of ICSICME-2020

Award Winner Participants of ICSICME-2020

Peer review statement

All papers published in this volume of *IOP Conference Series: Materials Science and Engineering* have been peer reviewed through processes administered by the proceedings Editors. Reviews were conducted by expert referees to the professional and scientific standards expected of a proceedings journal published by IOP Publishing.

IOP Conf. Series: Materials Science and Engineering **814** (2020) 012001 doi:10.1088/1757-899X/814/1/012001

Influence of Multi-walled carbon nanotubes on mechanical properties of cement concrete

Mavjot Kaur[1], Krishna Murari[2] and Gurpuneet Singh[3]

[1] UG Student, Guru Nanak Dev Engineering College, Ludhiana, Punjab, India
[2] PG Student, Guru Nanak Dev Engineering College, Ludhiana, Punjab, India
[3] Assistant Professor, Guru Nanak Dev Engineering College, Ludhiana, Punjab, India

E-mail: mavjotkaur@gmail.com

Abstract. The mechanical characteristic of cement concrete with incorporation of multi-walled carbon nanotubes (MWCNT) has been investigated in this paper. Disparate concrete mixes were prepared incorporating CNT-to-cement weight fractions of 0.075, 0.1, 0.125 wt.% in traditional concrete. The flexural and compressive strength were determined after 7 and 28 days of curing. The results of the experimental work showed the maximum increase in the strength was almost 1.5times than the conventional concrete which is believed to occur due to effective dispersion and nano-scale particle size. This was obtained in mix containing 0.075 wt.% of CNTs, making it the optimum content for improving strength parameters. Moreover, the post cracking strength of concrete after CNTs addition was found to be more than the conventional concrete with the reduction in the abrasion of the concrete.

1. Introduction

Plain concrete is the prominent material for construction around the globe [1] and having typical characteristics like quasi-brittle failure, low tensile strength and high crack propagation. To overcome these drawbacks, many approaches have been introduced, such as fibre incorporation [2], mineral admixtures [3–7] and many more. The Nano-materials, including Nano silica, Nano titanium, Nano alumina and, more lately, carbon nanotubes (CNTs), have been the focus of attention recently for the purpose of reinforcing cementitious matrix. Carbon nanotubes, discovered in 1991 by Iijima [8], are carbon allotropes and characterized as cylinders of graphite sheets of varying length from 1 to 100μm. [9]. The modulus of elasticity of CNTs is approximately 1TPa, that is, almost five-fold of steel, making it the quintessential material for reinforcement at Nano-scale [10-11].

Luo et al. [12] discussed the influence of the addition of CNTs varying between 0.1% and 1.0% by cement weight with flumed silica on properties of cement paste, obtained results, after 28 days curing for flexurual strength, showed highest improvements in 0.5 wt% batch. Nasibulina and coworkers [13] scrutinized the influence of using long functionalized CNTs of 0.02, 0.03, 0.05 and 0.09 wt% . The results show that the mix containing 0.03 wt% observed highest strength improvement while the lowest improvement was witnessed in 0.09 wt% mix. Xu et al [14] added 0.025, 0.05, 0.1 and 0.2% CNT-to-cement weight into cement paste and investigated their result on mechanical properties. The microstructurally supported results depicted slight improvement in both flexural and compression test with the gradational increase in content of CNTs.

IOP Conf. Series: Materials Science and Engineering **814** (2020) 012001 doi:10.1088/1757-899X/814/1/012001

So far, hardly any works regarding characterization of CNT-reinforced concrete have been published [15-16]. Therefore, the present work investigated the effect of incorporation of functionalized MWCNT of carboxyl group in pozzolana cement concrete on the compression, flexural strength as well as abrasion resistance.

2. Experimental methods and materials

2.1 Materials

The carbon nanotubes were purchased from United Nanotech Innovation Company (India). These were carboxyl group functionalized MWCNT in powder form. The properties are shown in Table 1 which has been taken from company's datasheet. Ethyl alcohol, as a surfactant, was used to aid CNT dispersion. For production of concrete, Portland Pozzolana cement (PPC) of grade 43 conforming to IS-1489, coarse sand (specific gravity 2.63, fineness modulus 2.94) was obtained by sieving normal river sand and crushed aggregate of 20mm size (specific gravity 2.68, fineness modulus 7.54) were selected.

Table 1. Properties of MWCNT

Parameters	Value
Density	2.1 gm/cm^3
Length (μm)	10-30
Purity (%)	>95%
w (ash) %	< 1.5
Poison ratio	0.1

2.2. MWCNT Dispersion

A physical and chemical process is involved in the Multi-walled CNT dispersion, which are in powder form: (a) magnetic stirring for deagglomeration of CNT, (b) to incorporate Ethanol for ensuring stability of the dispersion. The CNTs to dispersant mass ratio was 0.03:1. The suspension was subjected to magnetic stirring for 40mins. Subsequently, an aqueous suspension was prepared by mixing water and stirring magnetically for 40mins again.

2.3. Mix Compositions

As per IS 10262, M25 grade of was prepared. The proportions are shown in the Table 2.To investigate the effect of MWCNTs in cementitious concrete, four concretes were prepared with same composition but having varying amount of CNTs. The proportions of CNTs utilised in this study are 0%wt (control), 0.075%wt, 0.01%wt and 0.125%wt.

Table 2. Final Mix Proportions

Concrete Grade	Cement	FA	CA	w/c
M25	1	1.61	2.67	0.45

2.4. Mixing and Casting

The aggregates and cement are first dry mixed in rotating drum mixer for approximately 3 minutes which is followed by addition of aqueous suspension, having dispersed CNTs, in three equal parts. After achieving the required rheology of the concrete, the concrete was poured into the moulds, which were cleaned and lubricated with mineral oil, within 15minutes. The concrete was poured in the moulds in three equal layers. Then, compaction was carried out using vibrating table. The fresh concrete were casted into cube specimens ($150\times150\times150mm^3$) for compression test, prismatic beam specimen ($100\times100\times500mm^3$) for flexural test, as per IS 516: 1959 and cube specimen ($70\times70\times70mm^3$) for abrasion test. For each composition and curing period, three specimens were fabricated.

The specimens were demoulded after 24 hours and water cured in laboratory at temperature of $24\pm2^{\circ}C$.

2.5. Testing

The uniaxial compression test and flexural test on standard cubes was carried as per the provisions of IS-516:1959 after 7 and 28 days of curing. For the compression test, Universal testing machine with 2000KN capacity was utilised (Figure 1). Meanwhile, methodology for three-point flexural test was used for ultimate load and load-deflection analysis of the beam specimen. Abrasion test was only carried out on the optimal CNT (with respect to cement weight) content after 28 day curing. The specimen was loaded with 300N and the disk was given 22 revolutions at a speed of 30rpm and the loss in thickness is measured.

3. Experimental Results and Discussion

The results of the experiments conducted for Compressive strength after 7 days and 28 days of curing age and Flexure strength also abrasion resistance are shown in Figures (2 and 3) and Tables (3 and 4).

3.1. Compressive Strength

As expected, the incorporation of MWCNTs shows increase in the compressive strength. For this study cube were casted with concrete mixes having four distinct proportion of MWCNTs (0%, 0.075%, 0.1% and 0.125%) by weight of cement and compressive test was conducted on three samples of each mixes. Figure 1 represents the compressive strength after 7 days and 28 days of curing age, from which it is observed that the increase in strength is maximum for concrete mix with 0.075% of MWCNT by weight of cement.

Figure 1. Arrangement for Compression Test

IOP Conf. Series: Materials Science and Engineering **814** (2020) 012001 doi:10.1088/1757-899X/814/1/012001

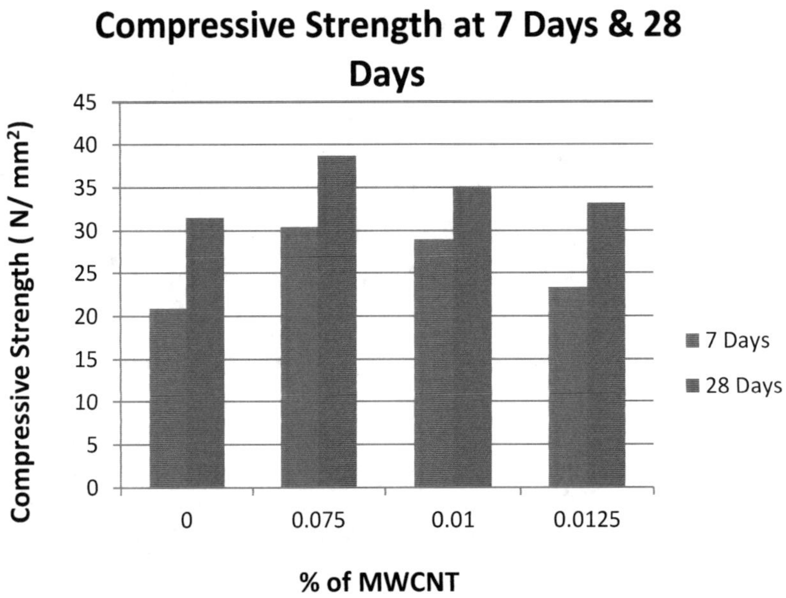

Figure 2. Compressive Strength at 7th and 28th Day for different proportion of MWCNT by weight of cement.

Mainly the reason for increase in strength is bridging effect and packing effect [14]. The small MWCNTs fills the spaces between the atoms in the crystal matrix of hydration product also combines the different components of hydration products in cement. Similar finding of filler effect were observed by other authors[14, 17]. As a consequence of agglomeration of CNTs, richer proportion of CNTs could not work effectively [11].

3.2. Flexural Strength
It can be seen in Table 1, there is increase in flexural strength due to addition of MWCNTs. Some of other authors [9,11,14] also found similar results. In the present study the flexural strength test was conducted for three beams casted with concrete mixes having distinct proportion (0%, 0.075% and 0.1%) of MWCNTs by weight of cement and the results are tabulated below.

Table 3. Ultimate Load Carrying Capacity of Beams of distinct samples

Beams specimen of distinct % of MWCNTs	Ultimate load carrying capacity (N)
0	7500
0.075	10500
0.1	8750

Results presented in Figure 2 shows Load v/s Deflection curve for 0%, 0.075 & 0.1% MWCNTs sample.
Figure 3 shows a typical flexural crack observed in the CNT reinforced concrete samples.

Figure 3. Typical crack pattern observed in the prism

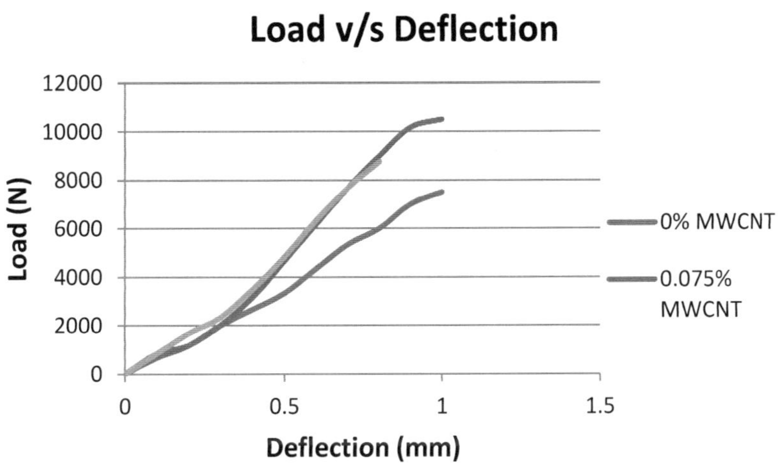

Figure 4. Comparision of Deflection for 0%, 0.075 and 0.1% MWCNTs sample.

3.3. Abrasion Resistance

The results of flexure test showed that 0.075 % by weight of cement is optimum proportion of MWCNTs in cement concrete. Hence specimen for abrasion test was casted using the cement concrete with this proportion only. The results of abrasion resistance values are shown in Table 4.

Table 4. % of MWCNT and abrasion resistance values

% of MWCNTs	Thickness loss in abrasion (mm)
0	1.77
0.075	1.05

IOP Conf. Series: Materials Science and Engineering **814** (2020) 012001 doi:10.1088/1757-899X/814/1/012001

4. Conclusion

In this paper, cement concretes, with CNTs as reinforcement in different proportions, were evaluated for their mechanical properties and the following conclusions have been drawn:

- Generally, MWCNTs with were effective in increasing the mechanical properties of the concrete, after 28 days curing, with an improvement of 23.35% and 40% in compressive strength and flexural strength respectively. These results were obtained for 0.075% CNTs, making it the optimum content.
- At early ages, MWCNTs were more effective in enhancing the compressive strength, i.e. 45.4%.
- Presence of MWCNTs also reduces the wear of concrete as it improves the abrasion resistance by 40.22%.Although, the improvement in mechanical strength is significant but it is not economically feasible to utilise it in general construction.

References

[1] Komloš, K., B. Babál, and T. Nürnbergerová. 1995 *Nuclear Engineering and Design* **156 (1–2)** 195–200.

[2] Hameed Rashid, Anaclet Turatsinze, Frédéric Duprat, and Alain Sellier. 2009 *J. Engineering and Applied Sciences* **4 (5)** 67–72.

[3] Bogas, J. Alexandre, and Augusto Gomes. 2014 *International Journal of Civil Engineering* **12 (2 A)** 268–78.

[4] Bogas, J. Alexandre, Rita Nogueira, and Nuno G. Almeida. 2014 *Materials and Design* **56** 1039–48.

[5] Mikhailova, Olesia, Grigory Yakovlev, Irina Maeva, and Sergey Senkov 2013 *Procedia Engineering* **57** 775–80.

[6] Ramezanianpour, A. A., and H. Bahrami Jovein. 2012. *Construction and Building Materials* **30** 470–79.

[7] Li, Zongjin, and Zhu Ding 2003 *Cement and Concrete Research* **33 (4)** 579–84.

[8] Tang, Z.R., Han, B., Han, C. and Xu, Y.J. 1991 *Nature* **354** 56–58.

[9] Jarolim, Tomas, Martin Labaj, Rudolf Hela, and Kamila Michnova. 2016 *Advances in Materials Science and Engineering* 1–7..

[10] Lefèvre, R., M. F. Goffman, V. Derycke, C. Miko, L. Forró, J. P. Bourgoin, and P. Hesto. 2005 *Physical Review Letters* **95 (18)** 1–4.

[11] Ahmed, Hawreen, José Alexandre Bogas, and Mafalda Guedes. 2018 *J. Materials in Civil Engineering* **30 (10)** 1–14 https://doi.org/10.1061/(ASCE)MT.1943-5533.0002470.

[12] Luo, Jianlin, Zhongdong Duan, Tiejun Zhao, and Qiuyi Li. 2011 *Advanced Materials Research* 146–147: 581–84.

[13] Nasibulina, Larisa I., Ilya V. Anoshkin, Albert G. Nasibulin, Andrzej Cwirzen, Vesa Penttala, and Esko I. Kauppinen. 2012 *Journal of Nanomaterial.*

[14] Xu, Shilang, Jintao Liu, and Qinghua Li. 2015 *Construction and Building Materials* **76** 16–23.

[15] Wille, Kay, and Kenneth J. Loh. 2010 *Transportation Research* Record, no. 2142: 119–26.

[16] Keriene, Jadvyga, Modestas Kligys, Antanas Laukaitis, Grigory Yakovlev, Algimantas Špokauskas, and Marius Aleknevičius 2013 *Construction and Building Materials* **49** 527–35.

[17] Carriço, A., J. A. Bogas, A. Hawreen, and M. Guedes. 2018 *Construction and Building Materials* **164** 121–33.

Experimental and Numerical study Behaviors on Single Cement Flyash Gravel pile under the combined effect of axial compression and Lateral load

N B Umravia[1] and Dr. C H Solanki[2]

[1] Ph. D, Scholar, Applied Mechanics Department, S.V.N.I.T. Surat, Gujarat, India
[2] Professor, Applied Mechanics Department, S.V.N.I.T. Surat, Gujarat, India

E-mail: nirav.umravia.gec@gmail.com

Abstract. A CFG Cement, Fly-ash and Gravel piles are one of the most popular approaches in geotechnical engineering for load transfer in soft soil and ground improvement in railway, road embankment, high rise building. Its behaviors are like as semi-rigid types of the pile. The analysis and computation of ultimate axial compression load and lateral load in soft soil experiments were conducted in a laboratory to obtaining the response of a single CFG pile. The experimental model of CFG pile is made up of cement fly ash and gravel cast in situ types ascertained using physical scaling fundamental, based on material size, properties, a grade of Mix and prototype of pile foundation. Moreover, the in-situ bored CFG piles have been cast in 70 mm and 100 mm external diameters for Laboratory tests. Hence the different CFG pile lengths are considered for variable L/D ratios. All the experiments were performed on soft clay under independent axial compression and lateral load test; it is also to investigate the combined effect of axial compression and lateral load. It has been observed that the behavior of CFG piles under independent individual loading is significantly different when compared with combined loading. It also derives the relationship between the ultimate load in axial compression and lateral loading as well as in the combined effect of loading. It observed that the ratio of ultimate lateral load and axial load in individual to combined loading are significantly increasing. However, the CFG pile vertical deflection /Lateral displacement at considering with its safe factor of lateral load/ axial compression was increasing effectively. In addition to determining the safe axial load / lateral load that has been considering in the present study may be required for design limiting displacement/ deflection failure criteria of the CFG pile effectively under combined load. It is noted that the CFG pile imposition with combined loading can substantially improve the soil properties and can lead to the economy in the design of foundations.

1. Introduction

Due to growth of urbanization and industrial development have been forced on infrastructures establishment, but the geotechnical engineer has faced the problem of unavailability of suitable ground for it. Therefore the ground improvement techniques are the use of the marginal site and weak soil construction of structure like railway embankment, high-rise tall building and others to ensure its stability, achieved required bearing capacity and reducing the settlement after construction. CFG pile composite foundations are used widely in china because after construction settlement and uneven settlement have been reducing. Many researchers had worked on optimization design of CFG pile [1-

IOP Conf. Series: Materials Science and Engineering **814** (2020) 012002 doi:10.1088/1757-899X/814/1/012002

2] for application in high-speed railway embankment, tall buildings [3-7], CFG piles combined with geogrid were adopted as an improvement method to reinforce the embankment built over Completely Decomposed Granite soils [8]; at site encountering as weak soil condition. However structure like Jetty structure, tall chimney, transmission tower and other structures which are subjected to a large lateral load, overturning moment due to the wind and ocean waves or combine effect of both. Many experimental works [9-11] and numerical analysis [12-13] studies have been conducted for investigate the behaviors of CFG pile supported embankment. Numerical analysis of lateral loading were study in Plaxis 3D followed based on [16].

2. Methodology

CFG piles are a new approach for the Indian environmentIn this present study a CFG piles models experiments were carried out in the laboratory to understanding the behaviors of CFG piles with influence with a diameter and its length. It was a significant help to measure the single CFG pile load versus settlement under a static axial compression load, lateral load,and combined load. Models were properly cast and cured 7 days under standard method, in this research paper focused to determine the primary function of lateral load transferring by applying the suspension lateral load. CFG pile can easily transfer the vertical load and lateral load. So the combined action of axial and lateral ground load can occur due to effect of an earthquake, heavy wind, slope failure and liquefaction lateral spared. Details methods of experimental setup and casting conducted are briefly discussion in the following Section.

2.1. Test set up for axial compression test

As shown in Figure 1 major equipment were used in all laboratory experiment are circular steel tank, loading frame, hydraulic jack, load cell with a display unit and 2 LVDT with a display unit.

- The inside dimension of the tank were fixed as 0.3 m (diameter) and 0.6 m (length) and the side and bottom of the tank were made up 8 mm thick steel sheet molded and made as a circular tank.
- The vertical loads were applied with a constant rate of 1.2 mm/ min in all experiments with help of hydraulic jack having maximum 100 kN capacity. Moreover the each load increment was carried out maintained constant rate of loading until CFG pile top head settlement uniformly. The applied load to the CFG pile was measured by using S type load cell having 100 kN capacity suspended from the spindle of the jack through an connecting rod and resting on a CFG pile with a ball bearing. Two LVDT (linear variable differential transformer) was used to measure settlement of loading plate. Minimum least count was 0.01 mm and maximum 50 mm.
- Display units were used to measure Load and settlement behaviors. The static compression loading tests are performed according to the (ASTM D1143–1994) for axial loading tests.

2.2. Laboratory setup for lateral loading

- The lateral loads were applied in a laboratory experiment by a manually suspended load, it was hangout under the gravitational load this load connected with high-level flexible wire and durable which maximum tensile load capacity 30 kN capacity. This was connected horizontally with CFG pile head collar.
- High flexible wire passed from a frictionless pulley. It was a special arrangement made and welds with steel tank as shown in figure 2. Wire one end is fixed with pile collar and while other end attached with a suspended load
- It wasvery difficult to measure the lateral load increment was maintained the constant until displacement was no significant change (i.e, 0.04 mm/min). For measuring the lateral displacement by LVDT were fixed horizontally.

2.3. Laboratory setup for combined loading

a. Figure 2 (c) shows also a photo of the combined loading experimental setup.

b. Axial compression load were applied on the CFG pile by hydraulic jack who have been connected with load frame and vertical displacement were measured by digital dial gauge.

c. Lateral load were measured as same method follow in lateral loading.

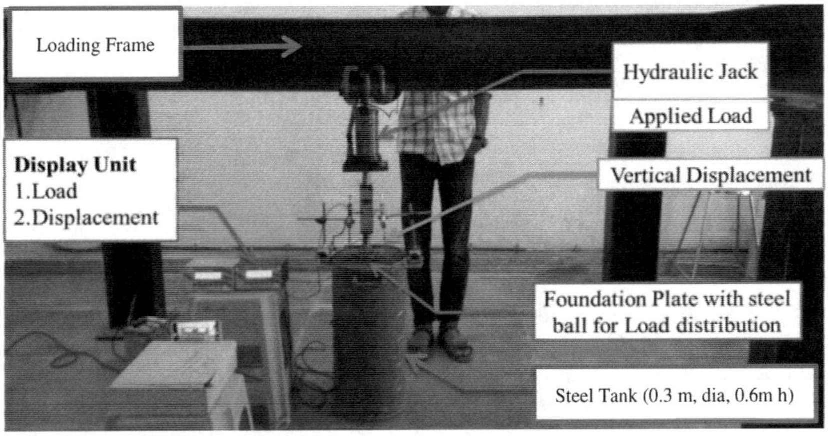

Figure 1. Photograph of experimental setup for axial loading.

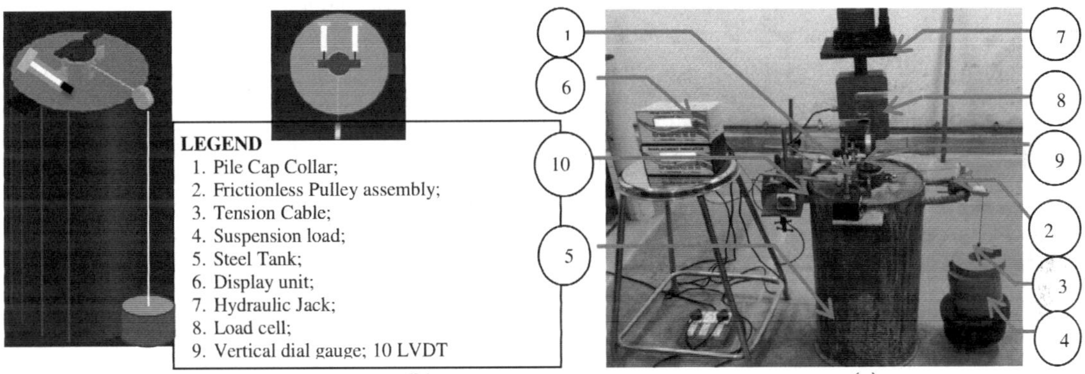

(a) (b) (c)

Figure 2. Experimental set up for Lateral Loading (a) 3D view of Experiment set up (b) Plane of Experimental setup (c) photograph of laboratory experimental setup

3. Model test

A Casted CFG pile tests were conducted in soft clay. The requirement of the materials and sample preparation are explained as per mention briefly in followed section.

3.1. Materials

In this study, casted CFG pile the cement, sand, and aggregate were used locally available from dahod city, India for casting bore pile in a laboratory. However, the standard sand 600 μ size according IS: 383-.2016 was also used to improve the capacity of the pile. The aggregate size was 6.3 mm less than and greater than 4.75 mm its coefficient of curvature Cc and coefficient of uniformity Cu was 2.86 and 0.38 respectively, the specific gravity was 2.55 and bulk unit weight was 16.2 kN/m^3.

3.1.1 Soil properties and sample preparation

The soil used for laboratory model tests it is clay. Test laboratory model end bearing CFG pile tank bottom 100 mm layer the sand used locally available river sand. The sand was cleaned and dry. Direct shear test were performed on this sample for laying a bed at 60 % relative density (RD) Friction angle 29.4% and Coefficient of curvature (Cc) and coefficient of uniformity Cu was to be 2.42 and 1.06 respectively. Preparation of soft clay a uniform moisture 23% added, which was determine results of unconfined compression test on clay. Clay mixture was kept 36 hrs. in five isolated plastic box for vaporization process. 11kg clay were filled in test tank and compacted uniformly by 9 kg weighted temper complete it for test tank in 10 layers 50 mm each layer achieve the bulk unit weight 18 kN/m^3.The experimental model of CFG piles were prepared cast in situ types bored pile made by cement,(C) sand (S), gravel (G), fly ash (F) and amount of water (W) with maintain as per design mix proposal (i.e, C:S:G:F:W ratio 1: 3.2 :4.13 :0.50: 0.61) a constant density of mix CFG piles were maintain for all tests to provide a similar test tendency in vertical axial load and lateral load. Test CFG piles installation has been followed by the procedure described by [14]. During the experimental, there are the two types of tests namely axial vertical compression test were consider IS 2911 Part 4 guideline on and other lateral pull types loading were follow according to ASTM-D3966-07. This experimental study, including laboratory model test of vertical, laterally loaded CFG pile and combined loading is an important measurement for comparison and validation of any design approach. where the vertical compression load specifically way as not induce the horizontal deflection and any friction resistance. So that's a factor that may be minimized using the plate and roller assemble according to ASTM A – 36. In this paper, a laboratory combined axial/ lateral load test compare with that of Finite element analysis by Plaxis 3D, this approach follows the [15]. The combined axial/lateral load test was conducted on a 70 mm diameter CFG piles with variable length 2.5 d, 3.0d, 3.5 d, and 4.0 d mm length. This test was executed in experimental work slightly based on ASTM D 3699 testing procedures. It is expected that laboratory piles are subjected to allowable load/ ultimate load only. It is a small sacle model test. The safe load is obtained by considering the different factors of safety (FOS = 1.3 to 2.5) according to many codes of practice. Hence tests were done in two-part; first part by providing constant 30, 50 and 70 % of ultimate axial load magnitude in conducting lateral load test and the second part of test conducting by keeping constant magnitude 30, 60 and 90% of ultimate lateral load. The overall test program is mention in Table 1.

Table 1. Overview of Test Program

CFG pile Dia.	Length (mm)	Axial compression load (Vu)		Lateral Load	
		Without Lateral Load	With lateral load (H/Hu)	Without axial load	With axial load (V/ Vu)
70 mm	175	Yes	0.3, 0.6, 09	Yes	0.3, 0.5, 0.7
	210	Yes	0.3, 0.6, 0.9	Yes	0.3, 0.5, 0.7
	245	Yes	0.3, 0.6, 0.9	Yes	0.3, 0.5, 0.7
	280	Yes	0.3, 06, 0.9	Yes	0.3, 0.5, 0.7
	315	Yes	0.3, 0.6, 0.9	Yes	0.3, 0.5, 0.7
	500End bearing	Yes	0.3, 0.6, 0.9	--	--

3.2. CFG pile subjected to Combined load

- A desire vertical load was applied in four increments and maintained for approximately one hour before applying a lateral force. The lateral force was applied initially by tension cable passed from the frictionless pulley and fixed with a suspended load. The loading should be applied in regular interval 5 percentage continuous at a rate sufficient constant rate of lateral displacement 0.01 mm per minute) this incremental repeated until failure of CFG piles or the specific deflection value of 13 mm. Vertical settlement of CFG pile measured by digital dial gauge which is position on the top of the pile Cap. The lateral load was reduced to every regular increment load every 30 minutes. Similerly test was then reduced to zero in five decrements. The remaining lateral force was removed over a three minutes period. End load reading was taken for 30 minutes.

IOP Conf. Series: Materials Science and Engineering **814** (2020) 012002 doi:10.1088/1757-899X/814/1/012002

4. Test Results and Discussion

Test were conducted on CFG pile at the constant bulk density and shear strength of soil the results corresponding to the CFG pile under consider axial, lateral and combination are presented in these sections. Figure 3 shows the load-displacement curve for the CFG pile at variable length under axial compression load. The ultimate load capacities of the CFG pile are interpreted using the tangential method for the analytical approach hence for validation of experimental work by Plaxis 3D modeling. It is seen in figure 3 (a) typical load – settlement behaviors are nonlinear and all the experimental CFG piles are short pile. However,the failure range of all pile were consider as nonlinear nature and its range maybe 12-20 mm vertical settlement. In the present study for axial compression laboratory, test settlement were measured up to 50 mm. To investigate the effect of end bearing CFG pile with hard strata and soft strata at the bottom it seems as per figure 3(b) it has been increasing 4.5 times.

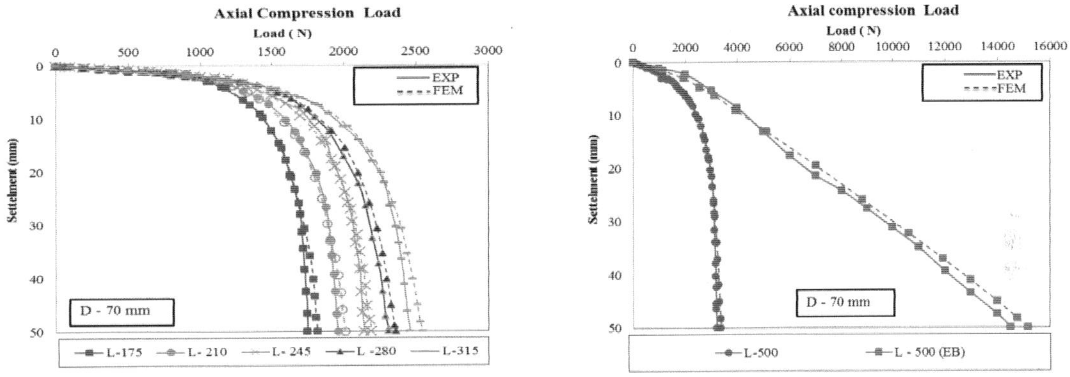

(a) (b)

Figure 3. Load -Settlement curve for (a) Floating Pile with Variable length (b) Comparative of Floating and end bearing pile

Figure 4 shows the lateral load-displacement curve of various lengths it seems that Different length of pile lateral behavious under lateral load figure 5 shows the relationship of different dia of CFG pile Lateral behavious. It is clearly observed 2.1 times more in 100 mm dia pile compare to 70 mm.

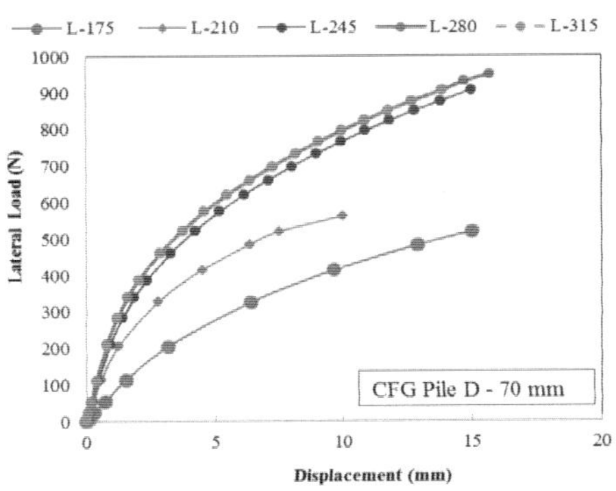

Figure 4. Lateral load vs displacement curve of various length of CFG pile

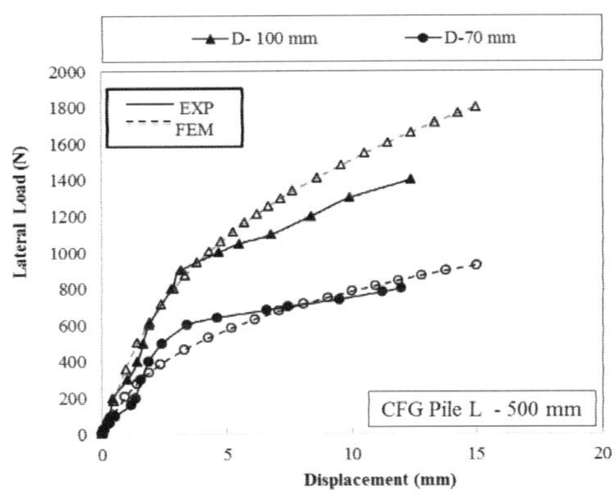

Figure 5. Typical Lateral load vs displacement curve of various diameters

IOP Conf. Series: Materials Science and Engineering **814** (2020) 012002 doi:10.1088/1757-899X/814/1/012002

Table 2. Ultimate Lateral load measured with and without constant ultimate axial load.

CFG pile dia.	Length (mm)	Ultimate Lateral Load Hu (N)			
70 mm		V = 0	V = 0.3 Vu	V = 0.5 Vu	V = 0.7 Vu
	175	440	460	510	580
	210	500	520	580	630
	245	580	620	670	780
	280	830	850	900	920
	315	850	840	900	940

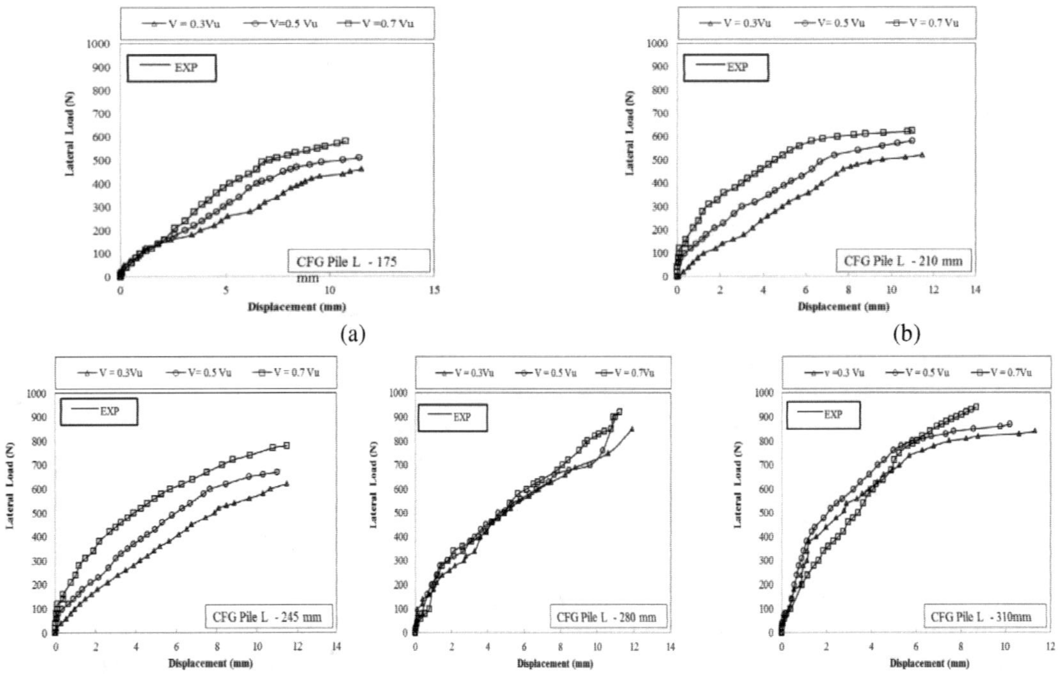

(a) (b) (c) (d) (e)

Figure 6. Variation of lateral load with pile top deflection measured under constant Axial load (a) 175 mm (b) 210 mm; L/D- 3.0 (c) 245 L/D- 3.5 (d) 280 L/D- 4.0 (e) 310 L/D- 4.5.

Table 3. Ultimate Axial load measured with and without lateral load 25 mm permissible settlements

CFG pile dia.	Length (mm)	Ultimate Axial compression Load Vu (N)			
		H = 0	H = 0.3 Hu	H = 0.6 Hu	H = 0.9 Hu
70 mm	175	1750	1600	1540	1440
	210	1960	1850	1740	1650
	245	2140	1910	1750	1610
	280	2310	1995	1860	1890
	315	2460	2240	2160	2030
	500	3110	2980	2880	2760
	500 EB	8820	8320	7850	7580

Figure 6 shows the Variation of lateral load with pile top deflection measured under constant axial load with a variable length of CFG pile the ultimate lateral load shows in table 3 during the combined loading the failure criteria consider 12 mm in lateral displacement and 25 mm in vertical displacement taken in all the laboratory test. It is observed those lateral loads were increasing with increasing L/D ratio. Figure 7 relatively of the axial load – settlement curve CFG pile deflection with 0.3 Hu, 0.6 Hu and 0.9 Hu lateral loads.

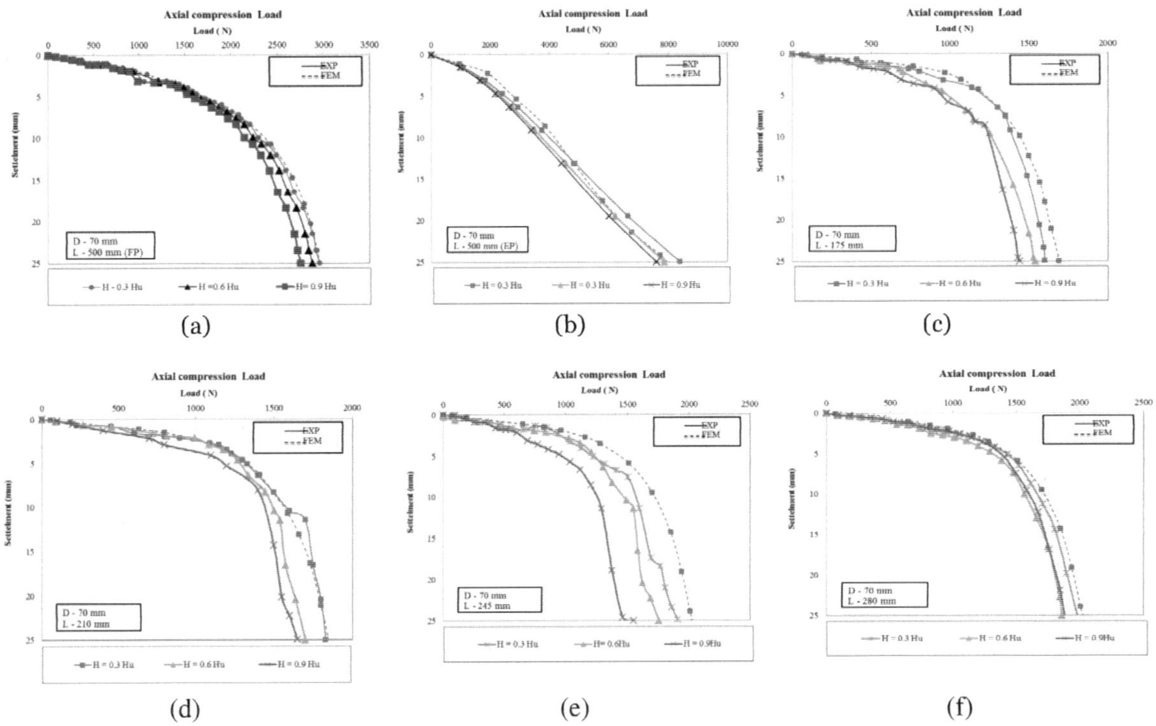

Figure 7. Variation of axial load with pile deflection measured under constant lateral load (a) L- 500 mm; (b) L -500 EB; (c) L – 175 mm; (d) L-210 mm; (e) L- 245mm; (f) L-280 mm ;

5. Conclusion

The study captures the behavior of single CFG pile subjected to axial, lateral and combined loading inSoft clay soil, the lateral deflection being limited to 12 mm. and vertical displacement being limit25 mm in combined loading the following conclusions can be made:

- The axial load test in single CFG pile load-carrying capacity significantly increasing a 6-9 % increase with an increasing (L/D) ratio.
- The ultimate lateral bearing capacity of CFG piles in clay significantly increasing with the increasing length of CFG pile. The lateral deflection was observed for the CFG pile subjected to only lateral load but in increasing, The axial l load-carrying capacity of a pile is decreasing 3-5 % in lateral load 0.3Hu; 6-8 % in lateral load 0.6 Hu and 9- 11 % in lateral load 0.9 Hu in combined load Due to lateral load distribution and relative moment generation.
- The presence of a vertical load had increased the lateral load capacity with increasing L/D ratios but the influence of vertical load decreases with an increasing slenderness ratio of CFG piles at all vertical load levels for both types of CFG piles. The lateral load-carrying capacity increased by 4-6 % 8- 10 and 11-13%, in 0.3 Vu, 0.5 Vu and 0.7 Vu respectively.

IOP Conf. Series: Materials Science and Engineering **814** (2020) 012002 doi:10.1088/1757-899X/814/1/012002

References

[1] H. Y. Cao and Y. F. Liu 2014 *Appl. Mech. Mater.* **578–579** 346–350
[2] Y. Shen and H. Wang 2016 *Transp. Infrastruct. Geotechnol.* **3** 1 3–20
[3] J. Zeng, J. Zhang, Y. Tu, and X. Tong 2009 *Second Int. Conf. Intell. Comput. Technol. Autom.* 638–640
[4] J. Feng, X. Y. Wu, and J. H. Zhang 2014 *Electron. J. Geotech. Eng.* **19** 6867–6878
[5] D. B. Zhang, Y. Zhang, T. Cheng, and J. Y. Yuan, *Int. J. Eng.* **30** 7 955–963
[6] L. C. Zhao 2013 *Appl. Mech. Mater.* **443** 123–128
[7] W. Li and X. Bian 2016 *Procedia Eng.* **143** Ictg 1059–1067
[8] L. Wu 2018 *Adv. Mater. Sci. Eng.* **2018** 1–11
[9] M. Zhang and X. Zhang 2017 **124** Isaeece 75–78
[10] A. Joseph, S. Chandrakaran and N. Sankar 2018 *Int. J. Geosynth. Gr. Eng.* **4** 4
[11] M. Hasan and N. K. Samadhiya 2017 *Comput. Geotech.* **87** 178–187
[12] D. Consolidation et al. 2017 *Brief Introduction of Application of Deep Mixing in China* 1–6
[13] K. Jennings 2013 *An investigation of geosynthetic reinforced piled embankments with particular reference to embankment extremities*
[14] G. Zheng, Y. Jiang, J. Han, and Y. F. Liu 2011 *Mar. Georesources Geotechnol.* **29** 2 145–161
[15] L. Zhang, Y. Wu, and M. Wei 2017 *IOP Conf. Ser. Mater. Sci. Eng.* **207** 1
[16] N. B. Umravia and C. H. Solanki 2018 **6** 2 37–44

Experimental investigation and effects of laser hardening process parameters on microhardness of En24 steel

Santoshkumar V. Wagh[1], Dhananjay V. Bhatt[2], Jyoti V. Menghani[3], Sujit S. Pardeshi[4] and Bhagyesh B. Deshmukh[5]

[1,4]Department of Mechanical Engineering, College of Engineering Pune, Pune-411005, India
[2,3]Department of Mechanical Engineering, Sardar Vallabhbhai National Institute of Technology, Surat-395007, India
[5]Department of Mechanical Engineering, Walchand Institute of Technology, Solapur-413006, India

E-mail: waghsv@gmail.com

Abstract. In presented paper through experimental investigation is carried out with fiber laser system so as to evaluate the effects of laser surface hardening process parameters on the microhardness (width and depth) of the laser treated En24 steel using a Taguchi method. The optical microscope and microhardness tester is used to examine the microstructure, microhardness (width and depth) of the laser hardened samples across the heat affected zone. The selected process parameters were laser beam power (210, 225 and 300 W), laser scanning speed (1.0, 5.5 and 10.0 mm/s) and standoff distance (175, 200 and 225 mm) respectively. The laser treated hardened surface layer has a microhardness of 728.1 $HV_{0.3}$ and reached a microhardness depth of 2471.44 μm, which is twice as high as untreated samples of En24 steel. From this study, it can be summarized that the microhardness of laser treated samples is maximized by controlling laser beam power, laser scanning speed and standoff distance, which improves the service life of laser-treated En24 steel components. The percentage error within the acceptable range was found to be small and was obtained by comparing the microhardness values obtained from the experimental results and predicted microhardness values obtained by using nonlinear regression equation.

1. Introduction

The laser hardening is a unique surface treatment process that has recently been used to enhance mechanical properties and wear resistance compared to conventional surface hardening. The benefit of this process is that only the selected area can be hardened which is required for the particular applications and the rest of the surface of bulk material remains unchanged in the form of mechanical properties. The laser hardening process was also used by the steel metal industry to boost the service life of the steel parts. In this study, the En24 steel material is used for research which is attributed to its uses in manufacturing of mechanical components like gears, kicker rods and bolts. Balasubramanian et al. [1] studied the effects of laser hardening process parameters like laser power and laser scanning speed on hardness, depth of hardness, microstructure change and wear for the materials from En8, En24, En36 and HCHCr. Zhang et al. [2] considered the impacts of laser hardening process to enhance wear and frictional properties of laser hardened En31 steel specimen. Rana et al. [3] investigated the microstructure and hardness of the different carbon composition laser-treated steel samples by weight percentages of 0.28, 0.40 and 0.46 respectively. The effects of process parameters such as laser power, traverse speed and spot size on the microstructure and microhardness were analyzed and the most important parameters which observed was traverse speed. Gisario et al. [4] evaluated the laser surface treatment capability tested on average roughness, the surface texture of laser hardened AISI 304 stainless steel with high power diode laser system. Wagh et al. [5] applied Taguchi's orthogonal array approach and investigated the impact of laser hardening process parameters on mechanical and wear properties of laser hardened CK45 steel. Visscher et al. [6] studied the impact of laser line hardening process on the wear action of AISI 1045 carbon steel compared to AISI 52100 ball bearing steel. It was investigated that wear resistance of a laser line hardened surface is minimum compared to the conventional carburization process. Grum and Sturm [7] evaluated that the laser hardening process is useful for increasing the hardness and wear resistance of the nodular iron 500-7 laser treated specimen. Pellizzari and De Flora [8] examined the impact of the laser hardening process parameters on the hardness of laser treated AF63CrMnMo6. The parent metal hardness was 300 HV and when the laser surface treatment is applied the hardness increased to 800 HV with the case depth from 1.2 - 2.0 mm. Zhang et al. [9] studied the effects of laser hardening on the microstructure, hardness and wear resistance of Cr_2Mo and 45 steel treated laser. The cross section of the area affected by heat was divided into three zones (overheated zone, the hardening zone and the transition zone). It was observed that the hardness within the overheated zone was lower than that of the hardening zone with the same values of laser process parameters. Prashanthi et al. [10] studied the effects of laser hardening process parameters to achieve maximum hardness for improving the working life of carbon steel, non-melleable cast iron and X20Cr13 materials. Babu et al. [11] investigated the results of laser hardening process parameters to improve the hardeness and wear resistance of laser treated EN25 steel materials using the methodology of the response surface. Mahapatra et al. [12] studied the effects of erosion wear process parameters such as impingement angle, fiber loading, erodent size and standoff distance on glass fiber reinforced polyester composites material using the Taguchi's approach. Zhang et al. [13] studied the impact of laser surface remelting on the performance of H21 steel microstructure, microhardness and thermal fatigue, observed microhardness values after laser surface melting was 780 HV and a microhardness depth of 650 μm achieved respectively.

2. Experimental details

2.1 Material used for experimentation

Nowadays, En24 steel is used for manufacturing of gear, kicker rods and bolts etc. Therefore, for the present experimental work, En24 steel is used as testing material. Table 1 shows the elemental composition in weight percent.

Table 1. Elemental composition of En24 steel in weight percentage

Elements	C	Si	S	P	Mn	Cr	Ni	Mo
Composition weight %	0.386	0.228	0.028	0.014	0.542	1.364	1.392	0.241

Figure 1 shows the SEM image of En24 steel specimens (a) parent metal and (b) after the laser hardening process respectively.

2.2 Experimental set-up and sample preparation

Figure 2 shows the image of fiber laser system 400 W continuous wave (CW) modes and 1070 nm laser wavelength. The experiments were conducted on the same laser system. The parameters of the laser hardening process used 210, 255 and 300 W laser beam power, 1.0, 5.5 and 10.0 mm/s laser scanning speed, standoff distance of 175, 200 and 225 mm as appeared in Table 2. The argon gas used as shielding gas at a flow rate of 10 liters/min and laser beam diameter of 2.1 mm. The specimens were cut to 30 mm × 8 mm × 8 mm by a wire electrical discharge machine. The standard specimen preparations technique was used to prepare and examine the specimen.

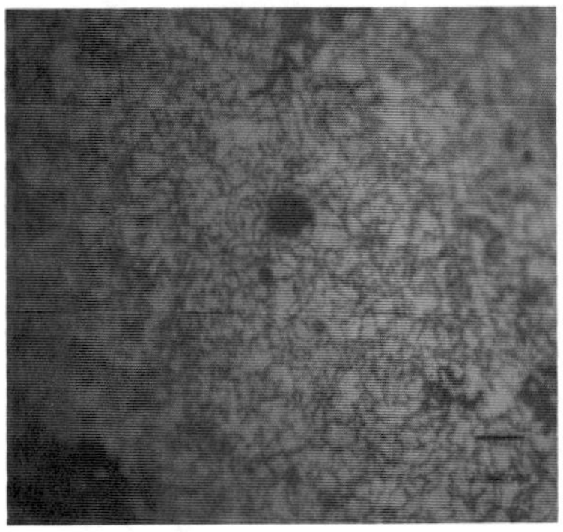

(a)

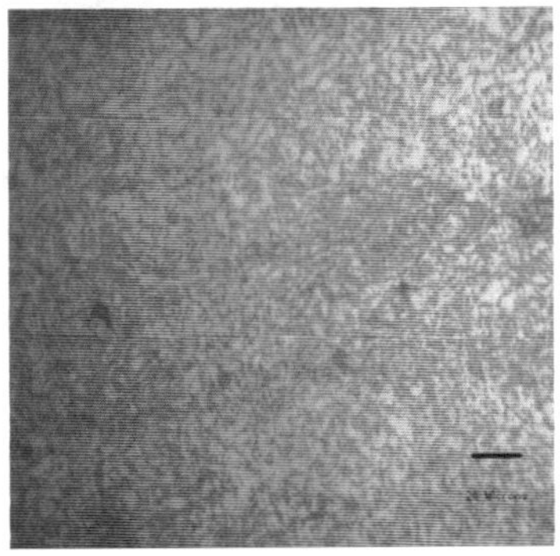

(b)

Figure 1. SEM image of En24 steel (a) base material (b) after laser hardening.

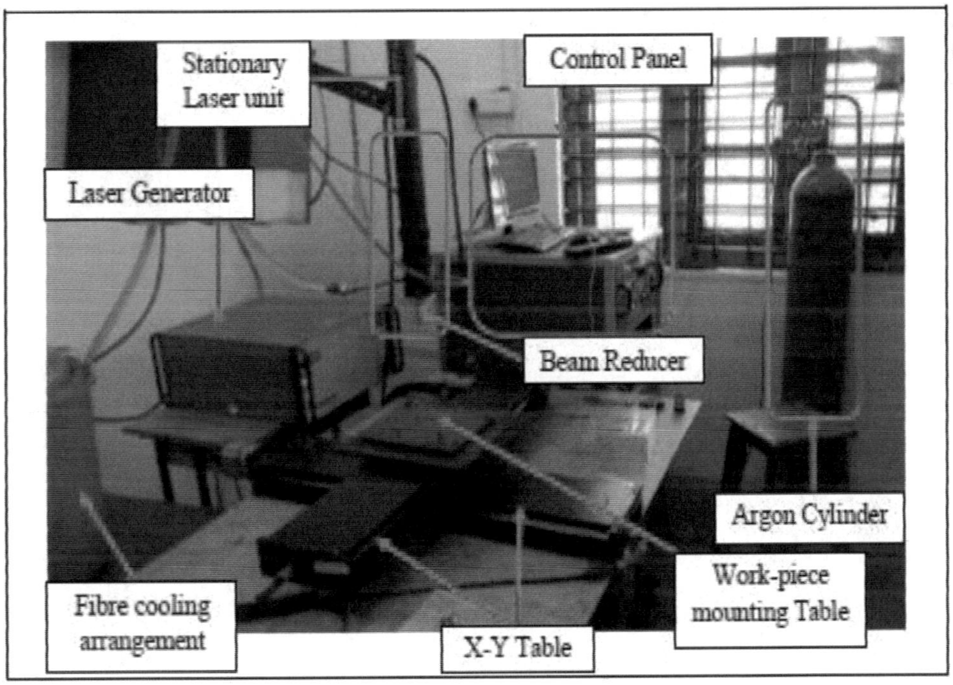

Figure 2. Fiber laser system 400 W

Table 2. Experimental testing parameters and their levels for laser hardening expt.

Sr. No.	Laser Parameters	Unit	Levels		
			1	2	3
1.	Laser beam power (P)	W	210	255	300
2.	Laser scan speed (v)	mm/s	1	5.5	10
3.	Standoff distance (f)	mm	175	200	225

2.3 Metallographic and microstructural analysis

Optimized parameters of laser hardening process were used to take a laser track on the En24 steel specimen. The laser hardened cross sectional area of the samples treated with laser was polished using emery papers of required grades to achieve mirror finish of the specimen. The microscopic examination is carried out to check the width and depth of microhardness and microstructure of the heat affected zone. The microhardness of the parent metal and laser-treated samples was checked by microhardness testing machine at a load of 300 gm with 10 seconds dual time.

2.4 Experimental Design

For the analysis of the influencing significant process parameters, the design of the experimental technique was used and based on that an output response model was developed. In the laser hardening process, three input parameters such as laser beam power, laser scanning speed and standoff distance are taken whereas microhardness was considerd as an output response. Table 2 shows the levels of process parameters used for the experiment. Table 3 displays the laser hardening process parameters and the output response of all runs, such as microhardness, microhardness width and microhardness depth. The output response values are translated to signal to noise ratio using Taguchi's L_9 orthogonal array method.

Table 3. Laser hardening process parameters, output response and S/N ratio

Sr. No.	Laser beam power (w)	Laser scan speed (mm/s)	Standoff distance (mm)	Depth of HAZ (μm)	Microhardness $HV_{0.3}$	S/N Ratio
1	210	1	175	1103.55	637.5	56.08
2	210	5.5	200	876.07	545.1	54.72
3	210	10	225	246.46	516.3	54.25
4	255	1	200	2237.30	689.5	56.77
5	255	5.5	225	989.30	656.6	56.34
6	255	10	175	444.79	618.3	55.82
7	300	1	225	2471.44	728.1	57.24
8	300	5.5	175	1063.85	687.8	56.74
9	300	10	200	535.01	651.7	56.28

For each variable rate combination, the signal to noise ratio was determined. The formula for the higher is better criteria as per Taguchi's signal to noise ratio is shown in Equation (1).

$$\text{Signal to noise ratio} = -10*\log\left(\Sigma\left(1/Y^2\right)/n\right) \quad (1)$$

Where, Y is the output response for the provided level combination and n is the number of factors levels combination output responses.

3. Results and Discussion

3.1 Microhardness analysis by Taguchi's experimental design

The results of the present experimental work are analyzed using MINITAB 14 software. Taguchi's orthogonal array L_9 design is used perform the experimental work. Table 3 shows details of the experimental parameters and respective results of S/N ratio for all the nine sets of experiments. The present experiment aims to increase the value of microhardness. Therefore, 'the larger the better' quality characteristic is selected and used for the present Taguchi's analysis. Table 4 shows the signal to noise response values for existing experimental conditions and the higher the characteristic of better quality.

In accodardance to the analysis shown in Table 4 it can be concluded that laser beam power is the most significant control parameter to achive surface microhardness. In affecting the surface microhardness of En24 steel, the remaining two parameters, namely laser scan speed and standoff distances, are rated in the second and third positions. The main effect plots of process parameters such as laser beam power, laser scan speed and standoff distances based on signal to noise ratio are shown in Figure 3. The optimum or maximum microhardness can be achieved at 300 W laser beam power, 1.0 mm/s laser scanning speed and standoff distance of 175 mm.

Table 4. S/N ratio response table using the larger the better characteristics

Level	Laser beam power	Laser scan speed	Standoff distance
I	55.03	56.70	56.22
II	56.31	55.94	55.93
III	56.76	55.45	55.95
Delta	1.73	1.25	0.29
Rank	1	2	3

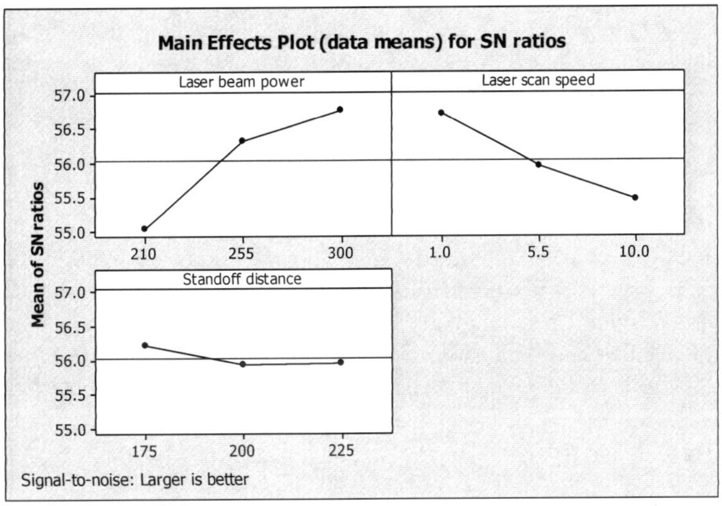

Figure 3. Main effects plot for signal to noise ratios.

3.2 Effect of laser power on microhardness

As the laser beam power increases from 210 to 300 W, the microhardness of the laser hardened sample also increases as shown in Figure 3. This is attributed due to the fact that at higher laser beam power maximum heating of surface takes place which further tends to form a martensitic structure. Moreover at higher laser power maximum heat energy concentrates on the available surface per unit time which allows maximum temperature difference for cooling and hence forming a microstructure tending towards martensitic nature.

3.3 Effects of laser scan speed on microhardness

From the main effect plots of Figure 3 the effect of laser scan speed can be simply illustrated from the nature. It is observed that that the microhardness of the surface decreses with increment in laser san speed, this is attributed due to the fact that as scanning speed increases the heat energy concentration per unit time decreases. Hence it is observed that the microhardness value decreases with the increase in the laser scan speed at constant laser beam power and variations in the stadoff distance.

3.4 Surface morphology using microstructure

Figure 4 displays the microstructure of the laser-treated En24 steel specimens (a) shows a laser treated specimen microstructure using 210 W laser beam power, 5.5 mm/s laser scanning speed and 200 mm standoff distance, (b) displays a laser treated specimen microstructure using laser beam power 255 W, laser scan speed 5.5 mm/s and standoff distance 225 mm and (c) shows microstructure of laser treated specimen using laser beam power 300 W, laser scanning speed 5.5 mm/s and standoff distance 175 mm. From this analysis, it is indicated that the formation of fine martensite with carbide dissolution may result increase the microhardness of the treated layers. The highly hardened area consisting of homogeneously dispersed martensite plates with a few pockets of undissolved carbides and austenite retained.

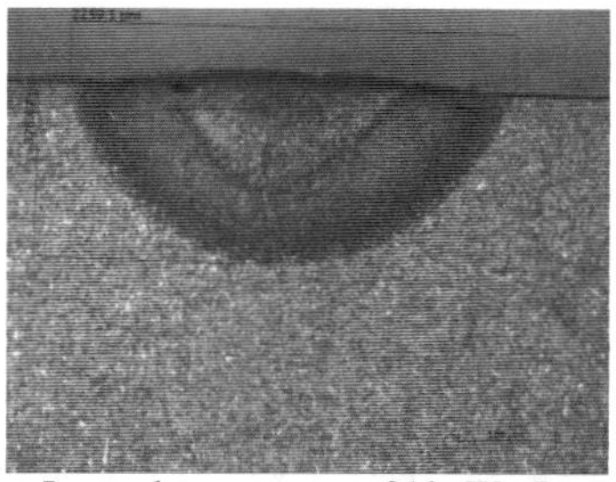
a. Laser beam power 210 W, Laser scan speed 5.5 mm/s and standoff distance 200 mm

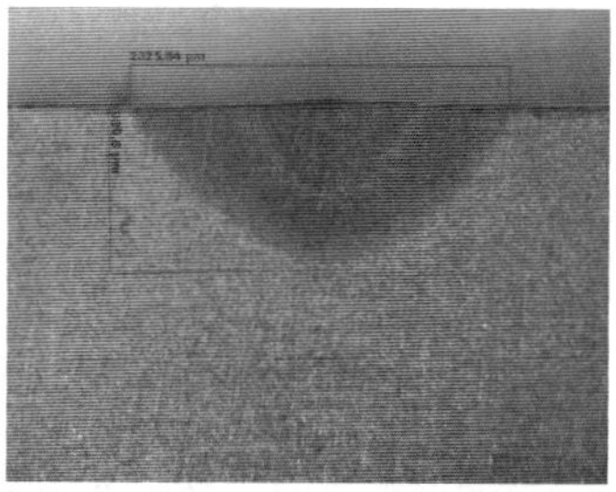
b. Laser beam power 255 W, Laser scan speed 5.5 mm/s and standoff distance 225 mm

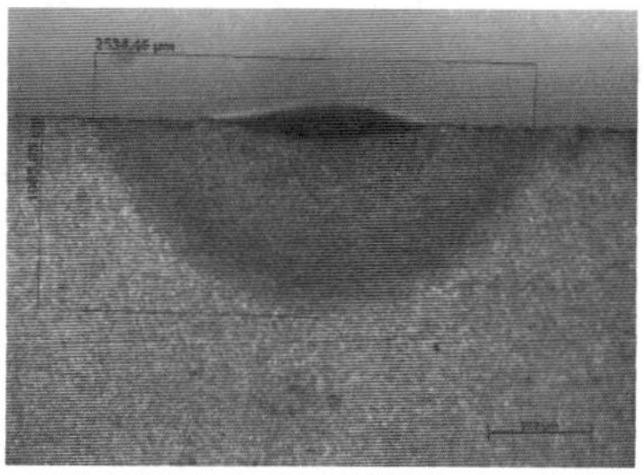
c. Laser beam power 300 W, Laser scan speed 5.5 mm/s and standoff distance 175 mm

Figure 4. Microstructure of the laser-treated En24 steel specimens

3.5 Hardness prediction using predictive equation

From the obtained values of microhardness as per the experimentation performed with reference to L_9 array, a regression equation is generated using MINITAB 14 software. This equation can be used to predict the values of microhardness at any levels of processing parameters. The generated equation is also tested by R^2 values and it is given a 93.1% which is accurate and hence useful to predict the responces at required levels. The generated regression equation is shown below in euation 2.

$$\text{Microhardness} = 400+1.37\,(P)-9.96\,(v) - 0.284\,(f) \qquad (2)$$

This equation is used to find the the microhardness at any given set of input process parameters. Here, the experimental results and the predicted results at each experimental condition are calculated and shown in Table 5.

Table 5. Comparison of experimental and predictive values for microhardness

Sr. No.	Results obtained from the experiments	Results obtained from the predictive equation	Percentage error (%)
1	637.5	628.0	1.50
2	545.1	576.1	5.38
3	516.3	524.2	1.50
4	689.5	682.5	1.01
5	656.6	630.6	4.11
6	618.3	600.1	3.04
7	728.1	737.1	1.22
8	687.8	706.5	2.65
9	651.7	654.6	0.44

A validation test was also performed on the En24 steel sample by considering optimal test parameters. The microhardness is obtained by using both nonlinear regression equation and confirmation experimental results are shown in Table 6. Approximately 3.42 % deviation was found which is to be acceptable by comparing the experimental results and analytical results. Therefore, the derived nonlinear regression equation with different control factors having a reasonable degree of approximation of the microhardness of the En24 steel material.

Table 6. Validation of experimental results for microhardness

Laser beam power (W)	Laser scan speed (mm/s)	Standoff distance (mm)	Hardness obtained from the experiments	Hardness obtained from the predictive equation	% error
300	1	175	725.6	751.34	3.42

4. Conclusions

The research study's major findings are summarized as follows:

- The optimum process parameters are found to be, laser beam power of 300 W, laser scanning speed of 1.0 mm/s and standoff distance of 175 mm to the En24 steel specimen.
- The base metal microhardness before laser hardening was observed 208.2 $HV_{0.3}$ and after laser surface hardening the microhardness of the treated laser specimens was observed from 516.3 $HV_{0.3}$ to 728.1 $HV_{0.3}$ respectively.
- The microhardness of the measured laser treated specimen increases twice in addition to the microhardness of the base metal without laser treated.
- The range of depth of the hardened surface layer is reached for the laser treated specimen from 246.46 to 2471.44 μm respectively.
- Laser beam power is the most influential laser hardening factor of En24 steel, followed laser scanning speed and standoff distance, respectively, from all three factors.

Acknowledgments

The authors are thankful to the Department of Mechanical Engineering of Sardar Vallabhbai National Institute of Technology, Surat, Gujarat, India. To carry out this research, authors have received the support from Department of Mechanical Engineering and the Department of Metallurgical Materials Science, COEP, Pune, Maharashtra, India.

References

[1] Balasubramanian S, Muthukumaran V and Sathyabalan P 2017 *Int J Civil Eng. Technol.* **8** 201
[2] Zhang X M, Man H C and Li H D 1997 *J. Mat. Proc. Technol.* **69** 162
[3] Rana J, Goswami G L, Jha S K, Mishra P K and Prasad B V 2007 *Optics Laser Techn.* **39** 385
[4] Gisario A, Boschetto A and Veniali F 2011 *Optics lasers Eng* **49** 41
[5] Wagh S V, Bhatt D V, Menghani J V and Shivnanda S S 2018 *Int. J Modern Manu. Technol.* **10** 86
[6] Visscher H, Rooij M B,Vroegop P H and Schipper D J 1995 *Wear* **181** 638
[7] Grum J and Sturm R 1998 *Surface Coatings Technol.* **100** 455
[8] Pellizzari M and De Flora M G *Wear* **271** 2402
[9] Zhang P L, Yan H, Xu P Q, Yu Z S, Li C G and Lu Q H 2014 *Lasers Eng.* **28** 135
[10] Prashanthi K and B. Ramakrishna B 2018 *Int. J. Sci. Res. Sci.Technol.* **4** 211
[11] Dinesh Babu P, Buvanashekaran G and Balasubramanian K R 2015 *J. Tribo. Trans.* **58** 602
[12] Mahapatra S S, Patnaik and Satapathy A 2008 *Wear* **265** 214
[13] Zhang Z, Lin P, Zhou H and Ren L 2013 *Appl. Surf. Sci.* **276** 62

IOP Conf. Series: Materials Science and Engineering **814** (2020) 012004 doi:10.1088/1757-899X/814/1/012004

Investigation of Hot Piercing method for manufacturing of conveyor chain link of bagasse carrier of the sugar industry to enhance the breaking load

B. D. Gaikwad[1], S. M. Khomane[2], S. S. Wangikar[3], C. C. Jadhav[4], T. A. Jagtap[5]

[1,3]Associate Professor, Department of Mechanical Engineering, SVERI's College of Engineering, Pandharpur-413304, Maharashtra, India
[2,4] Assistant Professor, Department of Mechanical Engineering, SVERI's College of Engineering, Pandharpur-413304, Maharashtra, India
[5]Engineer, DCPL, Gokul Shirgaon MIDC, Kolhapur-416234, Maharashtra, India

E-mail: bdgaikwad@coe.sveri.ac.in, smkhomane@coe.sveri.ac.in

Abstract. A chain is the most crucial element of sugar industry required for transmitting power and conveying of sugar cane, sugar & bagasse, etc. As these chains operate under various forces, failure of chain assembly is the major problem. The main cause of this failure is ineffective manufacturing processes like cold piercing method. The bearing length of the conveyor link hole is lesser (approx 33% straight and the remaining 67 % taper). The breaking strength is directly proportional to the bearing length of the conveyor chain link hole. Therefore, the bearing length is always desired higher. Only a few studies report in the literature related to the effect of bearing length on braking load. Thus, there is scope for work. This work is focused on hot piercing of the conveyor chain link of bagasse carrier to enhance the breaking load by increasing the bearing length of the pierced hole. It is carried out by varying heating temperature of chain link from 650°c to 850°c with an interval of100°c, by keeping variable clearance between punch and die. The bearing length is measured by using a Coordinate Measuring Machine (CMM). The comparative analysis is carried out for bearing lengths obtained at the cold condition and three hot conditions at temperatures650°c, 750°c, 850°c, respectively. The higher bearing length required for enhancing the breaking strength of the conveyor chain is predicted as an outcome of the study. This study is useful in Conveyor Manufacturing Industry.

1. Introduction

Failure of the conveyor chain can occur because of varieties of reasons. Still, majority of them can ultimately be attributed to defects of various stages of manufacturing the conveyor chain component [1]. Conveyor chain having several joint each joint subjected to tensile loading under working conditions. Each joint is an assembly of inner plate, outer plate, bush, pin and rollers. Breaking load of the conveyor chain usually is 8 times of working load. This braking load depends upon shear and bending stresses of the pin in the conveyor joint and minimum cross-section of the inner link plate of the conveyor. Pin in the joint has been fitted in outer link plate having holes which are pierced by press tool. So as per the basic principle of the piercing hole by press tool, the penetration of the punch 1/3 thickness of the conveyor link plate and after that remaining portion of the plates will get a fracture. This fracture length is depending upon clearance between punch and die. It is expected to make a straight fracture in the link plate to increase the bearing length of the hole. More bearing length more accurate pin can fit in the link plate. Because of this, there is less bending stress, and hence it will help to increase the breaking load of the conveyor chain. In cold working piercing, there is the optimum clearance between punch and die. If this clearance is reduced, then either punch or die will get cracks. In mass production, it is not affordable. So, it is expected with minimum clearance between punch and die. To get more bearing areas in the holes, it is only possible in hot piercing [2-5]. Various researchers have reported studies related to conveyor chain link. Sujata et al [1] reported a case study for failure analysis of conveyor chain link. The cause of the was determined as material processing.

From the above-reported literature, it is noted that very less work has been reported on the piercing process for carrier conveyor chain. So, there is a scope for employing a hot piercing method for fabrication of carrier conveyor chain. This work presents manufacturing of bagasse carrier conveyor chain for sugar industry application. The analysis of bearing length and breaking load is also performed.

2. Experimental Setup

2.1. Mechanical Power press for piercing operation
Need of mechanical power press, for piercing conveyor link plate by using press tools. This press machines having capacity 60 tons, 25 mm stroke length and 300 mm shut height. With proper table face for fitting press tools on the machine table.

2.2. Design and manufacturing of press tool
As per table 1, various sizes of press tool elements have been calculated, which are dependent upon cutting force required for piercing holes of conveyor chain link plate.

Table 1. Dimensions of Press tool for piercing operation

Parameter	Value/Size
Cutting force	59.66 tons
Clearance	0.6 mm
Punch Size	25 mm
Die Size	25.6 mm
Stripping Force	29.83 tons
Clamping bolt	M14x2 (4 no.)
Die thickness	40mm
Die Length	262 mm
Die Width	90mm
Punch Holder size	20x90x262
Stripper size	20x90x262
Backplate	10x90x262(2no.)
Bottom plate	60x190x362
top plate	60x190x362
Punch	Dia34x70

2.3. Material required for press tool
While selecting the press tool materials, there is a need to consider properties like dimensional stability, damping, wear-resistance and suitability of heat treatments [2-5]. Accordingly, suitable material has been selected, and details of the same have been shown in Table 2.

Table 2. Selected Material and Hardness for press tool

Sr. No.	Particular	Material	HRC
1.	Die	HCHCR	58-60
2.	Punch	HCHCR	58-60
3.	Stripper	OHNS	35-40
4.	Punch Holder	EN 31	35-40
5.	Back Plate	OHNS	30-35
6.	Top and bottom plate	M.S	Nil

2.4. Material selection for conveyor link plate.

While selecting the material for the conveyor link plate, there is a need to consider Mechanical Properties like tensile strength, excellent wear resistance, good shock-absorbing capacity [4]. By considering these properties, the material has been selected as per B. S. 970 which is shown in table 3

Table 3. Selected Material for conveyor link

Material	Tensile (kg/mm^2)	Yield (kg/mm^2)	Elongation (mm)	BHN	Composition
EN 24	75	57	20	270	C- 0.35-0.45, Si- 0.1-0.35 Si, Mn- 0.45-0.70, Ni- 1.32-1.8, Cr- 0.9-1.4, Mo- 0.2-0.35

2.5. Conveyor link data

Based on conveyor capacity and required breaking load of the conveyor chain, various specifications have been indicated in Table 4.

Table 4. Specifications of bagasse carrier chainplate

Sr. No.	Particular	Value
1	Plate size	225 x 65 x12 mm
2	Hole Size	Dia 25 mm
3	Pitch	150 mm
4	Material	EN24
5	Hardness	30-35 HRC

2.6. The cold piercing operation for conveyor chain link.

After design and manufacturing of press tool assembly, it has been fitted on mechanical power press having capacity 60 tons and stroke length 25 mm. And operation of piercing started on the conveyor link plate. As shown in figure. 1 (a, b) pierced plate holes has been inspected by the coordinate measuring machine. It has been observed that one side of the hole that is called punch entry the hole inside edge is straight up to 4 mm and after that remaining portion of the link plate is fractured that means one side hole size is diameter 25 mm, and the opposite side is 25.7 mm. so it is expected to reduce the size 25.7 mm and make it as possible as close to diameter 25 mm by decreasing the clearance between punch and die and optimum temperature for hot piercing.

IOP Conf. Series: Materials Science and Engineering **814** (2020) 012004 doi:10.1088/1757-899X/814/1/012004

Figure 1. (a) Cold piercing setup;

Figure 1. (b) Cold piercing plate

2.7. The hot piercing operation for conveyor chain link.

It is carried on the same mechanical power press with the same press tool by changing clearance between punch and die. For hot piercing, we have selected clearance between punch and die half of the cold piercing, i.e. 0.35 mm. That means punch size diameter 25 mm and die size is 25.35 mm. as shown in below Figure 2 (a, b). Like this way, it has been pierced 4 plates, and its hole sizes have been measured with different temperatures, and it will lead to the results.

Figure 2. (a) Hot piercing setup

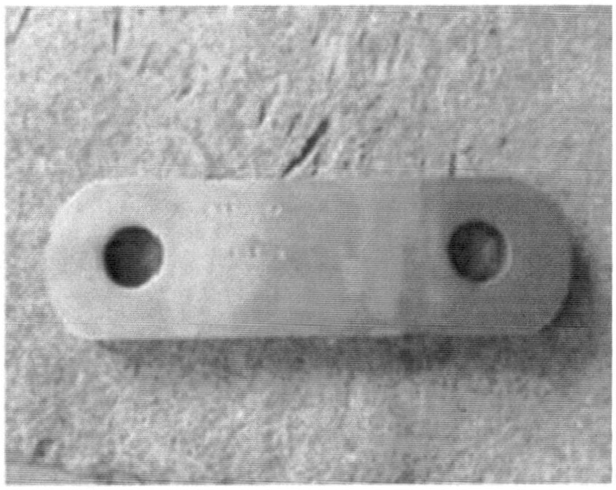

Figure 2. (b) Hot piercing plate

Table 5. Heat treatment cycle for bagasse carrier chainplate

Cycle Stages	Temperature (Degree)	Time (Min)
Preheating	650	60
Heating	850	Starting
Soaking	850	90
Oil quenching	20	30
Tempering	550	Starting
Soaking	550	90
Cooling	At Room Temp.	30

2.8. Heat treatment of the conveyor chain.

After piercing all sample, plates have been heat-treated by through hardening process. And pin bush rollers has been heat-treated by case hardening process [6]. Related brief data of heat treatment has been given in table 05.

3. Testing and measurements.

3.1. Estimation of bearing length of the hole.

This is performed by Coordinate measuring machine (CMM) as shown in below Figure 3 and the result has been noted in Table 6.

Figure 3. Testing of bearing length of link plate hole by using CMM

3.2. Checking of the breaking load of the conveyor chain.

Checking of the breaking load of conveyor chain assembly is done by ultimate tensile testing machine with special fixtures mounted on the two machine jaws as shown in figure.4 (a) having UTM Control unit as presented in figure 4(b) and reading has been noted in Table 6.

4. Result

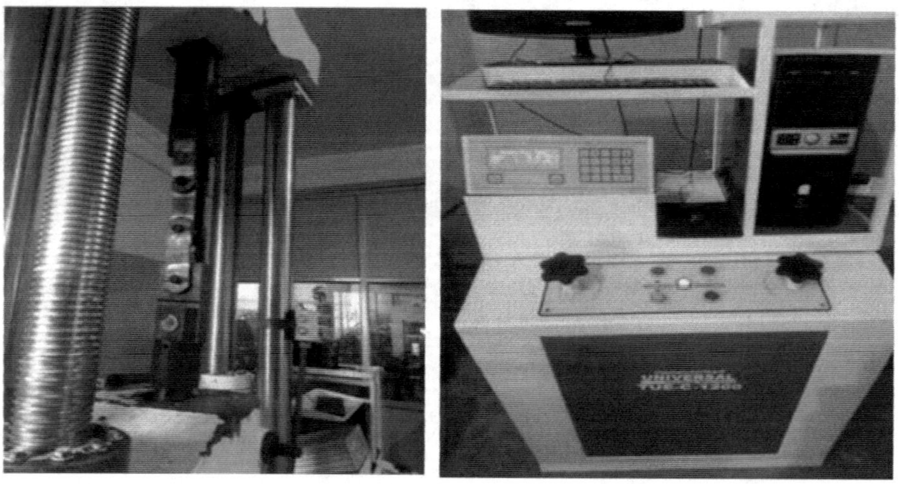

Figure 4. (a) UTM setup (b) UTM control unit

4.1 Comparison between Cold Piercing and Hot Piercing:

It is observed in Table 06, for cold piercing breaking load is lesser as compared to hot piercing.

Figure.5 and 6 depict the breaking load for the average length of bearing area for cold and hot piercing, respectively. The trend of the breaking load increase for cold piercing is in god agreement with that for hot piercing. Also, the length of the bearing area is observed to be lesser for cold piercing as compared to hot piercing.

By using Coordinate Measuring Machine (CMM), bearing length of the hole of the chain link plate has been measured. And after assembly of conveyor, the breaking load has been tested on UTM for the chain links manufactured by cold piercing and hot piercing.

Table 6. Breaking load and Average length of bearing area for cold and hot piercing

Operation	Conveyor link plate temperature (°C)	Breaking Load(KN)	Average Length of Bearing Area (mm)
Cold Piercing	Room Temp.	650	3
Hot Piercing	650	687	3.5
Hot Piercing	750	689	3.7
Hot Piercing	850	686	3.6

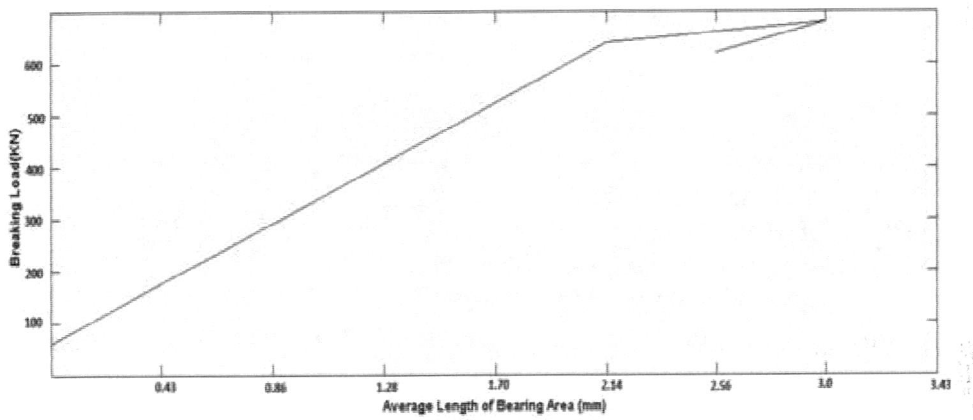

Figure 5. Average Length of Bearing Area (mm) Vs Breaking Load (KN) [Cold Piercing]

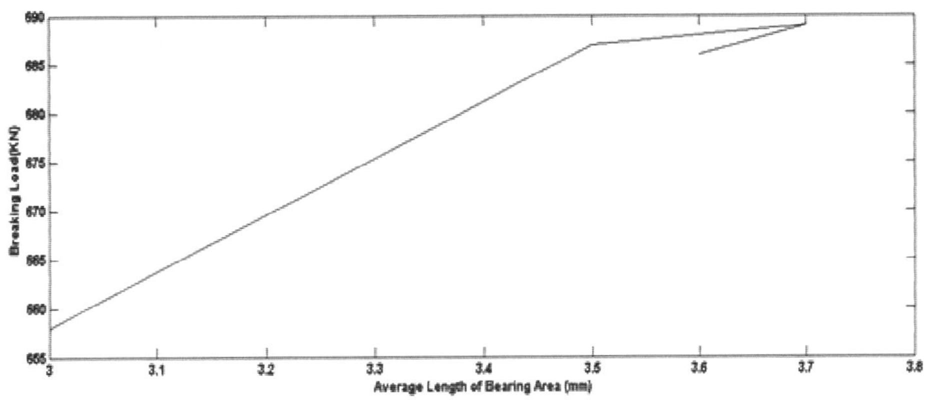

Figure 6. Average Length of Bearing Area (mm) Vs Breaking Load (KN) [Hot Piercing]

IOP Conf. Series: Materials Science and Engineering **814** (2020) 012004 doi:10.1088/1757-899X/814/1/012004

4.2 Effect of temperature in Hot Piercing:

From Table 6, it is observed that the breaking load increases with increase in temperature from 650 ºC to 750 ºC and then decreases to a further rise in temperature up to 850 ºC. The similar trend is also observed for the average length of the bearing area. This is presented in Figure 6.

5. Conclusion

The bagasse carrier conveyor chain link was manufactured by employing a hot piercing method to enhance the breaking load. The hot piercing was performed at three temperatures from 650°C to 850°C with an interval of 100°C, by keeping variable clearance between punch and die. The bearing length and the breaking load were analyzed. Comparative study for cold piercing and hot piercing is reported.Based on the work, the following conclusions are drawn:

- The breaking load and average length of bearing area are observed lesser for cold piercing as compared to hot piercing.
- The bearing length and the average length of bearing area increases with increase in temperature from 650°C to 750°C and further decreases with increase in temperature up to 850°C
- The maximum bearing length is noted as 3.7 mm and breaking load as 689 KN at the temperature of 750 ºC.

This study is useful for Conveyor Manufacturing Industry, especially for Sugar industry-related applications.

References

[1] Sujata M Venkataswamy M Parameswara M Bhaumik S 2006 *Eng. Fail. Anal.* **13** 914

[2] Sapate S and Didolkar V 2009 *Mater. Design.* **30** 2623

[3] Woelke S 1968 *J. Agr. Eng. Res.* **13** 168

[4] Low K H 1995 *Comput. Struct.* **55** 925

[5] Troedsson and Vedmar 2001 *J. Mech. Design.* **123** 395

IOP Conf. Series: Materials Science and Engineering **814** (2020) 012005 doi:10.1088/1757-899X/814/1/012005

Hard Turning Cutting Tool Materials used in Automotive and Bearing Manufacturing Applications – A review

Debabrata Samantaraya[1], Sanjay Lakade[2]

[1,2]Pimpri Chinchwad College of Engineering, Nigdi, Pradhikaran, Sector-26, Pune, Maharashtra, Pin-411044, India

E-mail: d.samantaraya@yahoo.co.in

Abstract. Hard turning for the machining of hardened steels by using super abrasive cutting tool materials is a comparatively newer manufacturing technology. Due to phenomenal advancements in super abrasive tool materials and machine tool technologies, hard turning becoming a potential alternate machining process to the traditional grinding. Hard turning offers many distinguished benefits over grinding process including fewer machining setups, shorter manufacturing time, lower manufacturing cost, achieving grinding comparable product quality. Fewer machining setups bring higher manufacturing flexibility, and lower the investment cost in machine tools in comparison to grinding. It is possible to carryout hard turning without coolant that eliminates the detrimental effect of coolants on the human health and environment. These benefits made hard turning a popular alternative for the automotive and bearing manufacturing industries for machining of cam shafts, gears, crank shafts, bearings, dies and mould. This paper reviews the researches that were performed on hard turning using cutting tool materials; cemented tungsten carbide (WC), poly-crystalline cubic-boron-nitride (PCBN) and ceramics. The research findings related to the effect of cutting parameters, tool geometry and surface coating on the tool life and tool wear phenomena of these hard turning tool materials in the hard turning of major steels used in automotive and bearing manufacturing industries have been discussed.

1. Introduction

Hard turning refers to the metal machining process for the various hardened steels with the help of single point super abrasive tool on modern CNC turning machine. The hardness of steels is usually ranges from 48 to 68 HRC [1-4]. Hard turning is applicable for the machining of various steel materials such as case-hardened steels, alloy steels, full hardened tool steels, super alloys, steels coated with hard chrome surface coatings, nitride steels, and heat-treatable sintered powder metal parts [5]. Finish machining of hardened steels, (e.g. through or full hardened steel AISI-52100, SAE-01 for bearing parts and case-hardened steel 20Mn5Cr5 or 16MnCr5 for automotive transmission parts) by hard turning process using super abrasive cutting tools (poly-crystalline cubic-boron-nitride - PCBN), ceramics, and cemented tungsten carbide were very well recognized in past by the automotive and bearing industries for the manufacturing of precision transmission parts [6]. Grinding was the only choice as a conventional or traditional machining process for the manufacturing of precession parts. Now-a-days, hard turning is establishing itself as a better alternate manufacturing methodology to grinding for the manufacturing of various dies, moulds and transmission parts (e.g. gears, crank shafts, bearings and cam shafts) due to some of its distinct features over traditional grinding such as; better flexibility as manufacturing of intricate shapes can possible on a single setup resulting into lower machining time. The machining feasibility in dry environment is one of the top most advantages of hard turning; it is possible to carryout machining without the environmentally harmful lubricants, coolants. All such benefits of hard turning resulted in substantial reduction in manufacturing time and cost [7-12]. There can be a reduction of 30% in overall machining costs with the application of hard turning process for the manufacture of the parts cited above. The resultant benefits of annual saving up to $6 billion has been realized by the United States industries by applying hard turning against grinding [13].

In the past, much experimentation was conducted on the hard turning of various alloy steels. Varaprasad et al [14] investigated the effects of cutting parameters on tool wear pattern in hard turning of AISI-D3 cold work tool steel with the help of ceramic inserts CC6050 by using RSM based DOE. Their study showed that the depth of cut has highest significance on tool flank wear. The feed rate and cutting speed have little influence. Diniz et al [15] experimented on the machining of SAE-01 steel (hardened to 58 ± 2 HRC). Their experiments were on the machining of continuous and interrupted features using two varieties of CBN inserts (high and low CBN content). They found that for the continuous and semi-interrupted feature machining, the low CBN content insert proved to be more suitable and high CBN content insert produced longer tool life in interrupted features machining. Khamel et al. [16] investigated the effects of machining parameters on cutting force, tool life and surface finish in hard turning of bearing steel AISI-52100 with CBN tool. They had used ANOVA and the composite desirability optimization technique based on RSM. The results state that both speed and feed are strong influential parameters that affect surface roughness and tool life and the cutting force is mostly affected by the depth of cut. Bartarya and Choudhury [17] have developed a force prediction model for hard turning of EN31 steel (hardened to 60 ± 2 HRC) using CBN insert with edge honning and without coating. They have used DOE (full factorial) to develop the regression models for surface roughness and cutting force. Their study showed that the three cutting forces are highly affected by the depth of cut followed by the feed rate and speed has little significance. Ozel et al [18] did the investigation to understand how the cutting forces and surface finish is affected by tool geometry, cutting conditions and workpiece hardness in hard turning of AISI-H13 steel. The results revealed that the cutting forces are affected by tool geometry, cutting conditions and workpiece surface hardness. However the surface finish is greately affected by tool geometry. Liu et al [19] presented an experimental study on the distribution of residual stress and how it is affected by insert nose radius and tool wear. They have experimented on bearing steel JIS-SUJ2. PCBN tools with three types of nose radius (1.2mm, 0.8mm and 0.4mm) were used. The results from the study showed that at the initial stage of machining, the distribution of residual stress on the machined workpiece surface is highly affected by insert nose radius. As the nose radius increases, the residual stress is converted to tensile stress alongwith the significant increase in residual compressive stress. Ren et al [20] studied on the hard turning of materials with high chromium hardfacing. They have found that the temperatures during machining is increased with the increase in speed and feed. However, the average temperatures were found in the range of 600 to 700 °C.

2. Superhard tool materials used in hard turnings

There is continuous evolution of various cutting tool materials with significant wear resistance properties for the machining of materials with higher hardness, toughness and chemically reactiveness. In view of this, the super abrasive tool materials such as CBN or PCBN, ceramics and tungsten carbides are the key enablers for hard turning. Therefore, the study on the mechanical, physical, and chemical properties of tool materials and their performances characteristics in machining are the subject of interest for both the cutting tool manufacturers and user industries [21].

Extreme pressure and temperature are exerted on the cutting tools during machining. Therefore, it is very important that the tools must have requisite properties to counter different wear phenomenon such as plastic deformation, fracture, chemical wear and should retain sharp cutting edge for a longer cutting time. The important criteria's that need to be considered while selecting or designing a tool are listed below.

- High hot hardness so that the tool hardness should not drop at elevated temperatures.
- High strength, resilience and stiffness so that they do not break and able to machine with positive geometries without breakage and edge chipping offs.
- High wear resistance to machine hard and abrasive job materials.
- Low frictional coefficient to counter built up edge.
- Better specific heat and thermal conductivity to carry heat away from the cutting interface.

The major superhard tool materials for hard turning are sintered or cemented tungsten carbides (WC), ceramics (e.g. Si3N4 or Al_2O_3 etc.), CBN or PCBN. The requirement of finished part quality parameters such as geometrical accuracy and surface roughness along with the machining process efficiency are the governing criteria for the selection of appropriate cutting tool material. There are different grades of tool materials with different cutting edge geometries available from various tool suppliers. Hence, the selection of most suitable grade of tool material and cutting edge geometry are two key factors in establishing a stable and efficient hard turning process. The different experiments conducted by the researchers using the above tool materials have been presented below:

2.1. Cemented tunsten carbide

Cemented carbide belongs to a class of hard and wear-resistant refractory materials. The hard tungsten carbide (WC) particles are cemented (bound) together with the help of a metal binder called as cobalt, at high temperature (1400-1500°C). Other alloying elements such as titanium, tantalum are also mixed in small proportion for different application [22].

Yigit et al [23] studied that carbide tools coated with CVD multi-layer coating enhances tool life and reduces the cutting force in comparison with uncoated tools. Ciftci [24] reported that the carbide tools coated with TiN multi-layer CVD coating exhibits lower coefficeint of friction than Al_2O_3 multi-layer CVD coated tools in hard turning of austenitic stainless steels. Shihab et al [25] conducted their experiments using TiN/ TiCN/ Al_2O_3/ TiN multi-layer coated carbide tools for establishing the relation between cutting parameters and the cutting temperature in the machining of AISI-52100 steel. They have found that speed and feed rate affects cutting temperature significantly. It is reported that the optimal temperature (566.593 C° and 592.028 C°) is attained at the speed of 100.12 m/min-250 m/min, feed rate of 0.13 mm/rev-0.22 mm/rev and depth of cut of 0.20 mm-0.85 mm. Dogra et al [26] reported the wear phenomenon of both carbide and CBN tools in hard turning of AISI-8620 steel (49–50 HRc). They have established that CBN tool produces the longest tool life. It was observed that the tool life of carbide tools is significantly affected by interrupted surfaces, while the tool life of CBN tool was not affected much by interruptions. However, the surface integrity achieved with both the tools were comparable with each other. There was no white layer created in interrupted and fully interrupted surfaces turning with both tools and the surface roughness observed below 1.6 μm with both tools. Selvam and Senthil [27] experimented on hardened medium carbon C45 steel with TiN coated carbide insert. The C45 steel is used as a raw material for transmission components (crankshafts, connecting rods, axle beams, push rods), shafts and machine spindles. They have concluded in their report that surface finish is significantly affected by tool nose radius followed by the feed rate, depth of cut and spindle speed. Mondal et al. [28] conducted the experiments on 16MnCrS5 steel (43 HRC) under wet and dry condition to judge the machining performance of TiC coated carbide inserts with wide groove and plain chip breaking geometry. They have found that the turning can be possible even at speed of 268 m/min with coolant. But the insert failed miserably when tried under dry condition. Cutting velocity was observed as an influential parameter on cutting forces in comparison to the dry or wet machining environment. They have concluded that the rate of tool wear is less in wet condition than dry condition for the both geometry inserts. The lesser wear rate in wet condition was due to reduction in friction and temperature at tool-part interface. Chinchanikar and Choudhury [29] studied the wear phenomenon of cemented tungsten carbide inserts in different conditions; uncoated, coated with PVD single-layer

TiAlN coating, and CVD multi-layer MT-TiCN/Al2O3/TiN coating in dry hard turning of 35 HRC AISI-4340 steel. Investigation results indicated that the cutting speed can be increased from 62 m/min to 200 m/min with a single layer coating on an uncoated insert. The cutting speed can be further increased to 300–350 m/min with the multi-layer coated inserts. The wear rate was lower while machining using inserts with single-layer TiAlN coating. The wear rate increases rapidly after removal of coating thin layer. Subsequently the wear rate of multi-layer coated inserts is dominated after the removal of upper thin layer. Similar trend was also seen for the cutting force components which increases rapidly with the occurrence of the tool failure. The prominent forms of tool wears were observed as the crater wear, flank wear and catastrophic failure.

2.2. Ceramics

Ceramics are non-metalic inorganic materials. The use of ceramic cutting tools are increasing due to the advent of ceramic-matrix composites and ceramic alloys. The metal cutting ceramic tools are based on either alumina (Al_2O_3) or silicon nitride (Si_3N_4). The three main Al_2O_3-base tool materials are Al_2O_3-TiC, Al_2O_3-ZrO3 and silicon carbide (SiC) whiskers reinforced Al_2O_3. Other Al_2O_3 base ceramics have additives of TiN, TiB, Ti(C,N) and Zr(C,N). Due to brittleness of ceramic tools, special edge preparation or treatment (edge honing) is done to prevent cutting edge chipping off [30].

Gaitonde et al [31] studied on AISI-D2 steel using ceramic insert CC650WG with wiper geometry. They have found that the increase in feed rate resulted in increase in power and decrease in specific cutting force, but the cutting force requirement was low at lower cutting time and feed rate values. The RSA also corobortes that the minimal surface roughness is attained with lower cutting time and feed rate and higher spindle speed values. The maximum wear of tool is observed at 150 m/min spindle speed with all feed rate values. Davim [32] studied on tool wear phenomena of ceramic tool in turning of 60 HRC AISI-D2 tool steel. The effect of speed was high than cutting time on the wear propagation. The resulted high flank wear on tool cutting edge deteriorated the surface roughness. The feed rate influences the specific cutting pressure, whereas surface finish is influenced by both cutting time and feed rate. It was possible to achieve surface finish below 0.8 μm and dimensional accuracy below IT7 with suitable adjustment in cutting parameters which can be comparable with grinding outcomes. Ferreira et al [33] in their research on hard turning of AISI-H13 steel with conventional and multi-radii ceramic tools. The surface finish is affected by tool edge geometry and feed rate significantly, whereas cutting speed has insignificant effect. Better surface roughness is attained with multi-radii geometry tool in comparison with conventional geometry tool. Drupal et al [34] carried out experimental investigation on AISI-4140 steel hard turning in dry condition using mixed ceramic insert Al2O3+TiCN with PVD-TiN coating. The results reveal that the surface roughness is affected by the feed rate, followed by cutting speed. However, cutting velocity in association with feed-depth of cut interaction affects the tool flank wear. Davoudinejad and Noordin [35] studied on 58 HRC DF-3 tool steel using mixed ceramic PVD-TiN coated inserts with chamfer and honed edge geometry with the help of ANOVA. The chamfered edge geometry produced longer tool life. The ceramic tool exhibited flank, crater and abrasive wear phenomena. The optimum values of tool life and material removal rate is attained at spindle speed of 155 m/min for both insert geometries. The combination of machining parameters (155–210 m/min cutting speed and 0.125–0.05 mm/rev feed rate) and with chamfer edge geometry, better surface finish is resulted. The cutting force had a decreasing relationship with cutting speed increase and affected considerably by tool wear. The higher values of cutting forces were observed with honed edge geometry.

2.3. Poly crystallıne cubic boron nitride (PCBN)/CBN

PCBN/CBN is the second hardest material (hardness 4500 HV) after PCD (hardness 9000 HV). It is synthesized by combining boron and nitrogen atoms to form a compound of hexagon structure called boron nitride (BN), which further converted into cubic boron nitride (CBN) at high temperature and pressure [36-37]. Individual CBN crystals are binded together (sintered) with the help of ceramic binders resulting into poly-crystalline cubic-boron-nitride (PCBN) [38].

Shalaby et al. [39] conducted experiments on 52 HRC D2 steel using Al2O3+TiC mixed ceramic, PCBN and PCBN with TiN coating inserts. They have reported that the prolonged tool life and lower cutting force is observed with mixed ceramic insert than PCBN insert. This is because of better thermal and chemical stability and formation of Cr-Ox lubricating films at high temperature during machining. However such characteristics was not exhibited by PCBN tools. The wear mechanism of PCBN tool was observed as combination of abrasion, adhesion, and diffusion, which is not improved even with TiN coating. They have concluded that Ceramic tools are the most suitable choice over PCBN for D2 steel cutting. Yallese et al. [40] experimented on 60HRC 100Cr6 steel hard turning using CBN tool to study the tool wear, temperature changes, cutting force and surface roughness. CBN tool exhibits better wear resistance characteristics. The maximum amount of generated heat was carried away with chips. At cutting speed of 280 m/min, heavy vibration and sparking were observed only after few minutes of cutting. Beyond this speed, stabilization of surface roughness is observed because of reduction in cutting forces at high speed, resulted in a stable machining condition. Optimal value of flank wear (below 0.4 mm) was observed at 120 m/min speed. Guddat et al. [41] conducted their experiments on 58-62 HRC AISI-52100 through hardened steel using PCBN inserts with wiper land. They studied the behavior of wiper geometry on cutting forces and surface integrity with the help of statistical DOE model. Their analysis reported that significant improvement in surface roughness and higher residual compressive stress were observed with the wiper geometry inserts in comparison with the conventional geometry inserts. Bushlya et al. [42] studied on continuous cut high speed machining (120 m/min – 180 m/min) of cold work tool steel to present a performance comparison of various superhard tools comprising of cBN-binderless, ceramic-bound and cBN-wBN materials. The experiments were conducted both in dry and wet condion. The performance of various tool material combination was studied in relation to surface roughness, cutting force and surface intigrity at sub-surface. For the above stated combined criteria, it was observed that low-cBN content tools with ceramic binder exhibits excellent tool performance.

3. Experıments on the tool lıfe and tool wear phenomena

Extreme temperature and pressure are exerted on cutting tools during machining due to the severe friction that occurs between chips and workpiece interface (metal-to-metal contact) resulting into tool wear. The tool wear predominantly affects the machined part dimensional accuracy and surface integrity [43]. Therefore, it is very much essential to study the tool wear phenomena in hard turning to address issues such as catastrophic tool failure, cutting edge breakage or chipping off, deterioration of workpiece surface integrity to establish a stable and economical machining condition [44-45].

In the past century, extensive research works were carried out by the researchers [46-50] to understand and predict the effect of various factors on tool wear behavior and mechanism. Diniz and Oliveira [51] investigated on 56 HRC AISI-4340 steel with continuous, semi interrupted and interrupted features using two grade of tools; high CBN content and low CBN content. The 7020 grade produced longer tool life irrespective of types of surface features. Based on the tool geometry it is observed that the tools with chamfered edge produced longer tool life in continuous cutting, whereas the tools with rounded edge exhibits better results in interrupted cutting. Oliveira et al. [52] studied the behavior of whisker-reinforced ceramic and PCBN tool in the machining of hardened steels with interrupted and continuous cutting features. In continuous cutting, the longest tool life was obtained with PCBN tool. For the interrupted cutting, similar longer tool life was found with both ceramic and PCBN tool. However, in view of surface roughness, PCBN tools exhibited better results for both continuous and interrupted cutting. Elmunafi et al. [53] studied on tool life of coated carbide tool in hard turning of AISI-420 steel hardened to 47-48 HRC using MQL system and castor oil as cooling media. The experiments were conducted under cutting velocity of 100, 135, 170 m/min. and feed rate of 0.16, 0.2, 0.24 mm/rev. It was reported that the tool life had a inversely proportional relationship with cutting velocity and feed rate. Chinchanikar and Choudhury [54] studied on the flank wear mechanism of coated tungsten carbide tools in relation with machining parameters, workpiece hardness and type of coating in hard turning of AISI-4340 steels of different hardness values. They have found that PVD coated tools shows crater wear as dominant wear phenomena and CVD coated tools shows flank wear as dominant wear phenomena. Also, they have reported that the foremost causes of tool wear are adhesion and abrasion in case of CVD coated tool, whereas abrasion, adhesion and diffusion are the main reason of wear for PVD tools. Dogra et al. [55] presented a comparative study on the tool flank wear of CBN inserts with respect to coated carbide inserts and cryogenically treated uncoated and coated carbide inserts in hard turning of 48-49 HRC AISI-H11 steel. They have reported that CBN insert shows lower flank wear values in comparison with other inserts.

4. Summary

Hard turning is becoming a popular machining process for hardened steels in bearing, automotive and other manufacturing industries. Grinding was used traditionally for machining of such materials till recent past. However, the technological evolution in cutting tool materials and machine tool technology made hard turning an alternate machining method to grinding. Hard turning offers distinct benefits over grinding, due to which it has created great interest among the research community. This paper has presented literature reviews on few of the past researches and focuses on hard turning of major steels and cutting tools materials used in bearings and automotive industries. Following are the conclusions drawn from the findings of the past researches which are reviewed and presented in this paper:

- The benefits of hard turning over traditional grinding includes substantial amount of reduction in manufacturing time and costs, achievement of grinding comparable surface quality.
- Hard turning is possible in dry condition, eliminating or reducing the use of hazardous coolant. This characteristic makes hard turning as an environment friendly green machining process.
- Hard turning encounters tremendous mechanical loads and high temperature. To sustain such high heat and load, the super hard cutting tool materials such as cemented tungsten carbide with surface coating, ceramic and PCBN are the suitable tool materials for hard turning.
- The component hardness and its features, machining parameters, tool material and geometry significantly affect the tool wear and tool life.
- The crucial information gained from the study of the tool wear phenomena in hard turning can be used as a basis for the cutting tools design, cutting parameters and tool wear optimization that in turn will help in establishing economical manufacturing strategies.

References

[1] Tönshoff H K, Arendt C and Amor R B 2000 *Ann. CIRP* **49** 547-566.
[2] Lima J G, Avila R F, Abrao A M, Faustino M and Davim J P 2005 *J. Mater. Process. Technol.* **169** 388-395.
[3] Singh D and Rao P V 2007 *Mater. Manuf. Proce.* **22** 15–21.
[4] Awadhesh Pal A, Choudhury S K and Chinchanikar S 2014 *Proc. Mater. Sci.* **6** 80-91.
[5] Shihab S K, Khan Z A, Mohammad A and Siddiquee A N 2014 *Prod. Manuf. Research* **2** 24-49.
[6] Davim J P 2011 *Machining of hard materials* (London:Springer-Verlag) 1-31 ISBN 978-1-84996-450-0.
[7] Molnar V and Kundrak J 2011 *Hungarian J. Industrial Chemistry* **39** 219-224.
[8] Mamalis A G, Kundrak J and Gyani K 2002 *Int. J. Adv. Manuf. Technol.* **19** 157-162.
[9] Kundrák J, Mamalis A G, Gyáni K and Markopoulos A 2006 *Mater. and Manuf. Processes* **21** 29-37.
[10] Klocke F, Brinksmeier E and Weinert K 2005 *Ann. CIRP* **54** 557-580.
[11] Kundrak J Mamalis A G and Markopoulos A 2004 *Mater. Manuf. Processes* **19** 979-993.
[12] Afteni M, Terecoasa I, Afteni C and Paunoiu V 2017 *Proc. Adv. Manuf. Engg. and Technol.* 95-111.
[13] Huang Y, Chou Y K and Liang S Y 2007 *Int. J. Adv. Manuf. Technol.* **35** 443-453.

[14] Varaprasad Bh, Srinivasa R Ch and Vinay P V 2014 *Global Congress on Manuf. and Management (GCMM) Proc. Engg.* **97** 338-345.
[15] Diniz A E, Gomes D M and Braghini A Jr. 2005 *J. Mater. Process. Technol.* **159** 240-248.
[16] Khamel S, Ouelaa N and Bouacha K 2012 *J. Mech. Sci. Technol.* **26** 3605-3616.
[17] Bartarya G and Choudhury S K 2012 *Proc. CIRP* **1** 651-656.
[18] Ozel T, Hsu T K and Zeren E 2005 *Int. J. Adv. Manuf. Technol.* **25** 262–269.
[19] Liu M, Takagi J and Tsukuda A 2004 *J. Mater. Process. Technol.* **150** 234-241
[20] Ren X J, Yang Q X, James R D and Wang L *J. Mater. Process. Technol.* **147** 38-44.
[21] Huang Y, Chou Y K and Liang S 2007 *Int. J. Adv. Manuf. Technol.* **35** 443-453.
[22] Santhanam A T and Tierney P *Metal handbook* 9th Ed. **16** 71-89.
[23] Yigit R, Celik E and Findik F 2009 *Turkish J. Engg. Environ. Sci.* **33** 147-57.
[24] Ciftci I 2006), *Tribol. Int.* **39** 565-569.
[25] Shihab S K, Khan Z A, Mohammad A and Siddiqueed A N 2014 *Int. Conf. on Mater. Process. and Character. (ICMPC) Proc. Mater. Sci.* **6** 1233-1242.
[26] Dogra M, Sharma V S, Sachdeva A and Suri N M 2012 *Mater. Manuf. Processes* **27** 523-530.
[27] Selvam M D and Senthil P 2015 *Australian J. Mech. Engg.* **14** 131-137.
[28] Mondal K, Das S, Mandal B and Sarkar D 2015 *Mater. Manuf. Processes* **31** 1770-1781.
[29] Chinchanikar S, Choudhury S K 2013 *J. Mech. Sci. Technol.* **27** 1451-1459.
[30] Komanduri R and Samanta S K Ceramics *Metal handbook* 9th Ed. **16** 71–89.
[31] Gaitonde V N, Karnik S R, Figueira L and Davim J P 2009 *Mater. Manuf. Processes* **24** 1373-1382.
[32] Davim J P and Figueira L 2007 *Mater. Des.* **28** 1186-1191.
[33] Ferreira R, Carou D, Lauro C H and Davim J P 2014 *Mater. Manuf. Processes* **31** 648-652.
[34] Das S R, Dhupal D and Kumar A 2015 *J. Mech. Sci. Technol.* **29** 4329-4340.
[35] Davoudinejad A, Noordin M Y 2014 *J. Mech. Sci. Technol.* **28** 4727-4736.
[36] Heath P J 1986 *Industrial Diamond Review* **46** 120-127.
[37] Huang Y, Chou Y K and Liang S.Y. 2007 *Int. Adv. Manuf. Technol.* **35** 443-453.
[38] Sumitomo Electric Carbide-CBN-Diamond 2001 *Product Catalog www.sumitool.com*
[39] Shalaby M A, El Hakim M A, Abdelhameed M M, Krzanowski J E, Veldhuis S C and Dosbaeva G K 2014 *Tribol. Int.* **70** 148-154.
[40] Yallese M A, Chaoui K, Zeghib N, Boulanouar L and Rigal J 2009 *J. Mater. Process. Technol.* **209** 1092-1104.
[41] Guddat J, Saoubi R M, Alm P and Meyer D 2011 *Conf. on Surface Integrity Proc. Engg.* **19** 118-124.
[42] Bushlya V M, Gutnichenko O A, Zhou J M, Ståhl J E and Gunnarsson S 2014 steel *J. Superhard Mater.* **36** 49-60.
[43] Poulachon G, Alberta A, Schluraff M and Jawahir I S 2005 *Int. J. Machine Tools Manuf.* **45** 211-218.
[44] Cora L, Gordon S and Phelan P 2007 *Robotics and Computer Integrated Manuf.* **23** 638-644.
[45] Arsecularatne J A, Zhang L C and Montross C 2006 *Int. J. Machine Tools Manuf.* **46** 482-91.
[46] Suresh R and Basavarajappa S 2014 *Proc. Mater. Sci.* **5** 1550-1559.
[47] Hooper R M, Shakib J I and Brookes C A 1998 *Mater. Sci. Engg.* **105-106** 429-433.
[48] König W, Neises A 1993 **162-164** 12-21.
[49] Kramer B M and Turkovich B F V 1986 *Ann. CIRP Manuf. Technol.* **35** 67-70.
[50] Diniz A E and Oliveira A J 2008 *J. Mater. Process. Technol.* **195** 275-281.
[51] Poulachon G, Moisan A and Jawahir I S 2001 *Wear* **250** 576–586.
[52] Oliveira A J, Diniz A E and Ursolino D J 2009 *J. Mater. Process. Technol.* **209** 5262–5270.
[53] Elmunafi M H S, Noordin M Y, Kurniawan D 2015 *Proc. Manuf.* **2** 563-567.
[54] Chinchanikar S, Choudhury S K 2013 *Int. J. Refract. Metals and Hard Mater.* **38** 124-133
[55] Dogra M, Sharma V S, Sachdeva A, Suri N M and Dureja J S 2011 *Int. J. Adv. Manuf. Technol.* **57** 541-553.

IOP Conf. Series: Materials Science and Engineering **814** (2020) 012006 doi:10.1088/1757-899X/814/1/012006

Naturally Available Materials and Manufacturing Processes for Development of Biodegradable Composite

Pratik Kulkarni[1]**, Swanand Kulkarni**[2]

[1]Dept. of Mech. Engg, PG Student, SKN Sinhgad College of Engineering, Maharashtra, India
[2]Dept of Mech. Engg, Associate Professor, SKN Sinhgad College of Engineering, Maharashtra, India

E-mail:pratikkulkarni019@gmail.com

Abstract. The hazardous effect of plastic is increasing day by day due to its usage in household and kitchen appliances. The researchers all over the world are looking for alternate material and Sustainable development without appropriate and biodegradable material is not possible. Use of natural resources in various material applications will give support for a rural development as well as it will help ecosystem. The present work provides information about availability of various natural substances which possesses great potential to be used in composite material. The present work also provides information of various techniques useful for manufacturing natural composites

1. Introduction

Presently use of plastic is increased tremendously in domestic household and kitchen appliances which have so many human health hazards. Also plastic is popular material for food packaging like polythene bags, foams, sheets etc which is non degradable, affecting on ecosystem and requires tremendous amount of man power as well as money for recycling. Though reason behind using the plastic material is its low weight and less cost, it has many health hazardous characteristics as it includes the chemical Bisphonol A (BPA) [1]. Worn Polycarbonate food and beverage products can lead to serious exposure to Bisphenol A [2] hence there is need to find alternative material which can efficiently replace the plastic. It will help in elimination of health hazards and disposal issues.

1.1 Motivation behind using natural ingredients

The hazardous effects of the plastic accelerated the rate of research in the field of natural fibre composites. Natural ingredients are likely to be harmless for the human health as well as eco system. Also, natural ingredients will degrade easily and thus eliminate issues regarding disposal. Moreover the use of natural ingredients will lead to the rural development by creating employment in agricultural sector.

The biodegradable composites have scope of application in kitchen appliances like food storage, containers, Tiffin boxes, vegetable cutting boards etc. Also it can further be used in packaging industries as replacement for polythene bags, foams, polypropylene sheets etc [3]. Xiaozhi Tang et. al. [3] identified the potential of starch Polyvinyl alcohol blend to efficiently replace LDPE film where good moisture barrier properties are not necessary.

Along with biodegradability the composite materials are cost competitive as compared to plastic materials. The matrix material like Polylacticacid (LPA) is little bit costlier than the polymeric matrices but the starch based matrices like corn starch and potato starch are cost competitive as compared to polymeric matrices. As all natural fibres like jute, coconut, sisal etc are naturally available; they are much cheaper than the conventional synthetic fibres like glass fibres and carbon fibres. Natural fibres have some advantages over polymeric matrices like low density, low cost, recyclability and availability [4].

2 Naturally available materials as reinforcements

Researchers have found potential in several natural materials to be used as reinforcement during natural composite manufacturing. Some commonly used natural reinforcements are as follows

2.1 Jute fibre

Jute fibre is one of the most affordable natural fibre and also available in adequate quantity. It has been found that around 19lakh tones of jute are being produced in India per annum. It has high toughness and aspect ratio as compared to other natural fibres [6]. However a Jute fibre possesses inconsistent mechanical and physical properties due to the factors such as growing conditions, fibre extraction technique and processing [7].

2.2 Coir fibre

Coir fibre is fibrous material found between the hard internal shell and outer coat of coconut and it has found various applications in floor mats, doormats, brushes [8]. It is also used for rope production over centuries due to its high strength [8].

2.3 Bamboo fibre

Bamboo is highly growing plant and possesses high strength with low density and low cost. Bamboo fibres are studied and used as reinforcement in a biodegradable Polylactic Acid (PLA) matrix to make ecofriendly composite [9].

2.4 Flax fibers

Flax fibre is soft, lustrous and flexible fibre extracted from the bust beneath the surface of stem of flax plant. It has been produced majority in Canada and being exported to Europe, US and Japan since 1994 [8]. Hamdy Ibrahim et al [10] illustrated the use of flax fibre and date palm fibre as reinforcement in a starch based matrix to make a biodegradable composite where the hybrid composite of flax and date palm with 25 wt% has shown good properties which in turn indicate that it can be used as competitive natural fibre as reinforcement.

2.5 Kenaf fibre

Kenaf is also known as Hibiscus Cannabinus and it is native to Southern Asia. Kenaf fibres are mostly preferred as reinforcement in polymeric matrices. Kenaf fibres require low processing cost as compared to other synthetic fibres along with comparable specific properties [11] hence they are preferred to be reinforced with polymeric matrices.

2.6 Coconut spathe fibres

Coconut spathe is covering of coconut inflorescence and though it has potential to be used as reinforcement in natural composite it is not rigorously studied as compared to other fibres from coconut plant [12].

Table1 gives the brief information about various types of natural materials which are used as reinforcement, manufacturing method and effect of reinforcement on mechanical as well as wear properties

Table 1. Natural materials used as reinforcement

Sr No	Name of Reinforcement	Manufacturing Method	Effect on various Properties	Ref.
1	Jute Fibre	Film stacking and compression moulding	It is observed that impact resistance increases when jute fibre is used as reinforcement	[6]
2	Bamboo fabric	Film stacking followed by compression moulding	Tensile, Flexural and impact properties have been increased when Polylactic acid is reinforced with bamboo fibres'	[9]
3	Date palm and Flax fibre	Hot pressing	Maximum Tensile strength is observed when 50wt% fibre content of Date palm and flax fibres is used as reinforcement	[10]
4	Coconut Spathe fibre	Hand lay up	Tensile strength of spathe fibres have been increased when reinforced in epoxy	[12]
5	Coir sheath	Hand layup followed by compression moulding	Decrease in wear rate is observed when coir sheath is used as reinforcement	[13]
6	Areca Frond Fibre	Hand layup followed by compression moulding	Flexural strength is increased when Areca Frond Fibres are used as reinforcement	[14]

3 Effect of chemical treatment of natural reinforcement on mechanical properties

Along with some advantages of natural fibres there is a main challenge of adhesion between the polymeric matrices and natural fibres. The natural fibres are likely to have hydrophilic nature (i.e natural fibres are water loving) and the polymeric matrices are hydrophobic in nature i.e. they repel water which causes adhesion quit difficult [5]. Adhesion problem can be resolved by treatment of the natural fibres which have been studied by several researchers. Table 2 represents the different practices conducted for chemical treatment of natural reinforcement along with its effect on properties.

4 Naturally available matrix materials

From last two decades the use of natural fibres has been trending as reinforcement in polymeric matrices in order to replace conventional synthetic fibres like glass and carbon fibres. There is need arising to make completely biodegradable material which accelerated the rate of research to find the biodegradable and efficient matrices as alternative to the conventional polymeric matrices. Following are some matrices which are identified and used to make green composite.

4.1 Starch based matrices

Starch is mostly used for producing biodegradable plastics. It is easily available, cheap and biodegradable. Starch in its original form lacks strength, water resistibility, process ability and thermal stability [3]. In order to eliminate these drawbacks it is blended with some synthetic polymer such as Poly (Є- capro lactone)(PCL), Polylactic acid (PLA), Polyvenyl alcohol (PVOH). Starch can be extracted from plants such as corn, wheat, rice and potatoes. Some ingredients are needed to be added in starch to make completely biodegradable matrix. Some binders are added in order to improve the bonding [14]. Kaith et al [17] used resorcinol formaldehyde cross linker as binder material in starch

Table 2. Different practices for chemical treatment of reinforcements

Sr. No	Type of treatment and Chemicals Used	Process carried	Effect of treatment on properties	Ref.
1	Alkali treatment with NaOH and Acetic acid	Fibres were washed several times with water and are dried at room temperature .dried coconut sheaths are then immersed in 5% NaOH solution for an hour and neutralized with 5% acetic acid solution	Cellulose percentage was found to be increased accompanied by decrease in hemicelluloses and lignin. Also treated composite shown higher tensile, flexural and impact strength	[4]
2	Alkali Treatment with NaOH and Acetic acid	Fibers were washed and dried for a week and treated to 105-110°C and treated with 5% aqueous solution for an hour and neutralized with acetic acid. Finally it is washed and dried with distilled water	Crystallinity of fibres increased with alkali treatment	[15]
3	Benzoyl peroxide treatment with acetone	Fibres were soaked in 6% benzoyl peroxide solution in acetone for 30 minutes and dried for 24 hours in air	Least weight loss in thermogravimetric analysis is observed	[22]
4	Stearic acid treatment	Solution of 1% Stearic acid in ethyl alcohol was added drop by drop to the fibres. These fibres were then dried at 80°C for about 45 minutes	More weight loss in thermogravimetric analysis is observed compared to fibres treated with Benzoyl peroxide	[22]
5	Alkali with NaOH	Fibers were soaked in a 1% NaOH soliution and washed thoroughly to remove excess amount of NaOH, it is then finally washed with acetic acid and dried	Alkali treated natural fiber composite shown better performance than potassium permanganate treated composite and poor performance as compared to stearic acid treated fibre composites	[22]
6	Potassium permanganate treatment with acetone	Fibres were soaked in 0.5% Potassium permanganate solution in acetone for 30 minutes and dried for 24 hours in air	Higher weight loss in thermogravimetric analysis is observed for fibres treated with Potassium permanganate	[22]

based biodegradable composite. Glycerol is used as plasticizer in starch based biodegradable matrix preparation which induces flexibility in the matrix resisting the cracking of starch based plastic film. Cinelli et al [20] illustrated the use of potato starch and corn fibres to produce foamed tray which can efficiently replace single use packaging material made of expanded Polysterene (EPS).

4.2 Polylacticacid as matrix material
Polylactic acid (PLA) is versatile thermoplastic produced from polymerization of heterocyclic dilactone (Lactide) monomer. Lactide is produced by fermentation of potato, corn, sugar beet, sugarcane. PLA shows properties which are quite similar to Polystyrene. PLA is little bit costlier than polystyrene but still used in industrial packaging and medical applications because of its degradability

5 Different Manufacturing techniques to manufacture biodegradable composite
In case of natural fibre reinforced polymeric composites various manufacturing techniques are available, but in case of biodegradable composites only few manufacturing techniques are available. Table 3 shows some commonly used techniques for manufacturing of biodegradable composite

Table 3. Manufacturing processes available for manufacturing of biodegradable composites

Sr. No	Name of Process	Reinforcement	Matrix	Maximum process temperature	Name of equipment	Ref
1	Extrusion process followed by compression moulding	Bamboo Fabric	PolyLactic Acid	210°C	Conventional compression moulding press (Dake press model 44-251)	[9]
2	Hot Pressing	Date Palm and Flax fibres	Corn starch	160±3°C	Carver Laboratory Press (Model C)	[10]
3	Hand layup followed by compaction	Areca Frond Fibres	Corn starch	140°C	Pneumatic Press	[14]
4	Hot pressing	Saccharumspontaneum L fibre	Corn starch	90°C	Carver Hydraulic hot press	[17]
5	Extrusion	Montmorillonite	Corn starch	120°C	Single Screw plastic extruder SJ25	[18]
6	compression/ explosion	-	Wheat starch. Corn starch and Potato starch	230°C	Hydraulic Press (Carver Model-C)	[19]

Porras et al [9] produced the laminate composite of bamboo fibres reinforced in polylactic acid matrix by using extrusion process followed by compression moulding. PLA films were prepared by extrusion and composite panels were produced by film stacking procedure followed by compression moulding. Hamdy Ibrahim et al [10] produced thermoplastic starch blend separately and then prepared composite by hot pressing the thermoplastic starch with chopped fibres. Srinivas Shenoy et al [14] prepared the biodegradable matrix by mixing base materials, binders and plasticizer followed by heating and mixing with fibres using hand layup process which is then compacted using pneumatic chamber and cured in a hot chamber.

Jun Young Jang et al [16] prepared coconut fibre reinforced PLA composites by using compression moulding technique. Kaith et al [17] employed hot pressing process for starch based composite preparation where as Mingfu Huang et al[18] prepared corn starch / Montmorillonite biodegradable composite in twosteps firstly granular cornstarch molecules were prepared using extrusion and mixed with montmorillonite powder which is further extruded to produce composite. G.M. Glenn et al [19] used compression/explosion process for making starch based foam whereas P.Cinelli et al [20] produced potato starch based foam tray by baking starch, corn fibres, magnesium stearate and PVA

6 Potential applications and Future prospectus

As a biodegradable composite is sustainable development it has great potential to contribute rural development along with employment in agricultural sector. Use of these composites will lead to save man power, machinery, and cost related to the waste management of existing single use plastic by replacing it with biodegradable composite. Starch-PVOH blends have potential to replace LDPE films in several applications [3] in packaging industry. Also starch based foam can be produced which has wide scope of applications in food and industrial product [20] which can replace single use expanded polystyrene (EPS) material significantly[19]

Biodegradable composites have stimulating potential to develop new material which can replace plastic material from kitchen and household appliances resulting in elimination of various hazards from plastics

7. Concluding Remarks

1. As jute fibres are produced in adequate quantity and also possesses good strength there is great potential in jute fibres to be used as reinforcement.
2. As corn starch is easily available and less costly, it will be the best alternative to the polymeric matrices.
3. Naturally available matrix material possesses great potential in case of impact resistance properties Manufacturing is quite simple using hand layup and compression moulding in case of biodegradable composites.
4. Biodegradable composites have potential to be used as replacement for Low density polyethylene and also possess future prospects to be used as replacement to plastic material in household and kitchen appliances.

References

[1] Erler, C., and Novak, J. 2010 *J. of pediatric nursing,* **25 (5)** 400-407.
[2] Vom Saal, F. S., and Hughes, C. 2005 *Environmental health perspectives* **113 (8)** 926-933.
[3] Tang, X., Alavi, S. 2011 *Carbohydrate polymers* **85 (1)** 7-16.
[4] Kumar, S. S., Duraibabu, D. A., Subramanian, K. 2014 *Materials and Design* **59** 63-69.
[5] Siva, I., Jappes, J. T. W., Suresha, B. 2012 *Polymer Composites* **33 (5)** 723-732.
[6] Russo, P., Simeoli, G., Lopresto, V., Langella, A., and Papa, I. 2018 *In Advances in Natural Fibre Composites Springer, Cham*pp. 119-126
[7] Acha, B. A., Marcovich, N. E., and Reboredo, M. M. 2005 *J. of Applied Polymer Science* **98 (2)** 639-650.
[8] Hao. L. C, Sapuan, S. M, Haassan, M. R., and Sheltami, R. M. 2018 *Composites Elsevier* 27-70
[9] Porras, A., and Maranon, A. 2012 *Composites Part B: Engg.* **43 (7)** 2782-2788.
[10] Ibrahim, H., Farag, M., Megahed, H., &Mehanny, S. 2014 *Carbohydrate polymers* **101** 11-19.
[11] Akil, H., Omar, M. F., Mazuki, A. A. M., Safiee, S. Z. A. M., Ishak, Z. M. and Bakar, A. A. 2011 *Kenaffiber reinforced composites: A review. Materials* and *Design* **32 (8-9)** 4107-4121.
[12] Sapuan, S. M., Zan, M. N. M., Zainudin, E. S., and Arora, P. R. (2006). *J. of Tropical* Agriculture **43** 63-65.

[13] Divya, G. S., Kakhandaki, A. and Suresha, B. 2014 *Int. J. of Innovative Research and Development* **3** 632-639.

[14] Heckadka, S. S., Kini, M. V., Ballambat, R. P., Beloor, S. S., Udupi, S. R. and Kini, U. A. 2014 *Int. J. of Manufacturing Engg.*

[15] Reddy, K. O., Reddy, G. S., Maheswari, C. U., Rajulu, A. V and Rao, K. M. 2010 *J. of Forestry Research* **21(1)** 53-58.

[16] Jang J. Y., Jeong T. K., Oh H. J., Youn, J. R. and Song Y. S. 2012 *Composites Part B: Engg.* **43** 2434-2438.

[17] Kaith, B. S., Jindal, R., Jana, A. K and Maiti, M. 2010 *Bioresource Technology* **101(17)** 6843-6851.

[18] Huang M. and Yu, J. 2006 *J. of Applied Polymer Science* **99(1)** 170-176.

[19] Glenn G. M. and Orts W. J. 2001 *Industrial Crops and Products* **13(2)** 135-143.

[20] Cinelli, P., Chiellini, E., Lawton, J. W. and Imam, S. H. 2006 *Polymer Degradation and Stability* **91(5)** 1147-1155.

[21] Jappes J. W., Siva I. and Rajini N. 2012 *Polymer-Plastics Technology and Engi* .**51** 419-424.

[22] Sreenivasan, V. S., Rajini, N., Alavudeen, A and Arumugaprabu, V. 2015 *Composites Part B: Engi.***69** 76-86.

[23] Siva I., Jappes J. T. W., Pitchipoo P., Amico S. C., Nagarajan, E. R and Azhagesan, N. 2013 *J. of Polymer Engi.* **33(7)** 665-672.

[24] Poddar, P., Dey, K., Ganguli, S., Sultana, S., Nur, H. P. and Chowdhury, A. S. 2018 *The J. of Pure and Applied Chemistry Research* **7(3)** 257

[25] Kalirasu, S., Rajini, N., WinowlinJappes, J. T., Uthayakumar, M., and Rajesh, S. 2015 *J. of Reinforced Plastics and Composites* **34(7)** 564-580.

[26] Naveen, J., Jawaid, M., Zainudin, E. S., Sultan, M. T., and Yahaya, R. 2019 *The Journal of the Textile Institute* **110 (8)** 1179-1189.

[27] Bennet, C., Rajini, N., Jappes, J. W., Siva, I., Sreenivasan, V. S. and Amico, S. C. *J. of Reinforced Plastics and Composites* **34 (4)** 293-306.

[28] Rosa, M. F., Chiou, B. S., Medeiros, E. S., Wood, D. F., Mattoso, L. H., Orts, W. J., & Imam, S. H. (2009). *J. of Applied Polymer Science* **111 (2)** 612-618.

[29] Ghori, W., Saba, N., Jawaid, M. and Asim, M. 2018. In *IOP Conference Series: Materials Science and Engineering IOP Publishing* **368 (1)** 012009.

IOP Conf. Series: Materials Science and Engineering **814** (2020) 012007 doi:10.1088/1757-899X/814/1/012007

Seismic Performance of Bridge Piers Reinforced with Shape Memory Alloys in Plastic Hinge Region

Kanan Thakkar[1] and Anant Parghi[2]

[1]Research Scholar, Department of Applied Mechanics, S. V. National Institute of Technology, Surat – 395 007, Gujarat, India
[2]Assistant Professor, Department of Applied Mechanics, S. V. National Institute of Technology, Surat – 395 007, Gujarat, India

Email: kananrthakkar@gmail.com, anantparghi@gmail.com

Abstract. The recent earthquake reconnaissance reports revealed that a large residual displacement had led to the devastation of bridge piers due to serviceability concerns. Shape Memory Alloys (SMA) is the distinctive category of smart materials, which can sustain enormous deformations and reappear to a parent shape after removal of loading or by removal of heat. Replacing the typical steel reinforcement with SMA reinforcement in the potential plastic hinge regions of bridge piers could reduce the residual displacement of a pier. This study is focused on the numerical investigation of circular bridge piers with SMA-based reinforcement in the plastic hinge regions to mitigate the residual displacement. The numerical model of a bridge pier is validated using the experimental results of SMA-reinforced bridge pier. Dynamic time history analyses are performed to compare the performance of SMA-reinforced bridge piers with steel-reinforced bridge piers under the effect of various earthquake ground motions. The results are represented in terms of displacement, base shear, and ductility. Moreover, the influence of several parameters on the performance of bridge piers is investigated, i.e., aspect ratio, axial load ratio, and compositions of SMA. The numerical results have indicated the effectiveness of bridge piers reinforced with SMA by reducing the residual displacement after major earthquake events.

1. Introduction

1.1. General
Bridges are the most important structures for the infrastructure network of a country. The damages of bridges could interrupt the traffic flow and cause substantial economic loss. The reinforced concrete (RC) bridges situated in the regions of high seismicity and designed according to the current seismic design guidelines are liable to significant damages during large earthquakes, which showed the large residual displacements {Formatting Citation}. The past earthquake investigation report showed that the bridge structures with high residual displacements were unserviceable for future use [2]. Subsequently, post-disaster respite and release operations were rigorously affected. Generally, it is presumed that the utmost seismic demand in the bridge piers focuses in a small zone, which has a supreme inelastic curvature and defined as the plastic hinge length [3]. To improve the performance of structures, numerous research has been carried out in the domain of smart materials, such as shape memory alloys (SMA).

SMA is a distinctive category of smart materials that can sustain enormous deformations and reappear to a parent shape after removal of loading (super-elasticity) or by removal of heat (shape memory effect) [4]. The re-centering phenomenon of SMA-RC structure is a unique and interesting feature for structural applications [5]. There are two crystal phases of SMA: high and low-temperature phase, named as austenite phase and martensite phase, respectively. SMA gets converted from one phase to another phase by applying/removing stresses/ temperature. SMA can deform up to six-to-eight percent strain limits [6]. There are various compositions of SMA. Among them, NiTi-SMA is proven to be the most suitable for applications in structural engineering.

1.2. Scope of previous research

The applications of super-elastic SMA are significantly expanded, and researchers have conducted extensive investigations exploring different structural applications and developing innovative devices making use of the distinctive characteristics of this smart material [2, 7]. Numerous innovative structural systems are developed incorporating NiTi-based and Cu-based SMA with an aim to dissipate earthquake forces, damping control, structural retrofit, etc. [8]. SMA offers numerous structural applications, namely, reinforcement in reinforced concrete, connectors in steel and concrete structures, damping devices, base isolation devices, seismic retrofit, serving as an active or passive confinement technique in terms of pre-stressing and post-tensioning [7, 9, 10]. There are various shapes in which SMA could be integrated into these structural systems, i.e., single and stranded wires, ribbons, strips, tubing, and bars.

Over the last decade, several applications of SMA in bridges are investigated for enhancing the dynamic performance of structures. Since the last decades, numerous researchers studied the behavior of super-elastic SMA-reinforced structures replacing conventional steel in plastic hinge locations to mitigate the seismic energy demand after severe seismic events.

The above literature shows that RC structures are susceptible to go through substantial deface during strong earthquakes with higher residual/permanent deformation, SMA could be used as the most suitable reinforcements at the plastic hinge area of RC structure. Furthermore, there is a limited investigation of the parametric study as an alternate of steel reinforcement at the plastic hinge location with super-elastic SMA-based reinforcement on the dynamic performance of the bridge piers. Recently, there is no code/guidelines available in the existing literature for the use of super-elastic SMA as an alternative reinforcement with conventional steel.

2. Numerical Investigation of Bridge Piers

This research is focused on the seismic performance of bridge piers with super-elastic SMA reinforcement in the plastic hinge region under dynamic loads. SMA is an expensive material, and large-scale structural elements require a massive quantity of reinforcing materials. As the maximum damage of a bridge pier is concentrated in a plastic hinge region, SMA reinforcement is provided only in the plastic hinge region of a bridge pier. The behaviour of super-elastic SMA-reinforced piers under the effect of various time histories is investigated and compared with conventional steel-reinforced piers using finite element program [11]. This section discusses the parameters for geometrical design and finite element analysis of a bridge pier. Additionally, a parametric investigation has been performed for bridge piers with SMA reinforcement in plastic hinge region under the effect of various ground motions. The parametric study has been conducted by selecting three variable parameters based on the existing literature, i.e. axial load ratio, aspect ratio, and the composition of SMA material.

2.1. Geometry of the bridge piers

In this research, a circular bridge pier has been considered for the nonlinear analysis [2]. The bridge pier is presumed to be situated in India (Seismic Zone-V) and seismically designed following the Indian Standard (IS) Codes and Indian Road Congress (IRC) guidelines. The geometry of the pier is used in this study, as shown in Figure 1. The diameter and height of the pier are 1.524 m and 7.62 m, respectively. The axial load ratio is used as 5% of the axial capacity of the pier, and the longitudinal reinforcement ratio (ρt) is used as 1%. Figure 1(a) illustrates the geometry of the pier in section A-A, and Figure 1(b) shows the vertical section of the pier with the longitudinal and lateral reinforcement details of bridge pier. It defines the steel rebar, SMA rebar, and their connection through a coupler, incorporated for this study.

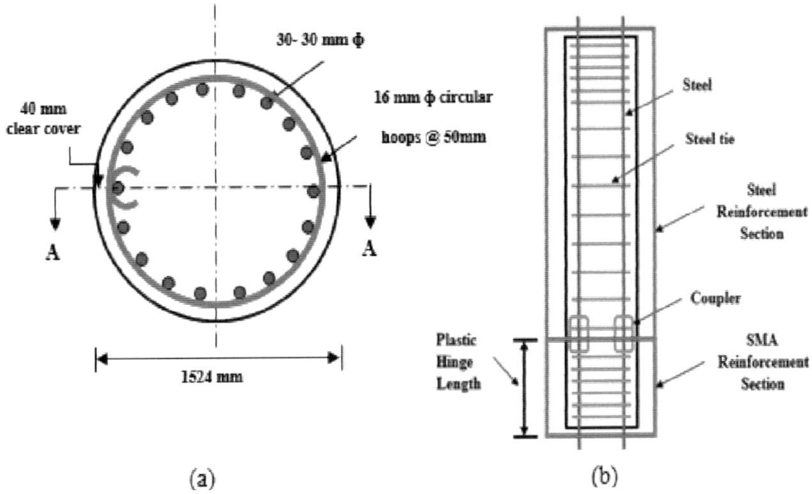

Figure 1. Geometry of SMA-reinforced bridge pier

2.2. Finite element modeling
To study the dynamic performance of SMA-RC and steel-RC bridge piers, a finite element (FE) tool is adopted [11]. This FE tool has the capability to predict both geometric nonlinearity and inelastic material behaviour of the structures under seismic load. To model a pier, a displacement-based nonlinear beam-column element has been chosen. The fiber section is used to discretize pier into three parts: core fiber, cover fiber, and steel/SMA fiber. The piers are modeled incorporating a force-based inelastic system owing to its accuracy. As shown in Figure 2(a) and Figure 2(b), there are 250 fibers, and 5 integration sections from the section are considered for mesh refinement. It has been discretized into 11 FE frames (Figure 2 (a)). The pier length is divided into a total of 11 sections, among which 3 sections represent plastic hinge length, and the remaining sections represent steel rebar.

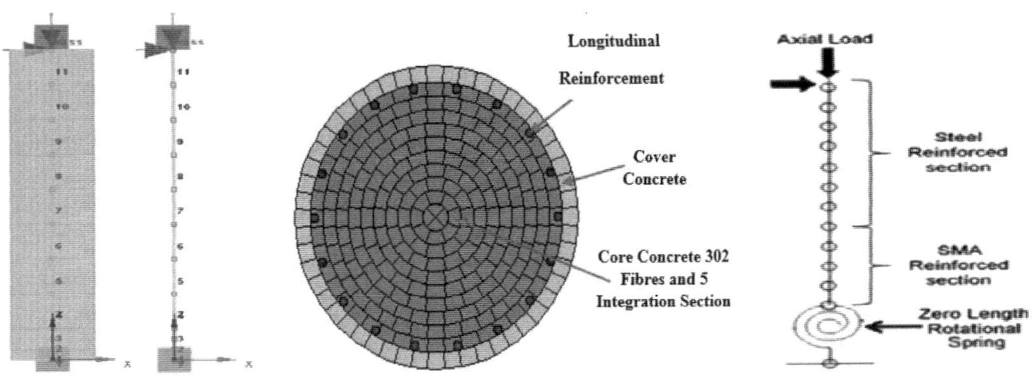

Figure 2. (a) Idealized numerical model (b) fiber discretization of cross-section and FE modeling of bridge piers

The link element has been introduced to connect two different types of nodes. The rotating spring was provided at the bottom of a bridge pier. Figure 2(b) describes the FE model of a bridge pier. As shown in Figure 2(b), dimension-less rotational spring is assigned at the lowermost part of a pier, which represents a mechanical coupler used to connect steel and SMA rebar. Axial load is applied on the top of a pier, and the footing is assumed to be fixed support.

2.3. Material constitutive models
While modeling the cross-section of a pier, concrete, and longitudinal steel/SMA reinforcement of each fiber element, their own constitutive stress-strain relationships are incorporated. In this study, three different material constitutive models are considered for the modeling of bridge piers: steel reinforcement, SMA reinforcement, and concrete.

Menegotto and Pinto [12] steel constitutive model is adopted to assign the properties of steel reinforcement in FE tool. NiTi (Nitinol) SMA having two metals (nickel and titanium), is chosen based on its performance. To represent NiTi reinforcement in the finite element model, Fugazza [12] model is implemented for this study. To predict the stress-strain behavior of confined concrete in this study, the stress-strain model by Madas [13] and the constitutive behavior by Mander et al. [14] are adopted. Table 1 depicts the material properties of steel-reinforced and SMA-reinforced bridge piers used for a numerical investigation.

Table 1. Materials properties used for steel-RC and SMA-RC bridge piers [15]

Material	Property	Value
Concrete	Compressive strength (MPa)	35
	Corresponding strain	0.0029
	Tensile strength (MPa)	3.5
	Elastic Modulus (GPa)	23.1
Steel	Elastic Modulus (GPa)	200
	Ultimate stress (MPa)	692
	Yield stress (MPa)	475
	Ultimate strain	0.14
	Plateau strain	0.016
SMA	Yield strength- f_y (MPa)	401
	Modulus of elasticity-E (GPa)	62.5
	Super-elastic plateau strain length- ε_s (%)	6
	A to M starting stress- f_y (MPa)	401.0
	A to M finishing stress- f_{P1} (MPa)	510
	M to A starting stress- f_{T1} (MPa)	370
	M to A finishing stress- f_{T2} (MPa)	130

Where A is the austenite phase, and M is the martensite phase.

2.4. Model validation
The precision of a proposed bridge pier model is validated through the experimental results by various researchers [16,17,18]. The proposed numerical model using a non-linear FE program [11] is compared with the experimental outcomes by Zadeh et al. [17] for the exactness of modeling techniques. They had conducted an experimental program on one-fifth scaled circular columns utilizing SMA in the plastic hinge region. The RNC (SMA reinforcement in the plastic hinge region) model is validated in this study. The structural displacement of RNC column in the experimental model and numerical model are 114 mm and 122 mm, respectively. The proposed numerical model is validated with the experimental results with an error of 6.56 %. Thus, the local responses of SMA-reinforced bridge piers under the effect of dynamic loads can be assessed reasonably with an implemented nonlinear FE modeling technique.

3. Parametric investigation of the bridge piers

In the previous section, a numerical FE model is generated for steel and SMA-reinforced circular concrete bridge pier using a fibre modeling approach [2]. Nine different pier models with three variable parameters and one conventional pier model are generated. These selected variable parameters are aspect ratio (3, 5, and 7), axial load ratio (5, 10, and 20 %), and compositions of SMA (NiTi, Fe, and Cu), as shown in Table 2. A total of ten numerical models are analyzed for the parametric investigation. The models are named as P-S, P-AR-3, P-AR-5, P-AR-7, P-AL-05, P-AL-10, P-AL-20, P-NiTi, P-Fe, and P-Cu. The details of these models are described in Table 3. The plastic length of the piers is estimated using Paule and Pristley [19] equation (1),

$$L_p = 0.08\ L + 0.022\ f_y \times d_b \quad (1)$$

Where, L_p is the plastic hinge length, L is the total height of a pier, f_y is the yield strength of rebar, and d_b is the diameter of rebar. A bridge pier's plastic hinge length depends on several parameters, i.e., axial load, aspect ratio, properties of SMA, reinforcements (longitudinal and transverse), and strength of confined concrete [3].

Table 2. Parameters of Steel-RC and SMA-RC bridge piers

Parameters	Values
Diameter (D) (m)	1.524
Height (L) (m)	7.62
Aspect ratio (L/D)	3, 5, 7
Axial load ratio P (%)	5, 10, 20
Longitudinal reinforcement ratio (ρ_t) (%)	1.55

Table 3. Details of parameters used for Steel-RC and SMA-RC bridge piers

Pier-ID	P-AR-3	P-AR-5	P-AR-7	P-AL-05	P-AL-10	P-AL-20	P-NiTi	P-Fe	P-Cu
L/D	3	5	7	5	5	5	5	5	5
$P/f_c'\times A_g$	0.05	0.05	0.05	0.05	0.10	0.20	0.05	0.05	0.05
P (kN)	3190.63	3190.63	3190.63	3190.63	6381.27	12762.5	3190.63	3190.63	3190.63
L (mm)	4572	7620	10668	7620	7620	7620	7620	7620	7620
L_p (mm)	741.96	985.8	1229.64	985.8	985.8	985.8	985.8	985.8	985.8
D (mm)	1524	1524	1524	1524	1524	1524	1524	1524	1524
$f_{y\text{-Steel}}$ (MPa)	475	475	475	475	475	475	475	475	475
$f_{y\text{-SMA}}$(MPa)	401	401	401	401	401	401	401	750	210
f_{p1} (MPa)	510	510	510	510	510	510	510	1200	275
f_{T1} (MPa)	370	370	370	370	370	370	370	300	200
f_{T2} (MPa)	130	130	130	130	130	130	130	200	150
E (MPa)	62500	62500	62500	62500	62500	62500	62500	46900	28000
ξ_s (%)	6	6	6	6	6	6	6	13.5	9
f_y/E (Strain)	0.0065	0.0065	0.0065	0.0065	0.0065	0.0065	0.0065	0.00159	0.0075

3.1. Time history analysis

A time-history analysis is a time-dependent analysis to predict the responses of a structure under the effect of dynamic loads. Dynamic time history analysis is performed on bridge pier models to predict the behavior of bridge piers under the effect of earthquake ground motions.

Table 4. Indian Major Earthquakes

Earthquake	Location	Occurrence	Magnitude	PGA (g)	Time (sec)	Seismic Zone
Bhuj	Kutch, Gujarat, India	January 26th, 2001	7.7	0.88	133.53	V
Chamoli	Chamoli, Uttar Pradesh, India	March 29th, 1999	6.8	0.55	87.05	V
Uttarkashi	Garhwal, Himalayas, Northern India	October 20th, 1991	6.8	0.14	21.34	V

The actual earthquake time-history accelerograms are chosen for the dynamic analysis of bridge piers in this study. Three different accelerograms are selected, such as they represent the actual responses of a structure for a particular geometric condition. Table 4 shows the details about Bhuj, Chamoli, and Uttarkashi earthquakes, respectively.

4. Results and Discussion

Figure 3 demonstrates the comparison of maximum top displacement of the piers with axial load ratios (5, 10, and 20 %) under the effect of Bhuj, Chamoli and Uttarkashi Earthquakes, respectively. As seen from these figures, the top displacement of piers upsurges with the raise in axial load ratio. The displacement was found to be a minimum for a 5% axial load ratio and maximum for a 20% axial load ratio. Further, base shear rises with the increment in axial load ratio.

Figure 4 describes the comparison of top displacement and base shear for the piers with aspect ratios (3, 5 and 7) under the effect of Bhuj, Chamoli, and Uttarkashi Earthquakes, respectively. It is evident from the figure that the top displacement of pier is minimum for aspect ratio 3 and maximum for aspect ratio 5. It can be stated that the lateral displacement and base shear increases with the surge in aspect ratio up to some extent. A similar observation has been made by Billah and Alam [3] also.

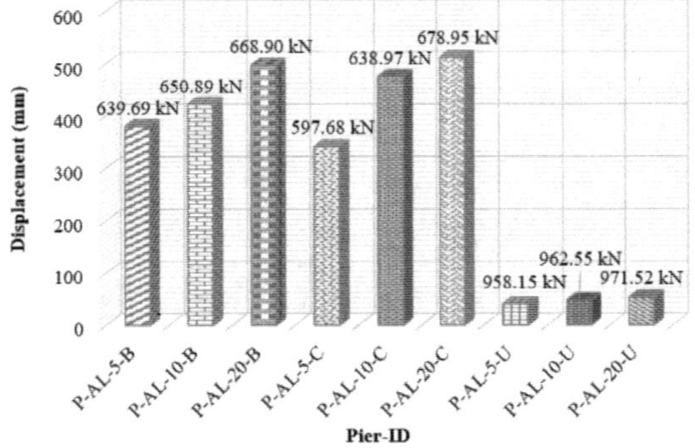

Figure 3. Base Shear and Displacement of Pier with various Axial Load Ratios under the effect of (a) Bhuj Earthquake, (b) Chamoli Earthquake, and (c) Uttarkashi Earthquake

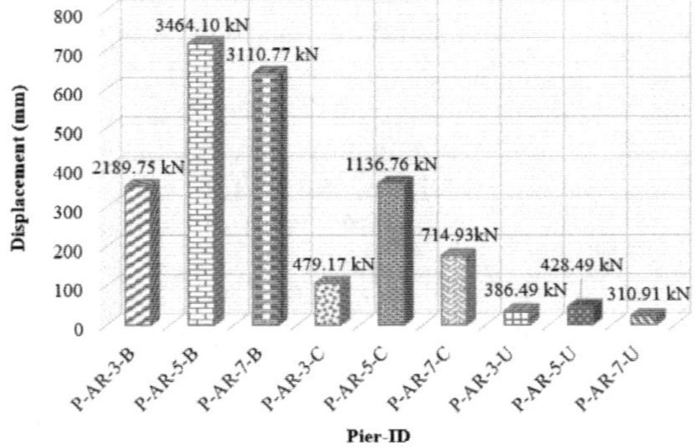

Figure 4. Base Shear and Displacement of Pier with various Aspect Ratio under the effect of (a) Bhuj Earthquake, (b) Chamoli Earthquake, and (c) Uttarkashi Earthquake

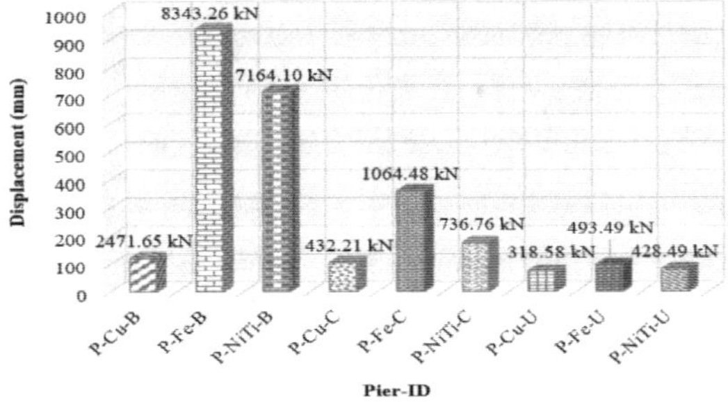

Figure 5. Base Shear and Displacement of Pier with various compositions of SMA under the effect of (a) Bhuj Earthquake, (b) Chamoli Earthquake, and (c) Uttarkashi Earthquake

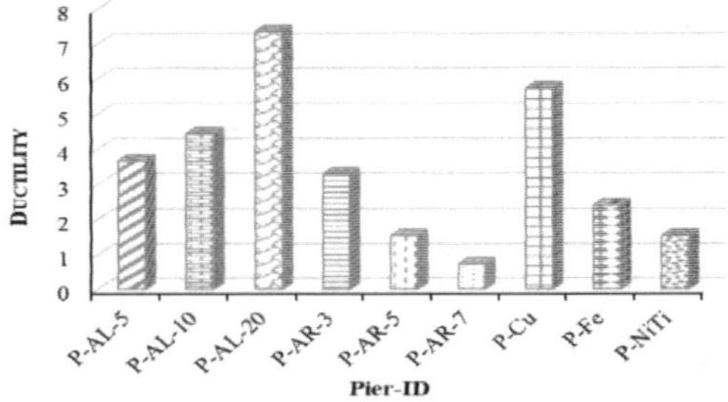

Figure 6. Ductility of piers for variable parameters

Figure 5 reveals the comparison of maximum top displacement and base shear of the piers with various compositions of SMA under the effect of Bhuj, Chamoli, and Uttarkashi Earthquakes, respectively. It is evident from the figure that the displacement and base shear are maximum for the piers with Fe-based SMA due to its higher super-elasticity. The top displacement and base shear of Cu-based piers are found to be less than Fe-based and NiTi-based piers, due to less super-elasticity of Cu-based SMA. These phenomena could be attributed that higher displacement is because of the super-elastic shape memory effect of SMA reinforcement. The higher yield strength of SMA leads to greater ductility of a pier. A similar study has also been done by Andrawes and Shin [20], and Dong et al. [21]. They have researched the improvement of ductility capacity of bridge piers. They determined that utilizing SMA in bridge piers upsurges the ductility, and reduces the residual displacement of a pier under the effect of earthquake ground motions.

Figure 6 shows the ductility of piers with three variable parameters. As shown in the figure, ductility rises with the increase in axial load ratio and reduces with the increase in aspect ratio. Ductility was found to be maximum for a pier with Cu-based SMA and minimum for a pier with NiTi-based SMA. Ductility of a pier with Fe-based SMA was less than the ductility of a pier with Cu-based SMA, and more than the ductility of a pier with NiTi-based SMA.

5. Conclusions

The numerical works were focussed on the consequence of several factors on the performance of circular bridge piers. By analyzing circular bridge piers for three aspect ratios, three axial load ratios and three compositions of SMA under the effect of three different ground motions, the following results were obtained:

Piers spectacles more deformation and higher base shear for larger values of axial load ratio. The pier deformation was most significant for 20% axial load ratio. Thus, larger the axial load ratio, more significant the deformation and base shear.

The top deformation and base shear of piers are found to be minimum for aspect ratio of 3. They get increased with the increase of aspect ratio up to certain limits. Then, the top deformation and base shear values decreased for further increase in aspect ratio.

Piers with Fe-based SMAs deformed more than other piers, owing to higher values of base shear and superior super-elasticity of Fe-based SMAs. Piers with Cu-based SMAs deformed minimum with comparison to other piers in arrears of less super-elasticity and less capacity to attract base shear of Cu-based SMAs. Piers with NiTi-based SMAs deformed to optimum level due to its optimal super-elasticity and ideal capacity to withstand the base shear.

Ductility of a pier was found to be in direct proportion with the axial load ratio and inverse proportion with the aspect ratio. Moreover, ductility of piers with Cu-based SMAs was the highest, and ductility of piers with NiTi-based SMAs was the lowest.

There is considerable potential for the utilization of super-elastic SMA in civil infrastructures. However, the high cost of the SMA is the main obscuring issue. Because of the large size of structural elements, the related forces are also large, which needs a tremendous quantity of reinforcing materials. Hence, it is a big issue for the application of expensive super-elastic SMA in the civil infrastructures. The cost of NiTi based alloys depends on the processing of the high-strength NiTi composites in specific shapes and forms. Thus, the custom-based shape/size and matching could make them expensive two to three times, although there is a considerable drop in the cost of the NiTi over the last decade. Thus, in order to initiate the large-scale applications on SMA, the development of low-cost SMA is mandatory for the reduction of overall cost of a structure.

Acknowledgements

The first author would like to acknowledge the financial support provided by Ministry of Human Resources Development (MHRD), Government of India, through the Ph.D. fellowship program. Further, the authors would like to acknowledge the online and offline library of SVNIT.

References

[1] Shrestha B and Hao H 2014 *23rd Australasian Conf. on the Mechanics of Structures and Materials (ACMSM23)* vol **I** (Byron Bay, NSW, Australia) p 375–80
[2] Billah M 2015 *Ph. D. Thesis* (University of British Columbia, Canada)
[3] Billah M and Alam M S 2016 *Eng. Struct.* **117** 321–31
[4] Boroschek R L, Farias G, Moroni O and Sarrazin M 2007 *J. Earthq. Eng.* **11** 326–42
[5] Abdulridha A, Palermo D, Foo S and Vecchio F J 2013 *Eng. Struct.* **49** 893–904
[6] Desroches R and Smith B 2004 *J. Earthq. Eng.* **8** 415–29
[7] Alam M S, Youssef M A and Nehdi M 2007 *Can. J. Civ. Eng.* **34** 1075–86
[8] Constantinou M C, Soong T T and Dargush G F 1997 *Passive energy dissipation systems for structural design and retrofit A report in Multidisciplinary Center for Earthquake Engineering Research (MCEER)* (University of Buffalo, Buffalo) p 322
[9] Ozbulut O E, Hurlebaus S and DesRoches R 2011 *J. Intell. Mater. Syst. Struct.* **22** 1531–49
[10] Menna C, Auricchio F and Asprone D 2015 *Applications of shape memory alloys in structural engineering Shape Memory Alloy Engineering* Elsevier pp 369–403
[11] SeismoSoft 2018 *SeismoStruct- A computer program for static and dynamic nonlinear analysis of framed structures*
[12] Menegotto M and Pinto P E 1973 *Method of analysis for cyclically loaded R.C. plane frames including changes in geometry and non-elastic behavior of elements under combined normal force and bending* A preliminary report in IABSE symposium on the resistance and ultimate deformability of structures acted on by well defined repeated loads (Lisbon) pp 15-22
[13] Madas P and Elnashai A S 1992 *Earthq. Eng. Struct. Dyn.* **21** 409–31
[14] Mander J B 1988 *J. Struct. Eng.* **114** 1804–26
[15] Billah M and Alam M S 2016 *Eng. Struct.* **117** 321–31
[16] Saiidi M S and Wang H 2006 *ACI Struct. J.* **103** 436–43
[17] Zadeh M S, Brien M O and Saiidi M S 2007 *A study of concrete bridge columns using innovative materials subjected to cyclic loading* A report by University of Nevada Reno and National Cooperative Highway Research Program (NCHRP) p 68
[18] Billah M and Alam M S 2012 *Constr. Build. Mater.* **28** 730–42
[19] Paulay T and Priestley M J N 1992 *Seismic design of reinforced concrete and masonry buildings* John Wiley & Sons, Inc. New York, USA
[20] Andrawes B and Shin M 2008 *Seismic retrofitting of bridge columns using shape memory alloys* SPIE - The International Society for Optical Engineering **6928** 69281K-1–9
[21] Dong J, Cai C S and Okeil A M 2011 *J. Bridg. Eng.* **16** 305–15

IOP Conf. Series: Materials Science and Engineering **814** (2020) 012008 doi:10.1088/1757-899X/814/1/012008

Enhancement of Properties of Geopolymer concrete with Crimped Steel Fibers

U.R.Kawade[1] and S.N.Daule[2]

[1]Prof. and Head, Civil Engg. Dept., DVVPCOE, Ahmednagar
[2]Asstt. Prof.,Civil Engg. Dept., DVVPCOE, Ahmednagar

Email:urmilaanagar@gmail.com, shashirdaule@gmail.com

Abstract: Geopolymer technology is nascent technology in the world of concrete. Indian standard code is not available for geopolymer concrete. However, at the initial stage, mix design of plain geopolymer concrete is done by testing various parameters. Calcium based, sodium based and potassium based solutions are available in the market. Sodium based alkaline solutions are used in the present research. Here, 100% replacement of cement is done with fly ash. Fly ash used is Pozzocrete 63. Geopolymer concrete with addition of crimped steel fibers are used in the research. Molarity is finalized as 13M and ratio of Na2SiO3/NaOH is finalized as 2.Alkaline solution to fly ash ratio is taken in the research. Crimped steel fibers of aspect ratio 70 and 100 are added in geopolymer concrete. Results obtained shows that as aspect ratio of crimped steel fibers increases there is increase in compressive strength, flexural strength and split tensile strength of concrete. It is observed that as the aspect ratio of fibers increased up to some extent, its compressive strength is also increased.

1.Introduction

Pollution free environment protects the life. In today's life apart from food, water and shelter, healthy environment is also important for everyone's life. The cement industry which is responsible for about 7% of CO2 emissions of total CO2 generation in environment, because for production of one ton of Portland cement it emits approximately one ton of CO2 into the atmosphere, as heating of lime stone releases CO2 directly. Among the total greenhouse gases, CO2 contributes about 65% of global warming. In Geopolymer technology, cement less concrete is possible by using any other base material which is rich in Silica and Alumina. GGBS, Rice husk ash, fly ash are the alternative materials to cement. Availability of fly ash is abundant as compared to these materials. In the present research, fly ash (pozzocrete63) is used as a base material to form geopolymer concrete. Disposal o fly ash is also a big problem. As per Ministry of Environment and Forest's (MoEF's) notification of 3rd November 2009, utilization of fly ash should be 100 %. But the highest utilization of fly ash is 62.60 % and that is in the year 2009.

2. Literature Review

Jain A.K. Technical Advisor, Ultra Tech cement Ltd.Has focused on cement production and distribution cost of cement .According to author there is great pressure on financial viability of cement business.Wallah S. E. and Rangan B. V.performed investigation on low calcium fly ash based geopolymer concreteon long term properties. The work considered two mixtures as mixture 1 contained 8 molar NaOH concentration and no extra water was added. Mixture 2 contained 14 molar NaOH concentration and with addition of extra water. Na2SiO3/NaOH ratio was 2.51 and Naphthalene superplastizser was used to achieve workability. Patankar S.V.et. al. (2015) studied on mix design procedure of fly ash based geopolymer concrete. Different parameters for selection of suitable ingredients of geopolymer concrete to achieve desire strength at required workability were studied.

3. Methodology

Here, molarity of NaOH is used as 13Molar and ratio of sodium silicate to sodium hydroxide is referred as 2. Steel crimped fibers having aspect ratios respectively 70 and 100 are added in geopolymer concrete. Fly ash, fine aggregates, coarse aggregates alkaline solution and water are ingredients of geopolymer concrete. Crimped steel fibers having aspect ratio 70 and 100 are added in geopolymer concrete. Fly ash is defined as fine, glass powder obtained from the gases of burning coal during the production of electricity. P63 pozzocrete is used in the present research procured from Dirk India Ltd., Nasik. As per the test results supplied from suppliers, fineness of fly ash is $435m^2/kg$. Locally available sand of fineness modulus 3.15 and specific gravity 2.65 is used. Coarse aggregates of size 20 mm and 12mm having sp. Gravity 2.84 and 2.79 respectively are used. Sodium hydroxide flakes are used for preparation of NaOH solution.

Quantity of materials obtained is as follows :

3.1. Selection of quantity of fly ash

Pozzocrete 63 is used in present research. As per the test results fineness of fly ash is $435m^2/kg$. Also, target mean strength of M_{30} grade of Concrete is 38.25 Mpa.

3.2. Alkaline Activators to Fly Ash Ratio:

Alkaline activator to fly ash ratio is fixed as 0.35. Mass of (Na_2SiO_3 + NaOH) / Fly Ash = 0.35. Mass of (Na_2SiO_3 + NaOH) = 0.35x420 = $147kg/m^3$. Na_2SiO_3/NaOH = 2.,NaOH = 49 kg/m^3, Na_2SiO_3 = 98 kg/ m^3

3.3 Solid content in Sodium Silicate solution:

It depends upon percentage of solids of Silicon Dioxide and Sodium Dioxide in Sodium Silicate solution. Here, Na_2O = 15.06 and SiO_2 = 34.01Total Solid = 49.07 (49.07/100) x 98 = 48.08 kg/ m^3. Solid Content of Sodium Hydroxide in Sodium Hydroxide Solution: 18.33 kg/m3.

3.4 To find water content in alkaline Solution and additional quantity of water required

Water content in alkaline solution = (Mass of Sodium Silicate + Sodium Hydroxide) – (Total Solid Content in Alkaline Solution) m Water content in Alkaline Solution = 147 – 60.93 = 86.07 kg/m^3, Additional quantity of water required = (Total quantity of water) – (Water content in Alkaline Solution)

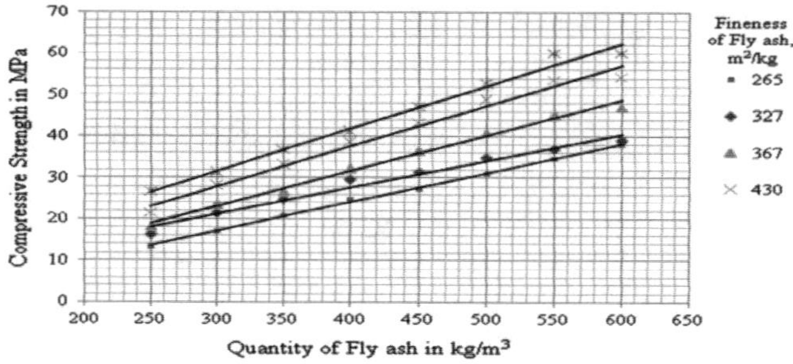

Figure1. Fineness of fly ash

From the figure 1, quantity of fly ash used is $420kg/m^3$.

If fineness of fly ash is in between 300 to 400 m^2/kg then quantity of water required is 110 kg/m^3
Additional quantity of Water required = 110 – 86.07 = 23.93 kg/m^3,
For fineness of fly ash 435m2/kg,density of concrete is obtained as: 2535 kg/m^3.

3.5. To find total aggregate content

To decide Fine to total aggregate Content in Percentage:
It depends upon fineness modulus of fine aggregate. Fineness modulus of sand is obtained 3.15.

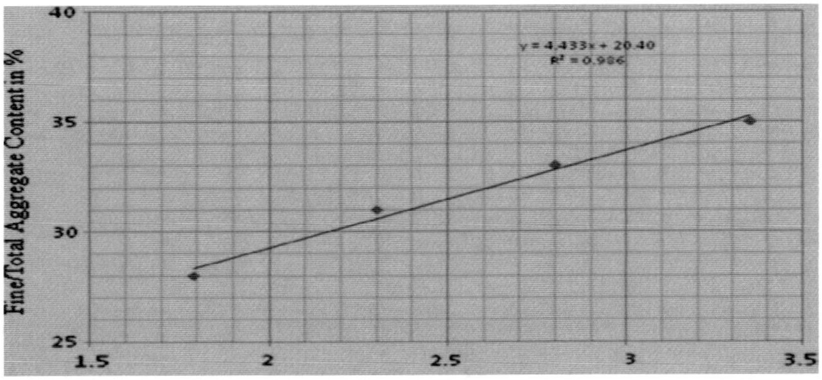

Figure 2. Fineness Modulus of aggregates

Fine aggregate / Total aggregate = 34.5 %=(34.5 / 100) x 1944.07 = 670.70 kg/m^3
Coarse Aggregate required = Total Aggregate – Fine Aggregate = 1944.07 – 670.70 = 1273.37 kg/m^3

Table 1. Materials required for 1m3 concrete

Fly ash kg/m^3	F A kg/m^3	C A m^3	NaOH kg/m^3	Na_2SiO_3 kg/m^3	Additional Water kg/m^3
420	668.81	1269.77	49	98	29.41

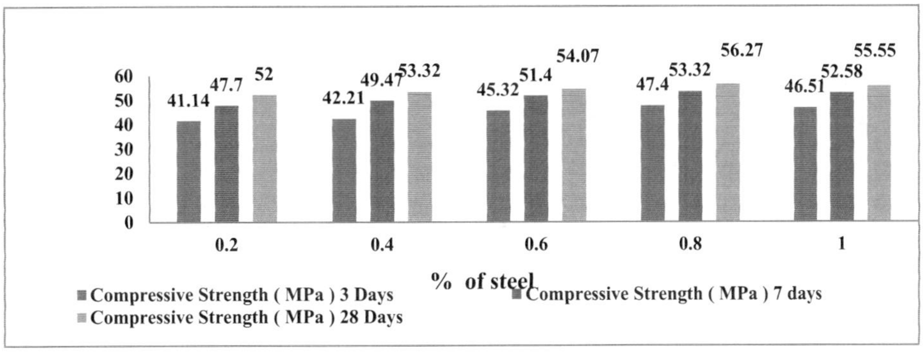

Figure 3. Compressive strength of GPC with Crimped Steel Fiber of aspect ratio 70

3.6. Quantity of Separate Fibers used in Geopolymer Concrete:

Alkaline solution to fly ash ratio is taken as 0.40. Percentage of fibers used in cube specimen and cylinder specimen is used as 0.2%, 0.4%, 0.6%, 0.8% and 1% by mass of concrete. Mechanical properties of specimens are checked. Crimped Steel Fiber 70:As addition of volume fraction of fiber increases its strength is increases up to 0.8 % and after that it reduces as flocculation offibers at one place takes place.

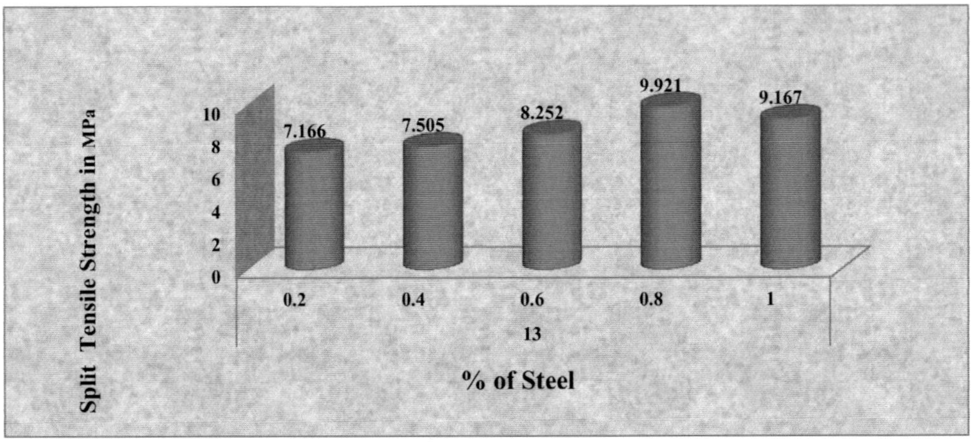

Figure 4. Split Tensile strength of GPC with Crimped Steel Fiber of aspect ratio 70

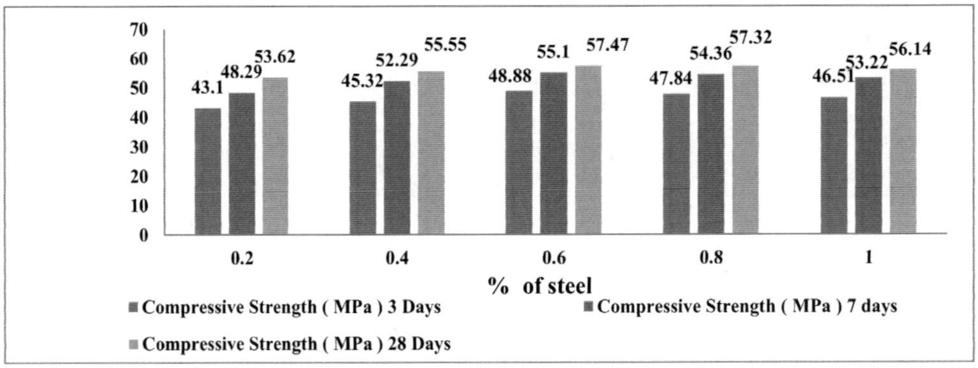

Figure 5. Compressive strength of GPC with Crimped Steel Fiber of aspect ratio 100

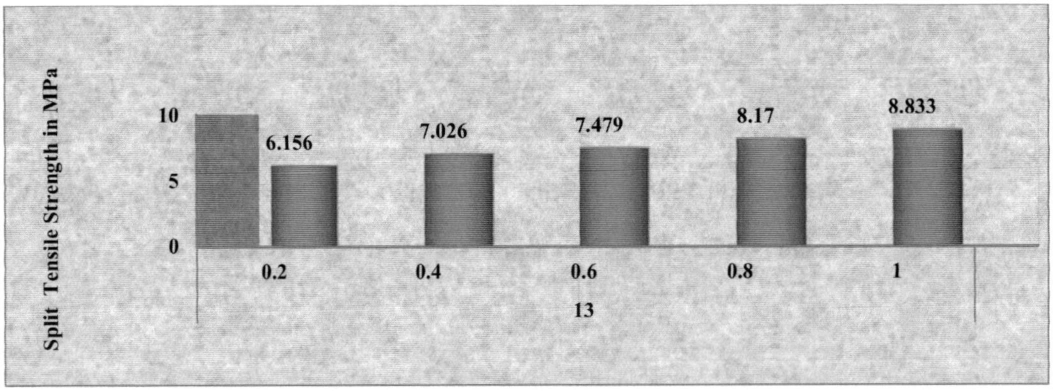

Figure 6. Split Tensile strength of GPC with Crimped Steel Fiber of aspect ratio 100

IOP Conf. Series: Materials Science and Engineering **814** (2020) 012008 doi:10.1088/1757-899X/814/1/012008

4. Conclusion:

For crimped steel fibers of 70 aspect ratio, maximum compressive strength is obtained for addition of fibers of 0.8 % volume fraction by mass of geopolymer concrete, whereas in case of 100 aspect ratio fibers, maximum compressive strength is obtained for addition of volume fraction of 0.6 %. Split tensile strength is observed maximum for 70aspect ratio at 0.8% addition of fibers and for 100 aspect ratio; it is maximum at 1% addition of fibers.

References

[1] Jain A K 2012 *Status of Availability, Utilization and Potential of Fly Ash Use in Construction* (A Report as a Technical Advisor Ultra Tech Cement Ltd.).
[2] Anuradha R, Sreevidya V, Venkatasubramani R and Rangan B V 2012 *Asian Journal of Civil Engineering Building and Housing* **13** 353.
[3] Patankar S V, Ghugal Y M and Jamkar S S2015 *Adv. Structural Engg. Springer* **161** 91633.
[4] PatankarS V,Jamkar S S and Ghugal Y M 2012 *Journal of Structural Engineering* **39** 7.
[5] Kawade U R,HussainMujahid 2018 *Int. J. Engineering Tech. Sci. and Research* **5** 3470.
[6] Kawade U R, Salunkhe P A and Kurhade S D 2014 *J. Innovative Res. Sci. Engg. Tech.* **3** 135.
[7] Abdul Aleem M I and Ammaira, P D 2012 *Int. J. Engg. Sci. Emerging Tech.***1** 118.
[8] Ali A Aliabdo, AbdElmoaty M, AbdElmoaty, Hazem A Salem 2016 *Construction and Building Materials, Elsevier* **121** 694.
[9] Antonio Caggiano, Serena Gambarelli, EnzoMartinelli, Nicola Nistico, Marco Pepe 2016 *Construction and Building Materials, Elsevier* **125** 1035.

IOP Conf. Series: Materials Science and Engineering **814** (2020) 012009 doi:10.1088/1757-899X/814/1/012009

Static buckling analysis of shape memory alloy reinforced composite laminated plate

Achchhe Lal[1] **and Kanif Markad** [2]

[1]Assistant Professor, Department of Mechanical Engineering, SVNIT, Surat
[2]Research Scholar, Department of Mechanical Engineering, SVNIT, Surat

E-mail: [1]lalachchhe@yahoo.co.in, [2]kmarkad13@gmail.com

Abstract. With high strength and stiffness composites are being widely used as structural materials. The properties of composites can further improve by reinforcing with advanced structural components like shape memory alloys (SMA). In the present study nonlinear buckling analysis of laminated composite plate with layer of shape memory alloys are done. Finite element governing equation based higher order shear deformation theory with principle of virtual work is considered. Properties of composite plates are formulated with micro-mechanical approach. A MATLAB program is developed with nine nodes isoperimetric element. Influence of volume fraction of SMA, recovery stress, with and without SMA layer, with temperature on critical buckling load is revealed. Convergence and validation study presented here shows the exactness and efficiency of the model.

1. Introduction

In the recent decades, laminated composites have found an increasing applications in various engineering fields especially in defence, marine, automobile, space and many more due to its high strength to weight ratio and stiffness. Reliability of the structure is still one of the major issue which can get fulfil with the utilisation of the smart material incorporation. Smart / intelligent materials, are those have one or more properties that can be considerably vary in a skilful fashion by peripheral stimuli, such as moisture, electric or light, stress, temperature, magnetic fields, pH, or chemical compounds. Piezoelectric materials, SMA and shape-memory polymers (SMP), Photovoltaic materials, Halochromic materials, Ferrofluids are the few examples of the smart materials. Generally when SMA wires are incorporated with layered composite materials, these are pre-strained, so that when it get utilised in high temperature application or environment, it regains its original shape. With the utilisation of this shape memory effect, in present paper, tried to show effect over nondimensional critical buckling temperature.

Initially the incorporation of SMA wire with various volume fractions were get utilised to regulate buckling of the composites. Rogers and Bakers [1] used SMA in laminated composites to improve in its natural frequency and control its vibration. Baz et al. [2] studied the control of SMA wires on the vibration and buckling performance on composite material. Abbas et al. [3] performed the study to show the influence of volume portion of composite fibers, plate dimensions, fiber orientation angle and SMA wires, on the stimulated plate natural frequencies. Prabhu and Dhanaraj [4] analysed the composite laminated plate against the thermal buckling load by first order shear deformation theory (FSDT) to find the effect of modulus ratio, fiber orientation, and aspect ratio over critical buckling load parameter. Ounis et al. [5] utilised the classical laminate plate theory with linear isoparametric element and a high accuracy rectangular Hermitian element, for the buckling analysis. In 2009, Zhang and Yang [6] studied the laminated composite plate with the advancement into the finite element analysis. From the available literatures it is found that, the study on the thermal buckling seems to be nominal related to the investigation on the mechanical buckling. Thus, the utmost works in the literature uses finite elements based on the first order (FSDT) [7–12]. Li et al. [15] presented numerical simulation of the post-buckling nature of composite shells embedded with SMA fibers under thermal loading. Investigated the effects of SMA fiber reinforcements on the characteristics of thermal buckling were carried out through numerical analysis in ABAQUS. Panda and Singh [16] estimated post-buckling behavior of composite laminated panel with and without SMA wire reinforcements through a mathematical model, established based on the HSDT taking Green–Lagrange nonlinearity. Xia and Shen [17] analysed the functionally graded material along with the piezo actuator with HSDT implementing Von Karman kinematics and under thermal atmosphere. Tawfik et al. [18] analysed SMA fibres reinforced laminated panels under the influence of combined thermal with aerodynamic loading through nonlinear FEM based on the von Karman relations.

When composite material with reinforcement of SMA wires subjected to high temperatures, then internal strain is get induced because of excessive recovered strain, which further generates the tension into the structural member and built its performance against thermal buckling and post buckling. From the available study, it is found that, limited study is focused on the implementation of smart shape memory wires embedded in laminated composites with the use of HSDT. Also, this study includes the convergence and validation study to shows the exactness and efficiency of the model. A MATLAB program is developed with nine nodes isoperimetric element. Influence of volume fraction of SMA, recovery stress, with and without SMA layer, with temperature on critical buckling load is revealed.

2. Mathematical Modelling

The displacement method shows distinct displacement field equation used, and fulfil the condition that, over the surface transverse shear stress zero and nonzero at additional places. Begin with the displacement field as, (Gadade et al. 2016),

$$\begin{aligned} U &= u_1(x,y)+(\Upsilon_1 z-\Upsilon_2 z^3)\psi_x(x,y)+(-\Upsilon_4 z^3)\theta_x(x,y) \\ V &= v_1(x,y)+(\Upsilon_1 z-\Upsilon_2 z^3)\psi_y(x,y)+(-\Upsilon_4 z^3)\theta_y(x,y) \\ W &= w_1(x,y) \end{aligned} \tag{1}$$

In Eq. (1), along x, y, z directions, the displacement occurred in plate given by U, V and W respectively, whereas midplane displacements are u_1, v_1, w_1. ψ_x and ψ_y are the rotations, their variation along the thickness of plate depends on the functions $\kappa_1(z)=(\Upsilon_1 z-\Upsilon_2 z^3)$ and $\kappa_2(z)=(-\Upsilon_4 z^3)$ respectively. Slope along x and y directions represented by θ_x and θ_y respectively, also $\Upsilon_1=1$; $\Upsilon_2=\Upsilon_4=4/3h^2$ in Eq. (1).

The representation of strain vector for the given displacement fields done by,

$$\{\varepsilon\}=\{\varepsilon_x,\varepsilon_y,\gamma_{xy},\gamma_{xz},\gamma_{yz}\}^T \tag{2}$$

The resultant strain vector denoted as follows,

$$\{\varepsilon\}=\{\varepsilon_l\}-\{\varepsilon_{nl}\}-\{\varepsilon_t\} \tag{3}$$

Where, $\{\varepsilon_t\}$ is the thermal strain vector, represented as, $\{\alpha_1\Delta T \quad \alpha_2\Delta T \quad 0 \quad 0 \quad 0\}$

For the kth lamina with composite SMA embedded wires, the relation between strain and stress with the HSDT for arbitrary fibre alignment angle θ, expressed as,

$$\{\sigma\}_k = \left[\bar{Q}\right]_k \{\varepsilon\}_k + \{\sigma_{\Delta r}\}_k Vs_k - \left\{\left[\bar{Q}\right]_m \{\alpha\}_m V_m\right\}_k \Delta T \tag{4}$$

Over the stress free state T, total stress $\{\sigma\}_k$ vector calculated. Due to temperature variation ΔT, the recovery stresses are generated into the SMA which is termed as $\{\sigma_{\Delta r}\}_k$ and for k[th] layer strain $\{\varepsilon\}_k$ vector denoted. In the expression $\left[\bar{Q}\right]_k$ and $\left[\bar{Q}\right]_m$ are the reduced transferred stiffness matrix of composite plate with SMA fiber and composite matrix respectively. $\{\sigma_{\Delta r}\}$ is the recovery stress generated in SMA due to temperature variation ΔT [14]. V_s and V_m are the volume fraction of SMA and matrix respectively.

The buckling equation of the smart composite laminated composite plate expressed as,

$$\left([K_s] - T_{cr}\left[K_g\right]\right)\{\delta\} = 0 \tag{5}$$

$[K_s]$ and $\left[K_g\right]$ are the global stiffness matrix and global geometrical stiffness matrix, and T_{cr} is the critical buckling temperature (CBT).

3. Result and discussions

In present study, along with the Figure 1, following boundary conditions are utilized for SMA embedded laminated composite plate. Figure 1 illustrates the geometry of the composite laminated plate, having length of plate '*a*', width '*b*' and '*h*' is the total height of plate.

Figure 2 shows the convergence study. Convergence of FEA has been be evaluated with the consideration of square laminated nine noded with seven degree of freedom composite plate of ply orientation (45/-45/45/-45) under the boundary condition of clamp. For the considered plate, it is observed that, from 3x3 graph gets converged and so, for further study 4x4 elements was used for the analysis.

Figure 3 shows the thermal critical buckling temperature (CBT) validation with the literature using layerwise theory. For the validation purpose square laminated composite plate with layer orientation of (45/-45/45/-45) under clamped and simply supported boundary conditions utilised. The obtained result shows the same nature of variation and close agreement with the previous available literature, and it is also observed that, T_{cr} is lowest under clamped condition, than the simply supported boundary condition.

Table 1 shows the effect of variation of SMA fiber volume fraction (Vs) over CBT. For the analysis square laminated composite plate [0/S/S/0] considered with clamped all edges. With the increment of SMA volume fraction, the CBT for the plate increases which is validated with the result [13] obtained by FSDT.

From the reference [14] - Fig. 1(a) and (b) show that the Young's modulus E_s and the recovery stress σ_r of SMA vary as the temperature with high nonlinearity. So, in present case Figure 4, with the increment in temperature, the corresponding modulus and recovery stress also get varied. The effect of variation of temperature over nondimensional CBT of clamped square laminated composite plate having aspect ratio of 50 is revealed in Figure 4. For the 5% SMA fiber volume portion, if temperature increases, the critical buckling temperature shows continuous increment of 244% between three layer and two layer, whereas, 20.5% between four layer and three layer respectively.

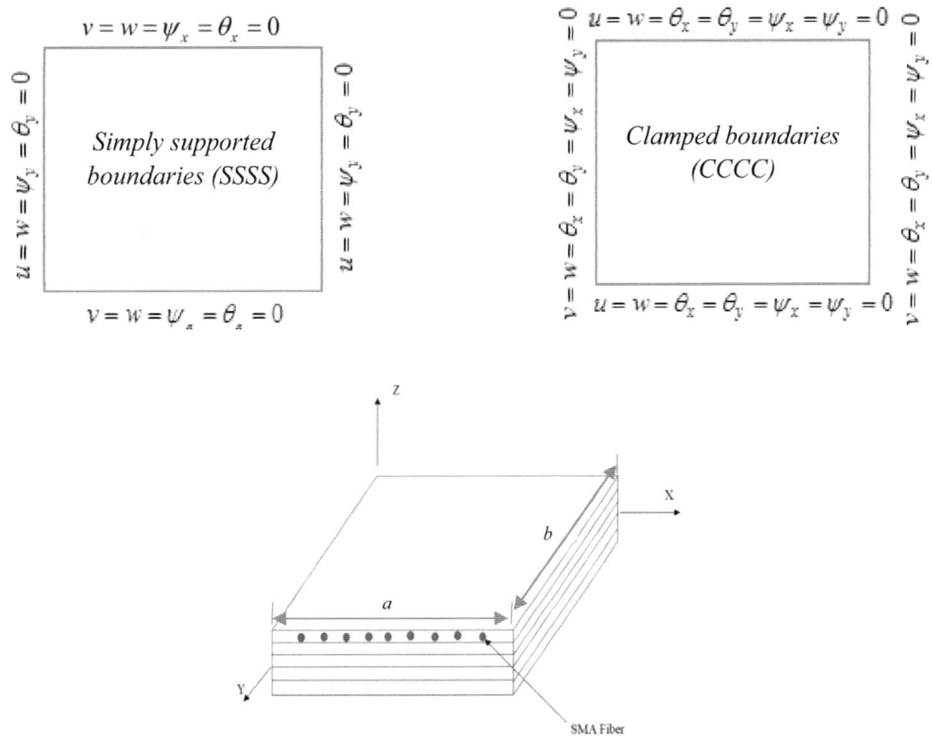

Figure 1. Schematic view of composite laminated plate with SMA reinforced layer

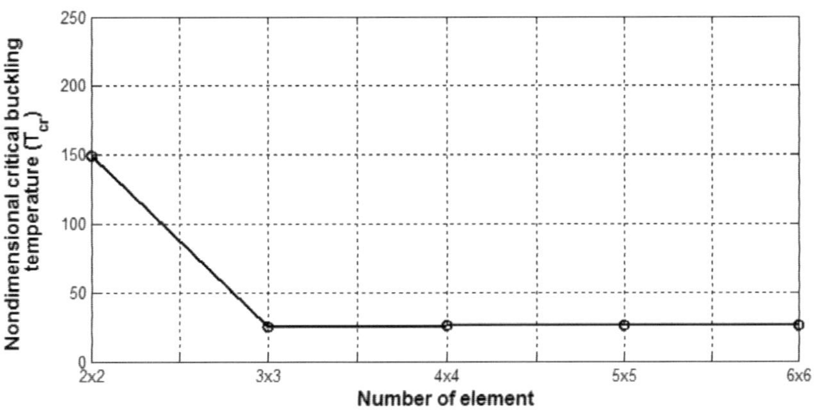

Figure 2. Convergence of square composite laminated plate with aspect ratio (a/h) of 50

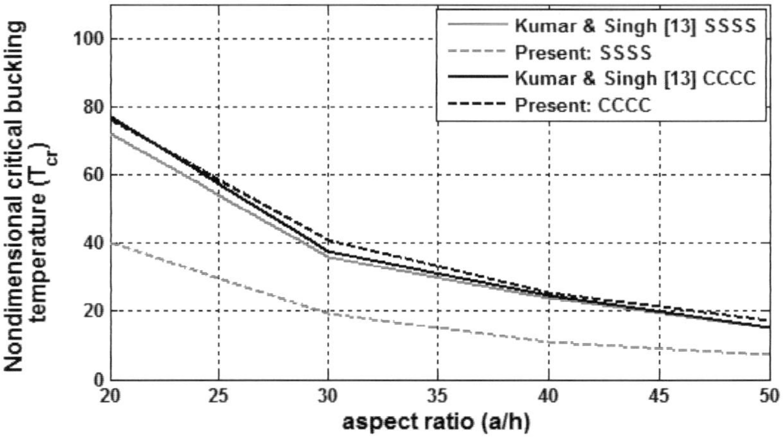

Figure 3. Critical buckling temperature variation with square composite plate aspect ratio

The effect of variation of (α_2/α_1) over CBT of clamped composite plate with stacking sequence of (0/90/90/0) and a/h= 50, shown in the Figure 5. With the increase in the ratio ($\alpha_2/\ \alpha_1$), critical buckling temperature reduces, but with the SMA fiber laminate (0/S/S/0), the value obtained for the CBT is minimum. Along with this the effect of number of layer also plays an vital role, as, these are reduces from four to two, critical buckling temperature is get decreased by 20% and 142% in (0/S/0) and (0/S), with reference to four SMA embedded layer.

Table 1. Effect of shape memory fiber over critical buckling temp with variation in SMA fiber volume fraction

a/h		SMA volume fraction (Vs)				
		0%	5%	10%	15%	20%
100	Present	7.1	9.5	10.5	11.5	12.6
	Ref.[13]	7	8.5	10	11	12
50	Present	25.6	32.6	37	41.1	44.5
	Ref.[13]	20	21.6	23.6	31	34
20	Present	106.5	127.1	153.35	178	203.3
	Ref.[13]	110	124	140	158	178

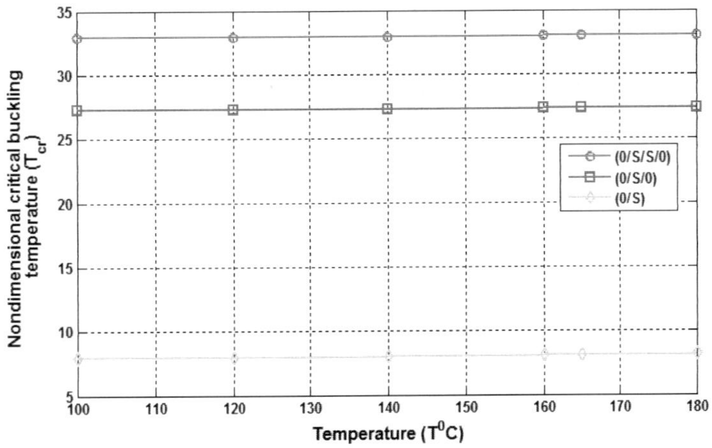

Figure 4. Effect of temperature over critical buckling temperature

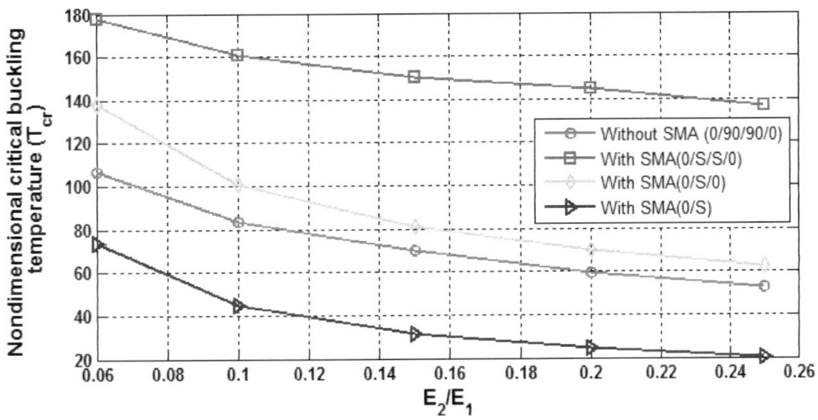

Figure 5. Effect of variation ($\alpha_2/\ \alpha_1$) over CBT of composite plate

4. Conclusions and future scope

Using FEM technique utilising HSDT the thermal analysis of composite laminated plate and composite sandwich plate means SMA embedded plate for buckling analysis is carried out. The effect of temperature variation, aspect ratio and ratio of coefficient of thermal expansion deviation is observed in the current work. From the study, it is witnessed that, as the SMA volume fraction added or increased in composites, the critical buckling load/temperature is get increased, in even temperature application. Based on the available literatures, it is clearly observe that, temperature variation is one of the important affecting factor in shape memory effect which is in alloys or in polymers. Based on that, researchers can simulate the shape memory effect in SMA as well as shape memory polymer simultaneously.

References

[1] Jeanette Epps and Ramesh Chandra 1997 *Smart Mater. Struct.* **6** 251
[2] Baz A, Poh S, Ro J, Mutua M, Gilheany J 1992 *Intelligent Structural Systems* **13** 169-212
[3] Abbas M K, Elshafei M A, Negm H M 2015 *International Conference on Applied Mechanics and Mechanical Engineering* **79** 1-30
[4] Prabhu M R, Dhanaraj R 1994 *Computers and structures* **53** (5) 1193-1204
[5] Ounis H, Tati A, Benchabane A 2014 *Front. Mech. Eng.* **9** 41
[6] Zhang Y X, Yang C H 2009 *Composite Structures* **88** 147
[7] Chen W J, Lin P D, Chen LW 1991 *Computers & Structures* **41** 637
[8] Chen W J, Lin P D, Chen LW 1991 *Composite Structures*, **18** (4) 379–397
[9] Huang N N, Tauchert T R 1992 *Thin-walled Structures* **13** (4) 259–273.
[10] Noor A K, Peters J M 1992 *Journal of Engineering Mechanics* **118** 351.
[11] Noor A K, Peters J M 1994 *Finite Elements in Analysis and Design* **15** 343.
[12] Noor A K, Starnes J H, Peters J M 1994 *AIAA Journal* **32** 1507
[13] Kumar C V, Singh B N 2009 *International Journal for Computational Methods in Engineering Science and Mechanics* **10** (6) 423-9
[14] Park J S, Kim J H, Moon S H 2005 *Compo. Part B* **36** 627-636
[15] Li S R, Yu W S, Batra R C 2010 *J Therm Stresses* **33** 79
[16] Panda S K, Singh B N 2009 *Composite Structures* **91** 366
[17] Xia X K, Shen H S 2008 *Engineering Structure* **30** 2420
[18] Tawfik M, Ro J J, Mei C 2002 *Smart Materials and Structures* **11** **(2)** 297
[19] Gadade A M, Lal A, Singh B N 2016 *Aerospace Science and Technology* **55** 227

IOP Conf. Series: Materials Science and Engineering **814** (2020) 012010 doi:10.1088/1757-899X/814/1/012010

Tribological characteristics of as-cast A356/10wt.% SiC_p functionally graded composite dry sliding against EN-31 steel under unidirectional and oscillatory modes

J Prem Kumar[1], V R Rajeev[1], K Krishnakumar[2] and M Hashim[3]

[1]College of Engineering Trivandrum, Kerala, India-695016
[2]Sree Vellappally Natesan College of Engineering, Kayamkulam, Kerala, India-690503
[3]Government Engineering College, Wayanad, Kerala, India-670644

E-mail: rajeevcet@gmail.com

Abstract. A comparison in tribological response characteristics of as-cast Al-SiCp functionally graded composite material (FGM) dry sliding against hardened EN-31 steel under unidirectional and oscillatory sliding modes was investigated for a sliding/oscillating velocity of 0.4m/s slid through 500m utilizing a pin-on-disc configuration. The A356/10 wt.% SiCp FGM was prepared using vertical centrifugal casting technique. The microstructure characterization was done by Leica optical microscope, and the microstructural analysis revealed the presence of different zones due to the radial gradient distribution of SiCp reinforcement. The hardness variation was found to have a good correlation with its microstructural features at different zones of FGM. Wear loss against load curve indicated an increase in trend for both sliding modes, whereas the coefficient of friction versus load plot showed a decreasing trend for both cases under an identical set of operating parameters. Further, oscillatory wear loss of FGM/EN-31 steel tribopair was observed to be higher in comparison with unidirectional sliding mode for the entire load range, (30-150N) investigated. At the same time, the coefficient of friction of tribopair in oscillatory sliding mode was found to lower than the unidirectional mode.

1. Introduction

The ceramic particulate strengthened functionally graded composite materials (FGM) has been the excellent contender for the automotive parts which include piston, piston rings, gears, and brake pads/rotors, because of its enhanced strength to weight ratio and wear resistance [1,2]. The use of composite materials with tailored properties known as FGM has nowadays become popular in automobile, aerospace and biomedical sectors. These materials are produced to attain gradient in properties such as strength, wear resistance, high-temperature resistance and corrosion across its geometry, thereby weight savings could be resulted compared to metal matrix composites. Most of the previous studies reported in the literature have shown that dry sliding wear tests for FGMs were investigated utilizing pin-on-disc and block on ring tribometers wherein the relative motion of tribopair is only in one direction (unidirectional). E Jayakumar et al. 2016 [3] investigated wear characteristics of A319-SiC_p FGM using DUCOM pin-on-disc tribometer and reported that the wear resistance of the particle rich zone (outer region) is higher than matrix rich (inner) zone. Zhang et al.

1997 [4] reported the transition from mild to severe wear characteristics of Al-Si alloys using block-on-ring tribotester. The authors noticed three different forms of transitions, namely, load, sliding velocity and sliding distance induced transitions. Recently, R Jojith et al. 2019 [5] investigated the friction and wear characteristics of heat-treated Al-7Si/B_4C FGM dry sliding against EN-8 steel using a pin on plate tribometer (DUCOM) as a function of load and sliding distance. The authors reported that the heat-treated outer zone of FGM having better wear resistance than as-cast composite. Studies reported on relative friction coefficient, and the wear resistance of these classes of materials in unidirectional and bidirectional/oscillating sliding modes are quite scanty [6-8]. P J Blau et al. 2000 [9] reported friction coefficent and wear characteristics of Mg alloy prepared by two different techniques tested using both unidirectional and reciprocating sliding contacts. The authors reported a higher friction coefficient for the sliding pair in unidirectional mode compared with reciprocating mode.

Etsuo Marui et al. 2001[10] reported the friction and wear characteristics in unidirectional and bidirectional sliding motion tests of Cu on steel. The authors pointed out a weak point in the pin-on-flat tribotester that sufficiently large velocity was difficult to achieve and could reach a maximum sliding speed of 1.11 cm/s only. In the present paper, it was planned to investigate the comparative study on the effect of friction and wear characteristics of as-cast A356/10wt.% SiCp functionally graded composite dry sliding against EN-31 steel under unidirectional and oscillatory modes using DUCOM make pin-on-disc tribometer with a sliding velocity of 0.4m/s.

2. Materials and methods

The material utilized in this investigation was hypo-eutectic Aluminium-Silicon alloy (A356) as the matrix for the preparation of functionally graded composite with 10wt.% SiC_p reinforcement. The composite melt was prepared through stir casting route followed by vertical centrifugal casting technique at a mould speed of 1300 rpm. The size of the prepared FGM disc was having an outer and inner diameters of 240mm and 60mm respectively with thickness 28mm. Microstructure characteristics of the prepared FGM were captured using a Leica optical microscope, and its correlation with hardness data was published elsewhere [11]. Wear test pins, 6 mm diameter whose axis was radial to the cast FGM disc to get particle rich and matrix rich zones at two ends of the pin, were cut from the disc using wire cut EDM machine. The pins were solution heat-treated at 500^0C for 6 hours, followed by water quenching at room temperature. Precipitation hardening was done at 200^0C for 12 hours, followed by furnace cooling to room temperature.

Comparison wear study of prepared FGM pin samples in unidirectional and oscillatory modes was performed under identical sliding conditions using DUCOM TR-20LE pin-on-disc tribometer. The same counter body disc of EN-31 steel was used for both unidirectional and oscillatory tribo testing modes since it is harder than the softer FGM pin. The steel disc was hardened to 55 HRC, ground and polished to a surface roughness value of 0.4μm (Ra). The pin was held stationary and pressed against the rotating/oscillating disc for conducting the wear test in both sliding modes. A sliding distance of 200 mm per cycle of the disc was selected for conducting wear test in oscillating tribotester so that a mean wear track radius calculated as 64mm used with a sector angle of 90^0 Figure 1. The same sliding distance per cycle was also fixed in unidirectional tribotester for comparing the wear characteristics of FGM in two different sliding modes. Accordingly wear track radius in unidirectional sliding condition was calculated as 32mm. Table 1 shows the test types and sliding conditions of FGM/EN-31steel tribopair. Wear tests were conducted in two sliding modes as a function of loads (30-150N) with a constant oscillating/sliding speed of 0.4m/s slid through 500m. The difference in weight of the wear test pin before and after testing was measured by using Shimadzu micro weighing balance having a resolution of 0.01 mg and converted to wear rate in mg/m. Bellows type load cell sensor attached with the tribotester was utilized for frictional force measurement.

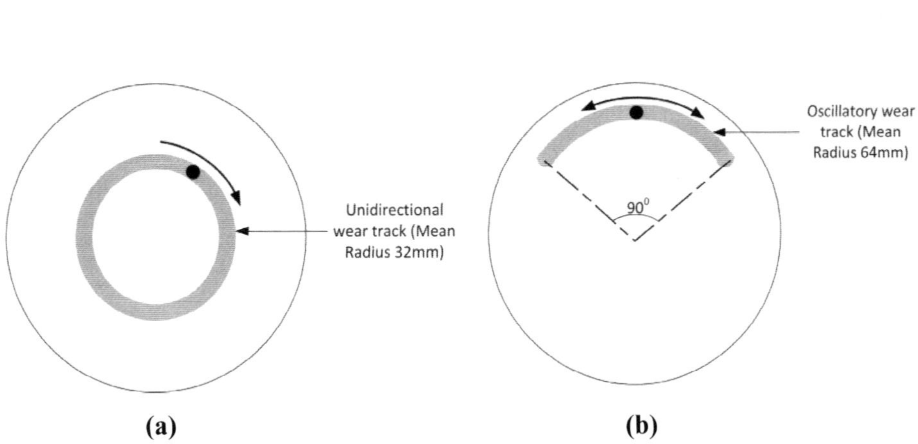

Figure 1. Schematic diagram of (a) Pin-on-rotating disc configuration (b) Pin-on-Oscillating disc configuration

Table 1. Test types and sliding conditions of FGM/EN-31steel tribopair

Operating parameter	Pin-on-Rotating disc	Pin-on-Oscillating disc
Speed/frequency	2 revolution/s	2 cycle/s
Sliding speed/Oscillating velocity	0.4 m/s	0.4 m/s
Mean Wear track radius	32 mm	64 mm
Sliding distance/cycle	200 mm	200mm

3. Results and discussion

A comparison study on dry friction characteristics of A356/10wt.%SiC_p FGM (outer zone cylindrical pin) dry sliding against EN-31 hardened steel disc was performed in unidirectional and oscillating sliding modes Figure 1, under similar conditions of operating parameters using DUCOM-TR20LE tribometer. The effect of load on the frictional characteristics of FGM was discussed as follows.

3.1 Effect of load on dry friction characteristics of FGM in two sliding modes

The Figure 2 showed the frictional behaviour of FGM against load with an oscillating/sliding velocity of 0.4m/s slid through 500m in oscillating and unidirectional sliding modes. It indicates that as the load increases friction coefficient of FGM was found to decrease in both sliding modes and is attributed to increased tribolayer thickness upon the increase of load causing an increase in wear rate, but the decrease in friction coefficient [12]. It was also noticed that the friction coefficient of FGM in unidirectional contact was found higher than oscillating contact in the selected conditions investigated with load in the range of 30-150N. The reason for the higher friction coefficient in unidirectional contact is due to the deposits of ploughed material ahead of the slider for a more extended period compared to that in oscillating contact. P J Blau et al. [9] also noted the same behaviour that friction coefficient of die-cast and Thixo moulded AZ91D alloys dry sliding against 440C stainless steel counter face in unidirectional motion test was found higher than in reciprocating motion test due to the build-up of leading-edge material ahead of the sliding pin for a more extended period. But in reciprocating test chances are there that this leading-edge material to expel at the ends of the reversing stroke. Figure 3(a-d) show continuous data of friction coefficient versus time plot was taken at low load (30N) and high load (150N) respectively in the case of unidirectional and oscillating contacts. It was clear from Figure 3(a, b) that the steady-state friction coefficient value is decreasing from 0.6 to

0.25 at low and high load, respectively, in the case of unidirectional contact. But in oscillating contact, it was found to decrease from 0.4 to 0.1 Figure 3(c, d).

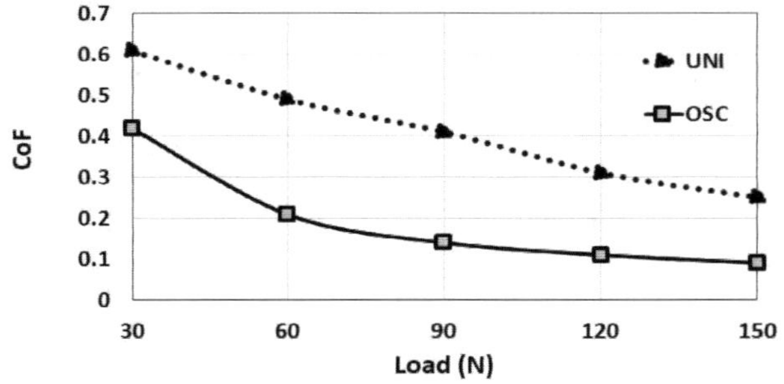

Figure 2. Load against Coefficient of friction (CoF) plot for unidirectional and oscillating wear tests under a test load range of 30-150N loads with a sliding speed/velocity of 0.4m/s slid through 500m.

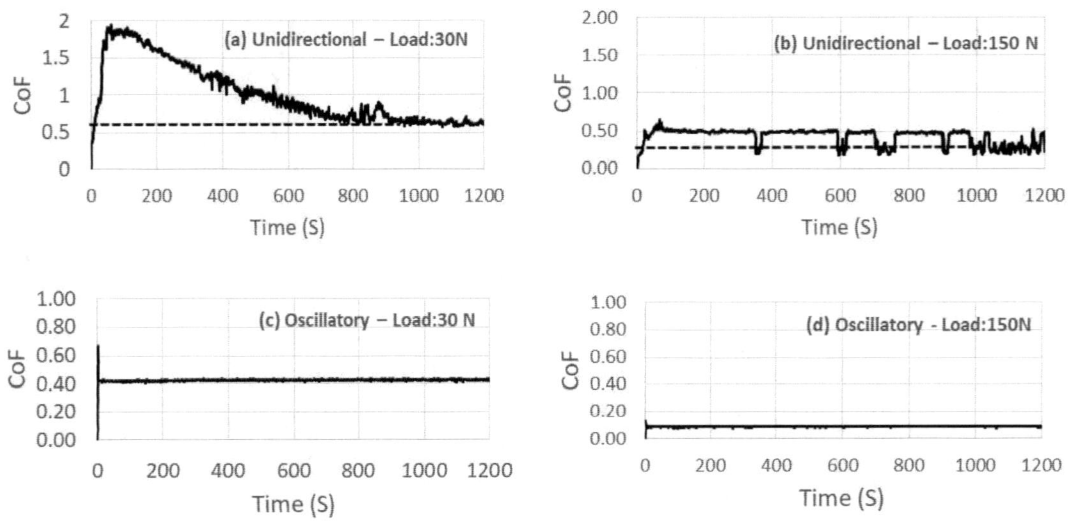

Figure 3. Continuous Coefficient of Friction plots for unidirectional and oscillating wear tests under 30N and 150 N loads with a sliding speed/velocity of 0.4m/s slid through 500m

3.2 Effect of load on dry wear characteristics of FGM in two sliding modes

Figure 4 shows wear characteristics of FGM against load with an oscillating/sliding velocity of 0.4m/s slid through 500m in oscillating and unidirectional sliding modes. It was observed from Figure 4 that as the load increases, the wear rate also increases in both sliding modes and is attributed to increased metallic intimacy due to higher load [13]. It was also noticed that the oscillating wear rate of FGM was found higher compared to unidirectional wear rate under similar operating parameters investigated in the study. This could be attributed to the fact that the additional probably possibilities of entrapment of wear particles within the oscillatory modes compared to the unidirectional mode of sliding that in turn successively roll and slide between the sliding interface and cause three-body abrasion. R Ward [6] also noted the same behaviour that wear rate of steel in reciprocating contacts was higher than in unidirectional contact. The author connected the explanation of it to two main differences; surface stressing and quantity of remaining wear particle within the wear track getting entrapped between the sliding interfaces throughout contact; each possibility of prevalence are comparatively high in reciprocal sliding.

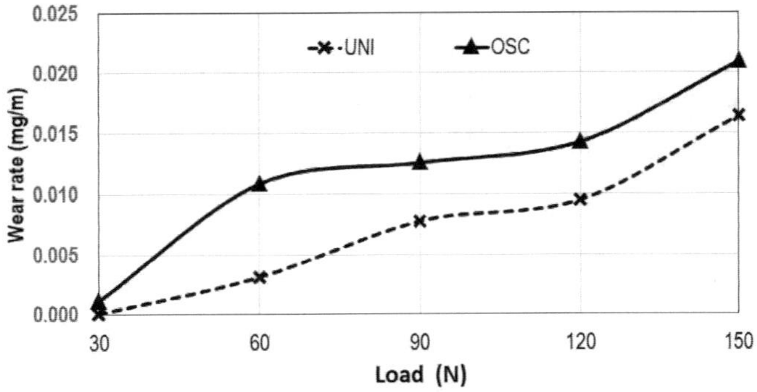

Figure 4. Load vs wear rate plots for unidirectional and oscillating wear tests under a test load range of 30-150 N with a sliding speed/velocity of 0.4m/s slid through 500m.

3.3 Worn out surface features of pin specimen by optical microscope

Figure 5(a-d) represent the stereo micrographs of worn-out surfaces of pin specimen after dry sliding in unidirectional and oscillating contacts with a sliding speed/velocity of 0.4m/s slid through 500m at a load of 30N and 150N respectively. It was observed from Figure 5(a, b) that as the load on the pin increases from 30N to 150N lip growth was found to occur at the trailing edges of the wearing surfaces proportional to the load in unidirectional contact. Since the leading and trailing edges of the sliding pin interchanging every cycle in oscillating contact, the lip growth likely to occur at both ends. However, the motion of the disc is oscillating, and hence the lip growth in the sliding direction is suppressed, and the same behaviour was also reported by Etsuo Marui et al. [10].

Figure 5. Stereo microscopic images of worn-out pin surfaces sliding in two different modes.
(a) Unidirectional-30N
(b) Unidirectional-150N
(c) Oscillatory-30N
(d) Oscillatory-150N

Figure 6 (a-d) shows the worn-out surface morphology of FGM outer pin specimen dry sliding in unidirectional and oscillating contacts at a load of 30N and 150N. It was observed from Figure 6(a, c) that at minimum load condition presence of fine, shallow scratches and grooves in sliding direction along with delamination at some locations happened in both sliding contacts. But at maximum load condition, SiC particle pullout along with heavy delamination was observed to be predominant in oscillating contact compared to unidirectional Figure 6(b, d).

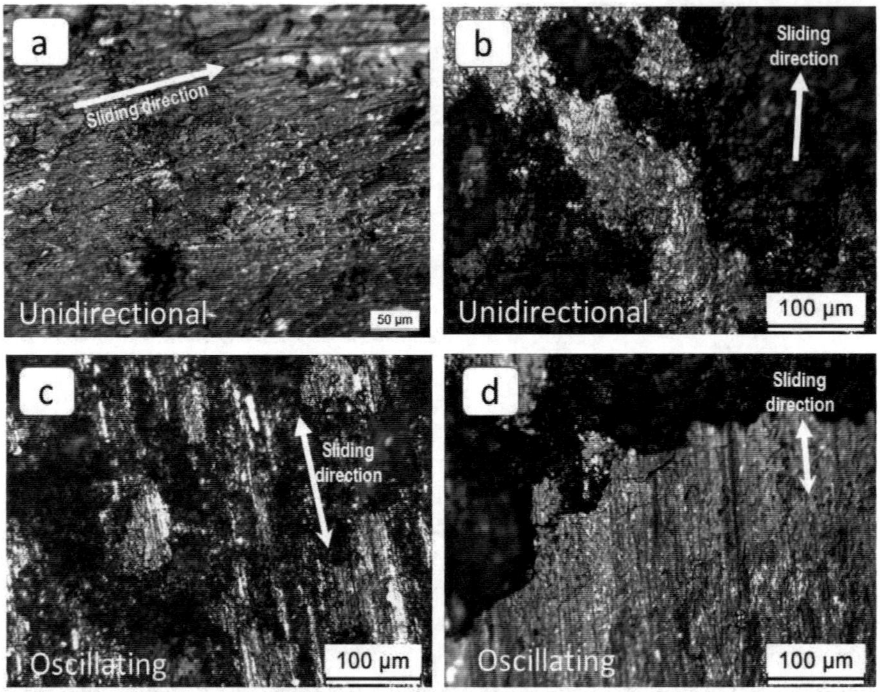

Figure 6. Optical images of worn-out pin surfaces captured at a sliding speed/velocity of 0.4m/s tested in two different sliding modes (a) Unidirectional-30N, (b) Unidirectional-150N, (c) Oscillatory-30N and (d) Oscillatory-150N

4. Conclusion

In the present study, the effect of sliding motion on the friction and wear characteristics of as-cast A356/10 wt.% SiC_p FGM was attempted using pin-on-disc tribometer in unidirectional and oscillating modes. Based on the investigation, the important findings are as follows.

The friction coefficient of as-cast A356/10 wt.% SiC_p FGM tribopair in unidirectional and oscillating sliding modes were found to decrease upon the increase of load from 30N to 150N. The reason is attributed to increased tribolayer thickness upon the increase of load and causing a decrease in friction coefficient.

The steady-state friction co-efficient captured in unidirectional mode was found to decrease from 0.6 at low load to 0.25 at high load. But in the oscillating mode, it was decreasing from 0.4 at low load to 0.1 at high load. Here the unidirectional friction coefficient of FGM was higher compared to that in oscillating mode within the load range investigated.

Lip growth was started to form during the wearing process of FGM tribopairs in both sliding modes except the difference is that it occurred at the leading and trailing edges of sliding pins of oscillating contact compared to that formed only at the leading edge of unidirectional contact.

The wear rate of FGM tribopair in both sliding modes found to increase upon normal load in the range of 30N to 150N. At low load a wear rate of 0.001mg/m while at high load 0.021mg/m noted in the oscillating sliding mode which was higher when compared with unidirectional sliding wherein at low load wear rate of 0.00004mg/m and high load 0.016 mg/m only noticed.

5. Acknowledgement

The authors acknowledge the fund received from the Centre for Engineering Research and Development (CERD), A P J Abdul Kalam Technological University, Govt. of Kerala, India.

References

[1] Vieira A C, Sequeira P D, Gomes J R and Rocha L A 2009 *Wear* **267** 585.
[2] Rajan T P D and Pai B C 2009 *Trans. of Indian Inst. of Metals* **62** 383.
[3] Jayakumar E, Jibin C J, Rajan T P D, Joseph M A and Pai B C 2016 *Metallurgical and Materials Trans.* **A 47(8)** 4306.
[4] Zhang J and Alpas A T 1997 *Acta Metallurgica* **45** 513.
[5] Jojith R and Radhika N 2019 *Silicon.* https://doi.org/10.1007/s12633-019-00294-3
[6] Ward R 1970 *Wear* **16**. 423.
[7] Odabas D and Su S 1997 *Wear* **208** 25.
[8] Marcia M M and Deniol K T 2006 *J. of the Braz. Soc. of Mech. Sci. & Eng:* **28** 278.
[9] Blau P J and Walukas M 2000 *Tribology International* **33** 573.
[10] Etsuo Marui and Hiroki Endo 2001 *Wear* **249** 582.
[11] Prem Kumar J, Rajeev V R, Krishnakumar K and Hashim M 2019 *10th Int. Conf. on Industrial Tribology (IndiaTrib 2019 Bangalore, India)* TSI 19219.
[12] Uyyuru R 2007 *Tribology International* **40** 365.
[13] Sannino A P and Rack H J 1995 *Wear* **189** 1.

IOP Conf. Series: Materials Science and Engineering **814** (2020) 012011 doi:10.1088/1757-899X/814/1/012011

Stress analysis of composite plate with cutout of various shape

Achchhe Lal[1], B.M Sutaria[2], Rahul Kumar[3]

[1]Assistant Professor, Mechanical Engineering Department, SVNIT, Surat
[2]Associate Professor, Mechanical Engineering Department, SVNIT, Surat
[3]Research Scholar, Mechanical Engineering Department, SVNIT, Surat

E-mail: lalachchhe@yahoo.co.in

Abstract. Composite plate found numerous applications in various structures due to its superior properties like high stiffness to weight and strength to weight ratio. Generally, cutouts in these structural elements are unavoidable for many purposes like access to internal parts, ventilation and sometimes to prevent resonance. These geometrical discontinuities in the form of cutouts may be of different shapes and arise nonuniform stress in the plate. Therefore, it becomes necessary to analyse stress concentration factor near cutouts and effect of shape of cutouts on axial deformation and stress distribution for better safety and stability of structure. In the present investigation a Carbon/epoxy composite plate with various shape of cutout subjected to in-plane tensile loading is considered. Finite element based plane stress analysis is used for mathematical modeling. A program for the present analysis is developed in MATLAB. Variation of in-plane displacement and stress concentration factor (SCF) with various shape of cutout is determined. The effect of the geometry of hole on the stress distribution around the hole is also studied. The results of various analyses have been presented with the help of tables, graphs, and nodal plots.

1. Introduction

Composite plates as a structural member finds its application in various fields like automobile, space vehicle, submarines and many others where high strength and stiffness to weight ratio is desired. Sometimes holes and cutouts of various shapes are unavoidable in these type of plates for their proper functioning. These structural members usually subjected to in-plane tensile loading. With geometrical discontinuities these composite plates with holes are more vulnerable due to rises of stress near discontinuities (hole) which results in SCF near hole. Therefore, analysis of stress near holes for better safety and to evaluate load bearing capacity of these composite plates are essential. Many research works are reported for investigation of SCF of composite plates with holes of various shapes.

Investigation for SCF of an isotropic rectangular plate with circular hole subjected to uniform tensile load is done analytically and using Finite element method is done by Babulal et al. [1]. Influence of hole diameter to plate width ratio (D/A) on stress distribution is studied. Approximate solution for normal stress distribution in the form of polynomial is presented by Konish et al. [2] and found good agreement for SCF. Saddique and Mirzana [3] presented a review investigation of stress analysis of composite and isotropic plate with circular hole with finite element simulation using Ansys and reveal the effect of fiber orientation on deformation, Von Mises Stress and SCF. Pan et al. [4] studied stress distribution and their influencing parameters in a finite plate with rectangular hole using Muskhelishvili's complex variable method and compared the results with those obtained from finite element simulation. Effect of geometry of elliptical hole and its aspect ratio on stress distribution in a plate is studied by Gunwant and Singh [5] using Ansys and deflection of plate with Von-Mises stress are determined. Mekalke et al. [6] used finite element method to investigate stress and displacement distribution in a rectangular plate with cylindrical hole subjected to uniform stress. Variation of results are studied using various meshes in Ansys. Classical laminated plate theory is used to determine stress distribution around an elliptical hole in a finite composite plate and effect of various parameters like hole size, stacking sequence and loading conditions on SCF are studied by Xiwu et al. [7]. Ogonowski [8] investigated the influence of finite geometry of isotropic and anisotropic plate with hole on stress distributions and compared with the known solution. Alhazmi et al. [9] studied stress distribution in a carbon fiber reinforced composite plate with open and bolt-loaded holes subjected to tensile and fatigue loading experimentally and numerically. SCF of orthotropic composite plates with a circular hole subjected to tensile loading is measured experimentally and effect of various parameter on strain and SCF is revealed by Tan and Kim [10]. Isotropic analysis, orthotropic analysis and finite element analysis is done by Park and kim [11] for comparative study of SCF in fiber reinforced plastic composite plate with central circular hole. Wu and Mu [12] found application of Scale Factors (SFs) to determine SCF of isotropic plate under biaxial loading and cylinder under uniaxial loading for suitable circular hole and compared results with that obtained with finite element simulation.

From the above literature it is observed that displacement and stress distribution of a composite plate with hole of different shapes are studied separately. To the best of author's knowledge there is no research work has been reported on comparative study of effect of different shapes of hole on displacement and SCF by keeping hole area constant. Therefore, in the present study stress analysis of carbon/epoxy composite plate with circular, square and rectangular hole by keeping cutout area constant subjected to uniaxial tensile loading is considered. Figure 1(a-c) shows the composite plate with hole of rectangular, square and circular shape under in-plane uniaxial tensile loading.

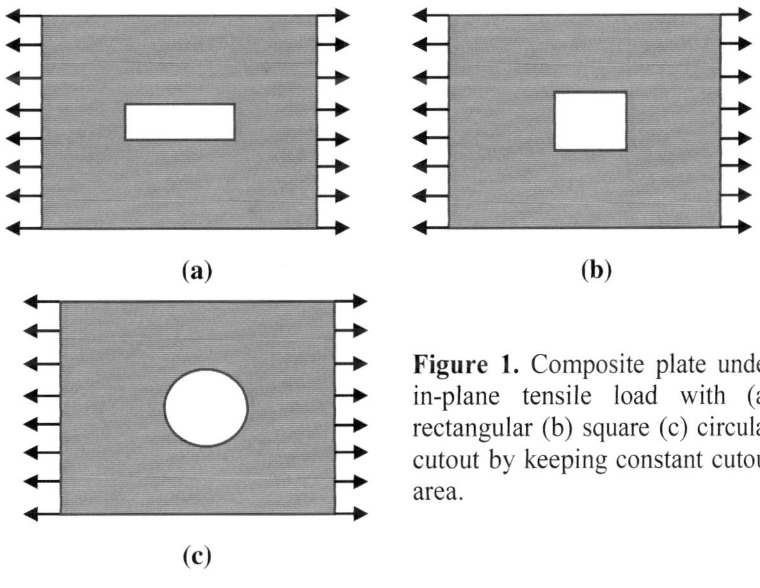

Figure 1. Composite plate under in-plane tensile load with (a) rectangular (b) square (c) circular cutout by keeping constant cutout area.

2. Mathematical Formulation

A MATLAB program is developed for plane stress analysis of carbon/epoxy composite plate with hole using four noded quadrilateral element with two degree of freedom. Material properties of composite plate considered for the present study is given in Table 1.

Table 1. Material properties of Carbon/epoxy

E_1 (GPa)	E_2 (GPa)	$\nu_{12}=\nu_{21}$	ν_{23}	G_{12} (GPa)
82.64	9.8034	0.3	0.0356	3.8545

Rectangular plate of 20 cm length, 10 cm width and 0.1 cm thickness with different shape of hole is considered unless otherwise specified for particular results. Uniformly distributed In-plane tensile load 5 KN is applied. The constant area of hole of different shape is $4\ \text{cm}^2$. Dimensions of holes are given in the Table 2.

Table 2. Dimensions of hole of different shape.

Shape of hole	Dimensions (in cm)	Area of Hole (in cm^2)
Circular	radius=1.128	4
Square	2x2	4
Rectangular	4x1	4

2.1. Displacement Field

The displacement field for the present study is given as

$$U_1 = u(x,y) \\ U_2 = v(x,y) \tag{1}$$

where u and v are displacements in x and y direction respectively.

2.2. Strain-Displacement Relations

Strain in respective direction can be determined by differentiating displacement field and can be written as,

$$\varepsilon_x = \frac{\partial u}{\partial x}, \qquad \varepsilon_y = \frac{\partial v}{\partial y}, \qquad \varepsilon_{xy} = \frac{\partial u}{\partial y} + \frac{\partial v}{\partial x} \tag{2}$$

2.3 Stress-Strain Relations

Stress in x-, y- direction and in xy-plane can be evaluated using reduced constitutive matrix and Eq.1 as,

$$\begin{bmatrix} \sigma_x \\ \sigma_y \\ \sigma_{xy} \end{bmatrix} = \begin{bmatrix} \overline{Q} \end{bmatrix} \begin{bmatrix} \varepsilon_x \\ \varepsilon_y \\ \varepsilon_{xy} \end{bmatrix} \tag{3}$$

Where $\begin{bmatrix} \overline{Q} \end{bmatrix}$ is reduced constitutive matrix.

Boundary condition applied in the present study is given as,

$u = 0 \qquad at \;\; x = 0$

2.4 Finite Element Analysis

Displacement vector can be given as,

$$U^e = \begin{bmatrix} u_1 & v_1 & u_2 & v_2 \dots u_n & v_n \end{bmatrix}^T \tag{4}$$

$$U_1 = \sum_{i=1}^{N} N_i^e u_i \;\; and \;\; U_1 = \sum_{i=1}^{N} N_i^e v_i \tag{5}$$

Where, N_i^e is elemental interpolation function.
Strain vector can be given as

$$\{\varepsilon\} = [B]\{U^e\} \tag{6}$$

Where, $[B]$ is strain displacement matrix.
Elemental stiffness matrix using 2 points Gauss integration method can be written as

$$[K^e] = \int_V t[B]^T[\bar{Q}][B]dV \tag{7}$$

where, t= thickness of plate.
Nodal force vector in the absence of traction forces can be written as,

$$\{f\} = \int_V t[N]^T P dV \tag{8}$$

Governing equation can be written as,

$$\{F\} = [K]\{U\} \tag{9}$$

Stress Concentration Factor (SCF): It is the ratio of stress of a point on the circumference of hole to the stress at same point when there is no hole and can be given as,

$$SCF = \frac{\text{Stress at a point on the circumference of hole}}{\text{stress at the same point when there is no hole}}$$

3. Results and Discussions

Using developed MATLAB program for the present analysis distribution of displacement and stress along x-direction is investigated. Maximum displacement and SCF are also evaluated for different fiber orientations and thickness of plate.

Table 3. Shows validation study for maximum displacement U_x and SCF. Good agreement of present results with already published results are observed.

Table 3. Validation of maximum displacement and SCF for carbon/epoxy plate with central circular hole for 0^0 fiber orientation

Parameters	Present Results	Published Results [1]
U_x (in mm)	2.43	2.71
SCF	2.1435	1.9115

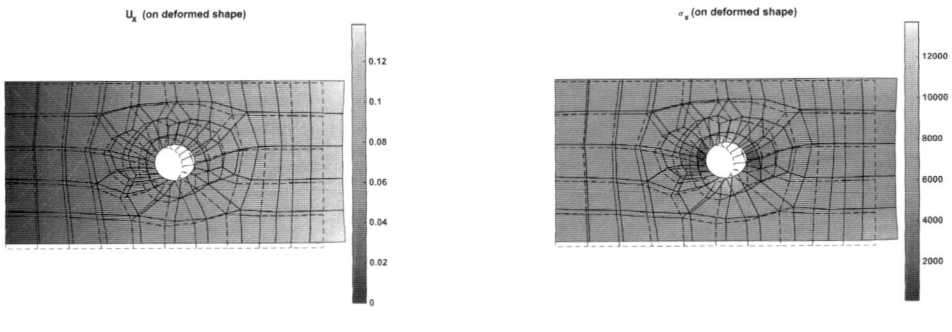

(a) (b)

Figure 2. Shows distribution of (a) displacement and (b) Stress on deformed shape of plate with 90^0 of fiber orientation for plate with circular hole.

IOP Conf. Series: Materials Science and Engineering **814** (2020) 012011 doi:10.1088/1757-899X/814/1/012011

It is well known fact that fibers have highest strength along its longitudinal direction which is observed from the Table 4 as orientation of fiber shifted from 0^0 to 90^0 maximum displacement increases. The variation of displacement with fiber orientation follows almost same trend in all shapes of hole due to equal cutout area. Maximum displacement is in case of plate with square hole followed by that of plate with circular hole. The maximum displacement in plate with square hole is approx. 1.64 % greater than that in case of plate with rectangular shape of hole.

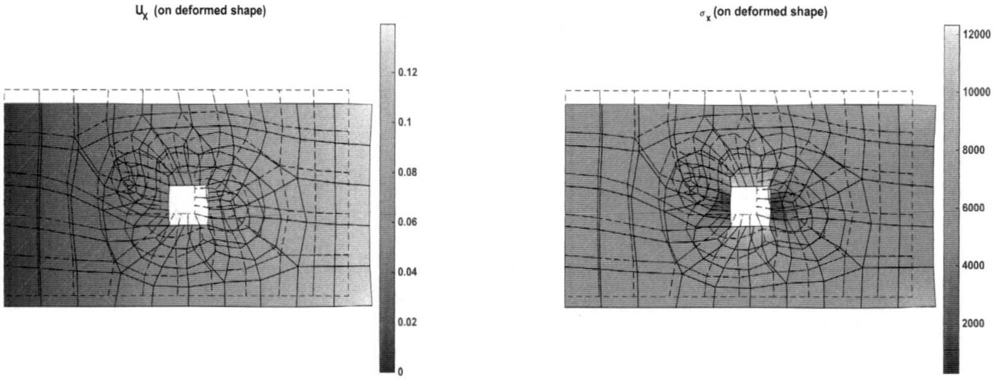

(a) (b)

Figure 3. Shows distribution of (a) displacement and (b) Stress on deformed shape of plate with 90^0 of fiber orientation for plate with square hole.

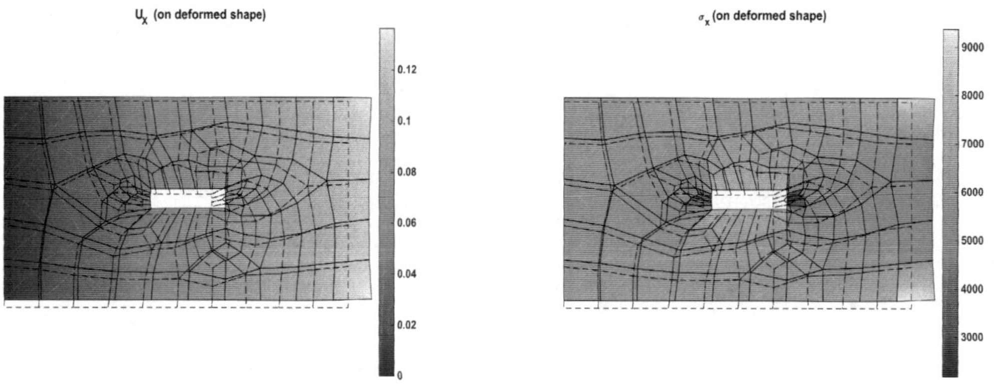

(a) (b)

Figure 4. Shows distribution of (a) displacement and (b) Stress on deformed shape of plate with 90^0 of fiber orientation for plate with rectangular hole.

In case of SCF the plate with circular hole shows maximum value and with the increase in the fiber orientation SCF decreases as resisting force per unit area of the plate decrease and which results in increment in displacement. The plate with circular hole shows 63.6 % more SCF than that of plate with rectangular hole.

Maximum stress along x-direction for plate with various shape of hole by keeping equal cutout area are shown in Table 5. It is observed that with the increase of fiber orientation stress decrease up to 45^0 of fiber angle while after that 90^0 it shows minimum stress which is due to weaker of fiber along transverse direction. Plate with circular hole shows maximum axial stress followed by plate with square hole. Plate with circular hole shows 63.73 % higher axial stress than that of plate with rectangular hole.

Table 4. Variation of deformation and stress concertation factor with fiber angle.

Types of cutout	$U_{x,max}$ 0^0	30^0	45^0	60^0	90^0	SCF 0^0	30^0	45^0	60^0	90^0
Plate with circular hole	0.0184	0.1327	0.1372	0.0485	0.1387	3.6487	2.7828	2.7208	3.8137	2.7392
Plate with square hole	0.0185	0.1332	0.1376	0.0486	0.1393	2.8601	2.4888	2.5716	3.3232	2.4642
Plate with rectangular hole	0.0182	0.1303	0.1350	0.0485	0.1364	2.2291	1.8646	1.9186	2.4329	1.8749

Table 5. Variation of maximum axial stress with fiber angle

Types of cutout	Maximum axial stress ($\sigma_{x,max}$ in N/m^2)				
	00	300	450	600	900
Plate with circular hole	1.824e+04	1.391e+04	1.360e+04	1.906e+04	1.369e+04
Plate with square hole	1.430e+04	1.244e+04	1.285e+04	1.661e+04	1.232e+04
Plate with rectangular hole	1.114e+04	9.323e+03	9.593e+03	1.216e+04	9.374e+03

Table 6. Variation of maximum displacement, maximum axial stress and SCF with thickness of plate with 0^0 fiber orientation.

Thicknes s of plate (in cm)	Plate with circular hole $U_{x,max}$ (in m)	$\sigma_{x,max}$ (N/m^2)	SCF	Plate with square hole $U_{x,max}$ (in m)	$\sigma_{x,max}$ (in N/m^2)	SCF	Plate with rectangular hole $U_{x,max}$ (in m)	$\sigma_{x,max}$ (in N/m^2)	SCF
0.1	0.0184	1.82e+04	3.6487	0.0185	1.430e+04	2.860	0.0182	0.018	2.229
0.2	0.0092	9.12e+03	3.6487	0.0092	7.150e+03	2.860	0.0091	5.572e+03	2.229
0.4	0.0046	4.56e+03	3.6487	0.0046	3.575e+03	2.860	0.0045	2.786e+03	2.229
0.6	0.0031	3.04e+03	3.6487	0.0031	2.383e+03	2.860	0.0030	1.857e+03	2.229
0.8	0.0023	2.28e+03	3.6487	0.0023	1.787e+03	2.860	0.0023	1.393e+03	2.229
1.0	0.0018	1.82e+03	3.6487	0.0018	1.430e+03	2.860	0.0018	1.114e+03	2.229

Variation of maximum axial displacement, stress and SCF with thickness of plate having various shapes of hole are shown in table 6. It is observed that with the increase in thickness of the plate axial displacement increases, maximum axial stress increases in the same proportion and there in no significant variation in SCF due to increment of corresponding nominal stresses.

4. Conclusions

In the present work analysis distribution of axial displacement and stress of composite plate with hole of various shape has been studied. The result of this analysis reveals the influence of fiber orientation and plate thickness on displacement, stress, and SCF when the plate is subjected to in-plane tensile loadings. Based on present analysis and results the following conclusion can be drafted,

1. Plate with square hole is more vulnerable and shows maximum displacement while cutout area of all shape of hole is kept equal.
2. Composite plate having any shape of hole shows minimum displacement with 0^0 fiber orientation when subjected to in-plane tensile loadings.

IOP Conf. Series: Materials Science and Engineering **814** (2020) 012011 doi:10.1088/1757-899X/814/1/012011

3. Plate with circular hole shows maximum SCF and plate with rectangular hole gives minimum SCF.
4. Composite plate with any shape of hole shows maximum value of SCF for 0^0 fiber orientation when subjected to in-plane tensile loadings.

References

[1] Babulal K S Vimalkannan I Pradeep A and Muthukumar P 2015 *IJMPE* **3** 114-117
[2] Konish H J and Whitney J M 1975 *J. Compos. Mater* **9** 157-166
[3] Shaik A S and Mirzana I M 2016 *IOSR-JMCE* **13** 1-5
[4] Pan Z Cheng Y and Liu J 2013 *Int j mech sci* **75** 265-277
[5] Gunwant D and Singh J P 2013 *IJEIT* **3** 387-392
[6] Mekalke G C Kavade M V and Deshpande S S 2015 *JMCE* 25-30
[7] Xiwu X Liangxin S and Xuqi F 1995 *Comput Struct* **57** 29-34
[8] Ogonowski J M (1980) *21st SDM Conf.* 694-698
[9] Alhazmi W H Elhady A A A Mousa S Atta M Sinna A A Bafakeeh O and Sallam H E M 2019 Procedia Structural Integrity **17** 292–299
[10] Tan S C and Kim R Y 1989 *SEM Spring Conference* 345-351
[11] Gao Z Park S J and Kim Y H 2016 Composite Research **29** 1-8
[12] Wu H C and Mu B 2003 *Compos Part B* **34** 127–134

IOP Conf. Series: Materials Science and Engineering **814** (2020) 012012 doi:10.1088/1757-899X/814/1/012012

Seismic investigation of longitudinally aligned shape memory alloy-stainless steel reinforced concrete column

Rahul Pardeshi[1], Bhairav Thakur[2], and Anant Parghi[3]

[1, 2] Research Scholar, Sardar Vallabhbhai National Institute of Technology, Surat, Gujarat, India.

[3] Assistant Professor, Sardar Vallabhbhai National Institute of Technology, Surat, Gujarat, India.

E-mail: rahultpardeshi@gmail.com

Abstract: Shape memory alloy (SMA) is a smart material sustain (8~13%) an enormous amount of strain and having an ability to regain parental shape after the removal of an external load and temperature due to its pseudo-elastic (SE) and shape memory effect (SME) properties. The SE and SME properties, lower modulus of elasticity, and higher elastic straining of SMA compared to conventional steel reinforcement makes it predictable for the design of the reinforced concrete (RC) structure. The present research is focused on the seismic investigation of RC columns reinforced with SMA rebar in the plastic hinge region along with stainless steel (SS) in the remaining portion of the column. A nonlinear static pushover seismic analysis is used for the numerical investigation of SMA-SS RC column. The numerical model is validated having 98% accuracy with existing literature results. The impact of various parameters on SMA-SS RC column is evaluated for SMA in the plastic hinge region. The results reveal that an aspect ratio, the yield strength of reinforcement, the compressive strength of concrete and axial load are significantly affecting the lateral strength and the ductility of the column under seismic loading.

1. Introduction

The national development generally depends on infrastructure facilities and the capacity to bear population load. But in the time of natural disasters like earthquakes or storms, this system may get disturb and its speedy repair necessary to make it original to get it into the flow. Generally, the structures have been cracked, buckled, or eventually collapse due to unforeseen huge energy accumulation subjected to earthquake or storm. The researchers are always trying to minimize this energy accumulation by introducing new innovations. The smart structural system through the utilization of smart materials like shape memory alloy (SMA) helps in controlling the structural response via, energy dissipation and optimization of systems. The response and characteristics of structures can be adjusted using a smart structural system through fortifying qualities like external pressure, moisture, magnetic field, and temperature, etc. In finding solutions to global challenges of sustainability of civil structures, the progress of responsive materials plays a vital role in system performance evaluation, restraining damage propagation, damage detection, and repair, etc. [1].

Shape memory alloy is a smart material sustains (8~13%) [2–5] enormous amount of strain and having an ability to regain parental shape after the removal of an external load due to its pseudo-elastic (SE) property [6]. Also, another distinct shape memory effect (SME) property provides a tendency to reappear an original shape while applying or removing heat. The SE and SME properties in SMA could be utilized to minimize the residual displacement of reinforced concrete (RC) structures subjected to seismic deformation. The behavioral change in the RC structure could be noticeable if the SMA is used as a reinforcement in structure. Lower modulus of elasticity and higher elastic straining with a distinctive yield point of the SMA compared to conventional steel reinforcement makes it predictable for the design of the RC structure.

Crystalline structures or phases of SMA like magneto-strictive or piezoelectric smart materials, changing its states by both external loading and thermal actuation [7]. Two phases of SMA, austenite phases (high temperature and solid material state), and marten site (cooler and softer material state) are known by high and low temperatures, respectively [6]. The reversible switching of phases of SMA and the related alterations (in thermal, mechanical and electric properties of the material) have been used for inventive applications and improvement in civil engineering structures for making them adaptive or smart structures [6]. Irrespective of unique properties, SMA has an initial high cost (US$20–30/kg [8], US $1250/m length of the bar [9]) restricted its utilization. In India, 20 mm, 1-meter length NiTi rod is net costing to Rs. 20,000/-. Generally, some researchers utilized the SMA in high moment concentrated area like plastic hinge region [10–12].

The RC columns are the most susceptible member of a structural system and agonize substantial deformation during an earthquake. The severe damage generally noticed in the plastic hinge region of the column since it experiences huge lateral deformation [13]. So, this area of the plastic hinge is important from the point of view of seismic performance assessment. The present research is focused on the seismic investigation of RC columns reinforced with SMA rebar in the plastic hinge region along with stainless steel (SS) in the remaining region of the column. The nonlinear seismic analysis was used for the numerical investigation of SMA-SS RC column. The numerical model was validated with existing literature results. The impact of various parameters on SMA-SS RC column in the plastic hinge region was evaluated. The parameters like an aspect ratio, the yield strength of reinforcement, the compressive strength of concrete and axial load affecting the lateral strength, failure mode, and the ductility of SMA-SS RC column under the seismic loading were evaluated.

2. The Geometry of Longitudinally Aligned SMA-SS RC Column

In this study, the longitudinal aligned SMA-SS rebar RC column was configured such as the SMA employed in the plastic hinge region and SS bar longitudinally aligned in remaining portion using a mechanical coupler. The geometry, reinforcement details, and material properties of SMA-SS rebar RC column were particular from Billah and Alam (2012) [14]. Figure 1 described the detailed geometry and rebar details having 450 mm × 450 mm cross-section with 8 numbers of 20.6 mm diameter SMA rebar and 8 numbers of 19.5 mm diameter SS rebar. Paulay and Priestley (1992) equation (Eq. 1) was used to calculate the length of the plastic hinge (L_P) region.

$$L_P = 0.08\, L + 0.022 f_y . d_b \tag{1}$$

Here, L is the column height (mm), f_y is the yield strength of the rebar (MPa), and d_b is the rebar diameter (mm). For the total height of the column (3200 mm), the L_P was calculated as 468 mm using the above equation. For this research study, particularly the NiTi SMA rebar was used. The SMA and SS were longitudinally aligned using a mechanical coupler.

3. Constitutive modelling of different materials

The constitutive modeling approach was adopted to describe the material properties and their unique nature in finite element analysis software [16]. Mander et al (1988) [17] nonlinear concrete model was adopted for the numerical modeling of concrete. The uniaxial bilinear stress-strain model was adopted for modeling of SS reinforcement. Whereas, Auricchio and Sacco's (1997) [18] super-elastic shape-memory alloys model [16] has been adopted for the numerical modeling of SMA rebar.

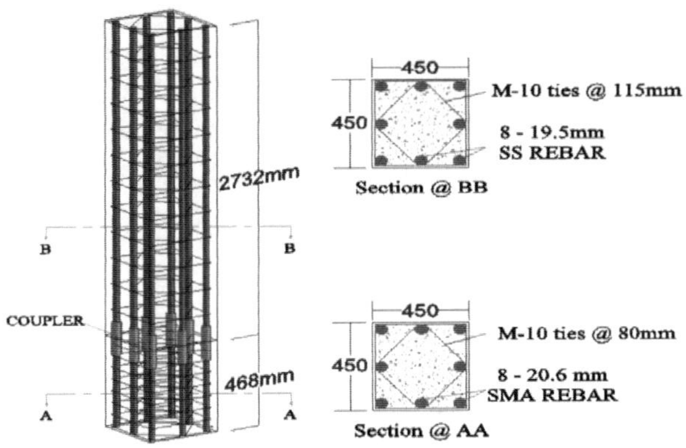

Figure 1. The geometry of a SMA-SS RC column reinforced.

4. Finite element analysis

The aim of research was to develop a fiber-based numerical model that should be proficient to predict the non-linear seismic behavior of SMA-SS RC column in terms of base shear force and displacement distribution. The Seismo struct fiber element-based software [16] was employed for modeling and non-linear static pushover seismic analysis. The rectangular columns were model as an inelastic beam-column element using a force-based modeling approach. The single barrel screw lock mechanical coupler developed by Alam et al. (2010) modified by Billah and Alam (2012) have been employed here to joint longitudinally aligned SMA and SS rebar. In modeling, it has been implemented at the base of the column by using a rotational spring zero length element as shown in Figure 2(c). The stress-slip behavior of the SMA bar inside the mechanical coupler has been used to model the bond-slip behaviour of rebar within the coupler. This bond-slip behaviour was used to define the joint element for modeling of rotational spring. The definition of joint element through moment rotation relationship has been incorporated in software through a modified Takeda hysteresis curve. The behaviour of curve has been characterized through five different parameters and their respective values have been defended through moment rotation relationship of a hybrid combination of SMA-SS rebar. The adapted FE model of the column is presented in Figure 2(a). A constant axial load of 400 kN was applied at top of the column to simulate the gravity load.

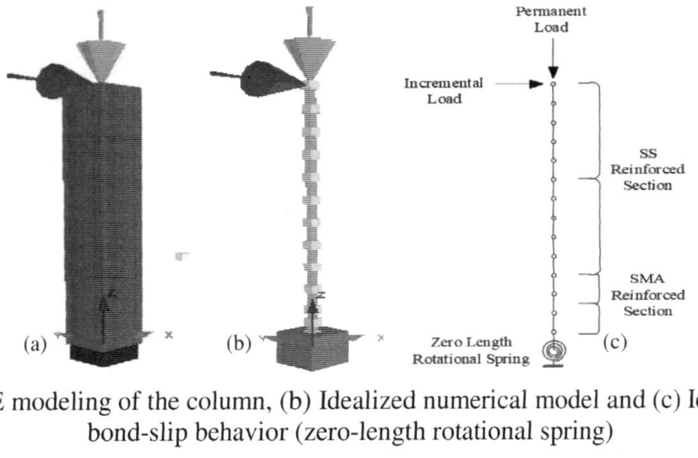

Figure 2. (a) FE modeling of the column, (b) Idealized numerical model and (c) Idealized form of bond-slip behavior (zero-length rotational spring)

5. Validation of Numerical Results

Billah and Alam (2012) carried experimental investigations through a pull-out test to ascertain the slippage of SS rebar and SMA rebar inside the mechanical coupler in the form of a stress-slip behavioural graph [14]. Where they used five sharp ended screws and mechanical coupler for connecting longitudinally aligned SMA and SS rebars. The stress-slip behavioural graph of the SMA bar inside the mechanical coupler has been used to model the bond-slip behaviour (rotational spring) of rebar within the coupler. The same modeling approach has been validated here with the analytical investigation of pushover analysis carried by Billah and Alam in 2012 with present research works result. Figure 3 and Table 1 show the performance results of nonlinear static pushover numerical investigation of the column and results of experimental-analytical investigation of Billah and Alam in 2012 [14]. The respective values of maximum base shear and its conforming displacement were calculated as 80.28 kN and 96 mm for SMA-SS RC column whereas, experimental-analytical result was 82 kN and 84 mm. The numerical results agree with the existing literature with an accuracy of about 98 %.

Table 1. The Comparative Performance of SMA-SS RC Columns

Column type	Concrete cracking		Steel yielding		Concrete crushing	
	Base shear (kN)	Δ (mm)	Base shear (kN)	Δ (mm)	Base shear (kN)	Δ (mm)
Numerical Investigation	23.32	3	76.34	57	77.46	168
Experimental-Analytical Investigation [14]	30.17	3	63.68	60	75.14	171

Note: Δ - Corresponding displacement in mm

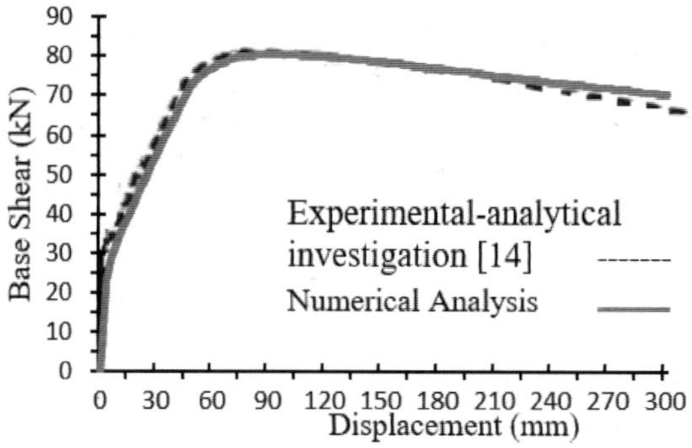

Figure 3. Base shear Vs. displacement - validation of numerical results

6. Parametric investigation of longitudinally aligned SMA-SS RC Column

In previous sections, numerical FE models have validated for SMA-SS reinforced concrete rectangular column using a fiber modelling approach with literature [14]. In this section, for parametric investigational study, total twelve different column models with four variable parameters have been used. The variable parameters are an aspect ratio (*l/d*), the compressive strength of concrete (f_c), the yield strength of (f_y) and axial load P (%) as shown in Table 2. The numerical models are named as P-1-5.5, P-2-7.5, P-3-9.5, P-1-35, P-2-45, P-3-55, P-1-250, P-2-450, P-3-650, P-1-5, P-2-7.5, P-3-12.5. For example, column P-1-5.5, P-2-7.5 & P-3-9.5 are defined as column models with aspect ratio 5.5, 7.5 and 9.5 respectively. Whereas column P-1-35, P-2-45, P-3-55 are defined as column models with the compressive strength of concrete (f_c) 35 MPa, 45 MPa and 55 MPa respectively. Yield strength of (f_y) of 250 MPa, 450 MPa and 650 MPa are considered for defining column P-1-250, P-2-450, P-3-650. Axial load P (%) with 5, 7.5 and 9.5 are selected in column P-1-5, P-2-7.5, and P-3-12.5 respectively.

Table 2. Parametric details of longitudinally aligned SMA-SS RC columns

Parameters	Column Id	Aspect ratio (*l/d*)	*P* (%)	*L* (mm)	f'_c (MPa)	f_y (MPa)	D (mm)
Aspect ratio (*l/d*)	P-1-5.5	5.5	10.0	2475	38.3	402	450
	P-2-7.5	7.5	10.0	3375	38.3	402	450
	P-3-9.5	9.5	10.0	4275	38.3	402	450
Compressive strength (f'_c)	P-1-35	7.1	10.0	3200	35	402	450
	P-2-45	7.1	10.0	3200	45	402	450
	P-3-55	7.1	10.0	3200	55	402	450
Yield strength (f_y)	P-1-250	7.1	10.0	3200	38.3	250	450
	P-2-450	7.1	10.0	3200	38.3	450	450
	P-3-650	7.1	10.0	3200	38.3	650	450
Axial load (*P*) %	P-1-5	7.1	5.0	3200	38.3	402	450
	P-2-7.5	7.1	7.5	3200	38.3	402	450
	P-3-12.5	7.1	12.5	3200	38.3	402	450

7. Results and discussion

The RC column incorporating SMAs rebar's in plastic hinge region along with SS rebar in the remaining portion has been modelled using the Seismo struct FE software [16].

Table 3. Parametric Investigation of Different Columns

Parameters	Column Id	First cracking of the concrete		First yielding of steel		First crushing of the concrete		Maximum base shear (kN)	Δ (mm)
		Base (mm)	Δ (mm)	Base (mm)	Δ (mm)	Base (mm)	Δ (mm)		
Aspect ratio (*l/d*)	P-1-5.5	35.58	2.5	105	37.5	112	100	113.55	220
	P-2-7.5	21.96	3	71.95	60	71.65	189	75.52	96
	P-3-9.5	21.19	8	52.94	88	43.79	324	54.79	128
Compressive strength (f'_c)	P-1-35	25.72	3	75.98	54	77.37	174	80.34	93
	P-2-45	23.43	3	76.93	54	79.05	153	81.53	78
	P-3-55	26.86	3	79.85	54	82.19	147	83.89	96
Yield strength (f_y)	P-1-250	25.90	3	60.68	33	56.92	249	68.68	78
	P-2-450	25.76	3	80.26	72	79.02	162	81.10	96
	P-3-650	28.00	5.2	80.35	109.2	78.19	169	80.49	101.4
Axial load (*P*) %	P-1-5	21.89	3	69.19	54	77.42	171	78.54	300
	P-2-7.5	21.95	3	72.93	54	77.85	168	78.72	120
	P-3-12.5	25.64	3	80.08	54	78.44	168	83.53	84

Note: Δ- Corresponding displacement in mm

A non-linear static pushover analysis has been performed for all models to compare the behaviour of columns. The value of base shear and displacement at which first cracking, first yielding, and first crushing occurs in pushover analysis for each model are considered for the brief study. Table 3 showing these values, and maximum value of base shear and corresponding displacement of above mentioned different configurational columns. Figure 4 to Figure 7 displayed the results of nonlinear seismic pushover analysis in terms of base shear vs. displacement conforming to different parameters namely, an aspect ratio, the compressive strength, the yield strength, and axial load, respectively. As described in Table 3, the value of base shear for column P-1-5.5 found to be maximum for the lower value of aspect ratio. Furthermore, as shown in Figure 4, base shear decreases with increases in aspect ratio. The compressive strength parameter does not make much difference in base shear for different values of f_c. The lower value of yield strength for column P-1-250 affects by 15% less value of base shear as compared to the higher yield strength column. Axial load, P (%) variation affect results by approximately 6% from lower value axial load to higher value axial load for column P-2-7.5 to P-3-12.5. Form Table 3 it is concluded that variation in aspect ratio makes much difference in the maximum value of base shear as compared to other parameters.

From Figure 4, it can be observed that the SMA-SS RC column flexural performance was significantly affected by the aspect ratio. Results from Table 3 reveals that the aspect ratio of 7.5 shows 38.28, 31.52 and 36.23% lower cracking, yielding and crushing base shear values, correspondingly revealed to those for an aspect ratio of a 5.5. For the aspect ratio of 7.5, crushing, yielding and cracking displacements increase by 47.09, 37.5 and 16.67%, correspondingly revealed to those having the aspect ratio of 5.5. According to the results presented in Figure 4 the aspect ratio of 5.5 expresses the lower deformability and higher stiffness respect to the aspect ratio of 7.5. The aspect ratio of 9.5 shows 3.5, 26.42 and 38.88% lower cracking, yielding and crushing base shear values, correspondingly revealed to those for an aspect ratio of 7.5 (see Table 3). For the aspect ratio of 9.5, crushing, yielding and cracking displacements raise by 41.67, 31.82 and 62.5% as shown in Table 3, correspondingly revealed to those for an aspect ratio of 7.5. According to the results presented in Figure 4, the aspect ratio of 7.5 expresses the lower deformability and higher stiffness respect to the aspect ratio of 9.5.

Results from Table 3 reveals that 45 MPa concrete raises pier yield and crushing base shear capacity by 1.23% and 2.15% respectively while cracking base shear capacity declines by 8.9% relative to 35 MPa concrete. Results also revealed a higher displacement for crushing by low strength concrete. There was no difference observed in 35, 45 and 55 MPa concrete yield and cracking displacement. The result shown in Figure 5demonstrates that the stiffness of all columns is quite identical until the concrete is cracked. The 35 MPa column shows comparatively lower stiffness than 45MPa concrete once concrete cracked and a similar observation was also seen in the case of 45 and 55MPa concrete. Results from Table 3 reveals that 55 MPa concrete raises 12.77, 3.65 and 3.82 % pier cracking, yielding and crushing base shear capacities, relative to 45 MPa concrete, respectively.

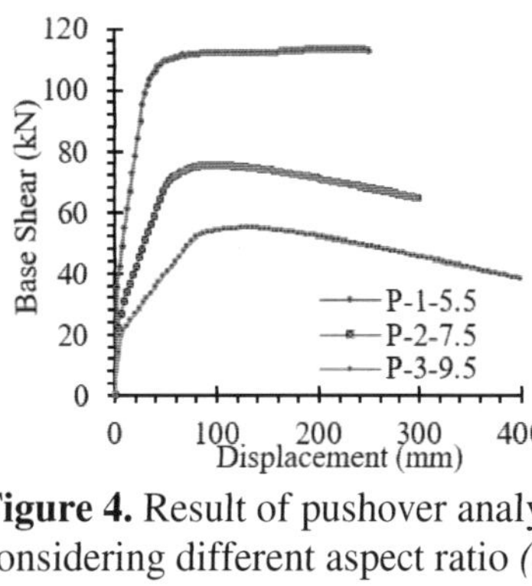

Figure 4. Result of pushover analysis considering different aspect ratio *(l/d)*

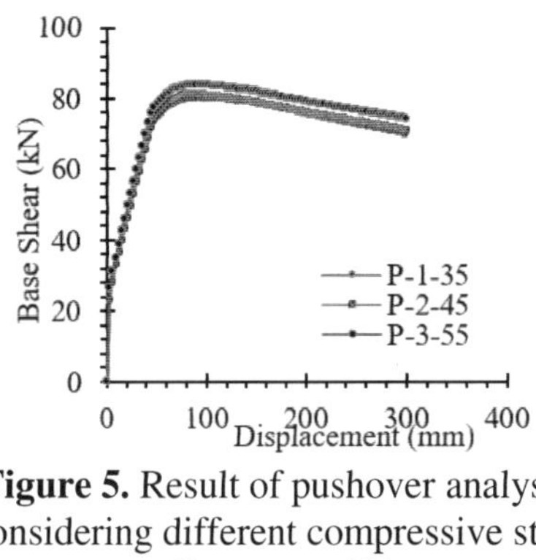

Figure 5. Result of pushover analysis considering different compressive strength of concrete (f'_c)

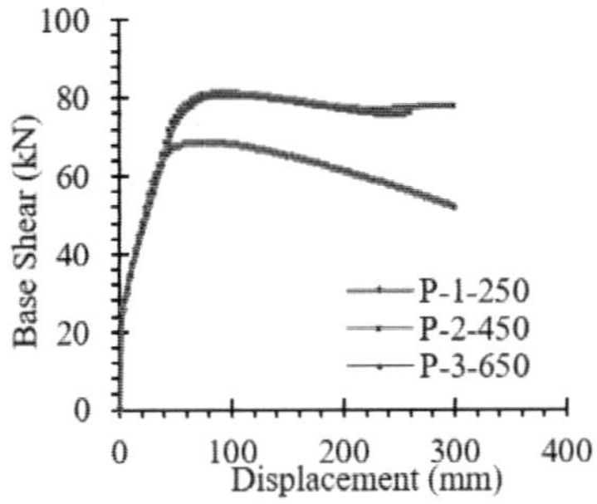

Figure 6. Result of pushover analysis considering different yield strength (f_y)

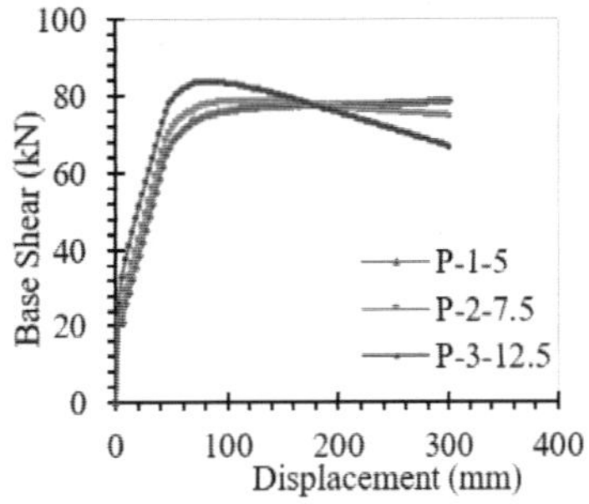

Figure 7. Result of pushover analysis considering different axial load *(P)*

The significant performance of SMA-SS RC column has been seen from Figure 6, due to the influential effect of yield strength of longitudinal reinforcement on the flexure. The high yield strength (450 MPa) rebar expresses the higher-crushing and yielding base shear of 27.96 and 24.39%, and larger yield displacement by 54.17% and lower crushing displacement by 34.94%, correspondingly associated to 250 MPa yield strength (see Table 3). There is no difference in the cracking and yield base shear for 450 MPa and 650 MPa yield strength. For the yield strength of 650 MPa, cracking, yielding and crushing displacements increase by 42.31, 34.06 and 4.14% as shown in Table 3, correspondingly associated with those for the 450 MPa yield strength.

It can be found from Table 3 that the yielding and crushing base shear rises by 5.13 and 0.5 % with an axial load of 7.5 % relative to that of 5 %. Figure 7 further indicates, the crushing displacement for 7.5% axial load decreases by 1.75% compared to 5% axial load, and with 7.5 and 5% axial loads, yielding and cracking displacements are the same. Table 3 further indicates that the cracking, yielding and crushing base shear rises by 14.39, 8.93 and 0.75 % relative to that of 7.5 % axial load at 12.5 % axial load and with 12.5 % and 7.5 % axial load, cracking yielding and crushing displacements are the same. It is clear from Figure 7 that the 5% axial load pier has a higher deformation relative to the column of 7.5, 12.5% axial load.

As shown in Figure 8, at the lower aspect ratio, the range of ductility is wider. The level of ductility decreases as the aspect ratio increases. Here, aspect ratio, yield strength, and axial load, are found to be significant on the ductility. Among them, the aspect ratio and yield strength have been responsive influential factors on the ductility of the longitudinally aligned SMA-SS RC column. With the axial load level, the impact of axial load on ductility is not important, the ductility decreases. Atalay and Penzien (1975), and Sheikh and Khoury (1993) also reported that increasing the axial load can reduce the displacement ductility of the column. The change in the compressive strength of concrete does not make any difference in the ductility of SMA-SS RC column.

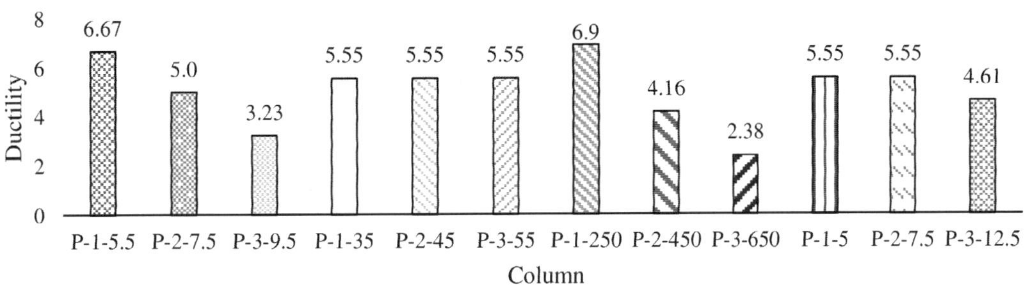

Figure 8. Ductility of different parametric configured columns

IOP Conf. Series: Materials Science and Engineering **814** (2020) 012012 doi:10.1088/1757-899X/814/1/012012

8. Conclusions

In this study, the inelastic pushover analysis was carried out on corrosion-free SMA-SS RC column for seismic investigation of RC column reinforced with SMA rebar in plastic hinge region along with SS in another region of the column. The results of the parametric analysis on SMA-SS RC column are demonstrated in terms of base shear and displacement. Based on the numerical investigation the resulting conclusions are extracted:

- The yield strength of rebar and axial load, the aspect ratio, and compressive strength have a significant impact on the crushing and yielding base shear of SMA-SS RC column.
- The compressive strength of concrete does not have a significant impact on the yield and crushing drift; However, it shows some contribution to the seismic performance of SMA-SS RC column in the yielding and crushing base shear
- The variation in aspect ratio makes much difference in the maximum value of base shear as compared to other parameters.
- The reinforced SMA columns demonstrated adequate ductility prior to failure.
- The aspect ratio, yield strength, and axial load, are found to be significant on the ductility. The change in the compressive strength of concrete does not make any difference in the ductility of SMA-SS RC column.

References

[1] Maghsoudi A A, Maghsoudi M and Haghighi H 2018 *Amirkabir J. Civ. Eng.* **49** 233–6
[2] Forrest B T and El-hacha R 2019 *Struct. Anal. Hist. Constr.* **18** 1931–9
[3] Maruyama T and Kubo H 2011 *Woodhead Publishing Limited* 141–59
[4] Tanaka Y, Himuro R, Sutou Y, Omori T and Ishida K 2010 *Science* **80** 327 4
[5] Logoudas D C 2008 *Shape memory alloys modeling and engineering applications* ed D C Logoudas (TX, USA: Springer International Publishing)
[6] Janke L, Czaderski C, Motavalli M and Ruth J 2005 *Mater. Struct.* **38** 578–92
[7] Callister Jr. W D 2007 *Materials Science and Engineering 7th Ed.: An Introduction* vol 26
[8] Chang W-S and Araki Y 2016 *Proc. Inst. Civ. Eng. - Civ. Eng.* **169** 87–95
[9] Pareek S, Suzuki Y, Araki Y, Youssef M A and Meshaly M 2018*Eng. Struct.* **175** 765–75
[10] Gencturk B and Hosseini S F 2014 *10th U.S. National Conference on Earthquake Engineering* (Earthquake Engineering Research Institute, Anchorage, AK)
[11] Nakashoji B A 2014 *Seismic performance of square Nickel-Titanium reinforced ECC columns with headed couplers* (University of Nevada, Reno)
[12] Saiidi M S, O'Brien M and Zadeh M S 2009 *ACI Struct. J.* **106** 69–77
[13] Bae S and Bayrak O 2008 *ACI Struct. J.* **105** 290–300
[14] Billah A H M M and Alam M S 2012 *Constr. Build. Mater.* **28** 730–42
[15] Paulay T and Priestley M J N 1992 *Seismic Design of Reinforced Concrete and Masonry Buildings, ISBN: 978-0-471-54915-4*
[16] SeismoStruct 2018 SeismoStruct - A computer program for static and dynamic nonlinear analysis of framed structures
[17] Mander J B, Priestley M J N and Park R 1988 *J. Struct. Eng.* **114** 1804–26
[18] Auricchio F and Sacco E 1997 *J. Intell. Mater. Syst. Struct.* **8** 489–501
[19] Alam M S, Youssef M A and Nehdi M L 2010 *Mater. Struct.* **43** 91–107
[20] Atalay M and Penzien J 1975 The seismic behavior of critical regions of reinforced concrete components as influenced by moment, shear and axial force
[21] Sheikh S A and Khoury S S 1993 *ACI Struct. J.* **90** 414

Performance Evaluation of Bridge Piers Reinforced with Shape Memory Alloys in Plastic Hinge Region: Part 1 – Statistical Analysis

Kanan Thakkar[1] and Anant Parghi[2]

[1]Research Scholar, Department of Applied Mechanics, S. V. National Institute of Technology, Surat-395 007, Gujarat, India
[2]Assistant Professor, Department of Applied Mechanics, S. V. National Institute of Technology, Surat-395 007, Gujarat, India

Email: kananrthakkar@gmail.com, anantparghi@gmail.com

Abstract. Bridge piers are the most perilous structural elements under the effect of enormous earthquake ground motions. The large residual deformation of a bridge pier turns out into the failure of an entire bridge structure. Thus, in the last decades, numerous novel materials were developed to reduce the residual displacement after the post-earthquake events. Shape Memory Alloys (SMA) is a special class of composite material, which can sustain enormous deformations and reappears to its parent state. The numerical results of SMA-reinforced bridge piers using non-linear finite element tools are compared with the experimental results. Two methods of statistical analysis are adopted for this study: One-Factor-at-a-Time (OFAT) Method, and Full-Factorial Design Method. Three variable material parameters, such as the compressive strength of concrete, the yield strength of steel, and the type of SMA composition are considered. Three levels for each factor are considered. The results of this research are presented in terms of drift values at various performance criteria, i.e., cracking, spalling, yielding, and crushing. Further, the regression analysis is performed to determine the relationship between material properties of SMA-reinforced bridge piers and the drift values at various performance criteria.

1. Introduction

1.1. *General*

Bridge piers are the most hazardous structural elements under the effect of enormous earthquake ground motions. The loss of infrastructure directly leads to the loss of the economy for any country. The maintenance and repair of bridges consume plenty of currency and time. Previous research indicates that the reinforced concrete (RC) bridges are susceptible to severe residual displacement because of tremendous earthquake vibrations [1]. This residual displacement causes permanent damage/collapse of RC bridges. The maximum damage of a bridge pier is concentrated in a small region, defined as a plastic hinge region [2]. Several innovative materials are invented in order to improve the performance of civil infrastructure, i.e., Shape Memory Alloys (SMA).

SMA is a novel composite material, exhibiting the self-cantering phenomena. SMA can sustain enormous deformations and reappear to a parent shape after removal of loading (super-elasticity) or by removal of heat (shape memory effect) [3]. This phenomenon of SMA is a unique feature for

IOP Conf. Series: Materials Science and Engineering **814** (2020) 012013 doi:10.1088/1757-899X/814/1/012013

applications in structural engineering. Austenite and marten site are two crystal phases of SMA. It gets converted from one phase to another phase by applying/removing stresses/ temperature. SMA can be composed of several compositions, i.e. nickel, titanium, copper, iron, manganese, etc. Nickel-titanium based SMA is proven to be the most suitable for the applications in structural elements.

1.2. *Previous* research

Over the last decade, super-elastic SMA reinforcements received significant attention to the researchers, which is reflected through an increasing number of researches conducted on SMA-equipped structural members and elements. SMA offers numerous structural applications, namely, reinforcement in reinforced concrete elements, connectors in steel and concrete structures, damping devices, base isolation devices, etc. [4]. Several applications of SMA in bridges have been explored to enhance the performance of bridge structures under the effect of a seismic event. Since last decades, numerous researchers studied the behaviour of SMA-reinforced bridge piers to mitigate seismic energy demand after severe seismic events [5].

Furthermore, there is a limited investigation of the parametric study as an alternate of steel reinforcement in the plastic hinge location with super-elastic SMA-based reinforcement on the dynamic performance of the bridge piers. Recently, there is no code/guidelines available in the existing literature for the use of super-elastic SMA as an alternative reinforcement with conventional steel.

2. Modeling of bridge piers

Following numerical analysis methods, material properties, geometrical data, material constitutive models, FE parameters, and modeling techniques are adopted for this study.

2.1. Static pushover analysis

Static pushover analysis is a nonlinear analysis methodology in which a monotonic lateral load is applied to the structure in an incremental manner. This monotonic loading is applied in such a way to determine the entire performance of a structure from the elastic phase to inelastic phase and failure. Figure 1(a) and illustrates the schematic diagram for typical pushover analysis of a pier. Figure 1(b) demonstrates the values of base shear, and displacement for flexural limit states, i.e., the values at which first cracking, spalling, yielding, crushing, buckling and fracture.

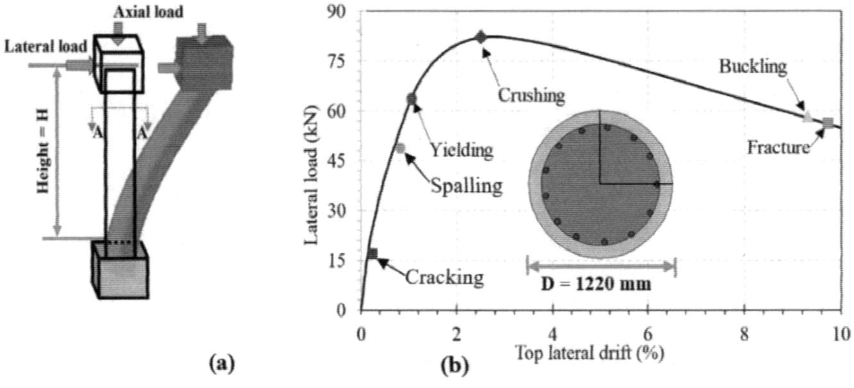

Figure 1. Schematic of a typical pushover analysis of pier (a) deformed and un-deformed shape, and (b) load vs. drift curve with limit states [6]

2.2. Flexural limit states

The performance points and their flexure limit states are essentials to be defined for the performance evaluation of SMA-reinforced bridge piers. Damages at several flexural limits are defined by the strain-based method [2]. The following flexural limits are taken into consideration to estimate the performance SMA-reinforced circular bridge piers, e.g., cover concrete's cracking and spalling, longitudinal SMA rebar's yielding, and core concrete's crushing.

2.3. The geometry of bridge piers
In this study, a 0.2 scaled circular bridge pier [7] has been considered for the nonlinear static pushover analysis. The geometry of the pier used in this study is, as shown in Figure 2.

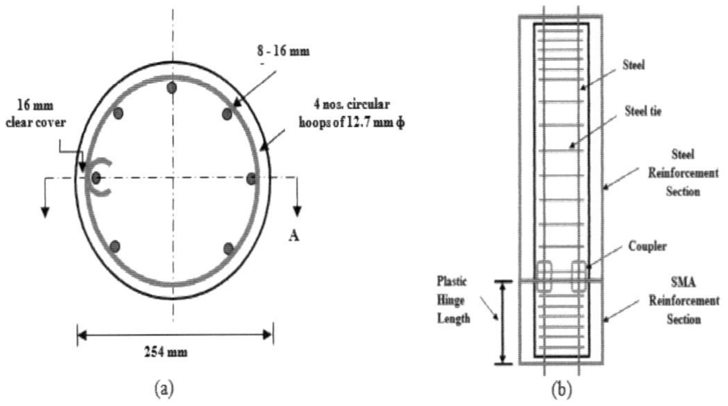

Figure 2. Reinforcement and geometrical details of SMA-reinforced bridge pier

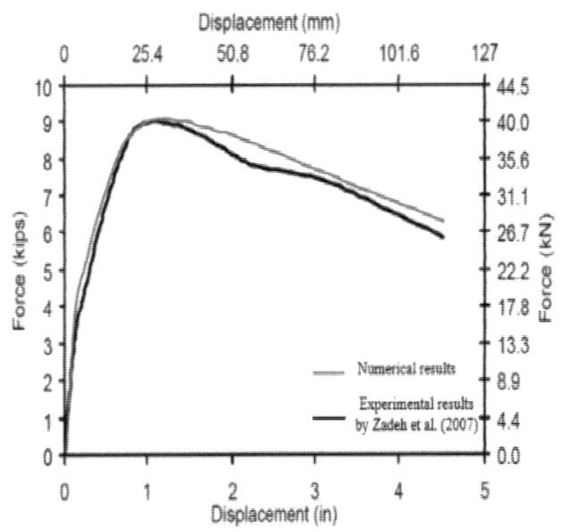

Figure 3. Validation of numerical model [7]

The diameter of a pier is 254 mm, and the height of a pier is 1270 mm. The height of a pier is calculated from the top of footing to the centreline of loading head, where the lateral load is applied. Axial load ratio is selected as 10% of the axial capacity of a pier, and a longitudinal reinforcement ratio (ρ_t) is used as 2%. The aspect ratio is selected as 5. Figure 2(a) illustrates the geometry of the pier, and section A-A in Figure 2(b) shows the vertical section of the pier with the longitudinal and lateral reinforcement details of the bridge pier. Plastic hinge length is calculated as 205.7 mm, according to Paule and Pristley [9] equation.

Table 1. Materials Properties used for piers reinforced with steel and SMA

Material	Property	Value	Reference
Confined concrete	Elastic Modulus (GPa)	33.23	-
	Compressive strength (MPa)	50	
	Tensile strength (MPa)	5	
	Corresponding strain	0.005	
Longitudinal steel	Elastic Modulus (GPa)	200	-
	Yield stress (MPa)	415	
	Fracture strain	0.10	
	Strain hardening parameter	0.05	
Super-elastic SMA	Yield strength- f_y (MPa)	210, 401, 750	[10] [11] [12]
	Modulus of elasticity-E (GPa)	28, 62.5, 46.9	
	Super-elastic plateau strain length- ε_s (%)	9, 6, 13.5	
	A to M starting stress- f_y (MPa)	210, 401, 750	
	A to M finishing stress- f_{P1} (MPa)	275, 510, 1200	
	M to A starting stress- f_{T1} (MPa)	200, 370, 300	
	M to A finishing stress- f_{T2} (MPa)	150, 130, 200	

Where A is the austenite phase, and M is the martensite phase.

2.4. Material constitutive models

In this study, three different material constitutive models are considered for the modeling of bridge piers: steel reinforcement, SMA reinforcement, and concrete. Menegotto and Pinto [13] steel constitutive model is adopted to assign the properties of steel reinforcement in FE tool. To represent NiTi reinforcement in a finite element model, Fugazza [14] model is implemented for this study. To predict the stress-strain behavior of confined concrete in this study, the stress-strain model by the constitutive behavior by Mander et al. [15] is adopted. Table 1 depicts the material properties of steel-reinforced and SMA-reinforced bridge piers used for a numerical investigation.

2.5. Finite element modeling

To study the performance of super-elastic SMA-RC and steel-reinforced bridge piers under the effect of a seismic event, a finite element (FE) program SeismoStruct [16] is adopted. The piers are modeled using the displacement-based inelastic system owing to its modeling requirements. There are 300 fibers considered across the section for the mesh refinement. The bar-slippage of super-elastic SMA rebar inside the coupler is neglected. The load is assigned as an axial load at the top of a pier.

2.6. Validation

The precision of the proposed model has been validated through experimental results proposed by Zadeh et al. [7] under the Highway Idea Project, University of Nevada, Reno (Figure 3). They conducted an experimental program on 0.2 scaled circular columns utilizing SMA in the plastic hinge zone under the effects of dynamic loads.

3. Design of experiment (DoE)

The design of experiments (DoE) is the orderly procedure to define a correlation between factors affecting a process and the output of that process. This research emphasis on the performance assessment of SMA-RC bridge piers under the effect of static pushover loads. Moreover, the effect of different materials properties, such as yield strength of steel, compressive strength of concrete, and composition of SMA along with their interactions are explored using full factorial design and Analysis of Variance (ANOVA). ANOVA is a statistical method used to test differences between two or more means. A significance level of 0.05 is selected. The results of SMA reinforced bridge piers are compared in terms of the performance criteria of limit state of design, e.g., cracking, yielding, spalling, and crushing.

3.1. Factors and Levels for DoE

Table 2. Factors and levels

Sr. No.	Factors	Levels		
		Low (-1)	Intermediate (0)	High (+1)
1.	Compressive Strength of Concrete, f_c (MPa)	35	50	65
2.	Yield Strength of Steel, f_y (MPa)	250	415	500
3.	Composition of SMA	CuAlMn	NiTi	FeNCATB

In this study, three factors, each having three levels have been defined. The no. of experiments to be performed in the one-factor-at-a-time (OFAT) method will be $3 \times 3 = 9$, and full factorial design will be $3^3 = 3 \times 3 \times 3 = 27$. The considered factors based on comprehensive literature and their levels are defined as shown in Table 2. Static pushover analysis is performed for 9 bridge pier models as per the OFAT method and 27 models as per the full factorial method of DoE (Table 2). These models are statistically analyzed using Minitab, and run orders are defined using Minitab.

4. Results and discussion

The non-linear static pushover analysis is performed for SMA-reinforced circular bridge piers as per the data described in previous sections. 9 models are prepared as per the OFAT method, and 27 models are prepared as per the full factorial method in the FE-software tool.

4.1. Capacity curves of SMA-RC bridge piers with OFAT method

By modelling 9 models (3 factors and each having 3 levels) under the OFAT method, the following results were derived:

Figure 4 shows the capacity curves of SMA-reinforced bridge piers for concrete compressive strengths, 35 MPa, 50 MPa, and 65 MPa. As seen from the figure, the capacity of a pier to attract base shear and displacement increases with the increment in concrete compressive strength until certain limits. Piers starts to behave inversely after that limit, reducing the capacity with the further surge in concrete compressive strength.

Figure 5 demonstrations the capacity curves of SMA-reinforced bridge piers for steel yield strengths 250 MPa, 415 MPa, and 500 Mpa. It can be observed from the figure that the capacity of a pier to attract base shear and displacement is almost the same for the different yield strengths of steel. Thus, there is no significant consequence of yield strength of steel on the seismic performance of SMA-reinforced circular bridge pier.

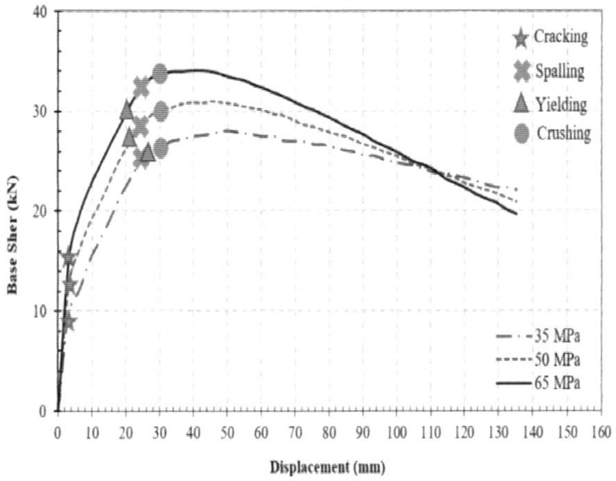

Figure 4. Capacity curves of SMA-reinforced bridge piers for variable concrete compressive strengths (f_c = 35 MPa, 50 MPa, and 65 MPa)

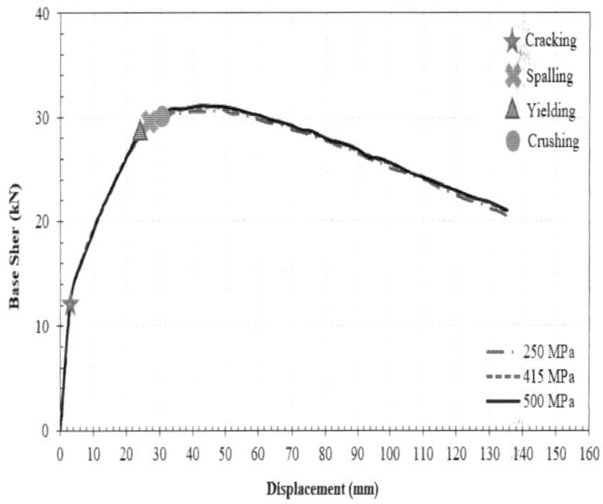

Figure 5. Capacity curves of SMA-reinforced bridge piers for variable steel yield strengths (f_y = 250 MPa, 415 MPa, and 500 MPa)

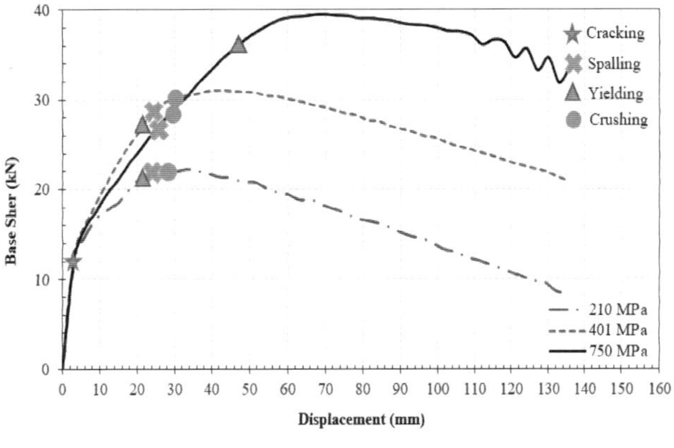

Figure 6. Capacity curves of SMA-reinforced bridge piers for variable SMA composites (CuAlMn, NiTi, and FeNCATB)

Figure 6 outlines the capacity curves of SMA-reinforced bridge piers for three types of SMA composites (CuAlMn, NiTi, and FeNCATB). As observed from the figure, Pier with Cu-based SMA reinforcement in the plastic hinge region has the minimum capacity to attract base shear and displacement owing to the less yield strength of Cu-based SMA. The capacity of a pier with NiTi-based SMA is found to be more than piers with Cu-based SMA, as the yield strength of NiTi-based SMA is more than Cu-based SMA. Pier with Fe-based SMA demonstrates the maximum capacity to attract base shear and displacement, because of the higher value of yield strength of Fe-based SMA.

4.2. Drift of SMA-RC bridge piers at various flexural limit states with full factorial design

The 27 pier models using a full factorial method of statistical analysis are modeled in a non-linear FE software tool. The outputs of those models are analyzed with the condition of "larger is better" for the drift-based outputs. Further, the main effects plots' and interaction plots for each output are determined. The slope in the main effects plot is directly proportional to the significance of a particular factor on the output. One-way ANOVA is performed for this study. The main effect plots and interaction plots for the distinct flexural limit states are presented in Figure 7 to Figure 16.

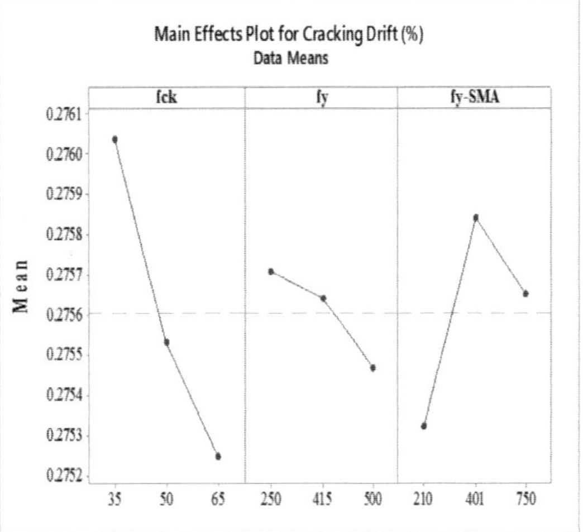

Figure 7. Main effects plot for cracking drift

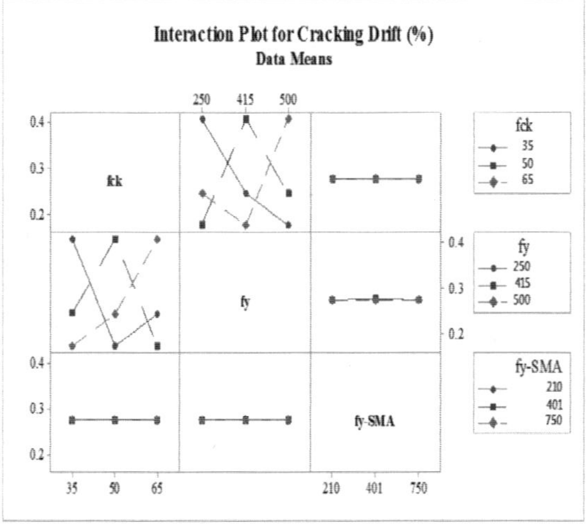

Figure 8. Interactions plot for cracking drift

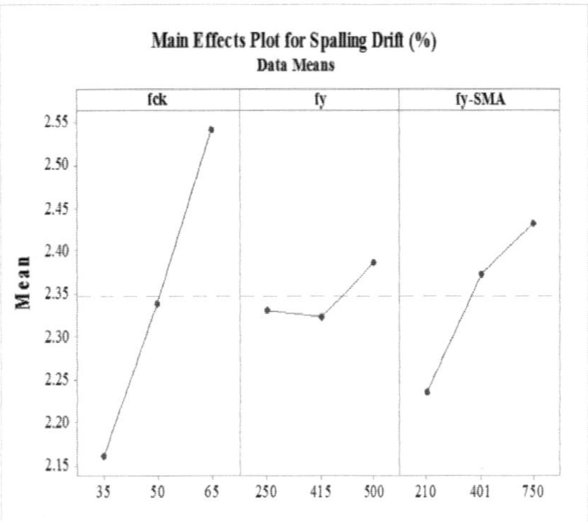

Figure 9. Main effects plot for spalling drift

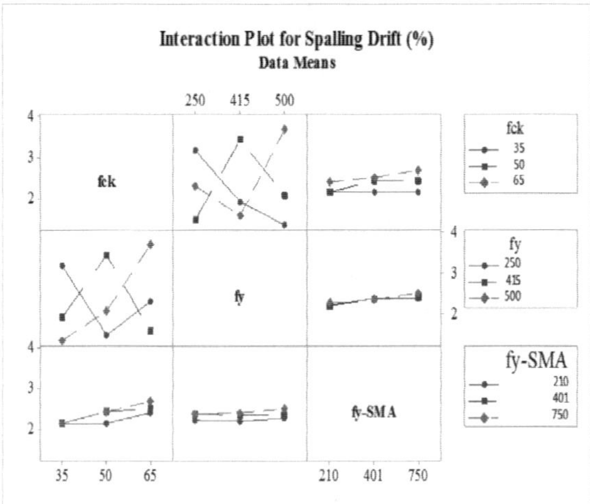

Figure 10. Interactions plot for spalling drift

IOP Conf. Series: Materials Science and Engineering **814** (2020) 012013 doi:10.1088/1757-899X/814/1/012013

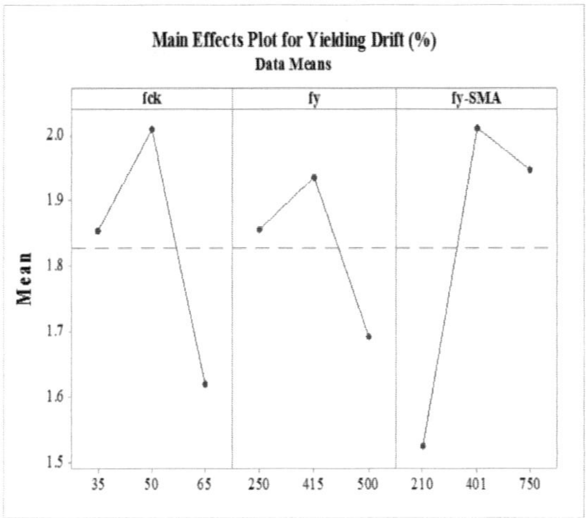

Figure 11. Main effects plot for yielding drift

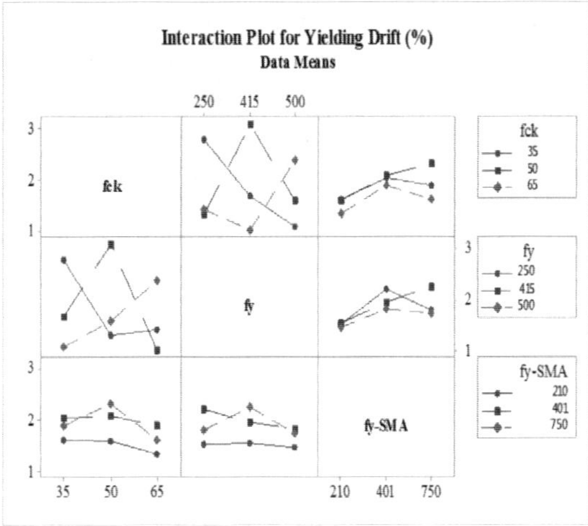

Figure 12. Interactions plot for yielding drift

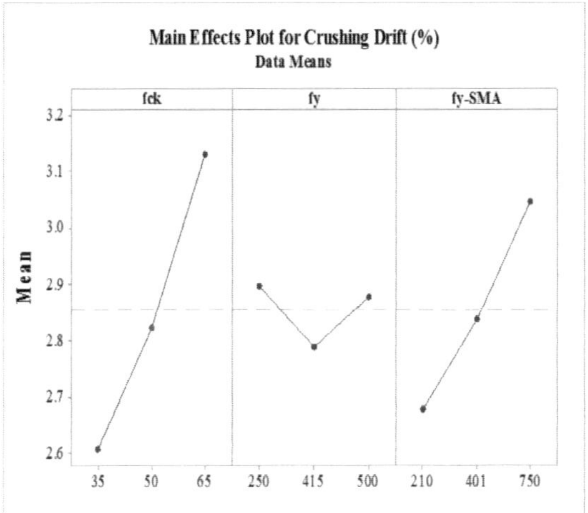

Figure 13. Main effects plot for crushing drift

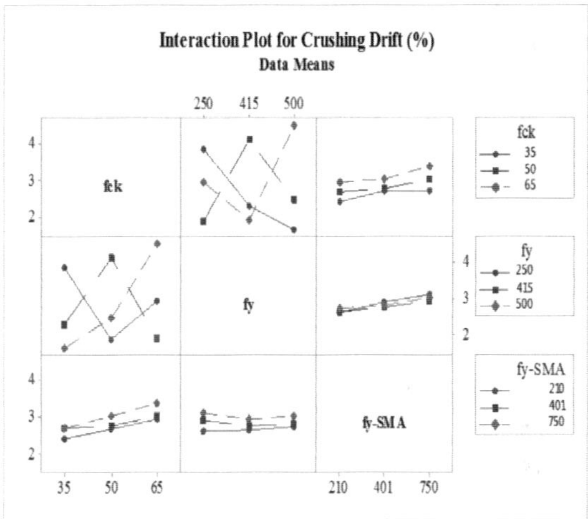

Figure 14. Interactions plot for crushing drift

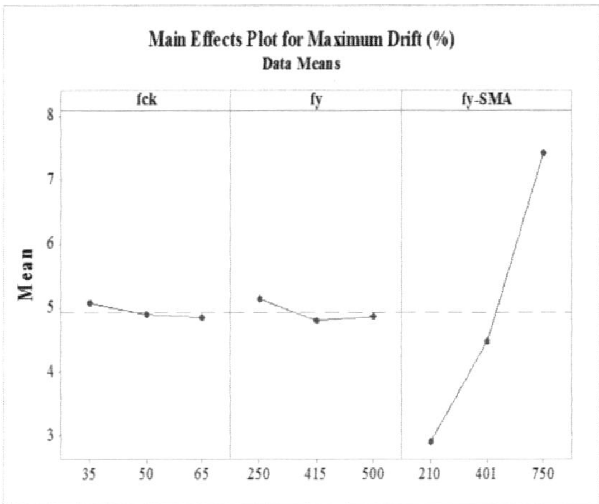

Figure 15. Main effects plot for maximum drift

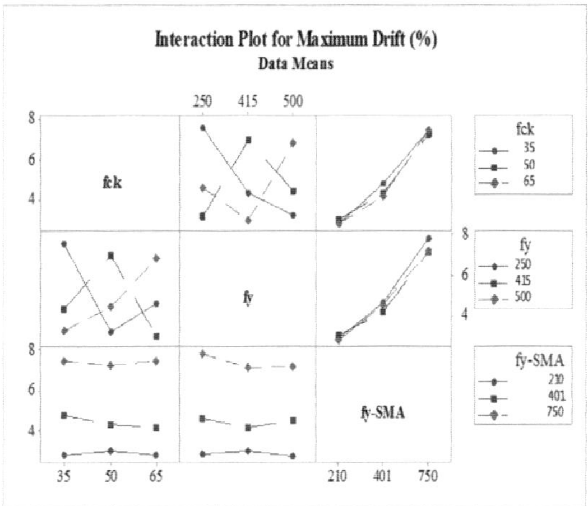

Figure 16. Interactions plot for maximum drift

Figure 7, Figure 9, Figure 11, Figure 13 and Figure 15 express the significance of the three factors on cracking, spalling, yielding, crushing, and maximum drift, respectively. As seen from these figures, f_c is the most significant factor for cracking, spalling, and crushing drift. Further, $f_{y\text{-}SMA}$ is the most significant factor for the yielding, and maximum drift. f_y is the least significant factor for each of the outputs. The results of main effect plots confirm with the results of OFAT method. Figure 8, Figure 10, Figure 12, Figure 14 and Figure 16 demonstrate the interaction effects for each factor with other factors on cracking, spalling, yielding, crushing, and maximum drift, respectively. As seen from these figures, f_c and f_y exhibit higher interaction effects with each other for each of the outputs. $f_{y\text{-}SMA}$ exhibits negligible interaction effects with f_c and f_y.

4.3. ANOVA
By performing ANOVA, f_c, and $f_{y\text{-}SMA}$ are the most significant factors for each of the outputs. Further, f_y does not exhibit any significance on the outputs. These results confirm the results of OFAT method and full factorial design method. The R^2 values for all the outputs in general linear models are more than 97 percent.

5. Conclusions
By performing a parametric investigation on RC bridge piers reinforced with SMA in the plastic hinge region, the following conclusions are drawn:

5.1. OFAT method
The capacity of a pier to attract base shear and displacement increases with the increment in concrete compressive strength until certain limits. There is no significant effect of yield strength of steel on the seismic performance of SMA-reinforced circular bridge pier. Pier with Cu-based SMA reinforcement in the plastic hinge region has the minimum capacity to attract base shear and displacement owing to the less yield strength of Cu-based SMA. The capacity of a pier with NiTi-based SMA is found to be more than piers with Cu-based SMA. Pier with Fe-based SMA demonstrates the maximum capacity to attract base shear and displacement.

5.2. Full factorial design
The compressive strength of concrete and the composition of SMA are the most significant factors for cracking, spalling, yielding, crushing, and maximum drift. The yield strength of steel does not exhibit any significance on the drift values of SMA-RC bridge piers at various flexural limits.

6. Acknowledgements
The first author would like to acknowledge the financial support provided by Ministry of Human Resources Development (MHRD), Government of India, through the Ph.D. fellowship program. Further, the authors would like to acknowledge the online and offline library of SVNIT.

References
[1] Shrestha B and Hao H 2014 *23rd Australasian Conference on the Mechanics of Structures and Materials vol I, ed S.T. Smith (Byron Bay, Australia)* 375–80
[2] Billah M 2015 *Performance-based seismic design and assessment of concrete bridge piers reinforced with shape memory alloy rebar* (Ph.D. Thesis, Univesity of British Columbia, Canada)
[3] Boroschek R L, Farias G, Moroni O and Sarrazin M 2007 *J. Earthq. Eng.* **11** 326–42
[4] Ozbulut O E, Hurlebaus S and Desroches R 2011 *J. Intell. Mater. Syst. Struct.* **22** 1531–49
[5] Paulay T and Priestley M J N 1992 *Seismic design of reinforced concrete and masonry buildings* New York, USA: John Wiley & Sons, Inc.
[6] Parghi A 2016 *Seismic performance evaluation of circular reinforced concrete bridge piers retrofitted with fibre reinforced polymer* PhD. Thesis, Univesity of British Columbia, Canada

IOP Conf. Series: Materials Science and Engineering **814** (2020) 012013 doi:10.1088/1757-899X/814/1/012013

[7] Zadeh M S, Brien M O and Saiidi M S 2007 *A study of concrete bridge columns using innovative materials subjected to cyclic loading* A report by University of Nevada Reno and National Cooperative Highway Research Program (NCHRP)
[8] Schoettler M J, Restrepo J I, Guerrini G and Duck D E 2015 *A full-scale, single-column bridge bent tested by shake-table excitation* Berkeley
[9] Paule T and Prisley M N J 1992 *Seismic design of reinforced concrete and masonry buildings* John Wiley & Sons, Inc.
[10] Shrestha K C, Araki Y, Nagae T, Koetaka Y, Suzuki Y, Omori T, Sutou Y, Kainuma R and Ishida K 2013 *Smart Mater. Struct* **22** 25025–12
[11] Youssef M A, Alam M S, Youssef M A and Nehdi M 2008 *Eng. Struct.* **30** 3399–411
[12] Tanaka Y, Himuro Y, Kainuma R, Sutou Y, Omori K and Ishida K 2010 *Science* **80** 327.
[13] Menegotto M and Pinto P E 1973 Method of analysis for cyclically loaded R. C. plane frames including changes in geometry and non-elastic behavior of elements under combined normal force and bending **13** (Zürich)
[14] Fugazza D 2003 *Eur. Sch. Adv. Stud. Reduct. Seism. Risk* **1** 141
[15] Mander J B 1988 *J. Struct. Eng.* **114** 1804–26
[16] SeismoStruct 2018 SeismoStruct- A computer program for static and dynamic nonlinear analysis of framed structures

IOP Conf. Series: Materials Science and Engineering **814** (2020) 012014 doi:10.1088/1757-899X/814/1/012014

Fracture analysis of single edge cracked functionally graded material plate under various loading conditions by extended finite element method

Achchhe Lal[1], B.M Sutaria[2] and Kundan Mishra[3]

[1] Assistant Professor, Department of Mechanical Engineering, SVNIT, Surat, India
[2]Associate Professor, Department of Mechanical Engineering, SVNIT, Surat, India
[3]Research Scholars, Department of Mechanical Engineering, SVNIT, Surat., India

E-mail: lalachchhe@yahoo.co.in

Abstract: Functionally graded material (FGM) is the perfect blend of two (or more) material phases of different materials and functional performance that vary in definite direction within the geometry. It was first introduced by group of scientist in 1984 for shielding and thermal barrier. In the present paper single edge cracked plate under different loadings are used to observe the behavior of FGM plate under different loading conditions by using extended finite element method (XFEM). The basic formulation is carried out in MATLAB environment by XFEM using interaction integral (*M*-integral) through partition of unity method.In the presentstudy the main focus is to utilize XFEM to find mixed mode stress intensity factor (MMSIF) of edge cracked FGM plate under the action of uniform tensile, shear, combined (tensile and shear) and point load.

1. Introduction

Functionally graded material (FGM) is a new class of composites, where material properties of constituent materials changes smoothly with desired gradation. It was first used for the thermal resisting materials and now it is being utilized for various engineering applications like furnace linear, artificial bones and other medical implants etc. In structural design FGM provides better resistance towards crack growth.

In spite of availability of advanced technology and manufacturing techniques, structural components contain various defects like cracks, which become very severe in various applications where, component with crack under complex loadings. So, it is essential to analyse the behavior of FGM plate with crack under various loadings. Many researchers are working and finding different FGM for better structural integrity.In this direction Kim and Paulino [1] evaluated MMSIF cracked FGM plate where variation of material is the function of geometry. In the present work MMSIF is calculated by utilizing various techniques incorporated with FEM and accuracy of present results are compared with theoretical and experimental results. Man et al. [2] found out the mode I and mixed mode crack solution of stress intensity factor (SIF) for the two-dimensional FGMwith crack by weight function method, where it is verified that proposed method is valid for the derivation of weight function in FGMs.In conventional FEM due to limitation in updating of mesh, many researchers have started working on XFEM, where fracture problem is solved by utilising enrichment functions without using re-meshing during evaluation of crack. Lal and Palekar [3] evaluated the stochastic MMSIF of cracked composite plate with random material properties. The proposed work is done by using XFEM combined different techniques and the present model is seen to be very accurate for uncertainty analysis. Mohammadi [4] Studied fracture mechanics by XFEM, where singular stress field is reproduced by using orthotropic crack tip enrichments. In the present work it is verified that orthotropic XFEM needs lesser DOFs as compared to conventional FEM. Mohammadi [5] presented fracture analysis by XFEM, where MMSIF of cracked FGM plate by using interaction integral through partition of unity method is discussed. Asadpoure et al [6] proposed XFEM with crack-tip displacement fields to evaluate MMSIF of cracked orthotropic plate. In the present study MMSIF is calculated by using interaction integral (M-integral) and the accuracy of the result is compared with analytical and other numerical methods. Kim and Paulino [7] presented interaction Integral (M-integral) method to calculate MMSIF of orthotic FGM plate with straight or curved crack under loading. The proposed work is accurately compared with available analytical and numerical solutions. Ebrahimi et al [8] investigated and modelled crack in orthotropic composite plate under uniaxial and bi-axial load by XFEM where re-meshing is not required while progress of crack. In the present study it is observed that crack tip domain size has not any effect on SIF when it reaches one third of crack length. Lal et al [9] investigated stochastic fracture analysis cracked composite beam under tensile, shear and combined loadings. The sensitivity of individual parameter on MMSIF is also investigated and here it is also confirmed that SIF is get reduced when orthotropic angle is above 400. So it is essential to keep orthotropic angle more than 400 for safety point of view. Khatri and Lal [10] presented stochastic fracture analysis of isotropic plate with emanating crack from hole under in plate loadings by utilizing XFEM. It is observed that the SIF for shorter crack length is also very high due to high concentration near crack tip.

In the present paper MMSIF of edge cracked FGM plate under the action of point load, uniformly distributed tensile, shear, combined is evaluated for different fracture parameter like crack length and crack angle. XFEM based fracture analysis is utilized in the present study and this approach is also validated with other numerical technique in the literature.

2. Problem formulation

Conventional FEM has limitations to handle fracture problem, where it is necessary to update mesh. To overcome this difficulty XFEM provides better approach to solve fracture problems. The displacement vector in XFEM framework is given below.

$$u^h(x) = \sum_{i=1}^{n} N_i(x)\overline{u_i} + \sum_{i=1}^{n_c} N_i(x)H(x)a_i + \sum_{i=1}^{n_{t1}} N_i(x)\sum_{\alpha=1}^{4}\Phi_\alpha^1(x)b_i^{\alpha 1} + \sum_{i=1}^{n_{t2}} N_i(x)\sum_{\alpha=1}^{4}\Phi_\alpha^2(x)b_i^{\alpha 2} \quad (1)$$

Where $\overline{u_i}$, a_i, $b_i^{\alpha 1}$ and $b_i^{\alpha 2}$ are the conventional degrees of freedom (dof), dof for enrichment function, crack tip and crack face asymptotic functions. Figure 1 shows a body of area Ω and its outer boundary Γ with a crack. The body is having volume / body loads b and the surface load at the boundary Γ_t. Boundary surface is represented as $\Gamma = \Gamma_u + \Gamma_t + \Gamma_c$ where u, t and c are displacement, traction and discontinuity. It is assumed that surface with crack is traction free.

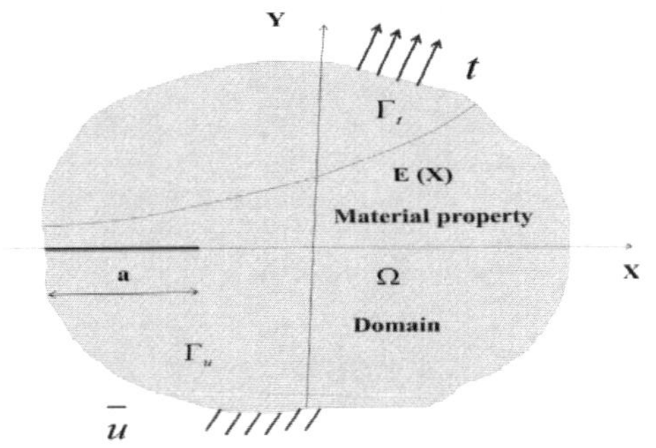

Figure 1. An arbitrary body with edge crack with displacement $\overline{u}$ and traction $\overline{t}$

IOP Conf. Series: Materials Science and Engineering **814** (2020) 012014 doi:10.1088/1757-899X/814/1/012014

$dofs = size(n) + size(n_c) + size(n_t)$ Where, $n_t = n_{t1} + n_{t2}$ (2)

Here, n, n_c and n_t are the nodes for FEM mesh, crack face and crack tip respectively. $H(x)$ denotes the Heaviside function for the enrichment of crack which values are +1 or -1. Crack tips asymptotic functions are represented by Φ_α^1 and Φ_α^2. The displacement function or asymptotic function near crack tip is represented as:

$$\Phi_\alpha^{t1} = \sqrt{r}\sin\left(\frac{\theta}{2}\right), \Phi_\alpha^{t2} = \sqrt{r}\cos\left(\frac{\theta}{2}\right) \quad (3)$$

$$\Phi_\alpha^{t3} = \sqrt{r}\sin\theta\cos\left(\frac{\theta}{2}\right), \Phi_\alpha^{t4} = \sqrt{r}\sin\theta\cos\left(\frac{\theta}{2}\right) \quad (4)$$

The relationship between SIF and J-integral in 2D can be represented as,

$$J = \frac{K_I^2 + K_{II}^2}{E_{eff}} \; Where, E_{eff} = \begin{cases} E & \text{for plane stress} \\ \frac{E}{1-\nu^2} & \text{for plane strain} \end{cases} \quad (5)$$

The J integral for the body with crack is given in equation (6), where Γ is the boundary surface, W is the strain energy, δ_{1j} is Kronecker delta, n_j is the outward unit normal to Γ for j^{th} component, σ_{ij} is stress tensor and u_i is displacement vector. A crack body is represented as two states and the sum of two stare is represented by equation (7).

$$J = \int_\Gamma \left(W\delta_{1j} - \sigma_{ij}\frac{\partial u_i}{\partial x_1} \right) n_j d\Gamma \quad (6)$$

$$J^{(1+2)} = \int_\Gamma \left[\frac{1}{2}\left(\sigma_{ij}^{(1)} + \sigma_{ij}^{(2)}\right)\left(\varepsilon_{ij}^{(1)} + \varepsilon_{ij}^{(2)}\right)\delta_{1j} - \left(\sigma_{ij}^{(1)} + \sigma_{ij}^{(2)}\right)\frac{\partial(u_i^{(1)} + u_i^{(2)})}{\partial x_i} \right] n_j d\Gamma \quad (7)$$

On further solving, we get,

$$J^{(1+2)} = J^{(1)} + J^{(2)} + \frac{2}{E_{eff}}(K_I^{(1)}K_I^{(2)} + K_{II}^{(1)}K_{II}^{(2)}) \quad (8)$$

Now, comparing the Equations. (5) and (8) we get,

$$I^{(1,2)} = \frac{2}{E_{eff}}(K_I^{(1)}K_I^{(2)} + K_{II}^{(1)}K_{II}^{(2)}) \quad (9)$$

The SIF in XFEM for two states $K_I^{(1)}$ and $K_{II}^{(1)}$ are evaluating by putting $K_I^{(2)} = 1$ and $K_{II}^{(2)} = 0$ and $K_I^{(2)} = 0$ and $K_{II}^{(2)} = 1$ in equation (9), we get,

$$K_I^{(1)} = \frac{M^{(1,Mode\ 1)}E_{eff}}{2} \; K_{II}^{(1)} = \frac{M^{(1,Mode\ II)}E_{eff}}{2} \quad (10)$$

Where, $I^{(1,\text{Mode I})}$ and $I^{(1,\text{Mode II})}$ are interaction integrals.

The following normalized mean MMSIF for first and second mode SIF under tensile loading are represented as

$$K_I = \bar{K}_I / \sigma\sqrt{\pi a} \; \text{and} \; K_{11} = \bar{K}_{II} / \sigma\sqrt{\pi a} \quad (11)$$

3. Result and Discussion

FGM plate with width W and length L = 4W under different loadings is shown. Material property of FGM plate is the function of distance along width. The material properties utilized for this present study are $E_1 = 1$, $v=0.3$ and unit uniform tensile, shear and point loads are considered. Figure 2 (a-d) shows the geometry of FGM plate with (a) single edge crack plate under uniaxial uniform tensile and shear stress, (b) crack enrichments, (c) deformation contour, and (d) stress contour.

$$E(x) = E_1 e^{\beta x}, \beta = \ln(E_2 / E_1), G_{12} = \frac{0.5\,E}{1+\upsilon} \tag{12}$$

The current approach is well validated from the literature, which is shown in figure 3. Figure 3(a) shows theFGM plate with uniform tensile load and Figure 3(b) shows the normalized K_I of edge cracked FGM plate under uniform tensile load.

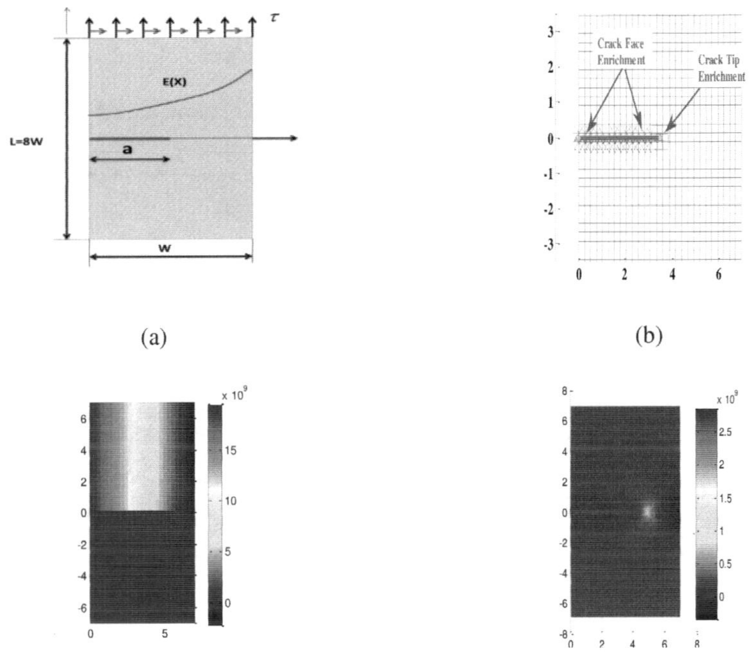

(a) (b)

(c) (d)

Figure 2. Geometry of FGMs plate with (a) edge crack under combined (Tensile and Shear) (b) crack enrichments, (c) deformation contour and (d) stress contour

Table 1 shows the effect of crack length on edge cracked FGM plate with various loadings. The present study is carried out for Crack angle (α) = 45^0 and $E_2/E_1=0.45$. It is observed that as crack length increases from 0.4 to 0.6 MMSIF $K_I$and K_{II} also increase for all loading conditions. For particular crack length maximum MMSIF K_I and K_{II} are observed for combined loading and minimum for uniform tensile loading. The percentage change in K_I when crack length varies from 0.4 to 0.6 for uniform tensile, shear, combined and point loadshare 77, 41, 43 and 75 respectively.

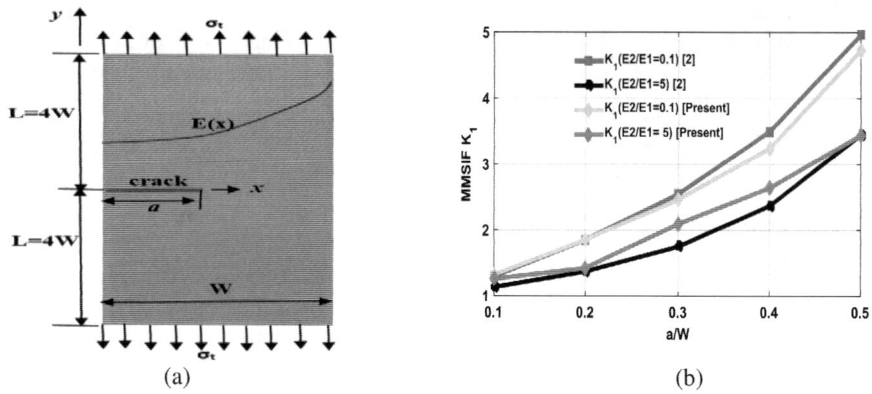

Figure 3. Geometry of FGMs plate with (a) edge crackunder uniaxial stress (b) MMSIF *(K_I)* with respect to *a*/W

Table 1. Effect of crack length on the normalized mean K_I and K_{II} of edge cracked FGM plate with different loadings.

a	SIF	Tensile (UDL)	Shear	Combined	Point Load
0.4	K_I	2.1490	32.4447	34.5937	2.3078
	K_{II}	0.8126	8.2275	9.0401	0.8593
0.5	K_I	2.7843	37.5495	40.3338	2.9681
	K_{II}	1.0191	8.8253	9.8445	1.0695
0.6	K_I	3.8150	45.8914	49.7064	4.0403
	K_{II}	1.2883	9.5875	10.8758	1.3435

Table 2. Effect of crack angle on the normalized mean of K_I and K_{II} of edge cracked FGM plate under various loadings.

α	SIF	Tensile (UDL)	Shear	Combined	Point Load
0	K_I	2.9506	38.7081	41.6587	3.1481
	K_{II}	0.0005	1.8327	1.8332	0.2164
15	K_I	2.8937	38.6665	41.5601	3.0882
	K_{II}	0.3554	2.0189	2.3743	0.3744
25	K_I	2.7932	37.9929	40.7861	2.9826
	K_{II}	0.5740	4.4792	5.0532	0.6048
45	K_I	2.4586	35.1594	37.6180	2.6306
	K_{II}	0.9259	8.7667	9.6926	0.9757

Table 2 shows the effect of crack angle on edge cracked FGM plate with various loadings. The present study is carried out for Crack length (*a*) = 0.45 and $E_2/E_1=0.45$. It is observed that as crack angle increases from 0^0 to 45^0 K_I decreases whereas K_{II} increases for all loading conditions. For particular crack angle maximum MMSIF $K_I$and K_{II} are observed for combined loading and minimum for uniform tensile loading. The percentage change in K_I when crack angle varies from 0 to 45 for uniform tensile, shear, combined and point loads are 16, 9.16, 9.69 and 16.43 respectively.

4. Conclusions

In the present study MMSIF of edge crack FGM plate with different loading conditions are presented. The mathematical formulation is carried out by utilizing XFEM. The conclusions from the present study are.

- MMSIF K_I and K_{II} are calculated for different crack length and crack angle and it is observed that effect of combined loading is maximum with respect to other loading conditions. This is due to combined effect of opening and shearing mode.
- It is observed that as crack length increases, MMSIF K_I and K_{II} also increase.
- Crack length is one of the most critical parameter for the fracture behavior of materials.
- On comparing uniform tensile and central point load, central point load is more critical because whole concentrated load is at center of plate, which is near to the crack whereas in uniform tensile same amount of load is distributed throughout the width of plate.

References

[1] Kim J H and Paulino G H 2002 *Int. J. Numer. Methods Eng.* **53** 1903–1935.
[2] Man S, Huaping W, Long L and Chai C 2014 *Int. J. Mech. Mater Des.* **10** 65–77.
[3] Lal A and Palekar S P 2015 *Int. J. Mech Mater Des.* **13-2** 195-228.
[4] Bayesteh H and Mohammadi S 2013 *Journal of Composites Part B* **44** 8-25.
[5] Mohammadi S 2012 *John Wiley and Sons Ltd. Publication*
[6] Asadpoure A, Mohammadi S and Vafai A 2006 *Thin walled Struc.* **44** 1031 – 1038.
[7] Kim J H and Paulino G H 2003 *International journal of solids and structures* **40** 3967-4001.
[8] Ebrahimi S H, Asadpoure A and Mohammadi S 2007 *Int. J. Civil Eng.* **6-3** 198-207.
[9] Lal A, Mulani S B and Kapania R K 2017 *journal of applied mechanics* **9-4** 1750061.
[10] Khatri K and Lal A 2018 *Mechanics of Advanced Materials and Structures* **25-9** 732-755.

IOP Conf. Series: Materials Science and Engineering **814** (2020) 012015 doi:10.1088/1757-899X/814/1/012015

Multi response optimization in WEDM of H13 steel using hybrid optimization approach

L Rathod[1], NS Poonawala[2], R Rudrapati[3]

[1]Ph.D. Scholar,Department of Mechanical Engineering, JJT University, Rajasthan, India
[2]Professor, Department of Mechanical Engineering, JJT University, Rajasthan, India
[3]Assistant Professor, Industrial Engineering Department, Bule Hora University, Ethiopia

E-mail: lakhanglb03@gmail.com

Abstract. Surface roughness and material removal rate (MRR) are important technological parameters which describe quality of machined surfaces and productivity of machining process. Surface roughness and MRR are significantly influenced by many interactive process parameters dynamically but those are difficult to quantify adequately in any machining process. Wire electrical discharge machining (WEDM) is one of the advanced machining processes which used thin wire as cutting tool for creating intricate features on machined parts. In WEDM, optimizing surface roughness(Ra) and material removal rate (MRR) combinedly by controlling process variables namely pulse on time, pulse off time, wire feed, voltage gap, etc, is difficult task and much needed area of research. In the present work, investigation is made to study and optimize the surface roughness of H13 steel in WEDM. L16 orthogonal array of Taguchi methodology has been used to conduct the experiments. Analysis of variance (ANOVA) has been applied on experimental data to determine the significance of input parameters on surface roughness and MRR. Mathematical relationships are developed to correlate the machining parameters and output responses: surface roughness and MRR. Contour plots have been drawn to illustrate the combined effects of process parameters on output responses. Multi-objective jaya optimization algorithm (MJOA) applied on developed mathematical equations to predict the multi-responses simultaneously. From the study, it is stated that hybridTaguchi method, RSM and MJOA is useful for studying, modeling and optimizing the multiple responses: surface roughness and MRR, simultaneously in WEDM of H13 steel.

1. Introduction

Electric discharge machining (EDM) is a non-conventionalmetal removal technologywhich uses an electricalspark to cut excess material from the work material immersed in a dielectric fluid [1]. In this process, electrical spark is created between the electrode and workpiece.EDM is used to cut, hard to cut materials and conductive with intricate shapes on it. Wire cut EDMis one of the types of EDM process which emerged as an economical substituteto create three dimensional or sharp angles on machined surfaces like blanking dies, extrusion dies, punches, etc. Since, WEDM is advanced machining process and useful to create complex features on machined parts, surface finish of WEDM'ed parts has greater influence on the qualities of job [2]. Surface finish in WEDM has been found to be affected by varying several factors: input parameters: discharge current, pulse duration, pulse frequency, wire speed, wire tension, dielectric flowrate[3], etc.,work material characteristics, use of cutting fluids, etc., WEDM is complicated machining process which required to control many interaction parameters to obtain the process efficiency. The systematic analysis is required to conduct the WEDM process economically and predictably.

Taguchi method is a robust design tool useful to understand the manufacturing processes / systems effectively to produce better output responses. Response surface methodology (RSM) is important techniques which are useful to postulate mathematical relations between parameters and responses [4]. Jaya optimization algorithm (JOA) is an advanced probabilistic technique used to solve engineering problems [5]. In the present work, Taguchi methodology, response surface methodology (RSM) and multi-jaya optimization algorithm (MJOA) is applied to analyze, model and optimize the control parameters in WEDM to optimize multi-responses: surface finish and material removal rate (MRR) of WEDM'edH13 tool steel material.

2. Literature Survey

Literature review is made in the present work to study the reported articles based on the various aspects of analysis and optimization of WEDM of H13 tool steel and other steel materials using Taguchi method and statistical analysis of variance. The details of literature review are given as follows:

Singh et al.[6] had been carried out experimental research work based on WEDM of H13 die tool steel material to analyze and optimize the surface roughness and MRR. Experimental plans were made by orthogonal array of Taguchi method by considering pulse on time, pulse off time, wire type, and peak current, as input variables. The effects of input parameters on output responses were determined by analysis of variance technique. Lodhi and Agarwal [7] had optimized the surface roughness of AISI D3 steel by controlling the pulse-on-time, pulse-off-time, peak current, and wire feed in WEDM using Taguchi method. L9 orthogonal array design was used to conduct experimental runs. They stated from the research that discharge current was most significant parameter for surface roughness. They were confirmed the predicted parametric condition by Taguchi validator test. Sharma et al. [8] had conducted a experimental work to study the factor effects on output performances of WEDM of high strength low alloy steel material. Authors used RSM to create mathematical relationships between the process variables and output responses and predict the multiple responses. They used ANOVA to determine the significant parameters on responses.

Neeraj et al.[9] were planned to investigate the significances of parameters on surface roughness and MRR in WEDM of high strength steel material using RSM approach. They observed from their study that both the responses were increasing with increase of pulse on time and decreasing with increase of pulse off time. Both responses were modeled and optimized by using RSM. Sunkara et al. [10]performed analysis on WEDM of aluminum alloy to model and optimize the multi responses using genetic algorithm. Agarwal et al. [11] had optimized the input parameters of EDM of D2 steel to predict multi-responses: surface roughness and MRR using RSM and jaya optimization algorithm. Again, Agarwal et al. [12] had used RSM along with JOA to predict radial over cut in EDM of titanium alloy. Rao et al. [13]had optimized the multi responses in EDM, plasma arc machining and micro-EDM using RSM along with multi-objective JOA and found improved responses.

IOP Conf. Series: Materials Science and Engineering **814** (2020) 012015 doi:10.1088/1757-899X/814/1/012015

3. Methodology

3.1 Taguchi method, RSM and jaya algorithm

Taguchi's methodology is an optimization technique which used to predict response through experimental analysis and to enhance the system performance without increasing production cost and time. The idea of orthogonal array of experiments in Taguchi technique is useful to determine the factors which most affect part quality with a reduced number of experiments, It uses statistical method of signal-to-noise (S/N) ratio to quantity performance of input parameters where signal is favourone and noise is unwanted value. S/N ratio in Taguchi method uses three different objective functions: larger-the-better, smaller-the-better and nominal-the better to analyze the response variable. The equation for smaller is the better and larger-the-better criterions are shown in Equations.1 and 2 respectively.

Smaller the better: $S/N = -10\log 1/n\left(\sum y^2\right)$ (1)

Larger-the-Better: S/N ratio = $-10\log_{10}(1/n\sum^{n}_{i=1}1/y_i^2)$ (2)

3.2 Response Surface Methodology

RSM is a statistical based mathematical modeling tool which is used to generate the relationships between the process parameters and performance variables. Second-order polynomial regression model is proposed and generated to correlate the independent process variables and dependent responses variable and given in Equation 3.

$$Y = \beta_0 + A\times\beta_1 + B\times\beta_2 + C\times\beta_3 + D\times\beta_4 + A_2\times\beta_{11} + B\times\beta_{22} + C_2\times\beta_{33} + D_2 \times\beta_{44} + A\times B\times\beta_{12} + A\times C\times\beta_{13} + A\times D\times\beta_{14} + B\times C\times\beta_{23} + B\times D \times\beta_{34} \quad (3)$$

Where, Y is output response, A, B, C and D are input parameters and β's are constants which are determined from experimental data by using least square method

3.3 Jaya algorithm

Let f(x) is the objective function to be optimized. In eachruni, 'm' number of design parameters(j=1,2…, m), and 'n' number of candidate solutions (population size, k=1,2,…,n) are selected. Let the best candidate best attains the best value of f(x) in the total candidate solutions and the worst candidate worst attains the worst value of f(x) in the entire solutions alternatives. If X_j,$k_i$is the value of jth variable for the kth candidate during the i^{th}repetition, then this value is alteredbased on Equation 4.

$$X'_{j,k,i} = X_{j,k,i} + r_{1,j,i}(X_{j,best,i-}\,|X_{j,k,i}|) - r_{2,j,i}(X_{j,worst,i-}\,|X_{j,k,i}|) \quad (4)$$

Where, $X_{j,best,i-}$is the value of the variable j for the best candidate and $X_{j,worst,i-}$ is the value of the variable j for the worst candidate. $X'_{j,k,i}$is the updated value of $X_{j,k,i}$and r1,j,i and r2,j,i are the two random numbers for the jth variable during the ith iteration in the range [0, 1]. The term "$r_{1,j,i}(X_{j,best,i-}\,|X_{j,k,i}|)$" indicates the tendency of the solution to move closer to the best solution and the term $r_{2,j,i}(X_{j,worst,i-}\,|X_{j,k,i}|)$" indicates the tendency of the solution to avoid the worst solution.

$X'_{j,k,i}$ is accepted if it gives better function value. All the accepted function values at the end of iteration are maintained and these values become the input to the next iteration.

4. Experimentation

The material used for this study is tool steel of H13 steel material. The experimental runs have been conducted on WEDM using L16 orthogonal array of Taguchi methods with four factors four levels each. The selected input variables and L16 orthogonal array designs for performing experiments are shown in Table 1. Experimental runs have been carried out on WEDM of H13 tool steel material.

5. Results and discussion

As mentioned earlier that experimental runs are performed on WEDM machine when machining of H13 steel material as per L_{16} orthogonal array of Taguchi method and response, surface roughness is measured and shown in Table 1. The data given in the Table 1 used to analyze, interpret and optimize using analysis of signal-to-noise ratio and Taguchi methodology.

Table 1. Full factorial design matrix and output response

	Input parameters				Output responses	
S. No	A	B	C	D	Surface roughness	MRR
1	17	3	116	60	2.652	11.625
2	17	4	118	57	2.640	15.500
3	17	5	120	54	2.690	19.375
4	17	6	122	51	3.220	23.250
5	18	3	118	54	2.576	11.625
6	18	4	116	51	2.399	15.500
7	18	5	122	60	2.771	19.375
8	18	6	120	57	2.474	22.250
9	19	3	120	51	2.639	13.625
10	19	4	122	54	2.772	15.500
11	19	5	116	57	2.343	19.375
12	19	6	118	60	2.611	23.250
13	20	3	122	57	2.838	11.625
14	20	4	120	60	2.579	15.500
15	20	5	118	51	2.929	19.375
16	20	6	116	54	2.629	23.250

5.1 Analysis of variance for output responses

The statistical tool analysis of signal-to-noise (AS/N) ratio from MINITAB 16.2 applied on experimental data of surface finish and MRR in WEDM'ed H13 steel. AS/N ratio tables for surface roughness and MRR are given in Tables 2 and 3 respectively. AS/N is very useful to determine the influences of processing variables on output performances [14]. AS/N results are assessed based on Probability (significant parameter, P) statistics [15]. From the Table 2, it is noted that process parameters like voltage gap (A) and pulse on time (C) have significant effects on surface finish (R_a), as its P values are less than 0.05. Pulse off time (D) have considerable influence on surface finish,

IOP Conf. Series: Materials Science and Engineering **814** (2020) 012015 doi:10.1088/1757-899X/814/1/012015

because its P value is very close to 0.05 as found from Table 2. From the A S/N ratio of MRR as shown in Table 3, it is observed that wire feed (B) is most influential for MRR, as its P value is less than 0.05 and very close to zero. Process parameters: voltage gap (A) and pulse off time (D) also have considerable influences on MRR, because itsvalues of P are closer to 0.05, as seen from Table 3.

Table 2. AS/N for surface roughness (Ra)

Source	DF	Seq SS	Adj SS	Adj MS	F	P
A	3	1.7138	1.7138	0.57125	15.39	0.025
B	3	0.351	0.351	0.11699	3.15	0.186
C	3	3.4934	3.4934	1.16446	31.37	0.009
D	3	0.9652	0.9652	0.32175	8.67	0.055
Residual Error	3	0.1114	0.1114	0.03712		
Total	15	6.6347	S = 0.1927; R-Sq = 98.3%; R-Sq(adj) = 91.6%			

Table 3. Analysis of variance for MRR

Source	DF	Seq SS	Adj SS	Adj MS	F	P
A	3	0.4496	0.4496	0.1499	1	0.5
B	3	70.2388	70.2388	23.4129	156.21	0.001
C	3	0.1864	0.1864	0.0621	0.41	0.756
D	3	0.4496	0.4496	0.1499	1	0.5
Residual Error	3	0.4496	0.4496	0.1499		
Total	15	71.7741	S = 0.3871; R-Sq = 99.4%; R-Sq(adj) = 96.9%			

5.2 Mathematical modelling and contour plots

The mathematical relationships between the WEDM control parameters and output responses: surface roughness and MRR are made by RSM approach by following the procedure given in Section 3.1. The obtained mathematical equations are given in Equations. 5 and 6.

$$Y_{Ra} = -421.564+6.23765\times A+15.7102\times B+6.21234\times C-1.48215\times D -0.147583\times A\times A +0.0322500\times B\times B-0.0232500\times C\times C)+0.0172593\times D\times D+0.0454633\times A\times B-0.00744812\times A\times C-0.00441134\times A\times D-0.103000\times B\times C-0.0823333\times B\times D \quad (5)$$

$$M_{RR} = (-(-4074.71+81.0380\times A+144.217\times B+56.0376\times C-11.1203\times D-1.56250\times A\times A +0.0625000\times B\times B-0.203125\times C\times C+0.125000\times D\times D-0.115708\times A\times B-0.189097\times A\times C-0.0272938\times A\times D-0.937500\times B\times C-0.500000\times B\times D)) \quad (6)$$

Contour plots are drawn from the mathematical equations of R_a(Equation 5) and MRR (Equation 6) using RSM applications in MINITAB 16.1 software. Contour plots for R_a are shown in Figure 1and for MRR are given in Figure 2.Contour plots are very useful for visualize the direct and combined effects of input parameters graphically on output responses [16, 17]. From the Figures 1 and 2, it is observed that process parameters are most significant for both the responses, because curved lines / bent lines / elliptical nature found in said plots.

5.3 Optimization by multi-jaya optimization algorithm (MJOA)

The weightage multi-objective JOA is applied on mathematical models of Ra and MRR of WEDM'ed H13 steel as given in Equations5 and 6 respectively to predict both the responses simultaneously with weightage of 22% for surface roughness and 78% weightage for MRR. And obtained parametric condition which corresponding to optimize both responses according to given weightage and corresponding response values are given in Table 4.

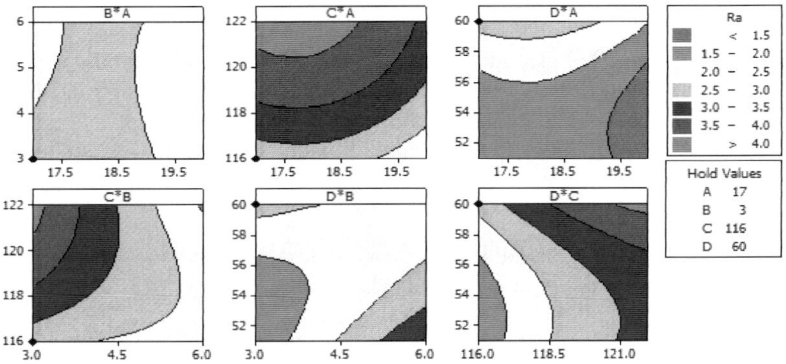

Figure 1 Contour plots showing effects of process parameters on R_a

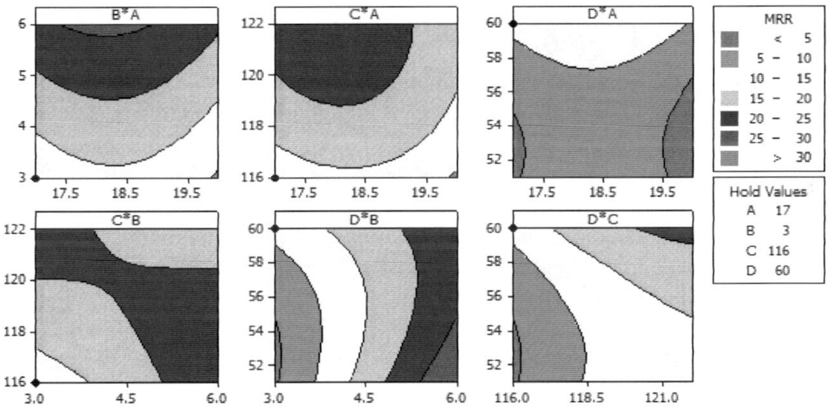

Figure 2 Contour plots showing effects of process parameters on MRR

Table 4. Multi-objective optimization by MJOA

Optimum condition by MJOA		
Parametric condition		Output response
Voltage (A)	17.98 volts	$Y_{Ra,min}$ = 2.56 μ
Wire feed (B)	6 mm/min	
Pulse on time (C)	116 μs	$Y_{MRR,max}$=25.82 mm^3/min
Pulse off time (D)	60 μs	

5.4 Confirmatory test

Confirmatory experiments are conducted for each response to validate the optimized WEDM parametric conditions obtained by MJAO for H13 material.From the confirmatory result, it is noticed that optimum parametric conditions provided are satisfactory results and those are good agreement with the experimental values given in Table 1.

6. Conclusion

Experimental investigation made to optimize the R_a and MRR in WEDM of AISI H13 tool Steel using Taguchi method, RSM and MJOA. The followings are the conclusions drawn from the study:

- From the Analysis of signal-to-noise ratio, it is found that voltage gap and pulse on time are significant for Raand wire feed is significant for MRR
- Mathematical models are developed for input parameters and output responsesby using RSMContour plots reveals that direct and interactions effects of process parameters are most significant for both the responses
- Multi-responses: surface roughness and MRR optimized simultaneously using multi-JOA
- Confirmatory test confirms the predicted WEDM condition for optimizing both the responses simultaneously
- From the resent investigation, it is stated that hybrid Taguchi method, RSM and MJOA is advantages for analyzing, modeling and optimizing multi responses in WEDM process.

References

[1] Rudrapati R 2018 *Stainless Steels and Alloys* (London: Intech Open)
[2] Huang JT, Liao YS and Hsue WJ 1999 *J Mater Proc Technol.* **87** 69
[3] Mahapatra S S and Amar P 2007 *Int. J. Adv. Manuf. Technol.* **34** 91
[4] Patil A and Rudrapati R 2019 *System Reliability, Quality Control, Safety, Maintenance and Management(Singapore: Springer Nature)*
[5] Rudrapati R and Patil A 2019 *Materials Sci. Forum* **969** 631
[6] Singh K, Goyal K and Goyal D K 2018 *Adv. Engi. Forum* **28** 55
[7] Lodhi B K and Agarwal S 2014 *Procedia CIRP* **14** 194
[8] Sharma N, Khanna R, Gupta R D and Sharma R 2013 *Int J Adv Manuf. Technl.* **67** 2269
[9] Neeraj S, Rajesh K and Gupta R 2013 *Procedia Engi.* **64** 710.
[10] Sunkara J K, Kayam S K,Monduru G K, Padaga K B 2019*Multiscale and Multidiscip Model Exp and Desdoi.org*/10.1007/s41939-019-00062-1.
[11] Agarwal N, Pradhan M K and Nitin S 2018 *Mater. Today: Proce.* **5** 23759-768.
[12] Agarwal N, Nitin S and Pradhan M K 2019 *Int. J. of Engi. And Adv. Technol.* **85** 274.
[13] Rao RV, Rai DP, Ramkumar J andBalic J 2016*Adv in Prod EngiManag* **11** 271.
[14] Ghosh N, Rudrapati R, Pal P K and Goutam N 2017 *Mater Today: Proc* **4** 2213.
[15] Rudrapati R, Bandyopadhyay A and Pal P K 2013 *Adv Mater Manuf. Charact* **3** 335.
[16] Rudrapati R, Pal P K and Bandyopadhyay A 2012 Int. J. of Mach. and Machina of Mater **12** 28.
[17] Rudrapati R, Sahoo P and Bandyopadhyay A 2016 *IOP Conf. Seri: Mater Sci. and Engi.* **149** 012039

Analysis, modeling and optimization of surface roughness in cylindrical traverse cut grinding using factorial design, RSM and simulated annealing algorithm

R Rudrapati[1], A Bandyopadhyay[2], P K Pal[3], L Rathod[4]

[1]Assistant Professor, Industrial Engineering Department, Bule Hora University, Ethiopia
[2, 3]Professor, Mechanical Engineering Department, Jadavpur University, Kolkata, India
[4] Ph.D. Scholar, Department of Mechanical Engineering, JJT University, Rajasthan, India

E-mail: rameshrudrapati@gmail.com

Abstract. In the present investigation, a new hybrid technique response surface methodology (RSM) based simulated annealing (SA) algorithm approach is proposed to predict surface roughness for stainless steel material in traverse cut cylindrical grinding process. Experiments are designed as per full factorial design, wherein infeed, longitudinal feed and work speed have been considered as the important input parameters. Analysis of variance and graphical main and interaction plots have been plotted for analyzing the experimental data for identifying the relationships between grinding parameters and surface roughness. The variation of the performance parameter (surface roughness) with grinding parameters has been mathematically modeled by RSM. SA algorithm has been employed for solving the obtained mathematical model. Finally, the validation exercise is performed with optimum levels of grinding parameters. The results confirm the efficiency of the approach employed for prediction of surface roughness in this study.

1. Introduction

Nowadays, a lot research work is carried out in metal cutting industries for optimizing control parameters for improving accuracy and surface qualities of the machined surface. Cylindrical grinder is an efficient and effective machining / finishing operations used in the mechanical industries to create the cylindrical parts with high profile accuracy as well as better surface roughness [1]. The complex machining structure in the cylindrical grinding process i.e. simultaneous rotation of grinding wheel and work speed and traverse movement of the work table along with other interactive parameters such as grinding wheel properties, work-piece and machine parameters are limiting the ability of grinding process to produce accurate and fine finish jobs [2]. Selection of process parameters seeking better machining economics in cylindrical grinding is important task. Wrong selection of parametric setting can have detrimental effects on quality of the ground surface and decreased efficiency grinding process.

To understand the significances of process conditions on output responses and developing relationships between the input and out parameters and optimize them is one of the important areas of research in traverse cut cylindrical grinding. Literature reveals that systematic optimization methodology based on design of experiments (DOE) could have ability to optimize the metal cutting operations (like grinding, turning, milling etc.) predictably, much research work is done so far in this respect. Full factorial design (FFD) is important methods of DOE, which was used by Baek et al. [3] for optimizing the grinding conditions and again, Thomas et al. [4] used same optimization methodology to analyze turning operation. Response surface methodology (RSM) is one of the widespreadtechniques of DOE. Kwak et al. [2] reported a study on RSM to optimize the surface finish and power spent during grinding in cylindrical external grinding operation. Lakshmi and Subbaiah [5] used integrated FFD along with RSM to optimize milling operations. Sahin and Riza [6] had been experimented and optimized the cutting conditions in turning using integrated factorial design and RSM approach. Rudrapati et al. [7] had made an experimental study to determine the factor effects and optimize the output responses by using RSM and genetic algorithm. Again, Rudrapati et al. [8] used RSM and TLBO to predict the performance characteristics in CNC turing operation. Adalarasan et al. [9] had been optimized the welding conditions in friction welding operation by Taguchi-SAA approach. Liao and Chen [10] utilized the applications of RSM along with SAA to analyze, model and optimize the process conditions in laser welding. Gohari et al. [11] had also optimized the milling process conditions to predict output responses by using RSM and SAA. Ganapathy [12] was analyzed turning operation to control turning parameters to predict output response by using regression analysis and SAA approach.

Present study is mainly focused to identify the effects of process settings and optimize them in traverse cut cylindrical grinding when machining of stainless steel. Experimental runs have been planned by factorial design and factor effects have been studied by analysis of variance (ANOVA) and graphical main and interaction plots. A new hybrid technique RSM based SA algorithm methodology is proposed and used in the present study to determine the global optimal conditions for traverse cut cylindrical grinding process.

2. Experimental planning and optimization techniques

Design of experiments (DOE): a procedure to attain a prognostic in formation of complex, multi-process conditions with minimum acceptable tests. Factorial design and RSM, which are two major approaches in DOE, those are used in the present study for examining and modelling the process parameters in cylindrical grinding operation.

A design in which each variable appears with every combination of every other variable is called full factorial design (FFD). In FFD, output performances are assessed at all settings of the input parameter levels. A three-factor three-level- FFD has been employed to plan the experiments. Factorial design allows for studying effects of each variable on the response characteristic and combined influences of parameters on the response(s) through graphical main and interaction effect plots. Significances of grinding variables on responses can be examined by using analysis of variance (ANOVA) method and it is comparatively simple to estimation of major effects of a parameter.

RSM is a set of mathematical procedures, employed in the present work to investigate the relationship between performance variables and a set of quantifiable experimental factors. RSM is often utilized to build the mathematical relation between controllable factors and output response [13]. Furthermore, developed mathematical equation can be utilized to determine the operating condition that produces improved response, satisfies process requirements and identifies optimum parametric condition which excepted produces enhanced component quality over the quality obtained [14]. Second order mathematical model developed by RSM is given as follows:

$$Y = \beta0+\beta1(A)+\beta2(B)+\beta3(C)+\beta11(A2)+\beta22(B2)+ \beta33(C2)+\beta12(A*B)+\beta13(A*C)+\beta23(B*C) \quad (1)$$

where, all β's are regression measurements; A, B, C are input process variables and Y is the performance characteristic which is needed to be enhanced.

2.1 Simulated annealing (SA) algorithm

In the present study, simulated annealing is proposed to solve the mathematical equation to optimize the surface finish. It was introduced in 1982 by Kirkpatrik et al [15]. It is a stochastic optimization technique that able to find global optima by using probability function. It works based on single point search method and then it resembles the coding procedure of molten metals by annealing process [16]. The atoms in the molten metal can move freely about to each other at higher temperature. If the temperature is slowly minimized and the passage of the atoms gets suppressed and starts to make well-ordered. And then crystals have been formed with least amount of energy levels. However, creation of crystal varies on cooling level(s). If the temperature is lowered drastically, the polycrystalline state has been formed, which may contains higher energy levels than crystalline state. So, in order to get the absolute lowest possible energy level / state, cooling has been done at slow rate. The method of slow cooling is termed as annealing. The simulated annealing process replicates this method of slow cooling of molten metal to obtain the minimum function value in optimization problems [17, 18]. While solving the second order response mathematical model by using SA in the present work, similar procedure is followed to optimize grinding conditions. In the present investigation, MATLAB software is utilized to optimize surface finish. SAA's basic operations are taken care of by SA optimization toolbox of MATLAB and finally, it produces the optimal combination for the desired surface roughness (Rq).

3. Experimental details

As stated earlier that present investigation is intended to analyze and optimize the control parameters in traverse cut cylindrical grinding. Three levels of the process variables are selected in the present work and shown in Table 1. Experimental runs have been conducted on stainless steel material on cylindrical grinding machine as per full factorial design as given in Table 1. After finishing the experimental runs, surface finish (Rq) has been evaluated using stylus-type profilometer. Surface roughness has been measured in three different places and the averaged. Surface roughness parameter Rq has been chosen for present study, because, it is important roughness parameters which is used to define the condition of the machine parts. The observed data are discussed and analyzed in the next section.

4. Results and discussion

As mentioned earlier, FFD experiments have been done and the corresponding output response is observed by measuring surface roughness. The output results along with full factorial design matrix are shown in Table 1. The data shown in the Table 1 has been used to analyze and optimize the cylindrical grinding process to enhance / minimize the surface finish by using ANOVA and RSM based SAA.

4.1 Parametric influence on surface roughness

Analysis of variance (ANOVA) technique is employed to determine the relative magnitude of the effect of each variable on Rq and to approximate the error variation. The larger the variable influence relative to the error variance may be assessed from the F-column. Higher the F-value, the more factor effect is compare to the error variance [19]. From Table 2, it is concluded that work speed is the variable which has greater influence on surface finish (Rq).

Table 1. FFD matrix and output response

S. No	Infeed (A)	Longitudinal feed (B)	Work speed (C)	Surface roughness (Rq)
1	0.04	70	160	0.995
2	0.06	90	80	0.744
3	0.06	90	160	0.874
4	0.05	90	80	0.961
5	0.06	80	112	1.246
6	0.05	90	112	1.503
7	0.04	80	112	1.33
8	0.05	90	160	1.077
9	0.04	80	80	1.39
10	0.04	70	112	1.077
11	0.04	90	80	0.858
12	0.06	70	80	1.08
13	0.05	70	112	1.253
14	0.04	90	160	0.995
15	0.06	80	80	0.979
16	0.05	80	80	0.968
17	0.04	80	160	1.11
18	0.05	70	80	0.924
19	0.05	80	160	1.003
20	0.06	80	160	1.41
21	0.04	70	80	1.22
22	0.06	70	112	1.193
23	0.06	90	112	1.313
24	0.05	80	112	1.147
25	0.04	90	112	0.818
26	0.06	70	160	1.123
27	0.05	70	160	0.984

The main (Figure 1) and interaction (Figure 2) effect plots for process parameters Vs. surface roughness (R_q) are illustrated by taking the mean values of R_q. These plots are very useful to identify the factor effects, individually as well as combinedly. In a main / interaction plot, when the lines are parallel, main / interaction effects are zero. The slopes of the lines are more, the more influence the main / interaction effect has on the response. From Figure 1, it is noted that longitudinal feed (B) has larger effect on Rq and next is work speed (C). Infeed is insignificant as the line is almost parallel, as found from the Figure 1. From Figure 2, it is concluded that the non-parallelism of the lines in a plot reveals that considerable amount of the interaction exists between the two parameters, although interconnecting lines are strong symptom of interaction effect on R_q.

Table 2. Analysis of variance for surface roughness (R_q)

Source	DF	Seq SS	Adj SS	Adj MS	F	P
A	2	0.00183	0.00183	0.00092	0.05	0.955
B	2	0.11521	0.11521	0.05761	2.92	0.111
C	2	0.18507	0.18507	0.09253	4.7	0.045
A*B	4	0.23062	0.23062	0.05765	2.93	0.092
A*C	4	0.19776	0.19776	0.04944	2.51	0.125
B*C	4	0.0674	0.0674	0.01685	0.86	0.529
Residual Error	8	0.15765	0.15765	0.01971		
Total	26	0.95555				

4.2 Mathematical modeling and optimization

As already mentioned above, RSM comprises of combination of mathematical and statistical techniques which is applied on experimental data and the mathematical equation is developed and shown in Equation 2.

$$Y_{Rq} = -5.303-37.027*A+0.172*B+0.012 *C+63.889*A2-0.001*B^2-0.0001*C^2+0.13*AB+0.18*AC \quad (2)$$

where, Y_{Rq} = output response, A = infeed, B = longitudinal feed and C = work speed. Mathematical equation (Equation 2) can be used to predict the optimum conditions for obtaining the better R_q value within the limits of process conditions. The simulated annealing algorithm is used in the present case, to solve the obtained mathematical model i.e. Equation 2. The steps are followed by using optimization toolbox in MATLAB software: 1. Select the objective function that is to be predicted, 2. Select the starting points of input parameters, 3. Fixing the lower and higher limits of input parameters, 4. Running the solver

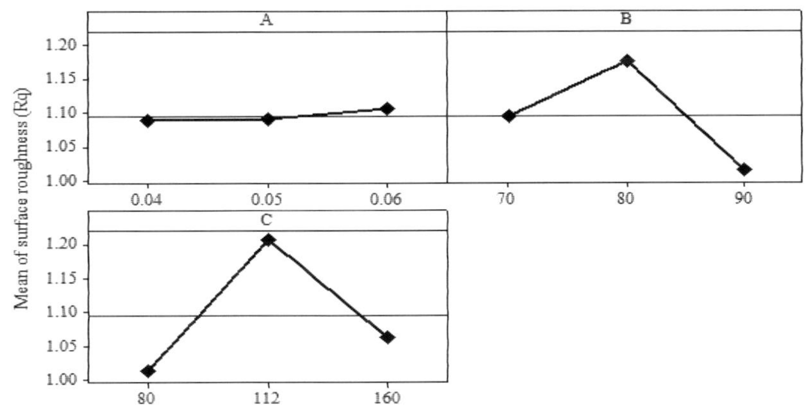

Figure 1. Main effect plots for Rq

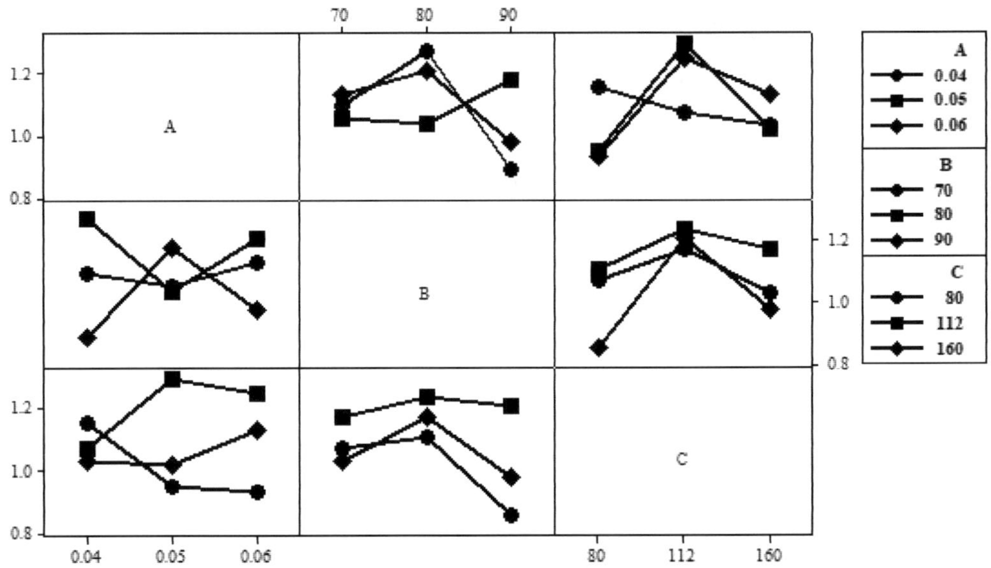

Figure 2. Interaction plots for Rq

At each iteration in the SA toolbox, a new combination of optimal parametric condition as well as output response variable is generated by the process. Optimum grinding condition is found from the SA toolbox is: infeed (A) = 0.06 mm/cycle, longitudinal feed (B) = 89.95 ≈90 mm/s and work speed (C) = 80.02 rpm and surface roughness (Rq) = 0.8599 μm. This setting is obtained in the range of the process variables selected in the work. Confirmatory experiment revels that the validity of the proposed hybrid optimization methodology for prediction of surface finish in stainless steel grinding.

5. Conclusion

Based on experimental results, optimal parametric setting for good surface roughness using RSM based SA algorithm using factorial design the following points can be concluded as listed below:

- From the ANOVA results, it is found that work speed is the factor that has more influence on surface roughness (Rq) than other factors as well as their interactions.
- From the main effect plots, it is noted that longitudinal feed and work speed are the significant factors for surface roughness
- Interaction effect plots revelled that all the interaction effects of input parameters have significant effect on output response (Rq).
- The second order mathematical equation is made by RSM to build relationships between input factors and output variable; this equation can be used for predicting Rq value for a given set of grinding parameters.
- The optimum grinding setting is obtained by using SAA is: infeed = 0.06 mm/cycle, longitudinal feed = 90 mm/s and work speed = 80 rpm. The optimum condition obtained from SA has been validated by confirmatory test.

References

[1] Rudrapati R, Pal K P, and Jaiswal N 2019 *J. of Prod. Research & Management* **6** 20-24.
[2] Kwak J S 2005 *Int. J. of machine Tools & Manufacture* **45** 327-34.
[3] Baek S K, Lee J K, Lee E S and Lee H D 2007 *Engg. materials* **329** 27-32.
[4] Thomas M, Beauchamp Y, Yossef Y A and Masounave J 1997 *Int. J. Quai Sci.* **2** 167-80.
[5] Lakshmi V K and Subbaiah K V 2012 *Int. J. of Engg. Research and Applications* **2** 674.
[6] Sahin Y and Riza A M 2005 *Materials and Design* **26** 321-26.
[7] Rudrapati R, Bandyopadhyay A and Pal P K 2018 *Iranian J. of Mechanical Engg.* **19** 34-62.
[8] Rudrapati R, Priya S and Bandyopadhyay A 2016 *IOP Conference Series Material Sci. and Engg.* **149** 1-13.
[9] Adalarasan R, Santhanakumar M, and Sundaram A S 2014 *Appl. Mechanics and Materials* **592** 595–599.
[10] Liao H T and Chen Z W 2013 *Int. J. of Advanced Manufacturing Technol.* **67** 1015–25.
[11] Gohari H, Hegab H A and Neamat G S A 2016 *American J. of Engg. Research* **5** 227-232.
[12] Ganapathy S 2016 *Int. J. of Multidisciplinary Research and Modern Education* **2** 1-11.
[13] Rudrapati R and Rathod L 2019 *Trans Tech Publications Ltd. Materials Sci. Forum* **969** 656-61.
[14] Das M, Rudrapati R, Ghosh N and Rathod L 2016 *Int. J. of Current Engg. and Technol.* **109** 12.
[15] Kirkpatrick S, Gelatt C D and Vecchi M P 1983 *Sci.* **220** 671-80
[16] Kolahan F, Abolbashari N H and Mohitzadeh S 2007 *Academy of Sci. Engg. and Technol.* **35** 326-329.
[17] Yang S H, Srinivas J, Mohan S, Lee D M and Balaji S 2009 *J. of Material Processing Technol.* **209** 4471-75.
[18] Mohd Z A, Haron H and Sharif S 2011 *Appl. soft computing* **11** 5350-59.
[19] Nixon K and Ravindra H V 2011 *Machining Sci. and Technol* **15** 47-75.

IOP Conf. Series: Materials Science and Engineering **814** (2020) 012017 doi:10.1088/1757-899X/814/1/012017

Numerical study of edge crack interaction with inclusion and void of an isotropic plate under different loadings by extended finite element method

Achchhe Lal[1], Manoj B. Vaghela[2] and Kundan Mishra[3]

[1]Assistant Professor, Department of Mechanical Engineering, SVNIT, Surat
[2,3]Research Scholar, Department of Mechanical Engineering, SVNIT, Surat

E-mail:lalachchhe@yahoo.co.in

Abstract: In the present work Extended Finite Element Method (XFEM) is utilized for interaction study of cracked isotropic plate under different loading conditions. XFEM is a powerful tool to handle fracture mechanics problems where, extra enrichment functions for the crack faces and tip are get utilized. In this present work main focus is interaction study of edge cracked isotropic plate with inclusion and void under tensile, shear, combined (tensile and shear) and exponential loading. Here, MATLAB environment is used to evaluate mixed mode stress intensity factor (MMSIF) of edge cracked isotropic plate when interacted with inclusion and void.

1. Introduction

In many practical applications in spite of using advanced production and operational methods it is very difficult to make defect freecomponent. Materials during casting, drilling, grinding etc. can have various unavoidable defects just as crack, inclusion, void etc. which degrade the material and finally affect the life. So, interaction study of cracked isotropic plate with inclusion and void under different loading conditions for higher safety reliability.

Structures are always subjected to various types of inplane loadings such as tensile, shear, combine (combination of tensile and shear) and exponential. Under the action of such loadings crack propagation behavior is abnormal. Hence the evaluation of crack behavior under such loading conditions is one of the areas of research for higher safety and reliability. The interaction between a crack and inclusion in the material is very important for understanding the fracture behavior and improving the fracture resistance. The crack tip in various engineering materials is affected by the geometry and the stiffness of near second phase inclusions and/or voids. The density and geometry of inclusion, the stiffness ratio between inclusion and voids and materials play a major role to enhance or reduce the stress at the crack tip. Crack-inclusion, interaction studies have great importance to understand the near crack-tip field for the fracture behavior of the material. In this direction Sih [1] proposed energy density based fracture mechanics, where "energy-density factor" (*S*) is measured similar to stress intensity factor (*K*). The fracture toughness of the material is measured from critical value of S and this proposed theory is also well suited for various mixed-mode crack problems.

Chienet et al [2] applied the concept of singularity at the crack tip with conventional FEM. In this present study SIF is calculated for edge cracked plate and compared the accuracy of these results. Sukumar and Prevost [3] presented an analysis of the 2-D crack of an isotropic and biomaterial plate by XFEM. In this present approach discontinuity is modelled by utilizing partition of unity in the standard finite element package. Li and Chen [4] studied the crack inclusion study based on transformation toughening theory for the prediction of stress intensity factors of cracked plate under loading. Sukumar et al [5] utilized XFEM with level set function for modelling discontinuities like hole and inclusion. The local enrichment functions are also incorporated for better accuracy of results. Limtrakarn and Dechaumphai [6] presented crack inclusion study of polycarbonate by adaptive FEM. In the present study SIF K_I and K_{II} are calculated and the accuracy of the present resultsis compared with results from photoelastic technique. Zhang et al. [7] Studied crack inclusion, interaction by distributing dislocation method with Gauss–Chebyshev quadrature and the resultobtained from the present study is compared with the result of FEM. Sharma [8] implemented XFEM based fracture analysis of a cracked plate with other discontinuities. In this present study,the interaction integral method is used to calculate SIF and conformed that SIF at the crack tip is reduced due to interaction of hard inclusion. Jiang et al [9] presented XFEM based fracture analysis of a plate with multiple discontinuities and the crack path is observed when interacted with hard and soft inclusions. Lal et al [10] presented stochastic fracture response of edge cracked composite beam under different loadings. In this present study randomness of various parameters is considered to check the severity of individual fracture parameter. The result obtained is well compared with other method in the literature. Khatri and Lal [11] investigated stochastic fracture analysis of the behavior of a hole emerging crack of a plate under in-plane loadings by XFEM. It is observed that the crack length of shorter length is also very sensitive because of very high stress concentration at the crack tip.

From the above literature review, it is observed that researchers are showing great interest in XFEM based fracture mechanics with the interaction of crack and inclusions/voids. In this present work main focus is numerical investigation of various discontinuities like crack, voids, soft and hard inclusions in an isotropic plate with an edge crack through implementing XFEM under different loading (Tensile, shear, combined and exponential).

2. Mathematical Formulation

Modelling of crack growth in finite element method is time-consuming and here it's necessary to update meshing at every step of crack growth. In XFEM remeshing is not required during crack growth and discontinuities are modelled by enrichment functions. Displacement vector in XFEM framework is given below.

$$u^h(x) = \sum_{i=1}^{n} N_i(x)\overline{u_i} + \sum_{i=1}^{n_c} N_i(x)H(x)a_i + \sum_{i=1}^{n_{t1}} N_i(x)\sum_{\alpha=1}^{4}\Phi_\alpha^1(x)b_i^{\alpha 1} + \sum_{i=1}^{n_{t2}} N_i(x)\sum_{\alpha=1}^{4}\Phi_\alpha^2(x)b_i^{\alpha 2} + \sum_{i=1}^{n_h} N_i(x)\chi(x)c_i \quad (1)$$

Where, $\overline{u_i}$ a_i, $b_i^{\alpha 1}, b_i^{\alpha 2}$ and c_i are the conventional degrees of freedom (dofs), and extra dofs is added for crack face, crack tip and for inclusion/void. A body is considered whose area is denoted by Ω and its outer boundary Γ containing a crack and inclusion indicated by Γ_c, as in Figure 1(a). The body is experiencing uniform volume/ body loads b, and the surface load or forces are applied at the boundary Γ_t. The parameter $\bar{u}$ is the displacement and $\bar{t}$ is the tractions. It is assumed that a surface with crack and inclusion is traction free. Figure 1 (b) represents an isotropic plate with edge crack and inclusion/void under tensile, shear and exponential loadings.

Here, a set of nodes for FEM mesh, enrichment due to fully cut the elements by crack, enrichment for the elements at the crack tip, and enrichment for the inclusion are denoted by n, n_c, n_t and n_h respectively. The total dofs is represented as

$$dofs = size(n) + size(n_c) + size(n_t) + size(n_h) \text{ Where, } n_t = n_{t1} + n_{t2} \quad (2)$$

The parameters $H(x)$ and $\chi(x)$ denote the Heaviside function for the enrichment of crack and inclusion/void which values are +1 or -1. Crack tips asymptotic functions are represented by Φ^1_α and Φ^2_α. The displacement function or asymptotic function near the crack tip is represented as:

$$\Phi^{t1}_\alpha = \sqrt{r}\sin\left(\frac{\theta}{2}\right), \Phi^{t2}_\alpha = \sqrt{r}\cos\left(\frac{\theta}{2}\right), \Phi^{t3}_\alpha = \sqrt{r}\sin\theta\cos\left(\frac{\theta}{2}\right), \Phi^{t4}_\alpha = \sqrt{r}\sin\theta\cos\left(\frac{\theta}{2}\right) \quad (2a)$$

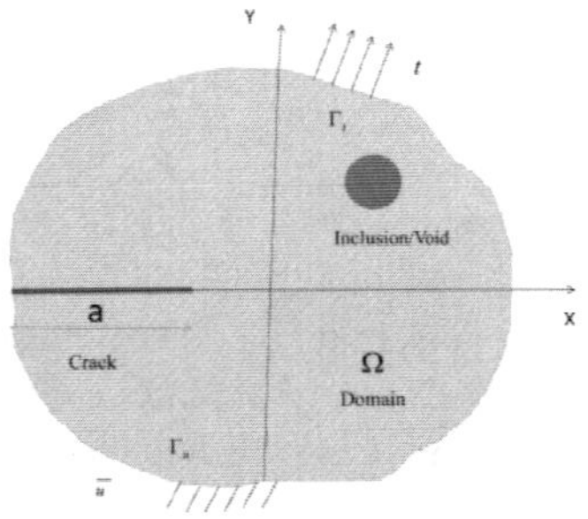

(a)

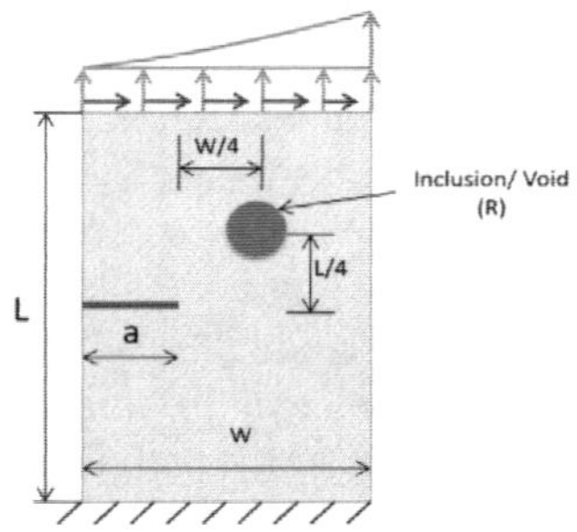

(b)

Figure 1. (a) An arbitrary body with edge crack and inclusion/void, subjected to traction $\bar{t}$ and displacement $\bar{u}$ (b) Edge cracked plate with inclusion under different loading conditions.

The relationship between MMSIFand J-integral for mixed-mode problems in 2D can be represented as,

$$J = \frac{K_I^2 + K_{II}^2}{E_{eff}} \quad Where, E_{eff} = \begin{cases} E & \text{for plane stress} \\ \frac{E}{1-\nu^2} & \text{for plane strain} \end{cases} \quad (3)$$

The J integral for the body with crack is given in equation (4), where Γ area inside contour W is the strain energy density, δ_{1j} is Kronecker delta, n_j is the outward unit normal to Γ for j^{th} component, σ_{ij} is stress tensor and u_i is displacement field vector. A crack body is denoted by two states and equation 5 represents combination of two states.

$$J = \int_\Gamma \left(W\delta_{1j} - \sigma_{ij}\frac{\partial u_i}{\partial x_1} \right) n_j d\Gamma \quad (4)$$

$$J^{(1+2)} = \int_\Gamma \left[\frac{1}{2}\left(\sigma_{ij}^{(1)} + \sigma_{ij}^{(2)}\right)\left(\varepsilon_{ij}^{(1)} + \varepsilon_{ij}^{(2)}\right)\delta_{1j} - \left(\sigma_{ij}^{(1)} + \sigma_{ij}^{(2)}\right)\frac{\partial(u_i^{(1)} + u_i^{(2)})}{\partial x_i} \right] n_j d\Gamma \quad (5)$$

On further solving, get

$$J^{(1+2)} = J^{(1)} + J^{(2)} + \frac{2}{E_{eff}}(K_I^{(1)}K_I^{(2)} + K_{II}^{(1)}K_{II}^{(2)}) \quad (6)$$

Now, from Equations. (3) and (6) get,

$$I^{(1,2)} = \frac{2}{E_{eff}}(K_I^{(1)}K_I^{(2)} + K_{II}^{(1)}K_{II}^{(2)}) \quad (7)$$

Where, I (1, Mode I) and I (1, Mode II) are interaction integrals
The SIFin XFEM for two states $K_I^{(1)}$ and $K_{II}^{(1)}$ are evaluating by putting $K_I^{(2)}=1$and $K_{II}^{(2)}=0$and$K_I^{(2)}=0$ and$K_{II}^{(2)}=1$in Equation. 7, we get

$$K_I^{(1)}=\frac{M^{(1,Mode\ I)}E_{eff}}{2}, \quad K_{II}^{(1)}=\frac{M^{(1,Mode\ II)}E_{eff}}{2} \tag{8}$$

3. Result and discussion.
An edge crack isotropic plate is considered for the validation study. The geometry of an edge crack isotropic plate under uniaxial tensile loading is shown in Figure 2(a). The geometry dimension of this plate is taken 72 mm x 36mm (L x W), crack length ratio (a/W) = 0.3, 0.4, 0.5, 0.6 ($10.8\leq a\leq 21.6$), polycarbonate material (PSI) is taken with E = 2.50 GPa, ν = 0.38, tensile stress σ = 1.1 MPa as taken from [6]

For validation purpose, a standard single edge crack isotropic plate under uniaxial tensile loading is considered. By applying XFEM Mode-1 K_I is determine for a/W = 0.3, 0.4, 0.5, 0.6 and compare with analytical result.Here, 2 (a) show isotropic plate with edge crack and inclusion / void under tensile loading and Analytically SIF can be calculated as:

$$K_1=\sigma\sqrt{(\pi a)}f(\alpha), \text{ Where, } \alpha=\frac{a}{W} \tag{9}$$

$$f(\alpha)=1.12-0.23\,\alpha+10.55\alpha^2-21.72\,\alpha^3+\ 30.39\alpha^4 \tag{10}$$

A validation graph 2 (b) shows, results of present work are very close to the analytical result. Some cases are solved by using this XFEM mixed-mode, numerical solution for an edge crack isotropic plate with inclusion/void under different loadings.

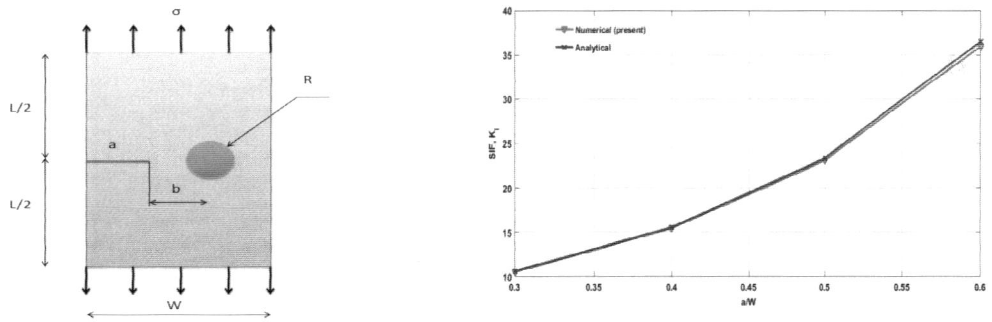

(a) (b)

Figure 2. (a) geometry of an edge crack isotropic plate with a single inclusion/void under different loading (b) Validation graph for K_I with respect to a/W

The geometry dimensions are taken 36 mm × 72 mm (W × L), crack length ratio a/W = 0.2, 0.3, 0.4, 0.5, 0.6 ($7.2\leq a\leq 21.6$), polycarbonate material (PSI) with E = 2.50 GPa, ν = 0.38. A circular void/soft and hard inclusionis considered with radius 6 mm. The position of a void/inclusion is taken at L/4 and W/4 for all cases as shown in Figure1 (b). The soft inclusion is of Teflon with E_{soft}/E_{PSI} ratio = 0.24 and hard inclusion taken of AZ61 with E_{hard}/E_{PSI} ratio = 18, tensile stress σ = 1.1 MPa [6] is considered. The shear stress τ = 1.1 MPa considered and in combine stress tensile and shear both are subjected on the plate. The exponential loading is considered with $p = \sigma \times e^{cx}$, $x(0 \leq x \leq W)$ and $c = 0.02778$.

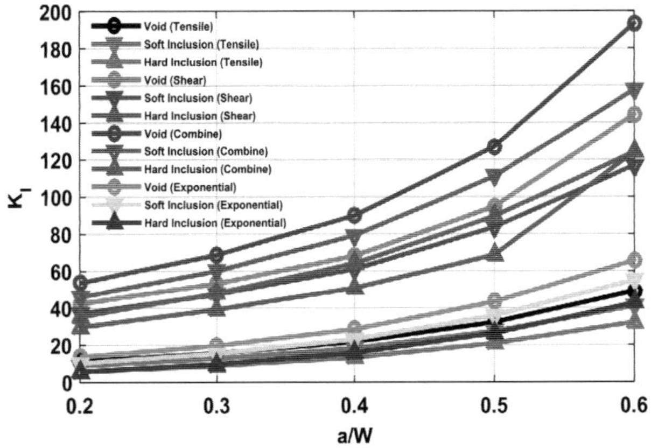

Figure 3. MMSIF K_I for an edge crack isotropic plate with a single inclusion/void under different loadings

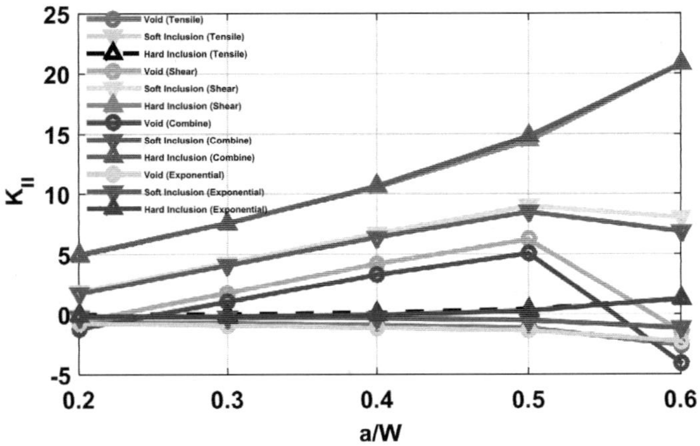

Figure 4. MMSIF K_{II} for an edge crack isotropic plate with a single inclusion/void under different loadings

The same meshing and elements are taken as above validation problem. MMSIF K_I and K_{II} are determined by XFEM method for various crack length ratios a/W = 0.2, 0.3, 0.4, 0.5, 0.6 as shown in Figure 3 and Figure 4, It is observed that MMSIF K_I and K_{II} increase by increasing a/W under different loadings. It is found that for void MMSIF K_I is increased up to 10% in exponential loading, around 3 times under shear loading and around 4 times under combine loading as compare to tensile loading. For soft inclusion, the MMSIF K_I is of the same nature and decrease near to 15%as compared to void under different loading. For hard inclusion, the MMSIF K_I is ofthe same nature and decreases near to 22% as compared to soft inclusion under different loading, which is due to the strength of the plate increase with the interaction with hard inclusion whereas decreases with the interaction with soft inclusion and void. The SIF K_{II} is nearly zero for tensile and exponential loadings whereas for shear and combined loadings are higher. It is concluded that maximum SIF K_I is found for void and

IOP Conf. Series: Materials Science and Engineering **814** (2020) 012017 doi:10.1088/1757-899X/814/1/012017

minimum in hard inclusion. So the plate strength is decreased in XFEM fracture behavior for void and it is increasing in XFEM fracture behavior for hard inclusion.

4. Conclusions

In the present work mixed mode SIF isotropic edge cracked plate with various discontinuities like crack, voids, soft and hard inclusions numerical are evaluated using XFEM under different loadings. The following observations are found during this numerical investigation study:

- With the increase of crack length MMSIF increases for voids and/or inclusions under different loadings.
- An isotropic plate with discontinuities like edge crack, void and inclusion are found more sensitive under shear and compound loading.
- The fracture behavior of the plate with a crack and voids/inclusions at near to crack tip is more sensitive.

References

[1] Sih G C 1973 *Eng. Fract. Mech.* **5** 1037-1040.
[2] Chien W Z, Xie Z C, Gu Q L, Yang Z F and Zhou C T 1982 *Eng. Fract. Mech.* **16** 95-103.
[3] Sukumar N and Prevost J H 2003*Int J Solids Struct.***40**7513-7537.
[4] LI Z and Chen Q 2002 *Int. J. Fract.***118** 29–40.
[5] Sukumar N, Chop D L, Moes N and Belytscko T 2001 *Apll. Mech.* **190** 6183-6200.
[6] Limtrakarn W and Dechaumphai P 2011 *T Can Soc. Mech. Eng.* **35** 355-368.
[7] Zhang J, Qu Z, Huang Q, Xie L and Xiong C 2013 *Arch. Appl. Mech.* **83** 861–873.
[8] Sharma K 2014 *J. Solid Mech.* **6-4** 410-421.
[9] Jiang S, Du C and Gu C 2014 *Struct. Eng. Mech.* **49-5** 597-618.
[10] Lal A, Mulani S B and Kapania R K 2017 *Int. J. Appl. Mech.* **9**1 750061- 1750094.
[11] Khatri K and Lal A 2018 *Mech. Adv. Mater. Struc.* **25-9** 732-755.

IOP Conf. Series: Materials Science and Engineering **814** (2020) 012018 doi:10.1088/1757-899X/814/1/012018

Computational analysis of novel cavity-based flameholder designs for supersonic combustion engines

Sanchayata Nayal[1], Sampada Lamb[2], Devabrata Sahoo[3], N. V. Raghavendra[4], S.A.I. Bellary[5]

[1-5]Annasaheb Dange College of Engineering and Technology, Ashta, Sangli, Maharashtra, India

E-mail: nayal.sanchayata@gmail.com

Abstract: Computational analysis for novel cavity-based flameholder designs with different fore-wall and aft-wall inclinations has been presented. The flameholder performance has been evaluated based on the following key parameters: pressure distribution, temperature distribution, recirculation zones and drag force. On comparing the different cavity designs, it has been found that both the fore-wall and aft-wall angles affect the flameholder performance. It was observed that an obtuse fore-wall angle gave favourable results numerically. Further studies can analyze cavities with different obtuse fore-wall ramp angles for optimum flameholder performance.

1. Introduction

The ever-increasing demand for faster flight vehicles has led to an active research in the area of propulsion mechanisms. Today, the fastest air-breathing engine is the SCRAMJET engine, which operates in supersonic conditions of Mach number greater than 5. The flow in the combustion chamber is supersonic, which results in extremely short residence time for injection, mixing and combustion of fuel with the incoming air. The high speed of air also affects the stability of the flame in the combustion chamber. Elevated temperatures of the combustor walls pose another technical challenge, which must be addressed to devise a plan for optimum use of heat-resistant materials [1].

Extensive research has been done on different types of flameholders, especially the cavity-based type, where multiple cavity configurations have been analyzed numerically as well as experimentally [2, 3]. A detailed analysis of different cavity configurations along with pylon-aided fuel injection and combustion studies has been done by Gruber et al [4-6], both experimentally and computationally. He concluded that smaller aft-wall ramp angles lead to lower fore-wall pressure, higher drag coefficient and smaller residence time. K H Yu et al [7] performed experimental study of rectangular and inclined cavities with varying aspect ratios (L/D), and found that short, inclined cavities and two tandem cavities give good flameholder performance and volumetric heating. Cavity-based flameholders are often combined with other types of flameholders like the strut-based flameholders [8, 9] and the pylon-based flameholders [5, 10] to develop larger recirculation regions and better fuel-air mixing characteristics for better combustor performance. A S Vishnu et al [11] conducted a detailed computational investigation on the effect of different heat fluxes on the cavity flow physics which was validated experimentally.

It was noted that general research based on cavity-type flameholder involved variation in the aft-wall ramp angle only. We specifically note that research has not been done to evaluate the effect of modification of fore-wall ramp angle on the flameholder performance.

The present work aims to provide parametric analysis of the effect of fore-wall ramp angle on the performance of a cavity-based flameholder. Computational simulations for supersonic flow over cavity configurations with fore-wall angle modifications have been performed. The analysis of results from these simulations can lead to better flameholder designs, and also help in material selection for the combustor walls.

1.1 Cavity-based Flame-holders

A common device to facilitate supersonic combustion is a *flameholder*, which introduces low-speed eddy-like recirculation zones in the incoming high-speed flow, as shown in Figure 1. These recirculation zones enable efficient mixing of fuel and air, resulting in continuous combustion (and prevent flame-out).

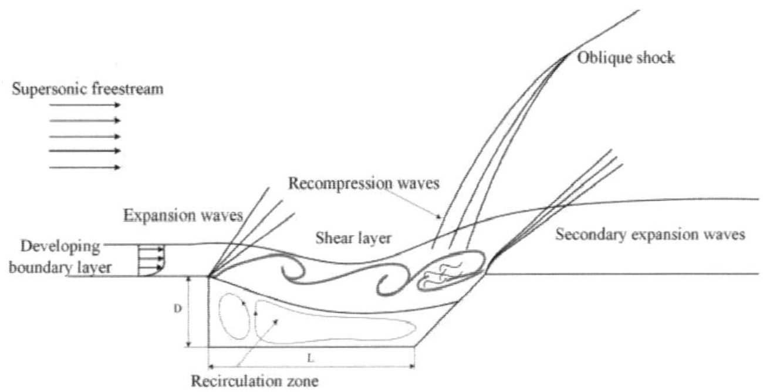

Figure 1. Flow physics of a cavity-based flameholder [11]

Extensive research has been done to understand the complex flow physics of cavity-based flameholders. This has led to progress in supersonic combustion and development of successful SCRAMJET prototypes.
The geometry of a typical cavity-based flameholder is shown in Figure 2.

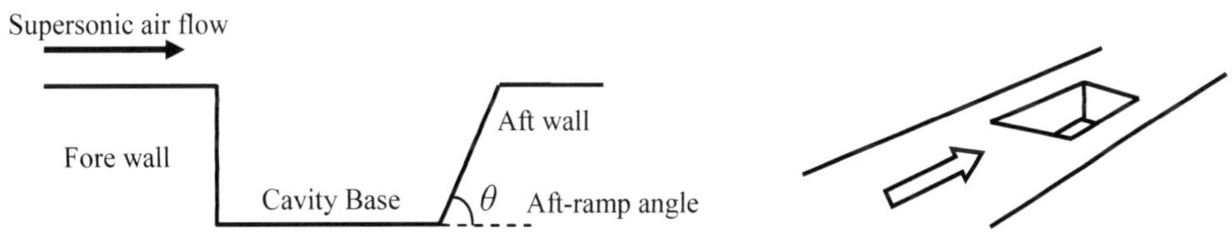

(a) Side view (b) Isometric view

Figure 2. Geometrical features of a cavity-based flameholder

The performance of a cavity-based flameholder design is evaluated based on the following broad parameters:

- Higher static pressure in cavity improves flameholder performance
- Higher temperatures in the cavity encourage better mixing of fuel and air, and also flame stability
- Large eddies or recirculation zones result in better mixing
- Low drag force on the cavity is desirable

In the present work, four novel cavity designs with fore-wall ramp angle modifications are proposed, and computational analysis of the above-mentioned parameters are performed.

1.2. Research objectives

Figure 3 depicts the definition of the fore-wall ramp angle θ_f, and aft-wall ramp angle θ_a of a cavity. To understand the effect of fore-wall and aft-wall ramp angles on cavity performance, different combinations of θ_f and θ_a are considered. Computational analysis of supersonic flow over these cavity designs is carried out and results are obtained for the various performance parameters.

Figure 3. Notations for fore-wall and aft-wall ramp angles of a cavity-based flameholder

2. Methodology

A total of six cavity designs are analyzed computationally. It may be noted that the first two designs with 90° fore-wall ramp angle correspond to the test cases studied in Gruber paper [4], and serve to validate our computational methodology. Remaining cavities are our novel designs with variation in fore-wall ramp angle.

For choosing the fore-wall ramp angle, there were two options: Acute angle and Obtuse angle. For the sake of uniformity, one acute angle and one obtuse angle were chosen. The mean difference from right angle of about 20 ° was kept constant for both the acute and obtuse fore-wall ramp angles. This resulted in fore-wall ramp angles of 70 ° and 110 ° calculated as per figure 3.

These two fore-wall ramp angles were combined with two different aft-wall ramp angles: 90 ° and 16°, which were taken from the reference of the Gruber paper [4]. This led to the creation of 4 novel cavities design with both acute and obtuse fore-wall ramp angles along with two standard aft-wall ramp angles. A list depicting the cavity geometries is provided in Table 1.

2.1. Computational Setup

Computational Fluid Dynamics (CFD) simulations are performed using ANSYS-Fluent® software, for steady supersonic viscous flow approaching the cavity. It is a two-dimensional, non-reacting, cold-flow investigation wherein only air is introduced in the cavity so we don't consider the actual combustion of fuel and air and the resulting thermodynamic effects. This is in accordance with our baseline reference paper [4].

For the Initial Flow conditions, the freestream conditions from the baseline paper [4] are used as the test conditions for our present work:

- Freestream Mach number: 3
- Stagnation pressure: 690 KPa
- Stagnation temperature: 300 K
- Working fluid: Air (considered as ideal gas)

IOP Conf. Series: Materials Science and Engineering **814** (2020) 012018 doi:10.1088/1757-899X/814/1/012018

Table 1. Cavity designs studied in present work

Sr. No.	Cavity Nomenclature	Cavity Design	Remarks
1	90_90	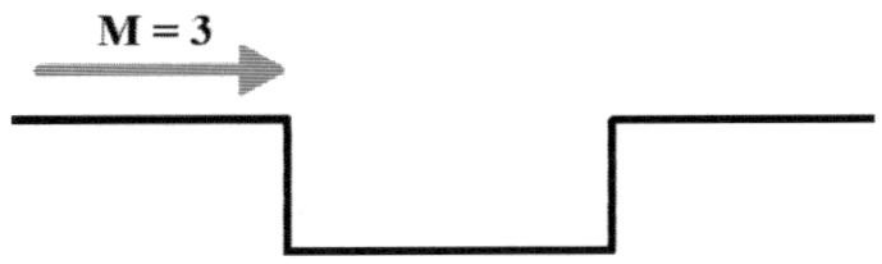	Validation Case $\theta_f = 90^{\circ}$ $\theta_a = 90^{\circ}$
2	90_16		Validation Case $\theta_f = 90^{\circ}$ $\theta_a = 16^{\circ}$
3	70_90	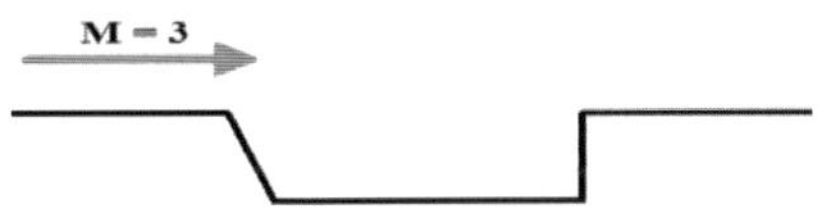	Novel design $\theta_f = 70^{\circ}$ $\theta_a = 90^{\circ}$
4	70_16		Novel design $\theta_f = 70^{\circ}$ $\theta_a = 16^{\circ}$
5	110_90		Novel design $\theta_f = 110^{\circ}$ $\theta_a = 90^{\circ}$
6	110_16		Novel design $\theta_f = 110^{\circ}$ $\theta_a = 16^{\circ}$

2.1.1. *Computational mesh*

A structured two-dimensional mesh is used as the computational domain, as shown in Figure 4. The height of the computational domain is about 20 times the cavity depth, to simulate far-field conditions at the top boundary. There are 250 x 200 cells in the cavity region and about 150 x 100 cells in the regions above. The smaller cell-size inside the cavity is used to capture the viscous effects of the flow like boundary layer, shear layer, shock formation and eddy formation. To get satisfactory results, y^+ is less than unity for all cavity grids. Further away from the cavity surface, the viscous effects become less important and hence the cell size is larger for quicker computations.

2.1.2. *Boundary conditions*

The boundary conditions for the numerical simulation are depicted in Figure 4. The lower part of the computational domain comprises of no-slip stationary wall condition which represents the combustor wall of the SCRAMJET engine. Inlet corresponds to the freestream boundary conditions mentioned in section 2.1. Both Outlet and Farfield boundary conditions correspond to Pressure-Farfield boundary type where the freestream Mach number is set as 3.

2.1.3. Flow solver setup

The flow solver is configured to use an explicit density-based solver, with Green-Gauss cell-based scheme for spatial discretization, and Roe-FDS scheme for the inviscid fluxes. Turbulence model is set to k-ω, as adopted in the reference paper [4]. The desired computational accuracy is configured to second order in both time and space.

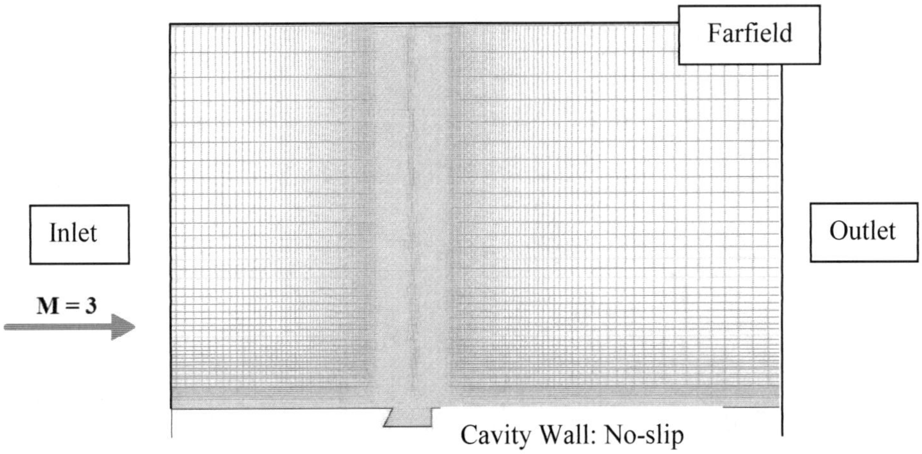

Figure 4. Computational domain

3. Results And Discussion

3.1. Validation of the computational methodology

Our computational methodology is first used to obtain results for baseline cavities presented in reference [4]. The results obtained are compared with experimental results presented in the reference [4]. As discussed below, there is a close agreement between the two sets of results. This is helps in validating our computational methodology. The same methodology is then used for computational analysis of the novel cavity designs presented in this work.

3.1.1. Normalized Pressure Plots

The baseline reference paper [4] provides the experimental data for the cavity designs 90_90 and 90_16. These designs are part of our parametric study as shown in Table 1. Figures 5 and 6 show the comparison of our present computational results (represented using solid lines) with the experimental results from Gruber *et al* [4] (represented using circles). The comparison shows a close agreement between the two results, thereby validating our computational method.

3.2 Performance parameters of novel cavity-based flameholder designs

CFD simulations are performed for Mach 3 supersonic flow approaching each of the four novel cavity designs: 70_90, 70_16, 110_90, 110_16; and results are obtained pertaining to the performance parameters – pressure distribution, temperature distribution, recirculation zones and drag force. Details of these results are presented below and discussed to evaluate the performance of each of these designs as a flame holder.

3.2.1 Pressure distribution

The results for normalized pressure distribution for the novel cavity designs are presented below. The pressure is normalized with respect to the inlet static pressure.

Figure 7 (a) shows the normalized pressure distribution for the cavity 70_90. It is seen that there is a slight increase in the pressure after the separation point, indicating a weak shock formation. The aft-wall corner point shows higher pressure than the rest of the cavity. When compared to the test cavity 90_90, we see more pressure variations with higher magnitude.

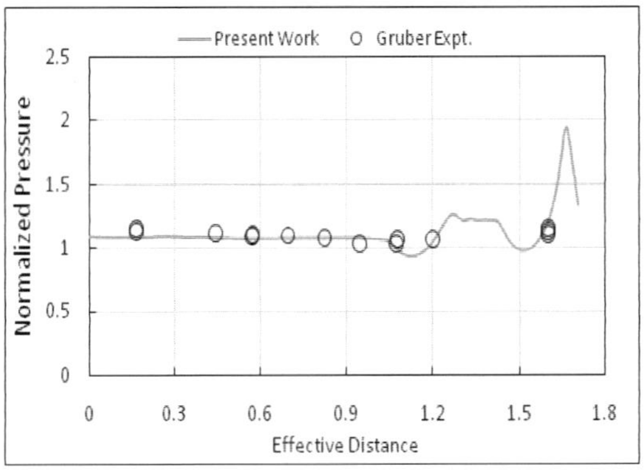

Figure 5. Normalized Pressure plot for cavity 90_90

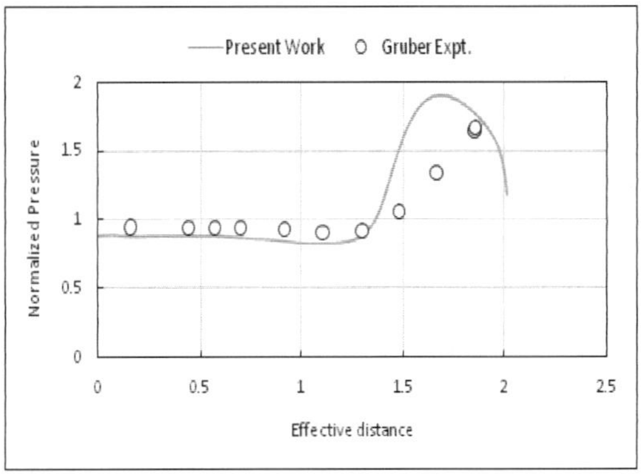

Figure 6. Normalized Pressure plot for cavity 90_16

Figure 7 (b) shows the normalized pressure distribution for the cavity 70_16. It is seen that the pressure decreases after the separation point, indicating a weak expansion wave formation. The aft-wall corner point shows higher pressure than the rest of the cavity. When compared to the test cavity 90_16, not much variation is seen.

Figure 7 (c) shows the normalized pressure distribution for the cavity 110_90. The pressure increase after the separation point is the highest when compared to the other cavities. The aft-wall corner point shows much higher pressure comparatively. When we compare our results for cavity 110_90 with those obtained for all the other cavities in Table 1, a higher-pressure distribution is observed for the novel cavity 110_90. Consequently, this design ranks best amongst the considered cavity designs for pressure distribution. Figure 7 (d) shows the normalized pressure distribution for the cavity 110_16. It is seen that the distribution is similar to that of cavity 70_16 and to the test cavity 90_16 also.

3.2.2 *Temperature distribution*

The results for normalized temperature distribution for the novel cavity designs are presented below. The temperature is normalized with respect to the inlet temperature.

Figure 8 (a) shows the normalized temperature distribution for the cavity 70_90. The shear layer is completely visible. It is seen that the temperature increases as we travel from free stream towards the cavity vertically downwards. The aft-wall corner point shows increase in temperature along the cavity line. When we compare the results from our computation for the test cavity 90_90, we see a lower temperature distribution for cavity 70_90.

Figures 8 (b) shows the normalized temperature distribution for cavity 70_16. On comparing with the results we obtained for the baseline test cavity 90_16, not much difference is seen, though temperature distribution along the aft-wall is relatively lower.

Figure 8 (c) shows the normalized temperature distribution for the cavity 110_90. The aft-wall corner point shows increase in temperature along the cavity line. The bottom corner of the aft-wall shows higher temperature than the interior of the cavity. The temperature variation also indicates the recirculation zones. When compared to all the remaining cavities, we see lower temperature distribution for 110_90 cavity, making it a better design choice.

Figure 8 (d) shows the normalized temperature distribution for the cavity 110_16. It is seen that the distribution is similar to that of cavity 70_16 and to the test cavity 90_16 also.

IOP Conf. Series: Materials Science and Engineering **814** (2020) 012018 doi:10.1088/1757-899X/814/1/012018

3.2.3 Recirculation zones in the cavity

Figure 9 depicts the results for the recirculation zones in the novel cavity designs. We see that the novel cavity design 110_90 shows stronger recirculation zones with contours of higher velocities, as compared to other cavities. This makes the cavity 110_90 a suitable candidate for flameholder design.

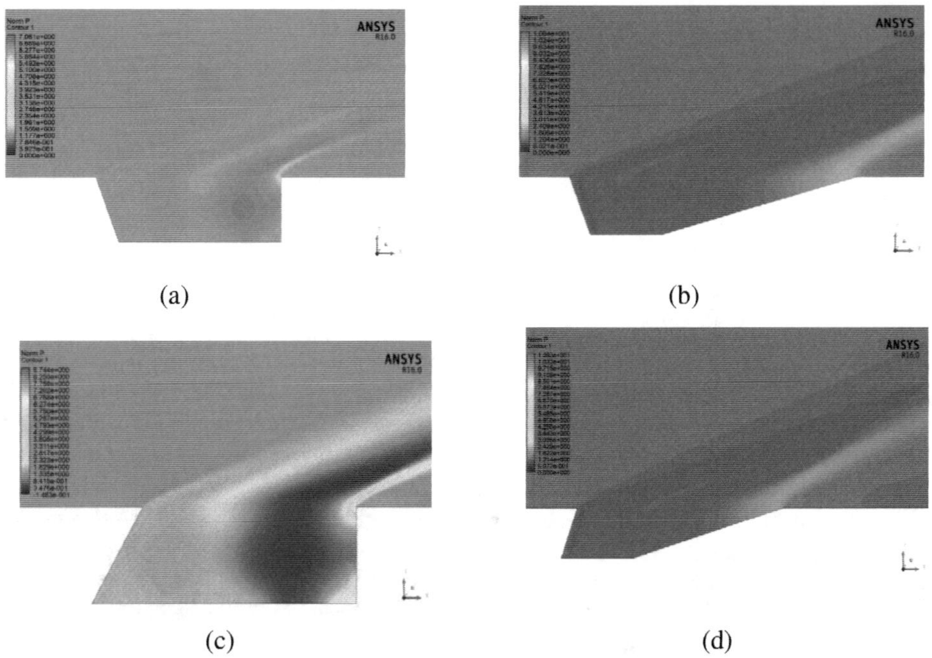

Figure 7: Normalized pressure contours for the novel cavity designs

3.2.4 *Drag force on the cavity*

The drag force obtained from ANSYS-Fluent® computations is normalized with respect to the drag force we had obtained for the baseline 90_90 cavity. Results from the computations show that cavity design 70_90 experiences the lowest drag amongst other cavities, making it a better cavity design with respect to the cavity drag. Results for the normalized drag force for the novel cavity designs are discussed in the next section 3.2.5.

3.2.5 *Discussion and comparison of results*

Based on each of the performance parameters, the four novel cavity designs are ranked as in Table 2. Cavity design which performs best for a parameter get a score of 4, while the one with relatively poorest performance get a score of 1.

The desirable performance parameters are:

- High pressure distribution
- High temperature distribution
- Large eddies or recirculation zones
- Low drag force on the cavity

IOP Conf. Series: Materials Science and Engineering **814** (2020) 012018 doi:10.1088/1757-899X/814/1/012018

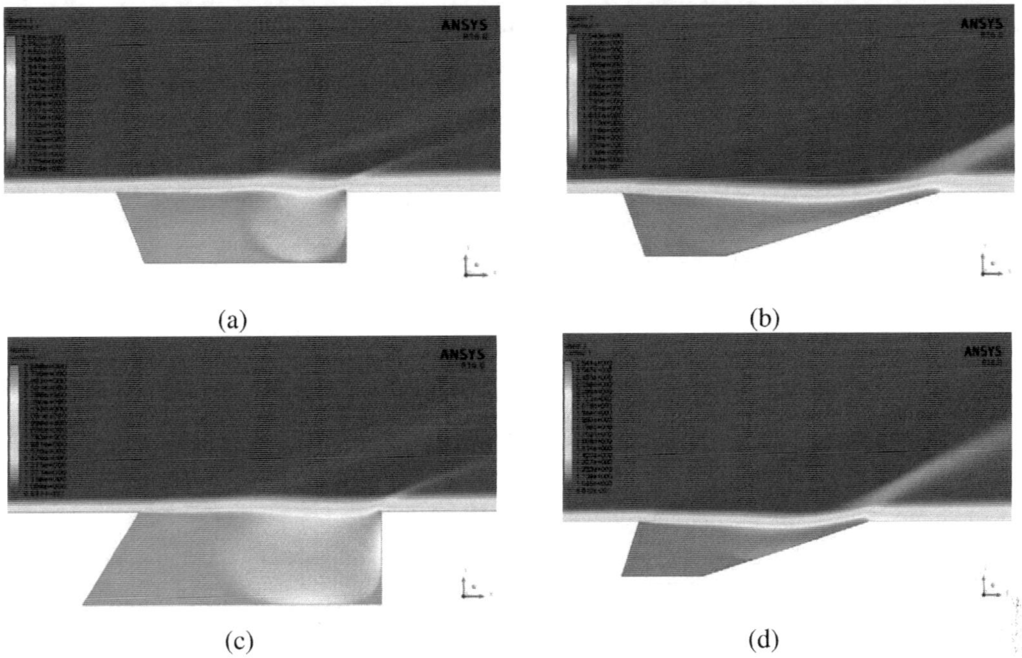

Figure 8: Normalized temperature contours for the novel cavity designs

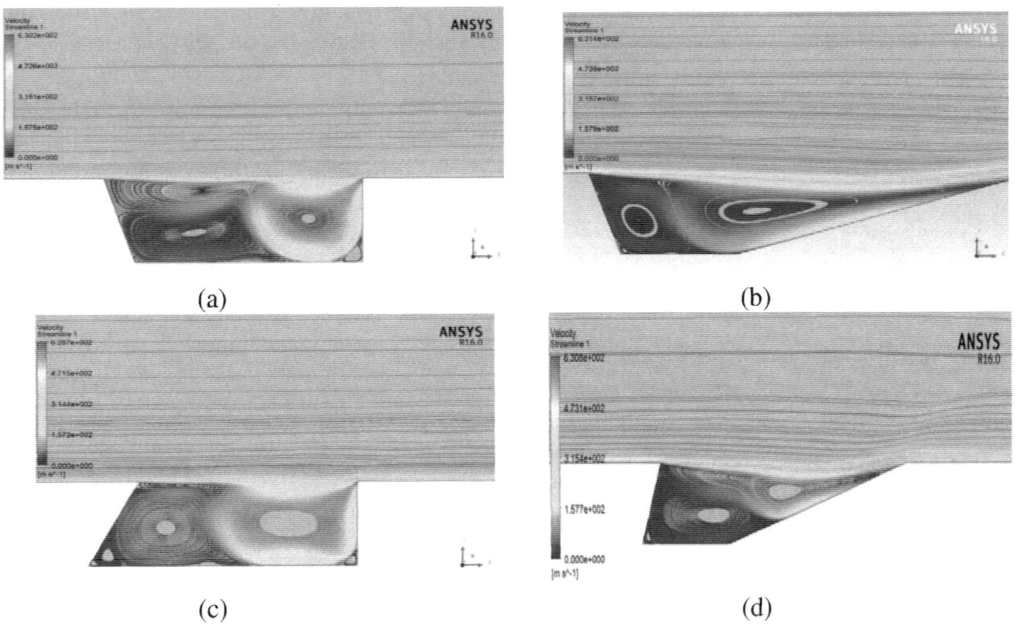

Figure 9: Recirculation zones for the novel cavity designs

Based on the relative performance scores in Table 2 and the discussions above, we see that cavity design 110_90 offers the best overall performance as a flameholder. In an incoming supersonic flow, this cavity develops good pressure distribution which promotes efficient combustion. This cavity lags in terms of high temperature distribution, which may slightly reduce fuel-air mixing. However, the strong recirculation zones in the cavity should bring about efficient mixing of fuel and air. The relatively lower temperatures at the cavity wall can be helpful in reducing damage to the material of the wall. Our finding of best overall performance with design 110_90 also corroborates with that of Gruber *et al* [4], who concluded that 90° for the aft-wall ramp angle works best for flameholder performance.

Table 2: Relative performance scores of the novel cavity designs

Cavity	Performance Parameters as a Flameholder				
	Pressure Distribution	Temperature Distribution	Recirculation Zones	Normalized Drag Force	Total Score
70_90	3	1	2	4	10
70_16	1	4	3	2	10
110_90	4	2	4	1	11
110_16	2	3	1	3	9

4. Conclusion

In this paper, computational analysis is done to qualitatively and quantitatively assess the performance of novel cavity designs based on key performance parameters of pressure distribution, temperature distribution, recirculation zones and drag force. The following conclusions are drawn:

- The fore-wall ramp angle of a cavity can affect its performance as a flameholder. Specifically, an obtuse fore-wall angle can result in stronger recirculation zones and efficient mixing of fuel and air.
- With respect to the aft-wall ramp angle, a setting of 90° works best for flameholder performance.
- Cavities with different obtuse fore-wall ramp angles can be analyzed for optimum flameholder performance.

References

[1] Yang W, Fu J, Ma X and Xing R 2018 *IOP Conference Series: Materials Science and Engineering* **408** 150.
[2] Ben-Yakar A and Hanson R K 2001 *Journal of Propulsion and Power* **17** 869–877.
[3] Santhosh K Gugulothu 2019 *Heat Transfer: Asian Research* **48** 3657-3681.
[4] Gruber M R, Baurle R A, Mathur T and Hsu K Y 2001 *Jr. Propulsion and Power* **17** 146–153.
[5] Gruber M R, Carter C D, Montes D R, Haubelt L C, King P I and Hsu K Y 2008 *Journal of Propulsion and Power* **24** 460–470.
[6] Gruber M R, Donbar J M, Carter C D and Hsu K Y 2004 *Journal of Propulsion and Power* **20** 769–777.
[7] Yu K H, Wilson K J and Schadow K C 2001 *Journal of Propulsion and Power* **17** 1287–1295.
[8] Hsu K Y, Carter C D, Gruber M R, Barhorst T, Smith S 2010 *Jr. Propulsion and Power* **26 6**.
[9] Choubey G, Devarajan Y, Huang W, Kulmani Mehar, Tiwari M and Pandey K M 2019 *International Journal of Hydrogen Energy* **44** 13895-13909.
[10] Freeborn A B, King P I and Gruber M R 2009 *Journal of Propulsion and Power* **25(3)**.
[11] Vishnu A S, Aravind G P, Deepu M and Sadanandan 2019 *International Journal of Heat and Mass Transfer* **141** 1140-1151.

IOP Conf. Series: Materials Science and Engineering **814** (2020) 012019 doi:10.1088/1757-899X/814/1/012019

Probabilistic design of composite leaf spring by using finite element method

S G Khiste[1], P Kulkarni[2], S D Katekar[3]

[1] Research Scholar, Department of Mechanical Engineering, SKN SCOE, Pandharpur
[2, 3] Asst. Professor, Department of Mechanical Engineering, SKN SCOE, Pandharpur

E-mail: shrigkhiste@gmail.com

Abstract. A parametric model is created in ANSYS workbench, FEA method is used to carry out probabilistic design in ANSYS workbench. Mono leaf spring is considered for analysis purpose and analysis is carried out. Composite material is applied to parametric model. Boundary conditions are applied to mono leaf spring, both ends are fixed and load (force) is applied to mono leaf spring. Meshing is carried out. The geometric parameters (thickness, width, etc.) are used as input parameter and Von-Mises stress considered as output parameter. For any further uncertainty in analysis, parametric co-relation function is used in ANSYS. The relation in geometric parameter and output parameter is achieved by correlation coefficients

1. Introduction

The leaf spring mainly provides flexibility between two mating parts subjected to fluctuated loads. Research shows when the composite material is used in leaf spring, its weight reduces and ultimately efficiency is increased. Most of the researchers focused on investigating the static behaviour of the leaf spring under loading condition with finite element analysis. The general process is to design spring in modelling software and then the model is exported to some neutral format and importing cad file for analysis in simulation software but this is a very tedious process. This work gives an idea about designing the parametric leaf spring in ANSYS workbench which reduces the time of designing.

The conventional material can be replaced by composite material so as to get better strength, many researchers are carried out analysis of leaf spring by using composite material to achieve better results as compared to conventional material [1-7]. By considering composite material for analysis the weight of vehicle is reduced and efficiency gets increased [1,3].The commercial vehicles leaf spring is used for analysis purpose to achieve the best solution for existing problems, the results calculated for the leaf spring are displacement, principal stress and energy [2-3].

It is observed from a literature survey that the probabilistic approach for the design of composite leaf spring by using FEA (ANSYS) is not taken into consideration. Also uncertainty in Von-Mises stress by considering various geometrical parameters is not focused by anyone in past years. Also, the results of probabilistic design are to be optimized. The result will provide a new approach.

2. Material selection

Carbon/epoxy composite materials have good strength as well as they have better absorption capacity and store more amount of energy. It has good modulus comparatively to conventional material.

IOP Conf. Series: Materials Science and Engineering **814** (2020) 012019 doi:10.1088/1757-899X/814/1/012019

Therefore, carbon/epoxy is used for designing of leaf spring and orthotropic material properties are considered [1].

3. Probabilistic Design

This technique is used to understand the effect of uncertain input and the assumptions on the model. This technique provides to determine the extent to which uncertainties in model effect on output parameter. An uncertainty random quantity of a parameter whose value is impossible to determine at a given point.

The uncertainty analysis of leaf spring can be classified into two parameters,

3.1 Input parameters

- Geometric parameters: chamber height (h), length of leaf spring (l), width (w) and thickness (t)
- Load parameters: Force (fy)

3.2. Output parameter

- Von-Mises stress (σMax)

Using these parameter, the probabilistic design is carried out in ANSYS to understand the sensitivity of input parameter on Von-Mises stress. Probabilistic design can be used to determine the effect of one or more these input variables variations on the outcome (Von-Mises Stress) of the analysis.

4. Finite element analysis

4.1 Parametric model

The parametric modelling of mono leaf spring is carried out in ANSYS workbench by taking various parameters length of leaf spring, width of leaf spring, thickness of leaf spring etc. Figure 1 shows the parametric model of mono leaf spring. The dimension commercial vehicle are considered [2].

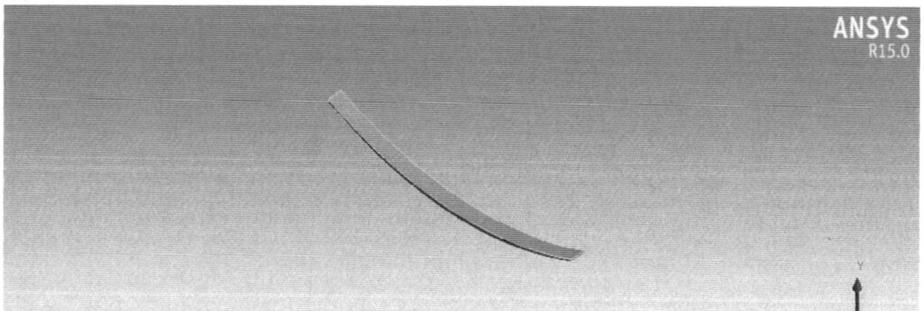

Figure 1. Parametric model of mono leaf spring

4.2 Meshing

Meshing is a process of dividing a model into a small number of pieces. The mono leaf spring has meshed, number of nodes 4577 and number of elements are 552 with element size 16 mm Figure 2.

IOP Conf. Series: Materials Science and Engineering **814** (2020) 012019 doi:10.1088/1757-899X/814/1/012019

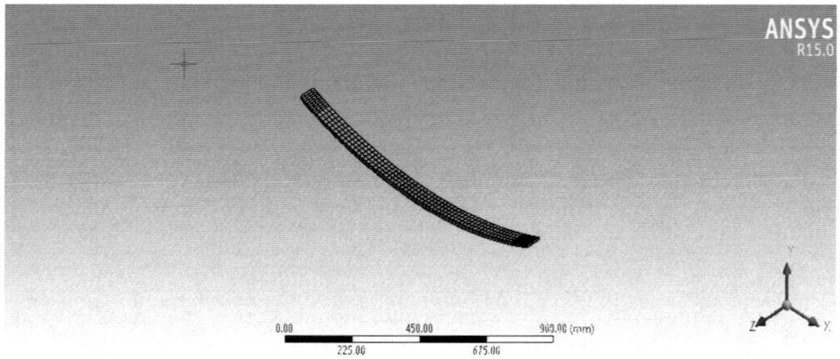

Figure 2. Meshed mono leaf spring

4.3 Boundary conditions

A finite element model with a segment of one mono leaf spring is considered for analysis. The mono leaf spring is considered as a fixed beam during analysis. Both ends of the leaf springs are fixed and a point load is applied on mono leaf spring Figure 3.

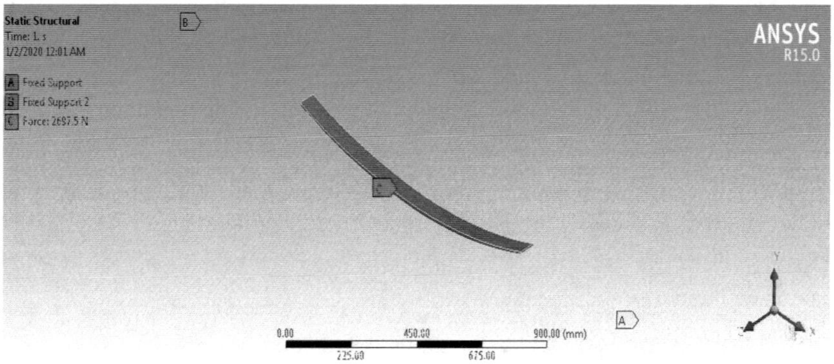

Figure 3. Mono leaf spring with boundary condition

4.4 Finite element analysis

After applying boundary condition to leaf spring, result is obtained as Von-Mises stress. The variation of the Von-Mises stress is shown in the Figure 4. The stress is maximum at fixed end and its minimum at a centre.

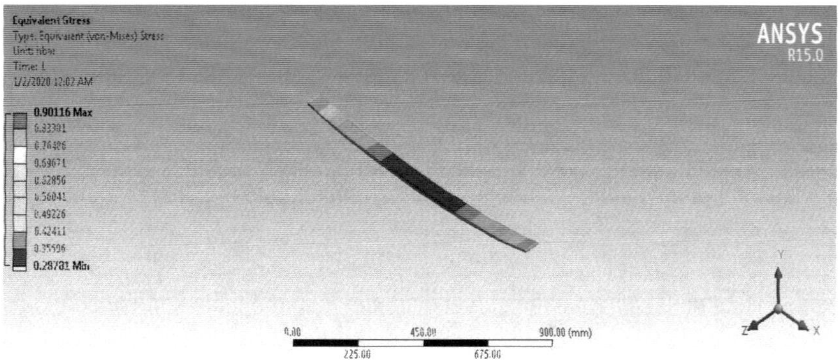

Figure 4. Representation of stress in ansys

Table shows various parameters used for probabilistic design and analysis of leaf spring.

Table 1. Dimensions for leaf spring

Length (mm) P1	Chamber height (mm) P2	Thickness (mm) P3	Width (mm) P4	Force (N) P5	Von-Mises Stress (hbar) P6
1072	95.4	8	60	2697.5	11.099

Each input parameter from its base dimension is varied by 5% in upper bound and lower bound to get the parameter set within that range Table 2.

5 Results

5.1 Probabilistic design

The probabilistic design of composite leaf spring is carried out and the results are obtained in Ansys as Von-Mises stress as objective function. From this study we can able to decide better parameter set to design the leaf spring, for parameter set in Table 2. From the study, the parameter set 15 is best amongst the all parameter set.

5.2 Correlation between input and output parameter

Correlation gives the process of design of experiments and it provides correlation in input and output parameter. Figure 5 shows about linear correlation between the input parameter and the output parameter. For the correlation matrix above twenty set of parameters are considered. The degree colour of the matrix shows the correlation between parameters. The correlation value for the input-output parameter is given in below Table 3.

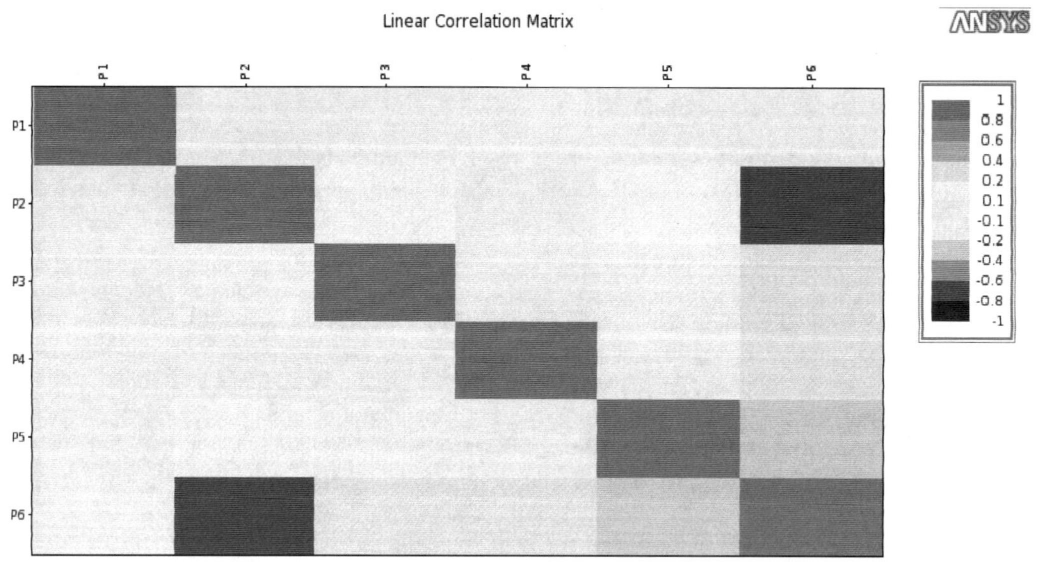

Figure 5. Correlation between input-output parameter

Table 2. Input and Output parameter optimisation

Sr. No	Width P1	Length P2	Chamber height P3	Thickness P4	Force P5	Von-Mises Stress P6
	W	L	H	t	F_Y	σ_{Max}
1	62.62	1114.95	97.87	7.68	2581.19	9.9717
2	61.69	1039.09	91.00	7.62	2793.10	12.7731
3	57.18	1062.29	90.43	7.98	2629.79	10.7803
4	57.61	1076.08	92.85	8.33	2785.50	10.8358
5	58.69	1111.09	92.31	8.19	2654.21	9.5721
6	62.92	1050.69	90.38	8.10	2678.28	10.0857
7	58.56	1069.90	95.78	7.91	2733.69	11.8230
8	57.43	1047.46	92.54	7.88	2595.69	12.4397
9	57.97	1119.11	94.38	7.82	2718.64	10.5433
10	59.37	1089.69	99.75	7.79	2586.29	11.8023
11	61.30	1092.27	93.69	8.07	2820.49	10.4405
12	60.01	1083.69	95.11	8.17	2738.77	10.6300
13	62.31	1123.99	91.96	8.39	2616.88	8.7722
14	60.95	1100.48	96.71	7.73	2767.83	11.0918
15	61.83	1040.93	93.43	7.92	2638.34	16.5849
16	59.11	1081.14	91.20	7.74	2705.00	10.7820
17	61.41	1066.91	96.31	7.80	2660.57	12.8250
18	62.62	1114.95	97.87	7.68	2581.19	9.9717
19	61.69	1039.09	91.00	7.62	2793.10	12.7731
20	57.18	1062.29	90.43	7.98	2629.79	10.7803

Table 3. Correlation matrix

Correlation Matrix	Width	Length	Chamber height	Thickness	Force	Von-Mises stress
Width	1.0000	0.0174	-0.0371	-0.1368	-0.0382	-0.0528
Length	0.0174	1.0000	0.1787	-0.0070	0.1658	-0.6043
Chamber height	-0.0371	0.1787	1.0000	-0.0576	-0.1192	0.2715
Thickness	-0.1368	-0.0070	-0.0576	1.0000	0.0197	0.3237
Force	-0.0382	0.1658	-0.1192	0.0197	1.0000	-0.2201
Von-Mises stress	-0.0528	-0.6043	0.2715	0.3237	-0.2201	1.0000

6. Conclusions

Based on the above study, the conclusions are made

- Parametric modelling and finite element analysis of leaf spring are successfully carried out in ANSYS Workbench.
- A new methodology for the design of leaf spring i.e. probabilistic design has been implemented
- Geometric parameters like Length of leaf spring, width, thickness etc. have a significant influence on Von-Mises stress
- Correlation Coefficients are obtained for the input and output parameters.

7. Acknowledgement

We express our sincere thanks to Dr. S. G. Kulkarni, Dean (R and D), Department of Mechanical Engineering, and Dr. S. G. Deshmukh, Dean (Publication), Department of Engineering Physics, SKN Sinhgad College of Engineering, Korti, Pandharpur, India for the encouragement and guidance.

References

[1] Trivedi A, Prof. Bhoraniya R. M. 2015 *IJSTE* **1** 151-156.
[2] Harshit, Kumar S and Verma A 2016 *IJIR* **2** 1421-1427.
[3] Singha H, Gurinder Singh Brar, 2018 *ICMPC* **5** 5857-5862.
[4] K Ashwini, Prof CV Mohan Rao 2018 *ICMPC* **5** 5716-5721.
[5] S. Rajesh, G. B. Bhaskar, J. Venkatachalam, k. Pazhanivek and Suresh Sagadevam 2016 *KSME* **30** 4291-4298.
[6] XIE Feng, ZHOU Yong and PENG sisi 2015 *Wuhan Univ. J. Nat. Sci* **20** 87-92.
[7] Krishan Kumar and M. L. Agrawal 2016 *ICMPC* **4** 1829-1836.

Fracture analysis of center cracked laminated composite plate subjected to bi-axial loading using extended finite element method

Achchhe Lal[1], Khushbu Jain[2] and S P Palekar[3]

[1]Assistant Professor, Department of Mechanical Engineering, SVNIT, Surat, GJ, India
[2]Research Scholar, Department of Mechanical Engineering, SVNIT, Surat, GJ, India
[3]Research Scholar, Department of Mechanical Engineering, SVNIT, Surat, GJ, India

E-mail: lalachchhe@yahoo.co.in

Abstract: Among various numerical method, finite element method (FEM) is extensively used but handling the fracture problem with conventional FEM is time-consuming and costly. So, extended finite element method (XFEM) which is based on partition of unity method and enrichment functions is used to handle material with discontinuities. In the present study, fracture analysis of center cracked layered composite plate under different biaxial loadings (tensile, shear and combined) using XFEM is investigated and mixed mode stress intensity factor (MMSIF) is calculated for different parameters.

1. Introduction

In recent year applications of fibre reinforced composites (FRCP) have increased in aerospace structure, automotive, marine, biomedical engineering due to its several advantages as high strength to weight ratio, low fabrication cost, better wear, corrosion, thermal resistance and high impact resistance. In most of the practical application, FRCP is subjected to complex state of stress and the presence of crack under complex loadings, accurate prediction of fracture parameter as stress intensity factor (SIF) is essential to understand the behaviour of fracture of materials. During manufacturing and fabrications, due to improper design of the constituent components, error in manufacturing and production techniques are the major cause of discontinuity like cracks, inclusions, voids and holes. The structural safety and reliability of FRCP structures can be increased by predicting fracture failure up to significant level.

When a crack exists in the structures, accurate predictions of fracture parameters like stress intensity factors (SIFs) and their effect on unstable fracture propagation under the action of biaxially acting tensile, shear or even the combination of both these stresses becomes necessary. In last decades many numerical methods are suggested and utilised by researchers. In conventional FEM every time remeshing is required when crack advances that leads to more computational time and cost. Significant improvement in conventional FEM are relised and now XFEM is used for modeling the discontinuities like crack, where extra enrichment functions are used for crack face and tip. In this direction Asadpoure and Mohammadi [1] proposed XFEM to understand the fracture analysis of orthotropic materials, where extra enrichment functions is used near crack tip. In this present paper, partition of unity based enrichment function is used to model crack geometry, where re-meshing is not required.

The accuracy of the present results is checked by other numerical methods. Meek and Ainsworth [2] presented finite element analysis with lower and upper bond theory of center cracked plate under biaxial loading. In this present study, by considering wide range of biaxial stress ratio limit load can be investigated. Lim et al. [3] presented fracture analysis of cracked orthotropic plate under biaxial loading. In the present study, it is confirmed that only singular term is not sufficient to predict the crack extension. Mostafavi et al. [4] carried out investigation based on experimental and FEM both to show the effect of stress multi-axiality on various fracture parameters of Aluminium alloy. Wang et al. [5] proposed a new approach called as modified dugdale model, which finds out the coupled effect at the crack tip, which was projected earlier for the crack tip opening displacement. Lal et al. [6] investigated XFEM based stochastic fracture analysis of composite plate with central crack under uniaxial loading, where, second-order perturbation technique (SOPT) and Monte Caro simulation (MCS) are used to evaluate MMSIF. The present results are compared with analytical results from literature. Tabarraei and Sukumar [7] introduced two-dimensional crack growth modelling by using Laplace interpolant and by adopting mean value co-ordinates for non-convex elements in the framework of XFEM. Yan [8] presented fracture analysis of square plate with emanating cracks from elliptical hole under bi-axial loadings by displacement discontinuity method. It is observed that the present numerical is quite efficient and accurate for the fracture analysis of complicated crack under biaxial loadings. Goldstein and Shifrin [9] solved first mode crack deviation of orthotropic plane under bi-axial loading where, crack geometry is represented as a thin elliptical hole. In this present work, condition for crack stability and crack deviation is obtained and the present results are compared with the results of the conventional model of the crack, where crack is considered as perfect cut. Lal and Palekar [10] presented XFEM based fracture analysis of laminated composite plate with through thickness cracks of different types like line, semi circular, semi elliptical and arbitrary curves under tensile and shear loading. The effect of randomness in different fracture parameters is studied by SOPT and MCS. It is observed that random variables like length of crack, crack depth, loading and angle of laminate with effect more on fracture behaviour of the plate. Lal et al. [11] presented stochastic fracture analysis of laminated composite plate with edge crack underneath tensile, shear and combined loadings. In this present work XFEM with M-interaction, SOPT and MCS are used to study the effect of variation of fracture parameter on MMSIF. The result obtained is also compared with other published results.

In this present work XFEM based fracture analysis of center cracked composite plate underneath biaxial tensile, shear and combined loading is investigated MATLAB environment. The effect of crack angle and orthotropic angle on MMSIF is also investigated.

2. Mathematical formulation

The stress- strain relationship by the form of Hooke's law for an orthotropic body is represented as

$$\sigma_i = R_{ij}\varepsilon_j \quad (i, j = 1,2,6) \tag{1}$$

where R_{ij} $(i, j = 1, 2, 6)$ are the compliance coefficients of the orthotropic material along the local Cartesian coordinates axes (x_1, y_1) as shown in figure 1.

At any point x displacement field located in the domain of cracked body by XFEM is represented as

$$u = u^{FEM} + u^{XFEM} \tag{2}$$

with

$$u^{XFEM} = u^{Crack\ Face} + u^{Crack\ tip} \tag{3}$$

Therefore, the Eq. (2) can take the form

$$u^h(x) = \sum_{j=1}^{n} L_j(x)u_j + \sum_{k=1}^{cr} L_k(x)\phi(x)a_k \tag{4}$$

where u_j is the nodal degrees of freedom (DOF) in conventional FEM, a_k the extra of DOF to CFEM model, L_j and L_k are the shape functions and $\phi(x)$ is the irregular enrichment function. The displacement field can be approximated by rearranging Eq. (4) as below

$$u^h(x)=\left[\sum_{j=1}^{n}L_j(x)u_j\right]+\left[\sum_{k=1}^{cf}L_k(x)\big(H(\xi(x))-H(\xi(x_k))\big)a_k\right]+\left[\begin{array}{l}\sum_{l=1}^{t_1}L_l(x)\left(\sum_{t=1}^{t_1 f}\big(F_t^1(x)-F_t^1(x_l)\big)b_l^{t_1}\right)\\+\sum_{m=1}^{t_2}L_m(x)\left(\sum_{t=1}^{t_2 f}\big(F_t^2(x)-F_t^2(x_m)\big)b_m^{t_2}\right)\end{array}\right] \tag{5}$$

Where, the three parentheses [ ] represent the linear, discontinuous and tip enrichment parts respectively. Nodes that belong to two crack tips are enriched with the two enrichment functions $F_t^1(x)$, $F_t^2(x)$ respectively for each crack tip, and *H (ξ)* is denoted by Heaviside function.

In Eq. (5) the *cf* are the set of nodes that have the crack face in their support domain, t_1 and t_2 are the sets of nodes for tips 1 and 2 respectively. u_j are the nodal displacements (standard degrees of freedom), $a_k, b_l^{t_1}, b_m^{t_2}$ are the vectors for extra DOF for the nodes located on the crack face and the two crack tips respectively. The functions F_t^1, F_t^2 are the crack tip enrichment functions for tip 1 and 2. *H (ξ(x))* is the Heaviside function is the step function for modelling strong discontinuity/crack face.

$$H(\xi)=\begin{cases}+1 & \xi\in\Omega^+\\-1 & \xi\in\Omega^-\end{cases} \tag{6}$$

Here, the value of +1 and -1 is considered for the positive side and other side of crack face.

In the XFEM the global form of linear equations for discrete system is written as

$$F=KU \tag{7}$$

Where *F* is the external force vector, K is the stiffness matrix and *U* is the DOF. *K* and F for each element are defined as

$$K_{ij}^e=\begin{bmatrix}K_{ij}^{uu} & K_{ij}^{ua} & K_{ij}^{ub}\\K_{ij}^{au} & K_{ij}^{aa} & K_{ij}^{ab}\\K_{ij}^{bu} & K_{ij}^{ba} & K_{ij}^{bb}\end{bmatrix} \tag{8}$$

Where

$$F_i^e=\left\{F_i^u\,F_i^a\,F_i^{b_{t1}}\,F_i^{b_{t2}}\,F_i^{b_{t3}}\,F_i^{b_{t4}}\right\} \tag{9}$$

$$U=\{u\ a\ b_{t1}\ b_{t2}\ b_{t3}\ b_{t4}\} \tag{10}$$

$$K_{ij}^{rs}=\int_{\Omega^e}(B_i^r)^T DB_j^s\,d\Omega(r,s=u,a,b) \tag{11}$$

Here, B and D are shape function and material stiffness matrix.

For orthotropic material

$$D=\begin{bmatrix}C_{11} & C_{12} & C_{13}\\C_{12} & C_{22} & C_{23}\\C_{13} & C_{23} & C_{33}\end{bmatrix}=\begin{bmatrix}\frac{E_{11}}{1-\upsilon_{12}\upsilon_{21}} & \frac{\upsilon_{21}E_{22}}{1-\upsilon_{12}\upsilon_{21}} & 0\\\frac{\upsilon_{21}E_{22}}{1-\upsilon_{12}\upsilon_{21}} & \frac{E_{22}}{1-\upsilon_{12}\upsilon_{21}} & 0\\0 & 0 & G_{12}\end{bmatrix} \tag{12}$$

(a) For laminated composite material [6]

IOP Conf. Series: Materials Science and Engineering **814** (2020) 012020 doi:10.1088/1757-899X/814/1/012020

$$D=\sum_{k=1}^{NL}\int_{z_{k-1}}^{k}\overline{Q_{ij}}\left(1\ z_k\right)dz \tag{13}$$

where the number of layers in layered composite is represented by *NL.*
The general form of *J*-integral method for the calculation of MMSIF can be written as

$$J=\int_A\left(\frac{1}{2}W-\sigma_{ij}\frac{\partial u_j}{\partial x_1}\right)n_j dA \tag{14}$$

Where A and W are the area surrounded by crack tip and strain energy density,

$$W=\frac{1}{2}\sigma_{ij}\varepsilon_{ij}=\frac{1}{2}R_{ijkl}\varepsilon_{kl}\varepsilon_{ij} \tag{15}$$

Using the equivalent domain integral, Eq. (14) can be transformed to

$$J=\int_A\left(\sigma_{ij}\frac{\partial u_j}{\partial x}-W\Delta_{1j}\right)\frac{\partial q}{\partial x_j}dA \tag{16}$$

Where, Δ_{1j} is the Kronecker delta and q is a linearly varying function, q=1 at the crack tip and q=0 at the outer boundary r as shown in figure 1. The *J*-integral is the combination of the actual and auxiliary states of J integrals as:

$$J=J^{act}+J^{aux}+M \tag{17}$$

Where M is the interaction integral and it is represented as

$$M=\int_A\left[\sigma_{ij}\frac{\partial u_i^{aux}}{\partial x}+\sigma_{ij}^{aux}\frac{\partial u_i}{\partial x}-W_{st}{}^{m}\Delta_{1J}\right]\frac{\partial q}{\partial x_j}dA \tag{18}$$

for linear elastic conditions

$$W^m=\frac{1}{2}(\sigma_{ij}\varepsilon_{ij}^{aux}+\sigma_{ij}^{aux}\varepsilon_{ij}) \tag{19}$$

After some arrangement the following equation is obtained [6]

$$M=2m_{11}K_IK_I^{aux}+m_{12}(K_IK_{II}^{aux}+K_I^{aux}K_{II})+2m_{22}K_{II}K_{II}^{aux} \tag{20}$$

By setting states as $\left(K_I^{aux}=1,\ K_{II}^{aux}=0\right)$ and $\left(K_{II}^{aux}=1,\ K_I^{aux}=0\right)$, MMSIF can be calculated
The MSIFs can be evaluated by calculating M and solving linear algebraic equations as.

$$M^{(I)}=2m_{11}K_I+m_{12} \tag{21}$$

$$M^{(II)}=m_{12}K_I+2m_{22}K_{II} \tag{22}$$

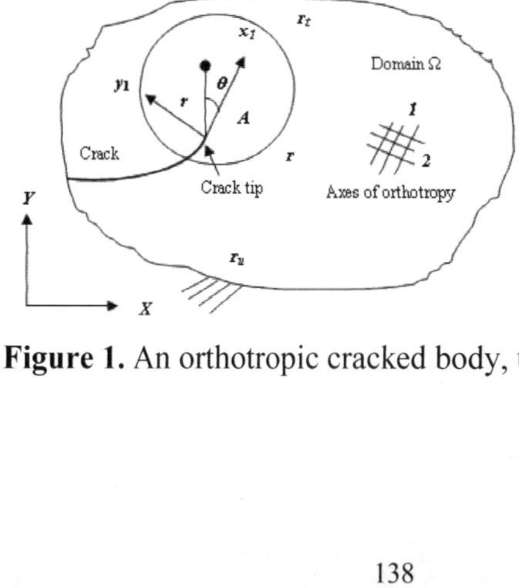

Figure 1. An orthotropic cracked body, underneath traction f_t

IOP Conf. Series: Materials Science and Engineering **814** (2020) 012020 doi:10.1088/1757-899X/814/1/012020

3. Result and discussion

A square plate with H=W =10 mm and unit thickness with inclined center crack of crack angle α and semi crack length (a=1mm) is considered. The plate is under uniform tensile stress and shear stress in the X and Y directions respectively is shown in figure 2(a). The factor K is the biaxial load factor, which is defined as the ratio of the load applied parallel to the crack to that applied perpendicular to it, where the crack is aligned with the X or Y coordinate axis. The meshing and enrichment of crack tip and face are shown in figure 2 (b-c). The material properties utilised for this study are E11 =144.80 Gpa, E22=11.70 Gpa, G12=9.66 Gpa and υ12 =0.21, the crack angle (α) = 450 and K = 2.

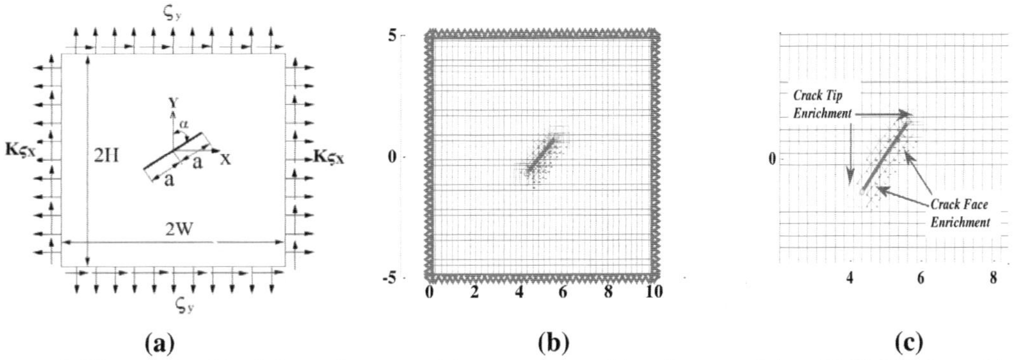

Figure 2. Plate under biaxial tensile stress (a) geometry and loading condition (b) X-FEM meshing of the plate with centre inclined crack (c) enrichment for crack tip and crack face by X-FEM.

Figure 3 shows the comparative study, where MMSIF of K_I and K_{II} are calculated for different crack angle and are compared with the result from literature. This comparative study shows that the current approach is quite efficient for this type of study.

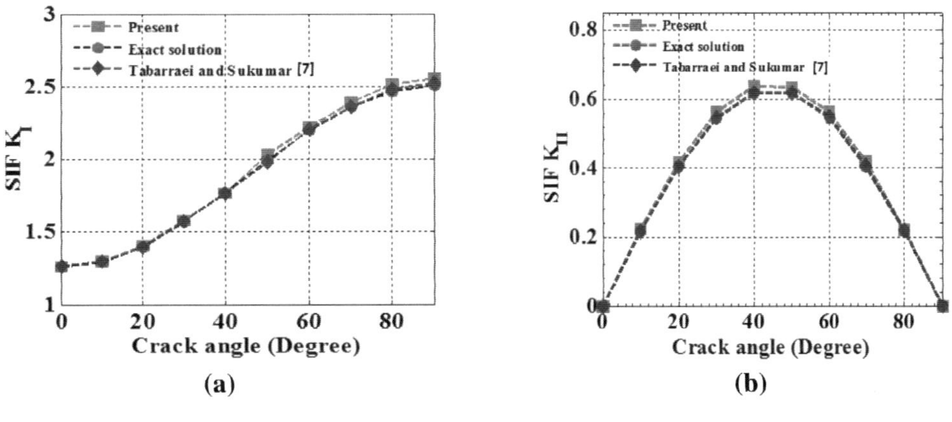

Figure 3. Comparison of present MSIFs results for K_I and K_{II}.

Figure 4 shows the effect of orthotropic angle on MMSIF of K_I and K_{II} for layered composite plate under the action of different loadings. From the figure 4 (a-b) it is observed that bi-axially applied tensile, shear and combined stresses follow the same trend for K_I and K_{II} The values of K_I are decreasing up to β = ±30°, and then they start to increase and are maximum at β = ±40° and for β = 0° and 90° the values are same. The values of K_{II} are very small as compared to K_I which represents there is no response of fibre angle on K_{II} which is also observed in uniaxial loading. The MSIFs K_I and K_{II} follow almost sinusoidal variation about 40° fibre angles for all types of biaxial loads considered in this study.

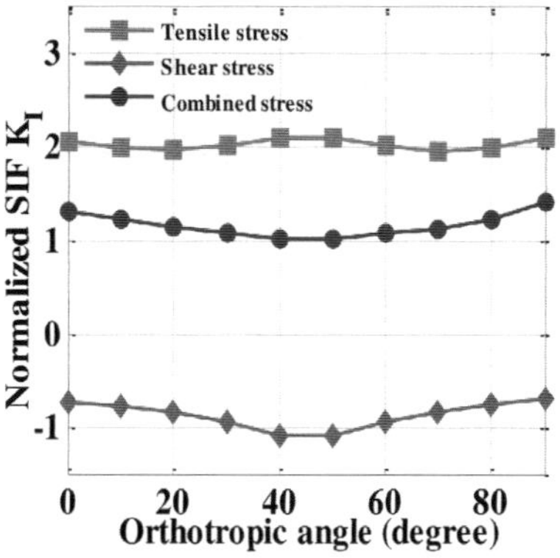

(a)

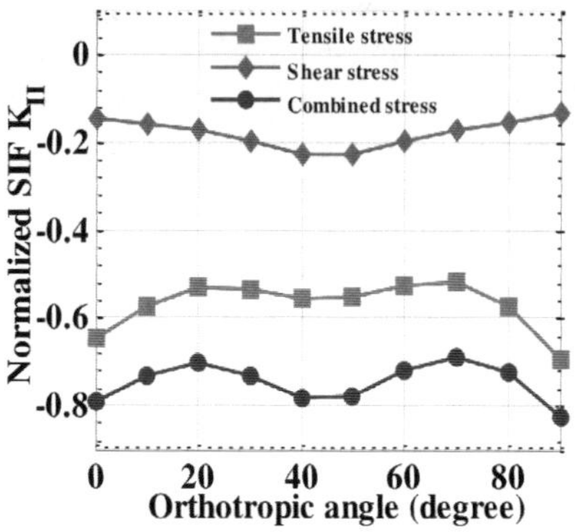

(b)

Figure 4. Effect of orthotropic axis ± β ° on the MSIFs of centrally inclined crack (α =45°) in an orthotropic plate under biaxial tensile, shear and combined stress with (K=2).

4. Conclusions

In the present work analysis of MSIFs of a central crack, inclined crack of orthotropic and symmetric angle-ply laminated composite plates under different biaxial loadings is presented. Present results show significant effects of variation of fibre angle of an inclined centre crack of laminated composite plates. For all types of stresses, the following conclusions are written

- There is a variation in the value of MSIFs with the increase in fibre orientation angle β°, which reveals that fibre orientation angle determines fracture axis in the case of orthotropic and laminated composite structures.
- Central inclined crack shows a mode-I failure under tensile stress, mode-II failure when subjected to shear stress and mixed mode failure under combine stress. While in the case of an eccentric crack, as the distance of crack from the plate edge increases the crack tips follow mixed mode fracture failure when subjected to shear and combine stress, and shows a mode-I failure under tensile stress.

References

[1] Asadpoure A, Mohammadi S and Vafaia A 2006 *Thin-Walled Struct.* **44** 1031–38.
[2] Meek C and Ainsworth R A 2015 *Eng. Fract. Mech.* **147** 306–17.
[3] LimW K, Choi S Y and Sankar B V 2001 *Eng. Fract. Mech.* **68** 403-16.
[4] Mostafavi M, Smith D J and Pavier M J 2011 *Eng. Fract. Mech.* **78** 1705–16.
[5] Wang J G, Ju D Y, Sun M J, and Li S L 2011 *Mater.Des.* **32** 664–70.
[6] Lal A, Palekar P S, Mulani S B and Kapania R K 2016 *Aerosp. Sci. and Tech.* **60** 131-51.
[7] Tabarraei A and Sukumar A 2007 *Comput.Methods in Appl. M.* **197** 425-38.
[8] Yan X 2004 *Eng. Fract. Mech.* **71** 615–23.
[9] Goldstein R V and Shifrin E I 2012 *Int.J. Eng.Sci.***61** 36–47.
[10] Lal A and Palekar S P 2015 *Int. J.Mech. Mater. Des.* **13-2** 195-228.
[11] Lal A, Mulani S B and Kapania R K 2017 *Int.J. Of Appl. Mech.* **9-4** 1750061- 94.

Performance Evaluation of Bridge Piers Reinforced with Shape Memory Alloys in Plastic Hinge Region: Part 2 – Optimization

Kanan Thakkar[1] and Anant Parghi[2]

[1]Research Scholar, Department of Applied Mechanics, S. V. National Institute of Technology, Surat – 395 007, Gujarat, India
[2]Assistant Professor, Department of Applied Mechanics, S. V. National Institute of Technology, Surat – 395 007, Gujarat, India

Email: kananrthakkar@gmail.com, anantparghi@gmail.com

Abstract. Shape Memory Alloy (SMA) rebars are emerging as a potential solution to mitigate the permanent/residual deformation of bridge piers in recent decades. The maximum damage of a bridge pier being concentrated in the plastic hinge region and SMA being the expensive material to be utilized in the large-scale structural elements, the bridge piers are reinforced with SMA in the plastic hinge regions only. The bridge piers are modeled with a non-linear finite element tool. Three variable material parameters are adopted for this study, such as the compressive strength of concrete, the yield strength of steel, and the type of SMA composition. The bridge pier models considering One-Factor-at-a-Time (OFAT) Method, and Full Factorial Design Method in the previous study (Part 1) are considered. In this paper, the results of full factorial design of the previous research are optimized using optimization techniques in order to determine the best suitable combination of these three material properties. The performance of SMA-reinforced bridge piers is evaluated based on the drift values at various performance criteria, i.e., cracking, spalling, yielding, and crushing. The results are presented in terms of the rank-wise significance for each possible combination of the considered material properties of bridge piers.

1. Introduction

1.1. *General*

The nation's development is directly proportional to its transportation network. Bridges are the backbones of a transportation network of any nation. Bridge pier is the most significant structural element in the entire bridge structure, as it transfers the lateral and horizontal loads to a foundation. It is observed from the previous research that a huge amount of displacement was detected in reinforced concrete (RC) bridges after the occurrence of massive earthquakes [1]. The maximum damage of a bridge pier is concentrated in a small region, defined as a plastic hinge region [2]. With an aim to improve the performance of structures, numerous innovative composite materials are invented. One such novel composite materials are Shape Memory Alloys (SMA).

SMA has two unique properties, named as super-elasticity and shape memory effect. SMA can sustain enormous deformations and reappear to a parent shape after removal of loading (super-elasticity) or by removal of heat (shape memory effect) [3]. SMA exhibits self-centering phenomena, which leads the structure to re-center its original position after removal of the loading. This phenomenon of SMA is a unique feature for applications in structural engineering. There are two crystal phases of SMA, austenite and martensite. SMA gets converted from austenite to martensite phase and martensite to austenite phase by applying/removing stresses/ temperature. There are various constituents of SMA, e.g., nickel, titanium, copper, iron, manganese, etc. Amongst these all SMA, nickel-titanium based SMA has been confirmed its potential for the applications in structural elements.

This research is focused on the performance assessment of bridge piers reinforced with SMA in the plastic hinge region with three variable parameters. The factors are varied altogether using a full factorial method for bridge pier models in the previous study. In this study, the best combination of the values for each factor is defined using several optimization methods.

Optimization is the process of defining the best utilization of a particular resource for the desired outcome. The purpose of optimization is to achieve the best design for a set of prioritized criteria or constraints. There are some functions associated with the demand of a specific optimization process, e.g., maximizing functions and minimizing functions. This decision-making process is known as optimization [4].

Multi-Criteria Decision Making (MCDM) is a category of methods existing for the decision making with more than one contradictory criterion. MCDM methods are classified primarily into two groups, such as Multiple Attribute Decision Making (MADM) and Multiple Objective Decision Making (MODM). In MODM methods, there are more than one goals to be achieved by the optimization process. Whereas in MADM methods, there are more than one criterion and only one goal to be achieved by the optimization process. There are numerous MADM methods, i.e., Analytic hierarchy process (AHP) method, Weighted product model (WPM) method, Simple Additive Weighting (SAW) method, Promethee method, Topsis method, etc. [4]. For this research, AHP, SAW, and WPM are the three methods selected for optimization.

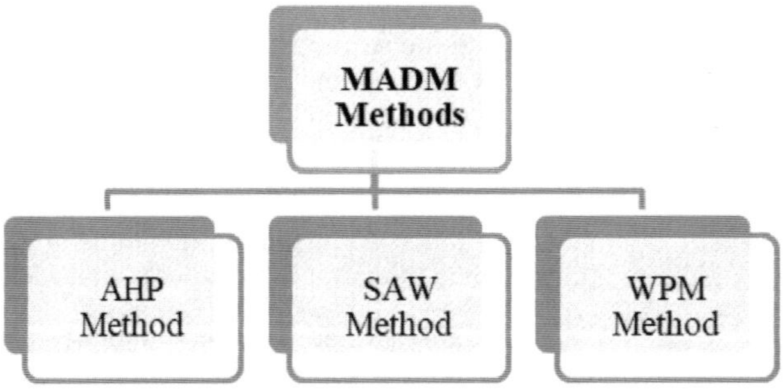

Figure 1. Selected methods of MADM

1.2. Scope of previous research

The applications of super-elastic SMA reinforcement in structural engineering is increasing hastily in recent decades. SMA offers numerous structural applications, namely, reinforcement in reinforced concrete elements, connectors in steel and concrete structures, damping devices, base isolation devices, etc. [5]. Numerous applications of SMA in the bridge structures are explored to improve the seismic performance of bridge structures. Since the last decades, several researchers have studied the behavior of SMA-reinforced bridge piers to mitigate the seismic energy after massive earthquake events [6].

Furthermore, there is a limited investigation of the parametric study as an alternate of steel reinforcement at the plastic hinge location with super-elastic SMA-based reinforcement on the dynamic performance of the bridge piers. Recently, there is no code/guidelines available in the existing literature for the use of super-elastic SMA as an alternative reinforcement with conventional steel.

2. Modeling of bridge piers

Following numerical analysis methods, material properties, geometrical data, material constitutive models, FE parameters, and modeling techniques are adopted for this study.

2.1. Static pushover analysis

Static pushover analysis is a nonlinear analysis methodology in which a monotonic lateral load is applied to the structure in an incremental manner. This monotonic loading is applied in such a way to determine the entire performance of a structure from the elastic phase to inelastic phase and failure. Figure 2(a) and illustrates the schematic diagram for typical pushover analysis of a pier. Figure 2(b) demonstrates the values of base shear, and displacement for flexural limit states, i.e., the values at which first cracking, spalling, yielding, crushing, buckling, and fracture.

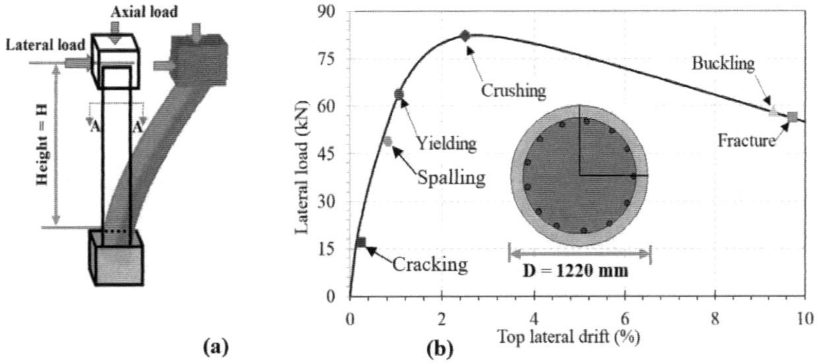

Figure 2. Schematic of a typical pushover analysis of pier (a) deformed and un-deformed shape, and (b) load vs. drift curve with limit states [7]

2.2. Flexural limit states

The performance points and their flexure limit states are essentials to be defined for the performance assessment of SMA reinforced bridge piers. Damages at several flexural limits are determined by strain-based method [2]. The following flexural limits are taken into consideration to estimate the performance SMA-reinforced circular bridge piers, e.g., cover concrete's cracking and spalling, longitudinal SMA rebar's yielding, and core concrete's crushing.

2.3. Geometry of bridge piers

In this study, a 0.2-scaled circular bridge pier [8] has been considered for the nonlinear static pushover analysis. The geometry of the pier used in this study is, as shown in Figure 3.

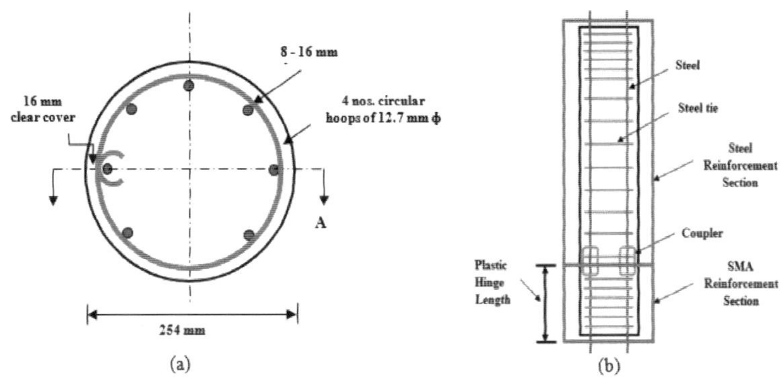

Figure 3. Reinforcement and geometrical details of SMA-RC bridge pier

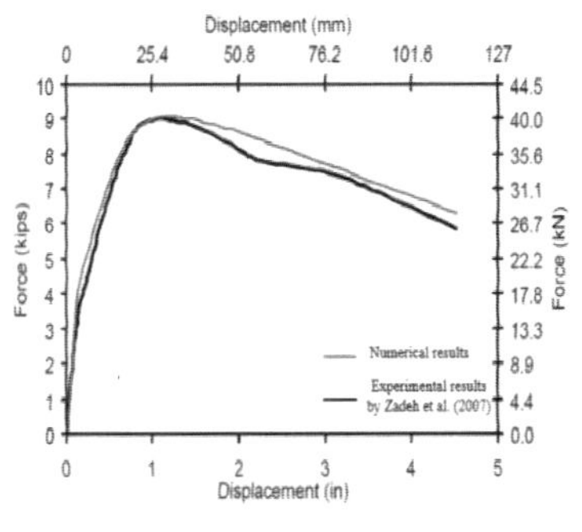

Figure 4. Validation of numerical model [8]

The diameter of a pier is 254 mm, and the height of a pier is 1270 mm. The height of a pier is calculated from the top of footing to the centerline of loading head, where the lateral load is applied. Axial load ratio is selected as 10% of the axial capacity of a pier, and longitudinal reinforcement ratio (ρ_t) is used as 2%. The aspect ratio is selected as 5. Figure 3(a) illustrates the geometry of the pier, and section A-A in Figure 3(b) shows the vertical section of the pier with the longitudinal and lateral reinforcement details of the bridge pier. Plastic hinge length is calculated as 205.7 mm, according to Paule and Pristley [6] equation.

2.4. *Material* constitutive models

Table 1. Materials properties used for Steel-RC and SMA-RC piers

Material	Property	Value	Reference
Confined concrete	Elastic Modulus (GPa)	33.23	-
	Compressive strength (MPa)	50	
	Tensile strength (MPa)	5	
	Corresponding strain	0.005	
Longitudinal steel	Elastic Modulus (GPa)	200	-
	Yield stress (MPa)	415	
	Fracture strain	0.10	
	Strain hardening parameter	0.05	
Super-elastic SMA	Yield strength- f_y (MPa)	210, 401, 750	[10] [11] [12]
	Modulus of elasticity-E (GPa)	28, 62.5, 46.9	
	Super-elastic plateau strain length- ε_s (%)	9, 6, 13.5	
	A to M starting stress- f_y (MPa)	210, 401, 750	
	A to M finishing stress- f_{P1} (MPa)	275, 510, 1200	
	M to A starting stress- f_{T1} (MPa)	200, 370, 300	
	M to A finishing stress- f_{T2} (MPa)	150, 130, 200	

Where A is the austenite phase, and M is the martensite phase.

In this study, three different material constitutive models are considered for the modelling of bridge piers: steel reinforcement, SMA reinforcement, and concrete. Menegotto and Pinto [13] steel constitutive model is adopted to assign the properties of steel reinforcement in FE tool. To represent NiTi reinforcement in the finite element model, Fugazza [14] model is implemented for this study. To predict the stress-strain behavior of confined concrete in this study, the stress-strain model constitutive behavior by Mander et al. [15] is adopted. Table 1depicts the material properties of steel-reinforced and SMA-reinforced bridge piers used for a numerical investigation.

2.5. *Finite element modeling*

To study the seismic performance of super-elastic SMA-RC and steel-reinforced bridge piers, a finite element (FE) program SeismoStruct [16] is adopted. The piers are modeled using the displacement-based inelastic system owing to its modeling requirements. There are 300 fibers considered across the section for the mesh refinement. The bar-slippage of super-elastic SMA rebar inside the coupler is neglected. Axial load is applied to the top of a pier.

2.6. *Validation*

The precision of the proposed model has been validated through experimental results proposed by Zadeh et al. [8] under the Highway Idea Project, University of Nevada, Reno (Figure 4). They conducted an experimental program on 0.2-scaled circular columns utilizing SMA in the plastic hinge zone under the effects of dynamic loads.

IOP Conf. Series: Materials Science and Engineering **814** (2020) 012021 doi:10.1088/1757-899X/814/1/012021

3. Optimization of bridge piers

This research emphasis on the performance assessment of SMA-RC bridge piers under the effect of static pushover loads. In this study, three factors, each having three levels, are defined. The considered factors based on comprehensive literature and their levels are defined as shown in Table 2. The no. of experiments performed in the previous study for one-factor-at-a-time (OFAT) method were $3 \times 3 = 9$, and full factorial design method were $3^3 = 3 \times 3 \times 3 = 27$. Static pushover analysis is performed for 27 bridge pier models as per the full factorial method of Design of Experiment (DoE), shown in Table 3.

Table 2. Factors and levels

Sr. No.	Factors	Levels		
		Low (-1)	Intermediate (0)	High (+1)
1.	Compressive Strength of Concrete, f_c (MPa)	35	50	65
2.	Yield Strength of Steel, f_y (MPa)	250	415	500
3.	Composition of SMA	CuAlMn (210 MPa)	NiTi (401 MPa)	FeNCATB (750 MPa)

Table 3. Factors and levels for each run order

Sr. No.	Run	Factor-1 f_c (MPa)	Factor-2 f_y (MPa)	Factor-3 Composition of SMA
1	1	35	250	CuAlMn (210 MPa)
2	2	35	250	NiTi (401 MPa)
3	3	35	250	FeNCATB (750 MPa)
4	4	35	415	CuAlMn (210 MPa)
5	5	35	415	NiTi (401 MPa)
6	6	35	415	FeNCATB (750 MPa)
7	7	35	500	CuAlMn (210 MPa)
8	8	35	500	NiTi (401 MPa)
9	9	35	500	FeNCATB (750 MPa)
10	10	50	250	CuAlMn (210 MPa)
11	11	50	250	NiTi (401 MPa)
12	12	50	250	FeNCATB (750 MPa)
13	13	50	415	CuAlMn (210 MPa)
14	14	50	415	NiTi (401 MPa)
15	15	50	415	FeNCATB (750 MPa)
16	16	50	500	CuAlMn (210 MPa)
17	17	50	500	NiTi (401 MPa)
18	18	50	500	FeNCATB (750 MPa)
19	19	65	250	CuAlMn (210 MPa)
20	20	65	250	NiTi (401 MPa)
21	21	65	250	FeNCATB (750 MPa)
22	22	65	415	CuAlMn (210 MPa)
23	23	65	415	NiTi (401 MPa)
24	24	65	415	FeNCATB (750 MPa)
25	25	65	500	CuAlMn (210 MPa)
26	26	65	500	NiTi (401 MPa)
27	27	65	500	FeNCATB (750 MPa)

The results of SMA-reinforced bridge piers are evaluated in terms of the performance criteria of limit state of design, e.g., cracking, yielding, spalling, and crushing. These five outputs are considered to be the maximizing functions. Furthermore, the effect of different materials properties, such as yield strength of steel, compressive strength of concrete, and the composition of SMA, are explored using three methods of optimization. The adopted methods of optimization are AHP method, WPM method, and SAW method. Moreover, the performance-wise rankings are assigned to a total of 27 bridge pier models.

4. Results and discussion

The non-linear static pushover analysis is performed for SMA-reinforced circular bridge piers as per the data described in previous sections. The results of these models in terms of cracking drift, spalling drift, yielding drift, crushing drift, and maximum drift, are adopted for the optimization process. These drift values are desirable to be maximum; hence, they are selected as maximizing functions. The results of the three optimization methods are discussed in the following section.

Table 4 demonstrates the significance-wise ranking of bridge piers as per AHP method. It can be observed from the table that run no. 3 (Rank no. 1) having f_c = 35 MPa, f_y = 250 MPa, and $f_{y\text{-}SMA}$ = 750 MPa gives the maximum values for drift at various flexural limit states, e.g., cracking drift, spalling drift, yielding drift, crushing drift, and maximum drift. It indicates that the low compressive strength of concrete, low yielding strength of steel, and high yielding strength of SMA lead to the higher drift values of a bridge pier. Further, the higher strain value of Fe-based SMA ($f_{y\text{-}SMA}$ = 750 MPa) makes a bridge pier more flexible to reduce permanent deformations. Further, run no. 11 having f_c = 50 MPa, f_y = 250 MPa, and $f_{y\text{-}SMA}$ = 401 MPa stand for rank no. 2. Run no. 23 having f_c = 65 MPa, f_y = 415 MPa, and $f_{y\text{-}SMA}$ = 401 MPa stand for rank no. 3.

Table **5** 5 expresses the significance-wise ranking as per SAW method. The results of AHP method and SAW method are similar to each other.

Table 4. Significance-wise ranking as per AHP method

Rank	Run	Input		
		Factor 1	Factor 2	Factor 3
		f_c	f_y	$f_{y\text{-}SMA}$
1	3	35	250	FeNCATB (750 MPa)
2	11	50	250	NiTi (401 MPa)
3	23	65	415	NiTi (401 MPa)
4	24	65	415	FeNCATB (750 MPa)
5	6	35	415	FeNCATB (750 MPa)

Table 5. Significance-wise ranking as per SAW method

Rank	Run	Input		
		Factor 1	Factor 2	Factor 3
		f_c	f_y	$f_{y\text{-}SMA}$
1	3	35	250	FeNCATB (750 MPa)
2	11	50	250	NiTi (401 MPa)
3	23	65	415	NiTi (401 MPa)
4	24	65	415	FeNCATB (750 MPa)
5	6	35	415	FeNCATB (750 MPa)

Table 6 demonstrates the significance-wise ranking of bridge piers as per WPM method. It can be observed from the table that run no. 3 (Rank no. 1) having f_c = 35 MPa, f_y = 250 MPa, and $f_{y\text{-}SMA}$ = 750 MPa gives the maximum values for drift at various flexural limit states, e.g., cracking drift, spalling drift, yielding drift, crushing drift, and maximum drift. It indicates that the low compressive strength of concrete, low yielding strength of steel, and high yielding strength of SMA lead to the higher drift values of a bridge pier. Further, the higher strain value of Fe-based SMA ($f_{y\text{-}SMA}$ = 750 MPa) makes a bridge pier more flexible to diminish permanent deformations. Further, run no. 11 having f_c = 50 MPa, f_y = 250 MPa, and $f_{y\text{-}SMA}$ = 401 MPa stand for rank no. 2. Run no. 6 having f_c = 65 MPa, f_y = 415 MPa, and $f_{y\text{-}SMA}$ = 401 MPa stand for rank no. 3.

Table 6. Significance-wise ranking as per WPM method

Rank	Run	Input		
		Factor 1	Factor 2	Factor 3
		f_c	f_y	$f_{y\text{-}SMA}$
1	3	35	250	FeNCATB (750 MPa)
2	11	50	250	NiTi (401 MPa)
3	6	35	415	FeNCATB (750 MPa)
4	19	65	250	CuAlMn (210 MPa)
5	17	50	500	NiTi (401 MPa)

5. Conclusions

By optimizing the results of parametric investigation of RC bridge piers with SMA reinforcement in the plastic hinge region using three methods of optimization (AHP method, SAW method, and WPM method), following conclusions are drawn:

- The significance-wise ranking is assigned based on the demand for outputs from static pushover analysis. The outputs of static pushover analysis on SMA-RC bridge piers (cracking drift, spalling drift, yielding drift, crushing drift, and maximum drift) are used as input factors for AHP method, SAW method, and WPM method. These drift values are designated as maximizing functions.
- AHP method and SAW method give the same results for the rank-wise significance of bridge pier models.
- The f_c as 35 MPa, f_y as 250 MPa, and $f_{y\text{-}SMA}$ as 750 MPa possess rank 1 for AHP method, SAW method, and WPM method of optimization.
- The f_c as 50 MPa, f_y as 250 MPa, and $f_{y\text{-}SMA}$ as 401 MPa possess rank 2 for AHP method, SAW method, and WPM method of optimization.
- The f_c as 50 MPa, f_y as 250 MPa, and $f_{y\text{-}SMA}$ as 401 MPa possess rank 2 for AHP method, SAW method, and WPM method of optimization.
- The f_c as 65 MPa, f_y as 415 MPa, and $f_{y\text{-}SMA}$ as 401 MPa possess rank 3 for AHP method, and SAW method of optimization.
- The f_c as 35 MPa, f_y as 415 MPa, and $f_{y\text{-}SMA}$ as 750 MPa possess rank 3 for WPM method of optimization.

Acknowledgements

The first author would like to acknowledge the financial support provided by Ministry of Human Resources Development (MHRD), Government of India, through the Ph.D. fellowship program. Further, the authors would like to acknowledge the online and offline library of SVNIT.

References

[1] Shrestha B and Hao H 2014 *23rd Australasian Conf. on the Mechanics of Structures and Materials (ACMSM23)* vol **I** (Byron Bay, NSW, Australia) p 375–80
[2] Billah M 2015 *Ph. D. Thesis* (University of British Columbia, Canada)
[3] Boroschek R L, Farias G, Moroni O and Sarrazin M 2007 *J. Earthq. Eng.* **11** 326–42
[4] Rao R V 2007 *Decision making in the manufacturing environment: Using graph theory and*

IOP Conf. Series: Materials Science and Engineering **814** (2020) 012021 doi:10.1088/1757-899X/814/1/012021

fuzzy multiple attribute decision making methods (Springer Science & Business Media) p 374
[5] Ozbulut O E, Hurlebaus S and DesRoches R 2011 *J. Intell. Mater. Syst. Struct.* **22** 1531–49
[6] Paulay T and Priestley M J N 1992 *Seismic design of reinforced concrete and masonry buildings* John Wiley & Sons, Inc. New York, USA
[7] Parghi A 2016 *Ph. D. Thesis* (University of British Columbia, Canada)
[8] Zadeh M S, Brien M O and Saiidi M S 2007 *A study of concrete bridge columns using innovative materials subjected to cyclic loading* A report by University of Nevada Reno and National Cooperative Highway Research Program (NCHRP) p 68
[9] Schoettler M J, Restrepo J I, Guerrini G and Duck D E 2015 *A full-scale, single-column bridge bent tested by shake-table excitation* A report by Pacific Earthquake Engineering Research Center (PEER) (Berkeley) p 153
[10] Shrestha K C, Araki Y, Nagae T, Koetaka Y, Suzuki Y, Omori T, Sutou Y, Kainuma R and Ishida K 2013 *Smart Mater. Struct.* **22** 25025–12
[11] Youssef M A, Alam M S, Youssef M A and Nehdi M 2008 *Eng. Struct.* **30** 3399–411
[12] Tanaka Y, Himuro Y, Kainuma R, Sutou Y, Omori K and Ishida K 2010 *Science* **327** 1488–90
[13] Menegotto M and Pinto P E 1973 *Method of analysis for cyclically loaded R.C. plane frames including changes in geometry and non-elastic behavior of elements under combined normal force and bending* A preliminary report in IABSE symposium on the resistance and ultimate deformability of structures acted on by well defined repeated loads (Lisbon) pp 15-22
[14] Fugazza D 2003 *M. Tech Thesis* (ROSE School, European School of Advanced Studies in Reduction of Seismic Risk)
[15] Mander J B 1988 *J. Struct. Eng.* **114** 1804–26
[16] SeismoSoft 2018 *SeismoStruct- A computer program for static and dynamic nonlinear analysis of framed structures*

IOP Conf. Series: Materials Science and Engineering **814** (2020) 012022 doi:10.1088/1757-899X/814/1/012022

Additive Manufacturing of Concrete: Challenges and opportunities

Avinash Nair[1*], Aditya S D[2], Adarsh R N[3], Nandan M[4], Manish S Dharek[5], B M Sreedhara[6] , Prashant C Sunagar[7], Sreekeshava K S[8]

[1,2,3,4,5] Civil Engg Dept, BMS Institute of Technology and Management (affiliated to VTU, Belagavi), Bengaluru, India
[6] Civil Engg Dept, Siddaganga Institute of Technology, Tumakuru
[7] Civil Engg Dept, Ramaiah Institute of Technology, Bengaluru, India
[8]Civil Engg Dept, Jyothy of Technology (affiliated to VTU, Belagavi), Bengaluru, India

*Corresponding Author: avinashunnikrishnan@gmail.com

Abstract. A modern incipient trend is being witnessed in the construction industry wherein concrete is pumped which flows through a nozzle connected to a robotic arm in successive layers in order to develop structural load bearing members generally referred as "Additive manufacturing of Concrete(AMOC)" or 3D Printing of Concrete (3DPC). Numerous challenges are being faced by the construction industry for implementation of Additive manufacturing of concrete to a large scale due to the scarcity of information available w.r.t this technology. This technology has opened up new opportunities which requires intensive research to be carried out to ensure that concrete gets pumped through the pipe(pumpability), extrudes through the nozzle (extrudability), retains its shape once this concrete is extruded from the nozzle(buildability). In addition to this, concrete has to be strong enough to withstand loads induced by upper layers without any deformations (shape retentivity). If the challenges related to pumpability, extrudability, buildability and shape retentivity can be addressed in a better manner, this technology of 3D concrete printing can be utilized to construct houses/ building components at a rapid rate and bring down the overall construction costs exponentially with optimal usage and minimal wastage of resources. This paper addresses the various challenges which are commonly encountered in the Additive manufacturing of Concrete (AMOC) summarizing the potential solutions to it discussing some of the case studies of projects which have used this technology.

1. Introduction

The earliest use of 3D Printing was to develop a tabletop scaled model. It was much economical option to erect these models compared to hand crafted model that had been and still in use. By the 1990s, many organizations were experimenting to use 3D Printing concept to build a full-scale project. By 2000s, the application was in full use and was getting a start to transform the entire construction industry. In 2006, Khoshnevis at the University of Southern California uncovered the Contour Crafting (CC) system [5]. It is a huge 3D Printer developed for building a whole building in place. It is similar to desktop 3D Printer but uses a crane to print and concrete as a medium to build the structure. In 2014, Dutch firm (DUS Architect) was able to showcase the use of 3D Printing by building a canal house in Amsterdam [7]. They used a giant crane printing arm called "Kamemaker" which mean

"Room Builder". In June 2016, a Chinese company Hua Shang Tenda announced that they had constructed an entire mansion in 45 Days [4]. The building frame was erected first followed by placing the plumbing and electrical fixtures in it and then printed the structures using a computer-controlled printer extruding concrete through it. The building comprised of 2 storeys with 4305 sq.ft carpet area. In April 2016, another Chinese company WinSun claimed to have constructed 10 houses in a span of 24hrs [6]. The walls of the houses were constructed using a mixture of recycled construction material and cement. They stated, this is both economical and environmentally friendly than the traditional practice. According to the United Nations world population prospects, the overall population is set to reach 9.6 billion people by 2050. A surge increase in population at this rate means requirement of more housing for people within a short period of time. With the traditional methods and current pace at which the construction industry is moving, it is a herculean challenge to provide housing to cater to this population. Hence, it is time to adopt to newer technologies which can increase the pace of construction industry in an efficient way with minimal wastage of materials and resources.3D printable concrete is an emerging trend in the construction industry which runs on the principle of additive manufacturing wherein concrete is designed in a way such that it is pumped through a nozzle which is connected to a gantry girder. This nozzle moves in the direction specified by the user which results in the formation of a 3D concrete model which when constructed on a larger scale can build houses and other buildings. Since this concrete is passed through a nozzle, there are minimal wastages taking place to construct a building when compared to traditional construction practices. This technology being an offspring in the construction industry, a lot of challenges are put in front due to the fact that not much research is being held in this field. This paper reviews the various challenges pertaining to 3D printable concrete. Possible solutions to these challenges which can be adopted are also discussed in this paper.

2. Challenges in 3D Printable Concrete(3DPC)

2.1 Pumpability

Concrete pumpability is defined as its ability to flow through a pipe using a pump and also the ability of confined concrete to flow under pressure while maintaining its initial properties [9]. The pumpability and buildability of 3D printable concrete is measured in terms of the pumpability index (the ratio of concrete flow rate to water flow rate of 2890 rpm), surface performance and maximum height printed [19]. Mixtures with low slump and slump-flow values tend to have a rough texture on the surface of the filaments with high persistence of form, large voids / cracks and poor pumpability index. [19]. Higher flowability (such as low 192 mm spread diameter during flow table testing) may also cause congestion and interruption of printing, poor printable material property can result in increased concrete filament voids [14].

2.2 Extrudability

Extrusion is a process in which a rigid paste is forced to pass through a rigid die with a high shear, resulting in a liquid behavior [17]. The size of the nozzle plays a crucial role in concrete extrudability. A larger diameter causes problems with the buildability because the layers could not support themselves while a smaller diameter posed problems with the separation of the concrete components [21]. In addition, lower flowability (such as low 192 mm spread diameter during flow table testing) could cause congestion and interruption of printing, poor printable material property can lead to an increase in concrete filament voids [14]. The sand-to-cement ratio (S / C) exceeding 1.5 significantly reduced the flowability that a nozzle cannot extrude.

2.3 Buildability

Buildability refers to the concrete layer's ability to hold the successive layers above it without subsiding. The concrete must also have a certain compressive strength [1]. Because of its relatively low stiffness and strength of the printed filament, concrete can pose buildability problems during its initial condition and another observation showed that the extruded filament temporarily suffers from cavity [1]. With a time gap of 19minutes, smaller deformations were measured as per layer settlement test results. Significant deformations (1.5mm) occurred after second layer was deposited for Poly-Propylene mixture with no time gap [2].

2.4 Strength

Panda et al. [13] studied the role of printing parameters such as the time gap between two successive layers, height from which the layer is deposited and its impact on the tensile bond strength. The study concluded that the surface moisture content at the interface between the layers played a vital role in regard to bond strength [16]. Le et al. [12] demonstrated the possibility of anisotropic 3D printable concrete components. Compressive and flexural strengths were tested in different directions on the printed specimen in the experiment and compared to the mold cast samples. High compressive and flexural strengths in the mould cast specimens was observed when compared with specimens which were 3D printed.

3. Solutions to Various issued related to 3D Printable Concrete (3DPC)

A square nozzle is generally used which improves buildability which is generally programmed so that the nozzle's orientation is always tangible to the tool direction. If the process is ignored, the filament tends to get twisted. Therefore, a rectangular nozzle is chosen for better performance efficiency [1]. A proper mix design requires high cement content or, more precisely, high paste content for adequate pumpability [10]. Addition of more water and 1% to 2% of super plasticizer makes the mix more flowable and hence improves the extrudability [12], use of polypropylene fibers reduces formation of cracks [1, 11] reported higher buildup rate of internal structure after shear induced breakdown for Nano-clay incorporated cementitious mixtures, especially at early stages. This could be a likely reason for enhanced buildability of Nano clay mixture [18] resolved the issue pertaining to cold joints by printing hollow structures so that the rebars can be placed afterwards. Increase in contact surface of layers resulted in increase in bonding strength [20].

4. Case Study

Construction of a Power distribution substation using 3D printable concrete technology [22].In this project, A mechanical column-type structure consisting of a four-column frame shown in figure 2 for 3D printing was set up. The 3D printer was divided into several structural parts in the frame: X-axis path, Y-axis path, Z-axis column path, print head, top stabilization system and ready-mixed concrete. This printer had a width of the X-axis and height of the Z-axis as 20m. The Y-axis length could be extended indefinitely by increasing the column number of Z axis. The Y-axis length could be extended indefinitely by increasing the column number of Z **axis**.

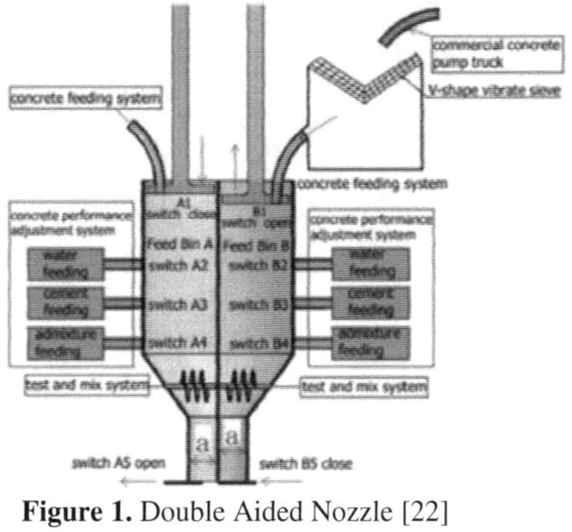

Figure 1. Double Aided Nozzle [22]

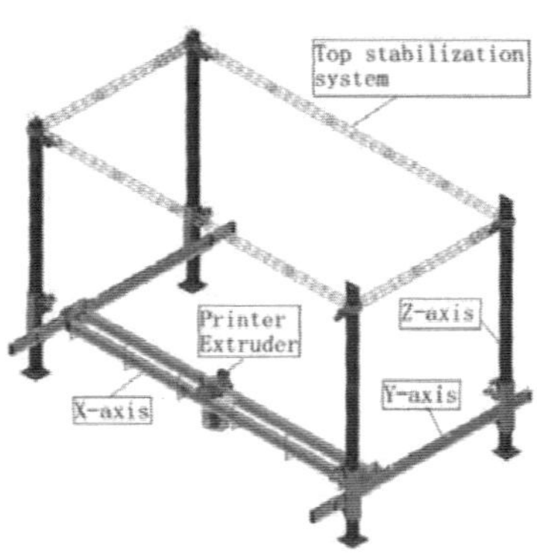

Figure 2. Four legged Column Frame [22]

A double aided nozzle was used in this project as shown in figure 1 that can extrude concrete with 15 mm diameter aggregates. The upper end of the concrete feeding system was constructed with a V-

IOP Conf. Series: Materials Science and Engineering **814** (2020) 012022 doi:10.1088/1757-899X/814/1/012022

shaped vibrating sieve. This sieve had a 15 mm pore spacing and can filter out aggregates with a diameter of more than 15 mm. The dual-assisted print head's working principle was that feed containers A and B can work together at the same time. The printing effect was highest when the concrete slump in the feed bin was between 120 and 130 mm. This nozzle was equipped in the concrete performance test system with a torque test screw. This ensures that the rough aggregate in the concrete does not exceed 15 mm in diameter. The concrete used had a strength of 20 MPa during the printing process of the power distribution substation, which met the structural design requirements. The concrete wall was first printed horizontally and the location of the building columns during the printing process was reserved. After the concrete printing wall was completed, construction columns were then constructed by supporting shaping, tying steel bars and casting concrete, as shown in the figure 3. After each printing layer reinforcement was given to strengthen the bond strength as well as the building's compressive and flexural strength. It should be noted that for the printing of the lintel above the door and window openings, it is still important to use the conventional formwork type, i.e. the prefabricated wooden formwork to support the lintel. The 3D design model was first performed, and the 3D printing slice technology was then used to slice the construction model, creating a layer of digitized code. The overall construction time of the power distribution substation project using conventional cast-in-place concrete was 54 days, while the construction period using 3D printing was less than 35 days, the construction time being reduced by 30% with concrete printing reduced the work needed but lead to increase in the cost of materials as a comparatively higher-class concrete was used.

Figure 3. (left) 3D printed structure before addition of columns,(Right) after addition of columns [22]

Figure 4. (Left) Addition of reinforcement after every printed layer. (Right) Final 3D printed structure [22]

5. Conclusions

In the above paper, various challenges faced by 3D printing concrete technology such as problems encountered have been summarized below

- Pumping of concrete from the source to the nozzle such as choking, changes in structural properties, segregation of concrete, etc. to which possible solutions were studied which include addition of superplasticizers, viscosity modifying agents and use of retarders.
- Extrudability of concrete from the nozzle which had issues in improper printing, formation of cavities, poor finishing, etc. These issues were solved by changing the section of the nozzle, varying the size of the nozzle, modifying the water cement ratio, etc.
- Buildability of concrete after extrusion which resulted in problems such as collapse of lower layers, deformations with respect to time gap between layers, formation of cold joints between layers, etc. These issues were resolved by use of Attapulgite clay or nano clay, creating hollow structures to avoid cold joints, increasing the cement paste content, etc.
- Variation in strength of concrete of printed specimen and mould casted specimen, weak bond strength between layers of concrete, low flexural strength, etc. These issues were overcome by using polypropylene fibers, introduction of steel reinforcement between layers of concrete, etc.

REFERENCES

[1] Bos, F., Wolfs, R., Ahmed, Z., and Salet, T. 2016 *Additive manufacturing of concrete in construction: potentials and challenges of 3D concrete printing.* Virtual And Physical Prototyping **11(3)** 209-225.

[2] Ali Kazemian,Xiao Yuan,Evan Cochran,Behrokh Khoshneivis,2017,Cementitious materials for construction scale 3D printing :Laboratory testing of fresh printing mixture,Construction and building materials, **145** 639-647

[3] Chen, Y. and Struble, L.J. and Paulino, G.H. 2006 *Extrudability of cement-based materials. American Ceramic Society Bulletin **85** X1-X5.*

[4] Chinese Construction Company 3D Prints an Entire Two-Story House On-Site in 45 Days - 3DPrint.com | The Voice of 3D Printing / Additive Manufacturing. 2019

[5] Contour Crafting Inventor Dr. Khoshnevis: Widespread 3D Printed Homes in 5 Years, High-Rises in 10 Years - 3DPrint.com | The Voice of 3D Printing / Additive Manufacturing [Internet]. 3DPrint.com | The Voice of 3D Printing / Additive Manufacturing. 2019 [cited 5 November 2019]. Available from: https://3dprint.com/53437/contour-crafting-dr-khoshnevis/

[6] Firm 3D prints 10 houses in 24 hours. (2019). Retrieved 5 November 2019, from https://www.bbc.com/news/blogs-news-from-elsewhere-27156775

[7] Frearson A. DUS Architects builds 3D-printed micro home in Amsterdam [Internet]. Dezeen. 2019 [cited 5 November 2019]. Available from: https://www.dezeen.com/2016/08/30/dus-architects-3d-printed-micro-home-amsterdam-cabin-bathtub

[8] G. Heirman, 2011 Modelling and Quantification of the effect of mineral Additions on the Rheology of Fresh Powder Tupe Self-compacting Concrete

[9] Jolin M., Chapdelaine F., Burns D., Gagnon F., Baupré D.(2006) Pumping concrete; a fundamental and practical approach, presentation Shotcrete for underground support X, **16** 9

[10] Jolin, Marc and Burns, Dennis and Bissonnette, Benoît and Gagnon, Frédéric and Bolduc, Louis-Samuel. (2009). UNDERSTANDING THE PUMPABILITY OF CONCRETE.

[11] Kawashima, S., Chaouche, M., Corr, D., and Shah, S. (2013). Rate of thixotropic rebuilding of cement pastes modified with highly purified attapulgite clays. Cement And Concrete Research, 53, 112-118. doi: 10.1016/j.cemconres.2013.05.019

[12] Le, T., Austin, S., Lim, S., Buswell, R., Law, R., Gibb, A., and Thorpe, T. (2012). Hardened properties of high-performance printing concrete. Cement And Concrete Research, 42(3), 558-566. doi: 10.1016/j.cemconres.2011.12.003

[13] Panda, B., Paul, S., Mohamed, N., Tay, Y., and Tan, M. (2018). Measurement of tensile bond strength of 3D printed geopolymer mortar. Measurement **113** 108-116.

[14] Rahul, A., Santhanam, M., Meena, H., and Ghani, Z. 2019 3D printable concrete: Mixture design and test methods. Cement And Concrete Composites **97** 13-23

[15] Rahul, A., Santhanam, M., Meena, H., and Ghani, Z. 2019 *Mechanical characterization of 3D printable concrete. Construction And Building Materials* **227** 116710.
[16] Sanjayan, J., Nematollahi, B., Xia, M., and Marchment, T. *2018 Effect of surface moisture on inter-layer strength of 3D printed concrete. Construction And Building Materials* **172** 468-475.
[17] Tay, Y., Qian, Y., and Tan, M. 2019 *Printability region for 3D concrete printing using slump and slump flow test. Composites Part B: Engineering, 174*
[18] Wangler, T., Lloret, E., Reiter, L., Hack, N., Gramazio, F., Kohler, M., Bernhard, M., Dillenburger, B., Buchli, J., Roussel, N. and Flatt, R. 2016 Digital Concrete: Opportunities and Challenges RILEM Technical Letter **10** 67-75.
[19] Yossef, M., and Chen, A. 2019. *Applicability and Limitations of 3D Printing for Civil Structures.* Retrieved 5
[20] Zareiyan, B., and Khoshnevis, B. (2017). *Effects of interlocking on interlayer adhesion and strength of structures in 3D printing of concrete*. Automation In Construction **83** 212-221
[21] Zeina Malaeb, Hussein Hachem, Adel Tourbah (2015). *3D Concrete Printing: Machine and Mix Design*, IJCIET **6(6)** 14-22.
[22] Ji, Guangchao and Ding, Tao and Xiao, Jianzhuang and Du, Shupeng and Li, Jun and Duan, Zhenhua (2019) A 3D Printed Ready-Mixed Concrete Power Distribution Substation: Materials and Construction Technology. Materials **12** 1540.

IOP Conf. Series: Materials Science and Engineering **814** (2020) 012023 doi:10.1088/1757-899X/814/1/012023

Techniques and various efficiency evaluation tests for self-healing cement-based matrix: State-of-the-art

Mavjot Kaur[1], Krishna Murari[2] and Inderpreet Kaur[3]

[1] Under Graduate Student, Guru Nanak Dev Engineering College, Ludhiana, India
[2] Post Graduate Student, Guru Nanak Dev Engineering College, Ludhiana, India
[3]Assistant Professor, Guru Nanak Dev Engineering College, Ludhiana, India

E-mail: mavjotkaur@gmail.com

Abstract. As the aging leads to dilapidation of reinforced concrete structures due to formation of cracks through which chlorides and other chemical agents promote corrosion. There is need of repair or maintenance of such existing structures. Therefore, self-healing, a leading-edge method for durability enhancement, has gained a lot of importance. In this paper, the various self-healing techniques for concrete, including intrinsic healing, capsule-based healing, vascular healing and also immobilization along with healing mechanism of some materials are reviewed with their comparisons. Moreover, a comprehensive study is also done on the different tests for assessment of self-healing in concrete at macro-scale, micro-scale and nano-scale to determine and characterize healing product and to evaluate their mechanical properties.

1. Introduction

In most of the infrastructures, concrete is used at large-scale as construction material having advantage of high characteristic compressive strength, durability and cost economy. However, under the action of tensile forces it is susceptible to crack formation at micro and macro levels. Initially micro-cracks are formed, which are unnoticeable, further leads to formation macro- cracks due to environmental and mechanical actions [1]. These cracks create a pathway for the oxygen, moisture and deleterious materials which causes steel corrosion adversely affecting the durability [2]. For this purpose, the inspection and maintenance techniques have gained increasing attention. Regular inspection and maintenance are difficult rather impossible for inaccessible large scale infrastructures o wing to capital investment required. Various other factors include labor and the location of the damage.

Grouting and coating materials are the temporary solutions for external cracks and has unpropitious impact on durability as 20% of the repairs fail after 5 years and 55% after 10years [3–5]. The repair of cracks by self-healing in concrete has had been great attraction as self-healing allows repairs in inaccessible regions without human intervention.

In the present paper, various techniques and evaluation tests of self-healing for cement based matrix are demonstrated.

2. Various techniques of self-healing cement based matrix

Self-healing to repair cracks in cement based matrix can be done by various techniques. Classification of self-healing for cement based matrix is given in figure 1.

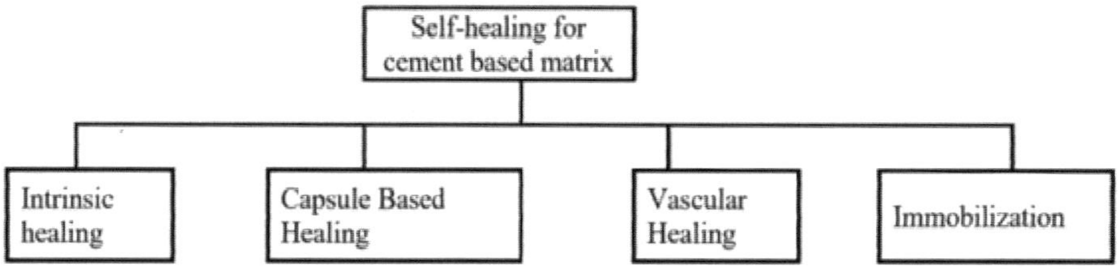

Figure 1. Flow Chart of various techniques of Self-healing

2.1 Intrinsic healing

Intrinsic healing is the capability to plug micro cracks by the concrete itself [2]. It is a crack healing phenomenon, which is effective for small crack width, involving chemical reactions in the presence of water. Reduction in crack size is also studied under intrinsic healing.

Summary of the early research works: the first investigation on water retaining structures, and pipelines by the French Academy of Science was done in 1836; formation of white crystalline precipitate in the cracks due to carbonization [6]. In 1926, Glanville executed an analysis of healing phenomenon [7] and followed by Soroker and Denson in 1926 and Brandeis in 1937 [8-9]. Different aspects of intrinsic healing have been studied by Jacobson and Sellevold [10], Clear [11] and Edvardsen [12]. Cracks inside the matrix can be naturally sealed by four mechanisms including [13]: Formation of $CaCO3$ from$Ca(OH)2$, blockage of cracks by fine particles, additional hydration of the unhydrated cementitious constituents, enlargement of the hydrated cementitious mix.

Major components of Portland concrete are C2S, C3S, C3Aand C4AF.After addition of water into the cement, C2S and C3S hydrates further produce CSH gel and CH and soon after mixing, the cement particles are covered with entringite layer leaving unreacted cement. On another note, Calcium silicate reacts with Carbon dioxide to produce calcium carbonate and silica gel, blocking cracks [14]. According to Neville [15], self-healing of early concrete is caused by hydration of the unhydrated cementitious materials. Divergently, at later stage it came to conclusion that calcium carbonate paves the way for crack filling [16]. Therefore, the involvement of water is significantly related to hydration of the un-hydrated components in the cementitious matrix. Recently, it was discovered that CSH gel accumulates on the crack surface whereas crystallization of portlandite occurs in the middle [17].

Li et al. used Engineered Cementitious Composite (ECC) with polyethylene fibers as reinforcement for limiting the width of crack [18]. In another study, crack width of 60μm was obtained by using Polyvinyl Alcohol (PVA) [19]. Several research studies started on the possibility of mixing Super Absorbent Polymers (SAP) in concrete mixture. SAPs are cross linked polymers which soak liquid from environment and swell. According to Minon et al. [20], SAPs are pH dependent. Kim et al. [21] used ECC in mortar mixture with PVA 2% and SAP content of 0.5% and 1% to cement weight. To develop cracks, Flexural test was conducted on pre-cracked specimens that were cured for 28days. The crack width of 15.7μm and 17.3μm were obtained for SAPs content and for the control mix 22. 3μm.Azarsa et al. [22] utilized crystalline admixtures (CA) by 2% of the cement content. After 28 days of air curing, standard crack-inducing jig was utilized to induce crack. The sample was sealed and exposed to constant water head on one side. The discharge through the specimen is measured. The sample with CA obtained healing ratio of 0.996 and 0.991 in control mix.

2.2 Capsule based healing

Capsule, usually micro-sized particle, is a shell which contains healing agent in solid, liquid or gas state [5]. These capsules provide mechanical resistance and releases the healing agent only when triggered by cracks, moisture and pH changes [23]. When cracks propagate in the concrete, the content present in the dispersed capsule is pulled out by capillary action (as shown in Figure 2) [5,23]. Low viscosity is the fundamental requirement of the healing agent for efficient release[4, 24]. Diameter of capsule, shell thickness and texture of surface are certain factors considered for encapsulation healing performance [25].

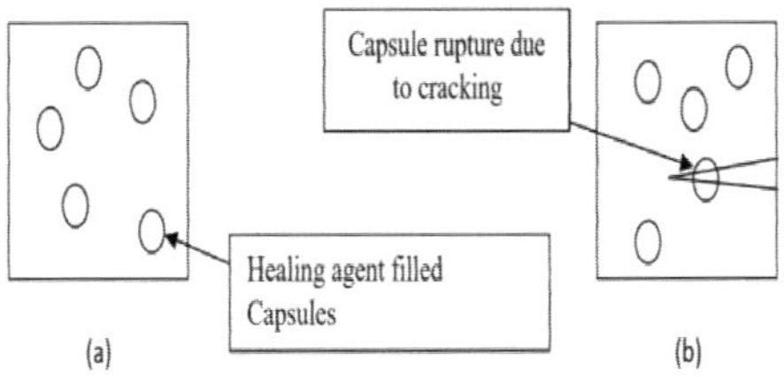

Figure 2. Capsule based Self-healing. (a). Before development of cracks (b). After development of cracks.

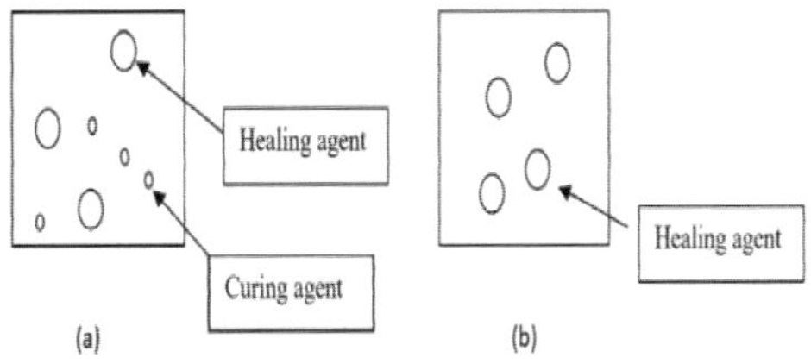

Figure 3. (a) Two Component micro capsules. (b) One component micro capsules.

In 1994, Dry, initiated a built-in capsule in concrete[24], utilized hollow porous polypropylene fibre capsule with methyl-methacrylate (healing agent), the improvement in flexural strength was reported [25]. Karaiskos and co-workers [26] embedded hollowed polyurethane-tabulated capsules in a grid and connected to steel bars. The preparation of capsules containing dyes was marked as the beginning of development of microencapsulation as they were included into papers replacing carbon paper[27].In 2001,White et al., introduced microcapsules utilization for polymers self-healing; used urea-formaldehydes (shell material) and dicyclopentadiene as healing agent [28]. Du et al. [3] fabricated toluene-di-isocyanate(healing material) in paraffin shell using met condensation method. In the work conducted by Wang et al.[29], melamine based microcapsules having inactive substance to protect the bacterial spores (encapsulated by poly-condensation reaction method) of size 5 μm was used. Milla et al [30] developed microcapsules with urea-formadehyde as shell material, using water-in-oil suspension polymerization technique, carrying calcium nitrate tetrahydrate healing agent and Sorbit anmono stearate surfactant was added to avoid acidification of the mixing water. Agitation rate of 800rpm and 1500rpm were utilized as it controls the average diameter of the microcapsule. The final microcapsule diameter was between 22 and 50μm.W. Du et al. prepared paraffin microcapsule containing toluene-di-isocyanate(TDI) as recovery agent and investigated microcapsule size distribution, under optimum preparation factors (one part of paraffin and two parts of TDI by mass with agitation rate 600rpm at temperature 75°C), was 30-300μm and one-tenth of their diameter was their thickness [3].

On the basis of healing agent reaction mechanism, microcapsules are classified into two categories, namely one-component and two-components (see figure 3). In the case of one-component, the healing material, which is encapsulated, is released, further reacts with air, water or curing agent scattered in the concrete mix which results in cracks filling whereas, in two-component, both the materials for healing and curing are encapsulated in separate microcapsules; after the rupture due to cracks, healing agent and curing agent reacts with each other to fill the cracks. To exemplify, Cailleux and Pollet [31] encapsulated bisphenol-F epoxy resin in gelatine based microcapsules of spherical shape and a hardener was scattered inside for the reaction to trigger. Yang et al. utilized microcapsules made of silica gel which encapsulates methyl methacrylate and triethylborane to synthesize two-component microcapsules [1,32].

2.3 Vascular healing

Vascular system is another approach to transfer healing agents into the concrete; as the name suggest, it is like a human body's vascular network consisting of single or multiple channels(hollow tubes) through which healing agent is applied from exterior of the structure[23,33] (refer figure 4).

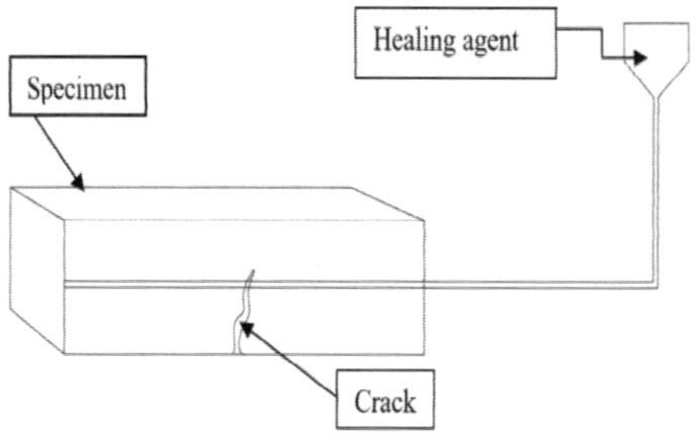

Figure 4. Schematic diagram of vascular healing.

In 1993, Dry, first to put forward vascular healing method, installed two fibre tubes and through them methyl methacrylate(healing agent) was drawn into the concrete[33]. Other structural elements such as concrete frames were also tested by Dry et al. by embedding glass tubes to heal cracks caused by dynamic loading[34-35]. Mihashi et al. incorporated two glass tubes, attached to external reservoirs. One component of epoxy glue was stored in one reservoir and second component of epoxy glue in the other one. During crack occurrence, both the tubes broke leading to polymerization reaction. However, the regain of strength was almost same as the components did not mix effectively[36].Joseph et al. made use of cyanoacrylate for healing by providing borosilicate glass tube as carrier in concrete[25, 37]. Sangadji and Schlangen [38–40] developed vascular network, made of porous concrete. Porous concrete cylinders were casted which were surrounded by PVA film that dissolves after casting of dense concrete around the porous matrix and creates possibility of exchange between interior and exterior layer. Installation of micro controllers is done which switchs on a pump, when a crack appears, to inject healing agent into porous concrete to densify it which further promotes crack healing.

2.4. Immobilization

Immobilization involves embedment of organic compounds into inorganic compounds into inorganic matrices. for immobilization application, disparate methods, as shown in figure 5. , have been adopted; [41]. In addition to high spectral transparency and controlled porosity, immobilization also offers stability mechanically and thermally [42].

Polyurethane (PU) has biochemically inert characteristics and is strong mechanically, making it a popular medium for immobilization [43]. B. Spaericus was immobilized by utilizing PU by some authors [44]. Bhasker et al. immobilized Sposarcina ureae and Sporosarcina pasteurii using zeolite [45]. Researchers also immobilized B. Cohnii in various carrier; Sierra-Beltran et al. used Light Weight Aggregate and Zhang et al. utilized expanded clay and perlite [46-47].

3. Materials

3.1. Bateria

The approach, biological repair technique of bacterial introduction in the concrete, dates back in 1990s in which environmentally friendly processes were suggested for crack repair [48]. Concrete is highly alkaline, the genus bacillus group of bacteria is capable of resisting to such alkaline conditions [49], [50]. A source of nutrient is provided to the bacteria which undergo metabolic process and precipitate of calcium carbonate is obtained [51-52]. In the precipitation pathway, urea -is converted into NH_4^+ and HCO_3^-[52].

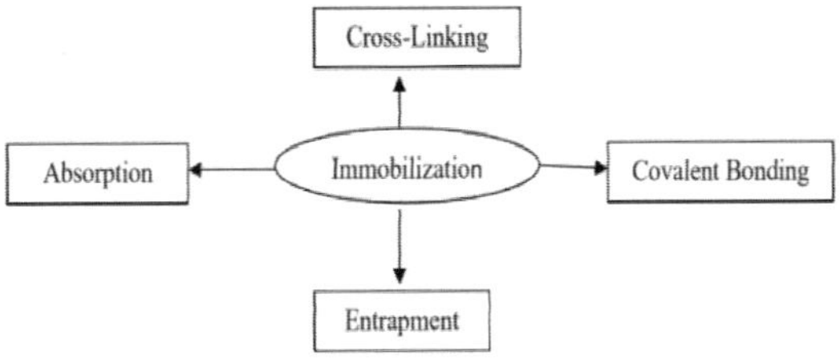

Figure 5. Different methods of Immobilization.

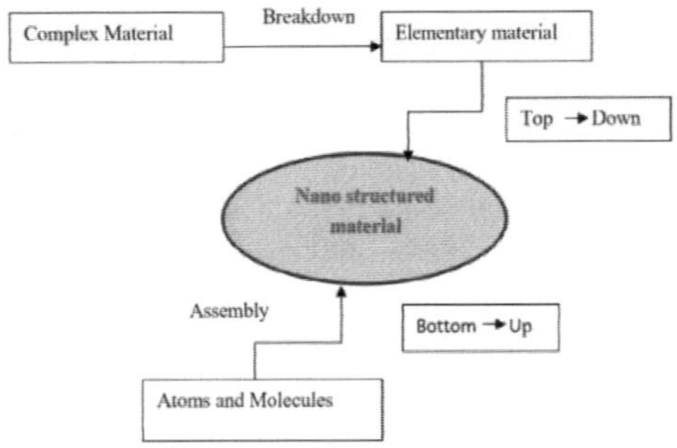

Figure 6. Routes for manufacture of nanomaterials.

In the starting of microbial urease mechanism, one mole of urea is hydrolyzed intracellularly to a mole of NH_3 and carbamate each. Further spontaneous hydrolysis of carbamate occurs leading to formation of NH_3mole and H_2CO_3 mole (Eqs. (1)-(2)) [49].

$$CO(NH_2)_2 + H_2O \rightarrow NH_3COOH + NH_3 \quad (1)$$

$$NH_2COOH + H_2O \rightarrow NH_3 + H_2CO_3 \quad (2)$$

Following that, these products further produce unimolecular bicarbonate and bimolecular ammonium and hydroxide ions each (Eqs. (3)-(4)). The rise in pH is observed which shifts bicarbonate equilibrium, and carbonate ion is formed (Eqs. (5)) [53].

$$H_2CO_3 \rightarrow HCO_3^- + H^+ \quad (3)$$

$$2NH_3 + 2H_2O \leftrightarrow 2NH_4^+ + 2OH^- \quad (4)$$

$$HCO_3^- + H^+ + 2NH_4^+ + 2\,OH^- \leftrightarrow CO_3^{2-} + 2NH_4^+ + 2H_2O \quad (5)$$

The formation of calcium carbonate precipitates due to availability of calcium ions after reaching a certain supersaturation level (Eq. (6)).

$$CO_3^{2-} + Ca^{2+} \rightarrow CaCO_3 \quad (6)$$

Negative charge of the cell wall leads to attraction of calcium ions which results in crystal formation on bacterial cell [54].

Another pathway for production of $CaCO_3$ is aerobic oxidation of organic acids, which shows less environmental effect in comparison to ureolytic pathway and through this metabolic pathway carbon dioxide is produced that leads to autogenous self-healing [55–58].

3.2 Admixtures

Self-healing can be imputed by reaction of admixtures in the cementitious matrix. Addition of these minerals is done during mixing of concrete mixture. After crack formation, these minerals react when water penetrates through them which leads to formation of products that imparts in crack healing [59]. They are classified into two categories: expansive additives and crystalline additives [60]. The creation of cementitious products in cracks such as Al2O3−Fe2O3-tri, Al2O3−Fe2O3-mono and Calcium carbonate and $Ca(OH)_2$ crystals in air voids were noticed [61]. Based on the assumption that re-crystallization takes place when these leached out products comes in content with water that is flowing through fractures, Kishi and co-workers formed a self-healing technique. Effects of several agents which includes geo-materials and chemical as well as expansive admixtures together with some of their combinations were evaluated and was concluded the proper dosage of CO3-2and expansive admixtures, can enhanced self recovery of cracks[61–64]. Jaroenratanapirom et al. also examined self-healing with different additives of varying proportions[65-66]. Researchers have utilized crystalline admixture which contains reactive silica and crystalline catalyst [22,60,67-68].

3.3 *Nanomaterials*

It is an irrefutable fact that nano particles are superior fillers than micro-sized materials. In recent years, nano nanomaterials had been used in concrete for various objectives, such as, to improve strength and durability, self-cleaning, self-healing and corrosion protection [69]. Nanomaterials can be converted from all materials by physical or chemical processes [70]. For the manufacture of nanomaterials, there are two ways : top-down method [71] and bottom-up method [72] (figure 6). For concrete, nanomaterials are mostly obtained from bottom-up approach [73].

Hardly any researches have been carried out for self -healing by incorporating nanomaterials in concrete. For construction usage, nanomaterials used, on the basis of accessibility, are nanosilica, nanoalumina, polycarboxylates and nanokaolin[74], [75]. In 2010, Qian et al. studied about nanoclay as water supply to promote hydration alongside the cracks in high performance concretes and showed considerable amount of improvement in recovery level by incorporation of nanoclay [76].Examination of capability of nanosilica to react with $Ca(OH)_2$in concrete matrix and production of Calcium silicate hydrate gel had been done [77–79].

The Pros and Cons of different self-healing techniques are discussed in Table 1.

4. Tests to assess and characterize Self-healing efficiency

The concrete self-healing evaluation has become an attraction because of the wide range of assessment methods available. These methods are utilized to form the criteria for quality of concrete which are of macro, micro or nano levels [91]. The macro level tests consist of mainly monitoring of concrete mechanical properties and durability tests while the micro level tests consist of optical and visual analysis tests. However, at nano level, there are few researchers on mechanical properties. Table 2 shows the various tests conducted. It is evident from the literature that all the research work is evaluated by conducting macro level tests.

4.1 *Macro-level tests*

The paramount consideration for healing efficiency in the concrete is the varieties of materials utilized for healing and their ability to regain the mechanical properties in varying circumstances [92]. The test included for mechanical properties recovery are flexural, compression, splitting, tensile and non-destructive tests. Bending tests that are mainly conducted for the evaluation of healing capacity are three-point bending and four-point bending. In three-point bending, a notch is fabricated in the center of the deflection section while in four-point bending, fabrication of the notch in not done for a practical approach [93]. This test has been used for cementitious material containing healing agent, such as bacteria [94], light-weight aggregate [95] and expansive agents [64]. Alghamri et al. utilized light-weight aggregate with sodium silicate (healing material) and showed 80% recovery in pre-cracked strength which is quintuple of the control specimen's recovery [95]. As defined by ASTM C39-03 [96], compression test is "applying a compressive axial load to mold cylinders or cores at a rate which is within a prescribed range until failure occurs". Gupta et al .[94]explored biochar, obtained from wood scrap, for carrying bacteria spores in cement matrix which precipitates carbonate to regain strength and permeability. The results showed an increase in strength by 38% and 65% decrease in permeability. Ultrasonic pulse velocity, a non- destructive test, has been conducted on several healing agents on hardened concrete [64,95,97].

To increase durability, diminution of porosity is the pre-dominant aspect which eventually effects the performance of concrete in long run [52]. Water and gas permeability are the measure often used. And moisture and gas affects the durability and their transportation into the micro-cracks enlarge the cracks [98]. Kanellopoulos et al. [84]performed gas permeability test used methanol as medium on cementitious matrices containing sodium silica, colloidal silica and tetraethyl orthosilicate which were injected manually in the cracks. The results obtained were 36% and 43% reduction in gas permeability in sodium silica and colloidal silica respectively. Various other tests are mentioned in the Table 2.

Table 1. Pros and Cons of different self-healing techniques

TECHNIQUE	BENEFITS	DRAWBACKS	REFERENCES
Vascular system	Healing agent discharge when required. Adjustable high quantity of healing material.	Casting difficulties. Strength parameters of cement matrix affected in presence of too many tubes.	[29]
Encapsulation	Healing agent discharge when required Medium quantity of healing agent Efficient under occurrence of multiple damage	Casting difficulties Strength parameters of cement matrix affected in presence of too many hollow fibers. Difficulty in release of healing agent may be possible	[80-84]
Microencapsul-ation	Healing agent discharged when required response to multiple cracked areas at same time	Complexity in preparation of capsules and casting Restricted amount of healing agent	[63,85]
Nanomaterials	High recovery in mechanical characteristics Outstanding healing efficiency Cost-effective and energy saving Applied internally and externally	High safety requirements	[86-87]
Bacteria	Natural and pollution less and biologically effective way	Bacteria protection measures should be taken Need to meet several prerequisites	[29,49,88]
Admixtures	Excellent efficiency in healing Healing products are more compatible with cement matrix	Under ill-treatment, undesirable expansion Uncertainty in generation of healing products.	[89,90]

4.2 Micro-level tests

The micro-level tests conducted to make the results more reliable. The purpose of these tests is the visualization of crack healing process in concrete and the morphology, identification and characterization of the accumulated matter post-healing process. The commonly used methods are scanning Electron Microscopy (SEM), X-ray diffraction (XRD) and Energy Dispersive Spectroscopy (EDS).

Some researchers used SEM to discover the mechanism involved in self-healing [99], while others conducted studies to observe to structure, texture and size of the material[100]. For the determination of the composition of healing product, XRD test is conducted. Using this technique, Wang and coworkers discovered that calcite was the major healing compound produced[100]. Similarly, Kalhori and Bagherpour also used XRD to recognize the healing products induced by bacteria, Bacillus subtilis[101]. It is interesting to know that, Wang and coworkers conducted X-ray tomography on products formed by using hydrogel immobilized bacteria and the results illustrated the minerals precipitated in the whole concrete specimen along with the cracks, thus, resulting in overall upsurge in the compressive strength [102]. Similarly, other tests have also been used to enhance the results as given in the Table 2.

Table 2. Tests for Evaluation of concrete self-healing

TESTS	EFFICIENT HEALING MEASUREMENT	AUTHORS
	Macro Level	
Bending test	Measure strength	[64,84,95,104-105]
Compression test	Measure strength	[44,47,88,94,106-107]
Tensile test	Measure strength	[102,108-109]
Splitting test	Measure strength	[110-111]
Ultrasonic Pulse Velocity	Measure tightness	[64,95,97,112]
Water permeability	Damage, healing efficiency	[29,113-115]
Gas permeability	Damage healing efficiency	[84,119-120]
Corrosion Test	Corrosion resistance	[117-118]
	Micro level	
Optical microscopy	Visualization of healed crack	[84,119-120]
Scanning Electron microscopy	Visualization of crystal deposition	[99-100,114,121]
X-ray diffraction	Identification of crystal matter	[84, 88,119]
Fourier transform infrared spectroscopy	Identification of crystal matter	[84, 95, 105]
Thermo-gravimetric analysis	identification of crystal matter	[64, 108, 122]
X-ray tomography	Visualization of healing	[102, 104, 123]
Neutron tomography	Visualization of healing	[124-125]
	Nano level	
Nano-indentation	Determination of Mechanical properties	[103]

4.3 Nano-level tests

These tests are conducted to further enhance the results by making them more reliable. Recently, Xu and Yao employed Nano indentation test, that is utilized for measurement of mechanical properties of small volumes (healing products), for evaluation of healing efficiency of bacteria-based concrete and discovered that the average Nano-mechanical values were 20% more of the transition zone in comparison to outer precipitates, that binds the deposited layer and matrix [103].

5. Conclusion

It is apparent that the mechanism of self-healing not only involves physical but also chemical process. In this review, several techniques of self -healing developed, including intrinsic, capsule-based, vascular and immobilization, as well as different healing materials have been discussed. Table 1 shows their reasons to support and oppose. Maximum research work has been reported using bacteria as it is proved to be comparatively more efficient in crack healing and the least by Nano-materials which has been recently discovered for self -healing in concrete.

For the examination of efficiency of several self- healing approaches, a series of test are conducted (Table 2); out of which macro-scale tests were adopted by most of the authors which showed contrasting results for the same healing agents, making them inconclusive. Determination and visualization of the healing product, by using micro-scale tests, increased reliability by measuring the reduction size of crack. Researchers opting for Nano-scale tests were a few. However, none of the authors reported the macro, micro as well as Nano-scale tests in one experiment. Hence, making it obvious that standardization of the methods to analyze self- healing efficiency is required.

6. Acknowledgement
We express our sincere thanks to Department of Civil Engineering, Guru Nanak Dev Engineering College, and Ludhiana for the encouragement and guidance.

References
[1] Yang Z, Hollar J, He X, Shi X 2010 *Transportation Research Record* **2142** 9-17.
[2] Algaifi H, Bakar S, Sam A, Abidin A, Shahir S, Al-Towayti W *Construction and Building Materials* **189** 816-24.
[3] Du W, Yu J, Gu Y, Li Y, Han X, and Liu Q 2019 *Constr. Build. Mater* **202**, 762–69.
[4] Van Belleghem B, Kessler S, Van den Heede P, Van Tittelboom K and De Belie N 2018 *Cement and Concrete Research* **113** 130–139.
[5] Al-tabbaa A, Litina C, Giannaros P, Kanellopoulos A and Souza L 2020 *Constr. Build. Mater.* **208** 669–85.
[6] Hearn N 1998 *Mat. Struct.* **31** 563.
[7] Glanville, W. H. (1931). *The permeability of Portland cement concrete.* (London: H.M. Stationery Off.)
[8] Soroker V and Denson A 1926 Autogenous healing of concrete *zement* **25** 76.
[9] Brandeis F 1937 Autogenous healing of concrete *Beton und Eisen*, **36**
[10] Jacobsen S and Sellevold E 1996 *Cement and Concrete Research* **26** 55–62.
[11] Clear, C. A. 1985. *The effects of autogenous healing upon the leakage of water through cracks in concrete*. (Wexham Springs, Slough: C and CA)
[12] Edvardsen C 1999 *Mater. J.* **96** 448–54.
[13] Talaiekhozan A 2014 *Journal of Environmental Treatment Techniques* **2** 1-11.
[14] Li G, Huang X, Lin J, Jiang X, and Zhang X 2019 *Constr. Build. Mater.* **200** 36–45.
[15] Neville A M 1995 *Properties of concrete* (Harlow, Essex : Longman)
[16] Neville A M 2002 *Concr. Int.* **24** 76–82.
[17] Jiang J, Zheng, Wu S,Liu Z and Zheng Q 2019 *Constr. Build. Mater.* **198** 696–709.
[18] Li V C, Mook Y, and Chan Y, 1998 *Composites Part B: Engineering* **8368** 819–27.
[19] Li V, Wang S, and Wu C 2001 *ACI Mater. J* **98** 483–92.
[20] Mignon A and Didier G G 2015 *Journal of Materials Science* 970-79.
[21] Kim JS and Schlangen Erik 2010 Super absorbent polymers to simulate self healing in ECC. *2nd International Symposium on Service Life Design for Infrastructure.* 849-58.
[22] Azarsa P, Gupta R, and Biparva A 2019 *Cem. Concr. Compos*. **99** 17–31.
[23] Souradeep G and Kua H W 2016 *Journal of Materials in Civil Engineering* **28** 1–15.
[24] Li W, Zhu X, Zhao N and Jiang Z 2016 *Materials (Basel)* **9(3)**
[25] Joseph C, Jefferson A D, Isaacs B, Lark R and Gardner D 2010 *Magazine of Concrete Research* **62** 831–43.
[26] Karaiskos G, Tsangouri E, Aggelis D G, Van Tittelboom K, De Belie N and Van Hemelrijck D, 2016, *Smart Mater. Struct* **25**.
[27] Wu M,Johannesson B, andGeiker M 2012 *Constr. Build. Mater* **28** 571–83.
[28] White S R et. al. *Nature* **409** 794–97.

[29] Wang J Y,Soens H, Verstraete W and De Belie N *Cem. Concr. Res* **56**, 139–152.
[30] Milla J, Hassan M M, Rupnow T and Daly W H 2019 *Constr. Build. Mater* **201** 526–38.
[31] Cailleux E and Pollet V 2009 *Proceedings of 2nd International Conference on Self Healing Materials* vol. **32** (Chicago) p 1
[32] Yang Z, Hollar J, He X, and Shi X, *Cem. Concr. Compos* **33** 506–12.
[33] Dry C 1994 *Smart Mater. Struct.* **31**18-23.
[34] Dry C 2001 Self-sensing and Self-repairing Materials for Engineering Applications *Proceedings of the SPIE* vol **4234**, p 23.
[35] Dry C, 1999 *Proceedings of the SPIE*, vol **3671**, p 253.
[36] Mihashi H, Kaneko Y, Nishiwaki T and Otsuka K 2000 *Concr. Res. Technol* **11** 21–28.
[37] Joseph C, Jefferson A D and Cantoni M B 2007 *Proceedings of the First International Conference on Self Healing Materials* (Noordwijk: Springer)
[38] S. Sangadji and E. Schlangen, 2011 **3rd International Conference on Self-Healing Materials** (Bath, UK)
[39] S. Senot, E. Schlangen, V. C. Li, and E. Herbert, 2012 *J of Advanced Concrete Technology* **10** 185–94.
[40] Sangadji S and Schlangen E 2011 *Int RILEM Conf on Advances in Construction Materials Through Science and Engineering* (RILEM Publications SARL) p 553.
[41] S. Datta and L. R. Christena 2012 *3 Biotech.* **3** 1-9.
[42] Soltmann U, Raff J, Universit T, Gmbu V, Schichten A F and Dresden D 2003 *Journal of Sol-Gel Science and Technology* **26** 1209–12.
[43] Bang S S, Galinat J K and Ramakrishnan V 2001 *Enzyme& Microbial Technology* 28 404.
[44] Wang J, Van Tittelboom K, De Belie N and Verstraete W 2012 *Constr. Build. Mater.* **26** 532–40.
[45] Bhaskar S, Anwar K M, Lachemi M, Wolfaardt G and Otini M 2017 *Cem. Concr. Compos.* **82** 23–33.
[46] Sierra-beltran M G, Jonkers H M and Schlangen E2014 *Constr. Build. Mater.* **67** 344–52.
[47] Zhang J et al. 2017 *Constr. Build. Mater.* **148** 610–17.
[48] Gollapudi U K, Knutson C L, Bang S S and Islam M R 1995 *Chemosphere* **30** 695–705.
[49] Van Tittelboom K, De Belie N, De Muynck W and Verstraete W2010 *Cem. Concr. Res.* **40** 157–66.
[50] Bravo F, Silva D, Boon N, Verstraete W and De Belie N, 2015 *Constr. Build. Mater* **88** 196
[51] Hammes F, Boon N, De Villiers J, Verstraete W and Siciliano S D 2003 *Applied Environmental Microbiology* **69** 4901–09.
[52] Vijay K, Murmu M and Deo S V 2017 *Constr. Build. Mater.* **152** 1008–14.
[53] Dick J et al 2006 *Biodegradation* **17** 357–67.
[54] De Muynck W, De Belie N and Verstraete W 2010 *Ecological Engineering,* **36** 118–36.
[55] Wiktor V and Jonkers H M 2011 *Cem. Concr. Compos.* **33** 763–70.
[56] Jonkers H M, Thijssen A, Muyzer G, Copuroglu O and Schlangen E, 2010 *Ecol. Eng.* **36** 230.
[57] Jonkers H M, 2008 Self healing concrete: a biological approach, *Self healing materials- An alternative approach to 20 centuries of materials science* (Springer) p 195.
[58] Jonkers H M and Schlangen E, 2007 *Proceedings of the First International Conference on Self Healing Materials* (Noordwijk: Springer).
[59] Huang H, Ye G, Qian C and Schlangen E 2016 *JMADE* **92** 499–511.
[60] Sisomphon K, Copuroglu O and Koenders E A B 2012 *Cem. Concr. Compos* **34** 566–74.
[61] Kishi T, Ahn T, Hosoda A, Suzuki S and Takaoka H 2007 *Proceedings of the First International Conference on Self Healing Materials* (Noordwijk: Springer) p 1.
[62] T. Ahn and T. Kishi, 2010 *Journal of Advanced Concrete Technology* **8** 171–186.
[63] Al-Ansari M, Abu-Taqa A G, Hassan M M, Senouci A and Milla J, 2017 *Constr. Build. Mater.* **149** 525–34.
[64] Sherir M A A, Hossain K M A and Lachemi M 2017 *Constr. Build. Mater.* **148** 789–810.

[65] Jaroenratanapirom D and Sahamitmongkol R 2011 *Annual Concrete Conference* 551-56.
[66] Jaroenratanapirom D and Sahamitmongkol R 2011 *Journal of Metals, Materials and Minerals* **21** 9–17.
[67] Roig-Flores M, Moscato S, Serna P and Ferrara L, 2015 *Constr. Build. Mater.* **86** 1–11.
[68] Jiang Z, Li W and Yuan Z, 2015 *Cem. Concr. Compos* **57** 116–127.
[69] Huseien G F, Shah K W and Sam A R M,2019 *J. Build. Eng* **23** 155–171.
[70] Sobolev K and Gutierrez M F 2005 *How Nanotechnology Can Change the Concrete World Progress in Nanotechnology* p 113-16.
[71] Abdoli H, Farnoush H R, Asgharzadeh H, Sadrnezhaad S K 2011 *Powder Metallurgy* **54** 24.
[72] E. Jankowska and W. Zatorski 2009 *Third International Conference on Quantum, Nano and Micro Technologies* (Cancun, Mexico: IEEE) p 147.
[73] Norhasri M S M, Hamidah M S, and Fadzil A M, 2017 *Constr. Build. Mater* **133** 91–97.
[74] Sanchez F and Sobolev K 2010 *Constr. Build. Mater.* **24** 2060–71.
[75] Shah K W, Sreethawong T, Liu S, Zhang S, Tan L S and Han M 2014 *Nanoscale* **6** 11273
[76] Qian S Z, Zhou J, and Schlangen E 2010 *Cem. Concr. Compos* **32** 686–93.
[77] Raki L, Beaudoin J, Alizadeh R, Makar J and Sato T,2010 *Materials (Basel)* **3** 918–942.
[78] Lopez-calvo H Z, Montes-garcia P, Bremner T W, Thomas M D A and Jiménez-quero V G 2012 *Cem. Concr. Compos* **34** 110–18.
[79] Wang L, Zheng D, Zhang S, Cui H and Li D, 2016 *Nanomaterials* **6** 10.3390/nano6120241.
[80] Pang J and Bond I P 2005 *Composites Part A Applied Science and Manufacturing* **36** 183.
[81] Pang J W C and Bond I P, 2005 *Composites Science and Technology* **65** 1791–99.
[82] Dry C M, 2001 *Cem. Concr. Res* **30** 1969–77.
[83] L. Lv et al., 2016 *Constr. Build. Mater* **105** 487–95.
[84] Kanellopoulos A and Qureshi T S, 2015 *Constr. Build. Mater* **98** 780–91.
[85] Zhu D Y, Rong M Z, and Zhang M Q, 2015 *Prog. Polym.* **49-50** 175-220.
[86] Lee M, Yoon S S and Yarin A L 2016 *ACS Applied Materials & Interfaces* **8** 10.1021/acsami.5b12358.
[87] Z Wang et al 2007 *Nanotechnology* **18** 455709
[88] Khaliq W and Ehsan M B, 2016 *Constr. Build. Mater* **102** 349–57.
[89] S. Nagataki and H. Gem,1998 *Cem. Concr. Compos* **20** 163–70.
[90] Ish-Shalom M and Bentur A 1974 *Cem. Concr. Res* **4** 519–32.
[91] Muhammad N Z et al., 2016 *Constr. Build. Mater.* **112** 1123–32.
[92] Sidiq A, Gravina R, and Giustozzi F, 2019 *Constr. Build. Mater* **205** 257–73.
[93] Wang X F et al., 2019 *Constr. Build. Mater.* **212** 400–21
[94] Gupta S, Wei H, and Dai S, 2018 *Cem. Concr. Compos* **86** 238–54.
[95] Alghamri R and Kanellopoulos A, 2016 *Constr. Build. Mater* **124** 910–21.
[96] ASTM C 39 2001 *Am. Soc. Test. Mater* 1–7.
[97] De Nardi C, Cecchi A, Ferrara L, Benedetti A and Cristofori D 2017 *Compos. Part B* **124** 144.
[98] Pijaudier-cabot G, Dufour F and Choinska M 2009 *Journal of Engg Mechanics* **135** 1022.
[99] Siad H, Lachemi M, Sahmaran M, Mesbah H A, and Anwar K, 2018 *Constr. Build. Mater* **176** 313–22
[100] Wang X, Fang C, Li D, Han N, and Xing F, 2018 *Cem. Concr. Compos* **92** 216–29
[101] Kalhori H and Bagherpour R, 2017 *Constr. Build. Mater* **148** 249–60.
[102] Wang J, Dewanckele J, Cnudde V, Van Vlierberghe S, Verstraete W and De Belie N 2014 *Cem. Concr. Compos* **53** 289–304
[103] Xu W, Yao J, 2014 *Cem. Concr. Res* **64** 1–10
[104] K. Van Tittelboom et al., 2016 *Constr. Build. Mater* **107** 125–37
[105] Qureshi T S and Kanellopoulos A, 2016 *Constr. Build. Mater* **121** 629–43
[106] Li W, Jiang Z, Yang Z, Zhao N, and Yuan W, 2013 *PLOS ONE* vol **8** 1–18
[107] Jo M and Douglas R, 2017 *Cement and Concrete Composites* **84** 167–74.

IOP Conf. Series: Materials Science and Engineering **814** (2020) 012023 doi:10.1088/1757-899X/814/1/012023

[108] Wang J Y, Snoeck D, Van Vlierberghe S, Verstraete W and De Belie N 2014 *Constr. Build. Mater* **68** 110–19

[109] Liu H, Zhang Q, Gu C, Su H, and Li V, 2017 *Constr. Build. Mater.* **153** 948–56

[110] Özbay E, Ahmaran M S, Lachemi M and Yücel H E, 2016 *Aci Mater. J.* **110** 33–43

[111] S. Choi, K. Wang, Z. Wen, and J. Chu, 2017 *Cem. Concr. Compos* **83** 209–221

[112] L. Ferrara, V. Krelani, and M. Carsana, 2014 *Constr. Build. Mater* **68** 535–51

[113] Araújo M, Van Tittelboom K, and Dubruel P, 2017 *Smart Materials and Structures* **26**

[114] Tziviloglou E, Wiktor V, Jonkers H M, and Schlangen E, 2016 *Constr. Build. Mater* **122** 118.

[115] Van Tittelboom K, Tsangouri E, Van Hemelrijck D, and De Belie N 2015 *Cem. Concr. Compos* **57** 142–52

[116] Qureshi T S, 2016 *Smart Mater. Struct* **25** 8 1–16

[117] Maes M, Van Tittelboom K, and De Belie N, 2014 *Constr. Build. Mater* **71** 528–37.

[118] N. Chahal, R. Siddique, and A. Rajor, 2012 *Constr. Build. Mater* **28** 351–56.

[119] Luo M, Qian C, and Li R, 2015 *Constr. Build. Mater* **87** 1–7.

[120] Hernandez-sanabria E, Boon N and De Belie N, 2016 *Cem. Concr. Compos* **70** 159–70.

[121] Sisomphon K, Copuroglu O and Koenders E A B 2013 *Constr. Build. Mater.* **42** 217–24.

[122] Bundur Z B, Amiri A, Ersan Y C, Boon N and De Belie N, 2017 *Cem. Concr. Res* **98** 44–9.

[123] Gilabert F A, Van Tittelboom K,Van Stappen J, Cnudde V, De Belie N and Van Paepegem W, 2017 *Cem. Concr. Compos.* **77** 68–80

[124] Zhang P, Dai Y, Ding X, Zhou C, Xue X and Zhao T 2018 *Constr. Build. Mater* **169** 705.

[125] Snoeck D, Van Den Heede P, Van Mullem T and De Belie N 2018 *Cem. Concr. Res.* **113** 8.

IOP Conf. Series: Materials Science and Engineering **814** (2020) 012024 doi:10.1088/1757-899X/814/1/012024

Seismic Assessment of Concrete Column Reinforced with Fiber Reinforced Polymer and Stainless Steel Rebars

Bhairav Thakur [1], Rahul Pardeshi [2], Mayank Patel [3] and Anant Parghi [4]

[1,2,3] Research Scholar, Applied Mechanics Department, SVNIT, Surat, Gujarat - 395007
[4]Assistant Professor, Applied Mechanics Department, SVNIT, Surat, Gujarat-395007

E-mail: bhairavthakur869@gmail.com, anantparghi@gmail.com

Abstract: The fiber reinforced polymer (FRP) rebar is utilized as a corrosion resistive reinforcement in the structural elements like a beam, column, and slab. The FRP rebar offers excellent mechanical and durability properties, like high strength to weight ratio, high stiffness, and corrosion resistance to structures. However, a lower ductility subjected to linear-elastic behaviour, and the brittle nature of FRP particularly in the case of column applications is getting attention among researchers. Whereas stainless steel (SS) having adequate ductility, low maintenance, and resistance to corrosion make it an ideal material for reinforcing applications. To improve the performance of FRP-reinforced elements and diminish the brittle behaviour, the present research is focused on the utilization of SS rebar in the plastic hinge region of the column along with FRP rebar in rest portion of a column. This hybrid configuration is adopted here to improve the ductility and corrosion resistance capacity of the RC column. The damage concentration of the hybrid column is analysed under the effect of nonlinear static loads. The parametric analysis is performed over the hybrid column to study the behaviour of different factors under pushover loading. The results indicate that the ductility of the SS-FRP RC column is achieved under seismic loads. The study reveals that axial load, the yield strength of steel and aspect ratio significantly affect the behaviour of the SS-FRP RC column.

1. Introduction

In the past few years, the seismic design guidelines are aiming at the performance-based design to forecast well-accomplished structural conditions and the post-earthquake functionality. The main goal of the present design concepts is to confirm that the failure of a structure is prohibited, and there is no need to limit the loss of post-seismic action retrieval. As per the modern philosophies of seismic design, a structure is predicated on suffering fatalities under the optimized predictable seismic force. However, the structure does not fail or collapse completely, and it will limit the substantial damages for post-seismic action retrieval [1].Hence, the current advancement in performance-based seismic assessment and design approach is emphasized to define and evaluate the permanent damages of a structure caused after the occurrence of massive earthquakes. During a seismic event, the longitudinal members of a column perform as a critical member of the frame structure to resist the seismic forces. In the recent earthquakes, the majority of structures were collapsed due to the poor performance of critical column sections. The maximum damage of a column is concentrated in the bottom region, defined is the plastic hinge region of a column.

IOP Conf. Series: Materials Science and Engineering **814** (2020) 012024 doi:10.1088/1757-899X/814/1/012024

Corrosion is the main factor that decreases the service life and long-term durability of RC structures. Several methods are adopted in order to minimize the corrosion and increase the endurance of the RC structures, such as fusion bonded epoxy (FBE) powder coatings and linings, cathodic and anodic coating, claddings, corrosion resistance alloys, and utilization of corrosion resistive SS composites. In RC structures, SS is utilized as a reinforcing material at seaside areas to prevent initial corrosion. In India, Europe, and North America, several bridges were built and repaired using SS for its corrosion resistance compared to conventional steel. The federal highway administration (FHWA)[2]has reported that SS bars are proficient at providing a complete service life of structures once the concrete becomes worsened. The field applications have unveiled that the utilization of SS rebar raises the total expenditure of a structure by 3% [3]. Additionally, SS rebar having adequate ductility, low maintenance, and resistance to corrosion make it an ideal material for the reinforcing applications. Hence, to diminish the overall cost and to increase the resistance from corrosion, FRP rebar is a suitable substitute for conventional reinforcement in the construction industry[12].

FRP is composed of mixing the fiber and resin and cured at room temperature afterward. In RC structures, rebar gets corroded due to de-salting, seawater, and natural factors like the freeze-thaw cycles[4]. FRP composite material can be utilized as a reinforcing material because of several characteristics of FRP, such as high stiffness and strength, corrosion-resistance, less weight, and easy to install[5]. Moreover, FRP rebar is utilized in civil engineering applications, e.g., RC buildings, bridges, etc. There are four main types of FRP, defined as glass (GFRP), basalt (BFRP), aramid (AFRP) and carbon (CFRP). Amongst these types, GFRP is the most utilized in civil engineering applications. The use of a polymeric matrix protects the surface and binds the fibers during manufacturing, transportation, handling, and throughout the service life of composite bars. Since polymeric matrix transfers the stresses to the fibers throughout the form, it should exist with the fibers in terms of its chemical and thermal properties. The commonly used resins are vinyl esters, epoxy, and polyester[6].

The main goal of this research is to examine the seismic behavior of joined RC columns reinforced with SS rebars in the plastic hinge region, and FRP rebars in the remaining part of a column. The performance of SS-FRP is evaluated in terms of base shear, displacement, and ductility.

2. Research Significance

In the present study, the effectiveness of SS-FRP hybrid reinforcement in RC columns is evaluated in order to improve the durability and seismic performance. Though FRP-reinforced structures can diminish the issues related to corrosion of steel, stiffness and compressive strength of FRP rebars make structures more susceptible to instability. ACI 440.1R-2006 [7] has recommended not to use the FRP bar as longitudinal reinforcement in flexural members (columns) because of the unavailability of adequate experimental support evidence. However, studies in the last few years reveal that the behavior of FRP-reinforced structures was found to be similar to the steel-reinforced structures [8]. While ductility can be incorporated in the design of FRP-RC, they will not only be extremely corrosion-resistant but also capable of dispersing significant energy during the earthquakes.

In this study, a hybrid SS-FRP configuration of reinforcement is incorporated to enhance the resistance to high corrosion in RC columns. Conventional steel is not used as an optimal solution to corrosion problems, whereas SS and FRP are likely to be considered to improve the performance of RC columns. Replacing SS rebar with conventional steel rebar in the plastic hinge regions increases the adequate ductility of columns, and utilizing FRP rebar in the remaining regions prevents the RC columns from corrosion. Generally, the unrecoverable deformation was observed in structural columns, as they are more vulnerable to severe damages subjected to enormous seismic energy. This study aims to indicate that the adequate ductility of the SS-FRP RC column is achieved under seismic loads.

3. Constitutive Material Model

The concrete is modeled utilizing a uniaxial nonlinear constant confinement model, developed by Madas [9], following the constituent framework suggested by Mander et al.[10] and the cyclic guidelines proposed by Martenize Rueda and Elnashai [11]. Mander et al. [10]proposed guidelines for the incorporation of confinement effects provided by the lateral transverse reinforcement, where the stress-strain range is assumed constant confining pressure is assumed throughout the entire stress-strain range. Five concrete material properties are used in Mander et al.[10]for the required calibration of a model. The input parameters and values are compressive strength (f'c= 38.30 MPa), mean tensile strength (ft=3.33 MPa), modulus of elasticity (Es=23.1 GPa) , Strain at peak stress (εc=0.0029 m/m), and specific weight (γ = 24 kN/m3) [12].

SS possesses a bilinear uniaxial stress-strain (σ-ε) model by kinematic strain hardening, and it remains constant throughout the elastic limit at the different loading stages. The kinematic hardening regulation for the surface of yield is predicated as a linear function of the plastic strain increase. This type of simple model is also characterized by easily distinguishable rectifying factors and through its computational efficacy. Four SS material properties are used in bilinear uniaxial σ-ε for the required calibration of a model. The parameters and properties of bilinear steel are elastic moduli (Es= 19 GPa), yield strength (fy=402MPa), specific weight (γ = 24 kN/m3) and strain hardening parameter (μ = 0.0125) are used in this model.

FRP was modeled as a generalized uniaxial elastic material system with a compression and tension specific elastic modulus. The different factors are utilized to determine the material in this model, e.g. elastic modulus (Es=52.2 GPa), ultimate compressive strength (f'cu=728 MPa), and ultimate tensile strength of FRP bar (f_{FRP},u = 364 MPa).

4. Design and Geometry of the SS-FRP RC Column

The detailed section, longitudinal reinforcement, and configuration of the SS-FRP RC column are adopted from Billah and Alam [12]. The detailed geometry of RC column is shown in Figure 1, shows the distribution of longitudinal reinforcement for SS-FRP columns, which has a square cross-section of 450 × 450 × 3200 mm, reinforced with M-20 grade concrete, 19.5 mm diameter of SS bar, 19.5mm diameter of FRP longitudinal rebars. The gross reinforcement ratio is 1.2 percent.

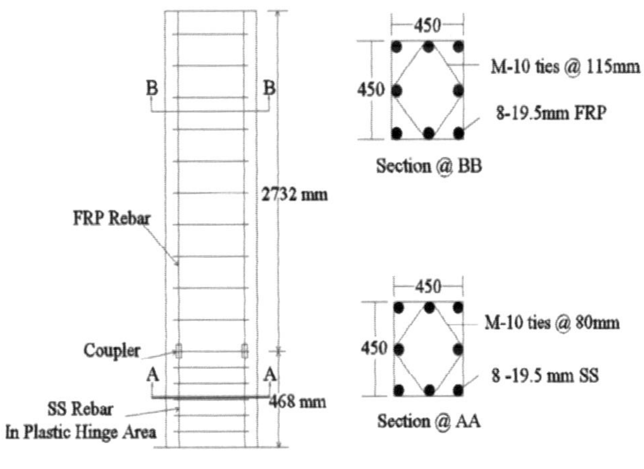

Figure 1. Geometry of Hybrid RC column

The investigators have recognized the effect of concrete strength on the plastic hinge length of RC members [12]. The equations of plastic hinge length given by most of the design codes and investigators do not consider the effect of the concrete strength. Though the influence of concrete strength is known only by the plastic hinge definition suggested by Billah and Alam [13], we have observed the increment in concrete strength as plastic hinge length decreases, as evident from their equation of plastic hinge length. The following equation1 is utilized to compute the plastic hinge length, according to Paulay and Priestley [14].

$$l_p = 0.08l + 0.022 d_b f_y \tag{1}$$

Where lp= plastic hinge length, db= diameter of rebar (mm), fy= yield strength of rebar (MPa)

5. Hybrid Rebar Connection

There are different categories of mechanical couplers utilized in the RC structures to connect smooth and deformed rebars. Yet modern splicing methodology is not appropriate for splicing varying genera of rebars. The connection of SS and FRP rebars involves economic, easy and effective techniques to empower the use of SS in sensitive positions of RC structures. For the splicing these rebars, screw lock-adhesive type couplers are used, as it does not require any assembly equipment and achieve speedy construction. Alam et al. 2010 [15] performed numerous experiments and developed a mechanical-adhesive style coupler. Using multi-row flat screws, the SS bar was connected to the coupler in one part, while the FRP bar was connected to a SS bar by epoxy on the other part.

6. Validation of Numerical Analysis

The column was designed and analytically tested under nonlinear static pushover loading defined by Billah and Alam [12]. The performance of the column applied nonlinear static pushover loading in relationships of tip load versus displacement. The applied load was displacement controlled along with 400 kN axial gravity loading. A numerical model in this investigation is validated using anon-linear FE software tool [16]. The hybrid column is modeled utilizing SS reinforcement in the plastic hinge region, and FRP reinforcement in the remaining length of a column. The columns are analyzed under the effect of nonlinear static pushover loads. Figure 2 portrays the comparison of predicted base shear versus displacement of the numerical model with experimental results by Billah and Alam [12]. It has been seen from Figure 2 that the numerical model with the adopted modeling techniques is verified with the actual empirical-numeric results with accuracy.

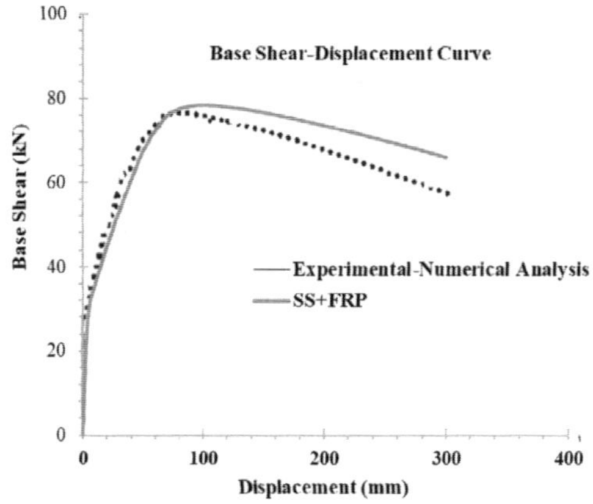

Figure 2. Validation of pushover analysis SS-FRP hybrid column with existing literature

7. Finite Element (FE) Modeling of SS-FRP RC Column

The finite element (FE) based program such as Seismo Struct [16], is adopted in the present study to analyse the seismic performance of the SS-FRP RC column. This program can predict large displacement behavior of the structure and is able to consider both material inelasticity and geometric non-linearity. The methodology of fiber modeling is used to describe the distribution of material nonlinearity along with the length of member and cross-section of the component. A bond-slip link is incorporated in the analytical model to join the hybrid RC column, which is signified by rotational spring at joint. A constant 400 KN axial load is applied at the top of a column to simulate the gravity loads.

A displacement-based nonlinear beam-column element is chosen for the RC column. The fiber section is used to discretize pier into three parts: core fiber for confined concrete, cover fiber for unconfined concrete, and steel fiber for reinforcement bars. The piers are modeled using a force-based inelastic system owing to its accuracy. The pier length is divided into a total of 11 sections, among which three sections represent plastic hinge length, and the remaining sections represent SS rebar[12].After the successful validation of a numerical model with literature, this study is extended for the performance analysis. Various factor-like axial load, compressive strength of concrete, yield strength of steel, and aspect ratio are considered for the parametric study. Details of parameter and corresponding terms are shown in Table 1.

Table 1. Details of parameter and corresponding terms of SS-FRP RC columns

Parameter	Column ID	Load (kN)	l/d	f'c (MPa)	Axial load (%)	fy (MPa)	d (mm)	l (mm)	lp (mm)
Aspect ratio,l/d	P-1-7.5	400	7.5	38.3	10	402	450	3375	442
	P-2-8.5	400	8.5	38.3	10	402	450	3825	478
	P-3-9.5	400	9.5	38.3	10	402	450	4275	514
Compressive strength,f'c	P-1-35	400	7.1	35.0	10	402	450	3200	428
	P-2-45	400	7.1	45.0	10	402	450	3200	428
	P-3-55	400	7.1	55.0	10	402	450	3200	428
Axial load, P	P-1-5.0	200	7.1	38.3	5	402	450	3200	428
	P-2-7.5	300	7.1	38.3	7.5	402	450	3200	428
	P-2-12.5	500	7.1	38.3	12.5	402	450	3200	428
Yield strength,fy	P-1-250	400	7.1	38.3	10	250	450	3200	363
	P-2-450	400	7.1	38.3	10	550	450	3200	449
	P-2-650	400	7.1	38.3	10	650	450	3200	535

8. Results and Discussion

The results of the numerical parametric analysis are described in Table 2. It can be observed that the aspect ratio (l/d), axial load (P), compressive strength of concrete (f'c), and yield strength of steel (fy) are significantly affecting the maximum base shear and top displacement of SS-FRP RC columns.

Table 2. Base shear and displacement at different limit states

Parameter	Column-ID	Cracking		Yielding		Crushing		Maximum	
		Base Shear (kN)	Disp. (mm)	Base Shear (kN)	Disp. (mm)	Base Shear (kN)	Disp. (mm)	Base Shear (kN)	Disp. (mm)
Aspect Ratio	P-1-7.5	29.45	6.0	43.41	24	69.72	186	73.01	108
	P-2-8.5	25.86	8.0	36.04	28	55.35	244	61.12	128
	P-3-9.5	22.22	8.0	31.65	36	42.73	316	51.54	148
Axial Load	P-1-5	28.09	3.0	33.63	18	75.15	171	78.54	300
	P-2-7.5	25.22	3.0	41.07	21	75.59	168	76.22	123
	P-3-12.5	35.31	6.0	50.82	21	77.01	159	81.16	90
Compressive strength	P-1-35	25.02	3.0	45.62	21	69.98	255	77.95	99
	P-2-45	23.43	3.0	76.93	21	79.05	153	79.18	102
	P-3-55	26.86	3.0	79.85	18	82.19	147	81.08	102
Yield strength of steel	P-1-250	24.99	3.0	45.03	21	56.90	153	59.52	75
	P-2-450	25.11	3.0	46.36	21	82.32	168	81.09	96
	P-3-650	25.23	3.0	47.50	21	107.66	183	75.72	300

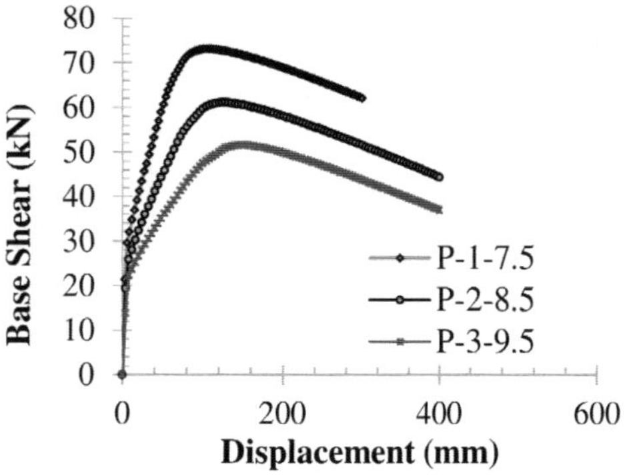

Figure 3. Result of pushover analysis (aspect load ratio)

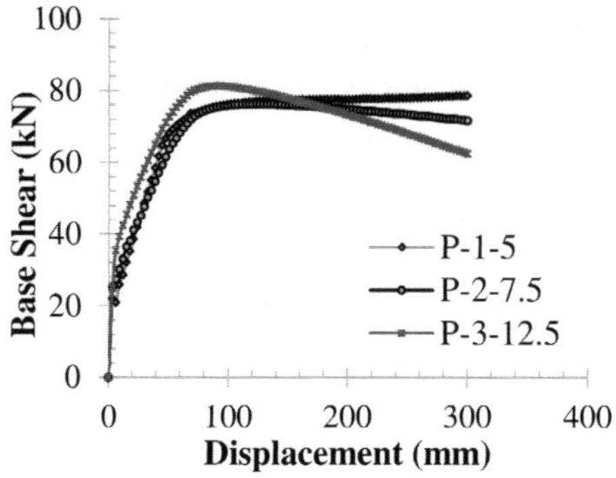

Figure 4. Result of pushover analysis (axial load ratio)

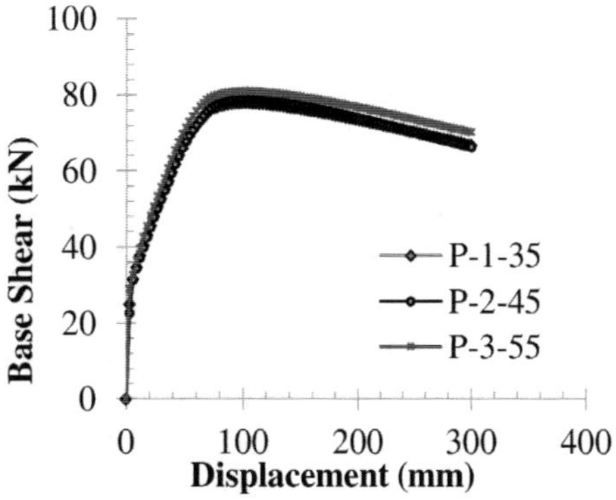

Figure 5. Result of pushover analysis (compressive strength of concrete)

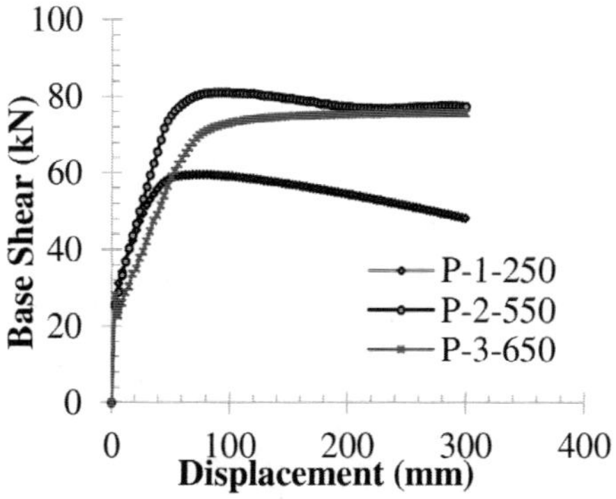

Figure 6. Result of pushover analysis (yield strength of steel)

The l/d ratio substantially decreases the value of maximum base shear as the aspect ratio increases from 7.5 to 9.5. Corresponding top displacements increase with an increase in the l/d ratio from 7.5 to 9.5. The axial load significantly decreases the value of maximum base shear as axial load increases from 5 to 7.5. Further, increase the amount of maximum base shear with increase the axial load. Corresponding top displacements decrease with an increase in axial load from 5 to 7.5. The f'c is significantly increasing the value of maximum base shear as f'c increases from 35 to 55. Corresponding top displacements increase with an increase in f'c from 35 to 55. The fy of steel firstly decreases the value of maximum base shear as the fy of steel at 250. Further reduces the amount of maximum base shear with increase the fy of steel. Corresponding top displacements increase with an increase in fy of steel from 250 to 650. As described in Table 2 and Figure 3, the aspect ratio (l/d) of 7.5 shows higher base shear values for cracking, yielding and crushing as compare to an aspect ratio of 8.5 by 12.19, 16.98 and 20.61 %, respectively. The aspect ratio of 7.5 shows lower values of displacement for cracking, yielding and crushing as compared to an aspect ratio of 8.5 by 25, 14.28 and 22.54 %, respectively. Also, the l/d ratio of 8.5 shows the higher base shear value for cracking, yielding and crushing as compare to an l/d ratio of 9.5 by 14.07, 12.18 and 22.80 %, respectively. The l/d ratio of 8.5 shows lower values of displacement for cracking, yielding and crushing as compared to an l/d ratio 9.5 by 0, 22.22 and 22.78 %, respectively.

As described in Table 2 and Figure 4, the axial load (P) of 5 shows the higher base shear value for cracking compared to the axial load of 7.5 by 10.22. The axial load of 5 shows the lower base shear value for yielding and crushing as compare to the axial load of 7.5 by 18.12 and 0.58, respectively.

IOP Conf. Series: Materials Science and Engineering **814** (2020) 012024 doi:10.1088/1757-899X/814/1/012024

The axial load of 5 shows the same, lower and higher value of displacement for cracking, yielding and crushing as compared to axial load of 7.5 by 0.00, 14.29 and 1.75 %, respectively. Also, the axial load of 7.5 shows the higher base shear value for cracking, yielding and crushing as compare to an axial load of 12.5 by 28.58, 19.19 and 01.84 %, respectively. The axial load of 7.5 shows lower, same and higher value of displacement for cracking, yielding and crushing as compare to an axial load of 12.5 by 50, 0.0 and 5.06 %, respectively.

As described in Table 2andFigure 5,the compressive strength of concrete, (f'c)of 55 shows higher base shear values for cracking compared to the fc of 45 by 6.35. The f'c of 35 shows lower base shear values for yielding and crushing as compared to the f'c of 45 by 40.70 and 11.47, respectively. The low fc strength of 35 shows the same, lower and higher values of displacement for cracking, yielding and crushing as compare to f'c of 45 by 0, 61.11 and 40.00 %, respectively. Also, the effect fcof 45 shows lower base shear values for cracking, yielding and crushing as compared to f'c of 55 by 12.77,3.66 and3.82 %, respectively. The f'c of 45 shows the same, same and higher values of displacement for cracking, yielding and crushing as compared to the f'c of 55 by 0.00, 0.00 and 03.92%, respectively.

As described in Table 2andFigure 6, The yield strength of steel (fy)250 shows higher base shear values for cracking, yielding and crushing as compared to fy of 450 by 0.47, 2.85 and 30.88 %, respectively. The fy of 250 shows the same, same and lower values of displacement for cracking, yielding and crushing as compare to fy of 450 by 0.0, 0.0 and 8.93 %, respectively. Also, the fy of 450 shows the lower base shear value for cracking, yielding and crushing as compared to fy of 650 by 0.476, 2.40 and 23.54 %, respectively. The fy of 450 shows the same, same and lower values of displacement for cracking, yielding and crushing as compare to fy of 650 by0.00, 0.00 and 8.20 %, respectively.

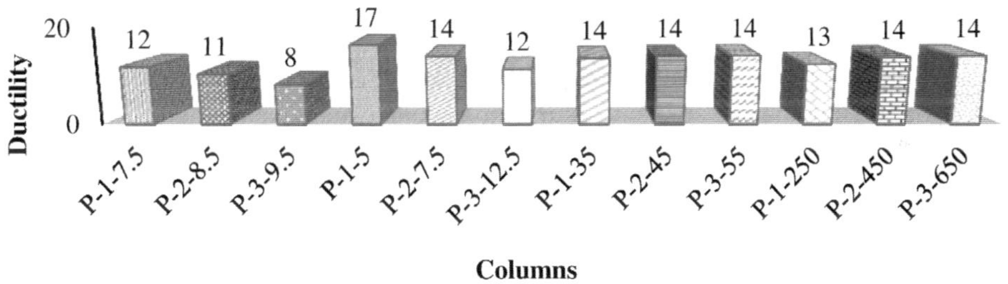

Figure 7. Ductility of FRP-SS columns

In terms of ductility, the performances of different types of columns with variable parameters are also compared. In evaluating performance and evaluating seismic design forces during an earthquake event, structural ductility plays an important role. As well as ductility is also a contributing factor for structures to evaluate the capacity of energy dissipation. Ductility is defined as the ratio of ultimate displacement (μd) to yield displacement, which is a dimensionless variable. The displacement of yield corresponds to the first yielding of tension reinforcement. Various investigators have revealed the final movement in a different way. Numerous researchers and nearly all the design guidelines (FEMA440) [17]have identified the ultimate displacement as the maximum displacement corresponding to optimum base shear. The final movement in this study is the displacement corresponding to a load falling from the peak load. Figure 7shows the variation in ductility for various hybrid RC columns. The ductility is decreased from 12 to 8, with an increase in the aspect ratio from 7.5 to 9.5. Similarly, the same observation was observed in case of axial load, where 5 % axial load possesses the ductility value of 17, and 12.5 % axial load possesses the ductility value of 12, and 7.5 % axial load ratio possesses the ductility value of 14.

9. Conclusions

The study was carried out through inelastic pushover analysis on corrosion-free SS-FRP RC columns for seismic investigation of the RC column, where stainless steel is used in the plastic hinge region and FRP in another area of the column. The results of the SS-FRP RC column in terms of base shear vs. displacement are validated with existing literature. Based on the numerical study, under the seismic excitations, using SS in the plastic hinge zone improved the seismic behaviour of the columns significantly. Based on the numerical parametric analysis are significantly affecting the maximum base shear and displacement of SS-FRP RC column, the following result can be obtained.

- It is observed that with increases in the aspect ratio substantially decreases the value of maximum base shear.
- The aspect ratio (l/d) of 7.5 shows the higher base shear value for cracking, yielding and crushing as compare to an aspect ratio of 8.5 by 12.19, 16.98 and 20.61 %, respectively.
- The axial load ratio of 5 shows a higher base shear value for cracking compare to an axial load ratio of 7.5 by 10.22.
- The higher compressive strength of concrete (f'c) of 55 show higher base shear value for cracking compare to the lower compressive strength of concrete of 45 by 6.35.
- The lower yield strength of steel (fy) 250 shows the higher base shear value for cracking, yielding and crushing as compare to higher yield strength of steel of 450 by 0.47, 2.85 and 30.88 %, respectively.
- The aspect ratio and axial load are the most significant factors for the ductility of the SS-FRP RC column.
- The results indicate adequate ductility of the SS-FRP RC column is achieved under seismic loads.
- The study reveals that the axial load (P), yield strength of steel (fy), and aspect ratio (l/d) significantly affect the behavior of the SS-FRP RC column.

References

[1] Pettinga J D, Pampanin S, Christopoulos C and Priestley M J N 2006 *First European Conference on Earthquake Engineering and Seismology* pp 1–10
[2] Virmani P, McDonald D B, Pfeifer D W, and Sherman M R 1998 FHWA-RD-98
[3] Association I M 2008 Stainless Steel Reinforcement *Int. Molybdenum Assoc.*June 20
[4] Aydin F 2018 *Constr. Build. Mater.***163** 1–8
[5] Benmokrane B and Ali A H 2018 *Int. Congr. Polym. Concr.* 49–59
[6] Parghi A M 2016 Seismic performance evaluation of circular reinforced concrete bridge piers retrofitted with fiber reinforced polymer (University of British Columbia)
[7] ACI Committee 440 2006 ACI 440.1R-06: Guide for the Design and Construction of Structural Concrete Reinforced with FRP Bars *Am. Concr. Inst.* 44
[8] De Luca A, Matta F and Nanni A 2010 *ACI Struct. J.***107** 589–96
[9] Madas P J 1993 *Advanced modelling of composite frames subjected to earthquake loading* (London)
[10] Mander J B, Priestley M J N, and Park R 1988 *J. Struct. Eng.***114** 1804–26
[11] Enrique Martínez-Rueda J and Elnashai A S 1997 *Mater. Struct. Constr.***30** 139–47
[12] Muntasir Billah A H M and Shahria Alam M 2012 *Constr. Build. Mater.***28** 730–42
[13] Billah A H M M and Shahria Alam M 2016 *Eng. Struct.***117** 321–31
[14] Paulay T and Priestley M J N 1992 *Seismic design of reinforced concrete and masonry buildings* ed I John Wiley & Sons (New York)
[15] Alam M S, Youssef M A, and Nehdi M L 2010 *Mater. Struct.***43** 91–107
[16] SeismoSoft S 2018 SeismoStruct - A computer program for static and dynamic nonlinear analysis of framed structures *http//www. Seism. com*
[17] FEMA 440 A T C (ATC-55 P 2005) *Improvement of nonlinear static seismic analysis procedures*.

IOP Conf. Series: Materials Science and Engineering **814** (2020) 012025 doi:10.1088/1757-899X/814/1/012025

Review of Strengthened FRP-confined RC column with various aspect ratios under axial load and high-temperature

Mayank Patel [1], Rahul Pardeshi [2], Bhairav Thakur [3], Anant Parghi [4]

[1,2,3] Research Scholar, Applied Mechanics Department, SVNIT, Surat, Gujarat – 395007
[4]Assistant Professor, Applied Mechanics Department, SVNIT, Surat, Gujarat – 395007

E-mail: mayankpg09@gmail.com, anantparghi@gmail.com

Abstract. In civil and structural engineering, building structures with vigorous stability and strength utilizing economical materials is challenging. Stability of structures during their lifespan is a very demanding endeavor in civil engineering systems. Recent trends are highly focused on high strength materials, strong corrosion-resistance in structural elements, slender structure development, broad span provision, and load reduction. in order to achieve these conditions, composite materials have proved to be a successful aspirant. The fiber-reinforced polymer (FRP) possesses novel properties that encourage the researchers to strengthen or restore the structural degradation of the reinforced concrete (RC) columns via confinement. The present study highlighting the different aspects of (FRP) confined (RC) column having different aspect ratios, the axial load, and the high temperature under extensive literature review. The FRP confinement is much more effective in the case of circular columns than sharp-edged rectangular columns. The variation of the cross-sectional aspect ratio (section depth to width ratios) of RC columns plays a vital role in the evaluation of the efficiency of strengthening techniques. In spite of the clear and proven advantages of utilizing FRPs over conventional materials, awareness of the behavior of such composite materials after exposure to high temperature is noticeable and requires more research.

1. Introduction

In India, many existing structures do not reach its load-carrying capacity level due to the deterioration generated by a combination of decay, improper design, or unforeseen activity like earthquakes, storm, etc., and not capable to sustain cautiously the same [1,2]. Dismantling a structure and rebuilt it again is not a practical solution. Because of time and economic aspects, the most desirable choice is to immediately overhaul the damaged part of the structure. Confinement of RC columns using FRP sheets is considered to be a very effective technique. It improves both axial compressive load carrying capacity and structural ductility of RC columns [3–6]. The previous study has confirmed that the repairing of damaged concrete by means of external confinement is successful in restoring the original strength of RC columns [7–10]. The researchers noticed that confinement was a better option for refurbishing the ductility and strengthening of damaged concrete [11, 12]. The FRP is a combination of high strength fibres (glass, carbon, aramid or basalt) and polymer matrix (polyester, vinyl ester or epoxy)[13]. The FRP has excellent properties having outstanding corrosion resistance, low specific gravity at high mechanical strength, less effect of weather, good insulation qualities, high strength to weight ratio as compared to steel, and innate protection to decay [14]. The FRP confinement is more effective in the circular columns due to uniformity in stress distribution than in the rectangular columns [6, 15–17]. The confining pressure is different in the corners and the flat side in the case of the rectangular column. Most research and evaluation work focused on characterizing the performance of FRP confined circular concrete columns [18–23]. Nevertheless, very few tests and research are performed on the rectangular column because of the complex distribution [8, 20, 24–27]. The efficacy of FRP confinement in RC columns depends on different influencing factors like the type of fiber, the thickness of jacketing, orientation of fiber, bond between core, concrete strength, and shape of cross-sectional, etc [6]. The cross-sectional aspect ratio affects the confinement efficiency of FRP composites and stress-strain behaviour. The maximum value of the cross-sectional aspect ratio was limited up to 1.5 [28, 29] and it was revised to the value of 2.0 [30].

The FRP has been widely used in a practice due to its various advantages. However, utilizing the FRP in the repair and retrofit of structures, it's poor performance under the effect of high temperature or fire exposure was noticeable [31]. Moreover, designing a structural element, it needs to design a satisfactory level of fire safety and minimized risk from heat and smoke. The utilization of FRP materials is a major challenge owing to the lack of standard codes and complex behaviour of FRP materials at various temperatures.

2. Background

In the 1930s, united states have shown interest in the utilization of FRP based confinement in concrete structures. Nevertheless, the actual practice to utilize these materials for retrofitting was stared in the 1980s after the initiative of the National Science Foundation (NSF) and Federal Highway Administration (FHWA). Retrofitting the concrete structures using FRP was testified in the year 1978 in Germany [32]. Similar kinds of investigations for the retrofitting of concrete structures were done in Japan and Europe in the 1980s. In the mid of 1980, externally bonded FRP systems were being used universally. Externally bonded FRP system was first utilized in Switzerland for the flexural strengthening of bridges [33]. In Europe, FRP method was being used as an alternative method to steel plate bonding. In Japan, the FRP system was being used in the 1980s to improve the confinement [34]. The utilization of FRP systems has been increased in the last decade from a few projects to several thousand projects around the world [14]. Europe, Japan, Canada, and the United States have developed the design rules and guidelines in the area of applications of externally bonded FRP systems. Preceding research and field applications for the rehabilitation and strengthening using FRP are described in (ACI 440.2R-17) [35]. Canadian Standard Association (CSA) has approved the code, "Design and construction of building a component with Fiber Reinforced Polymers" [36].

3. Type of confining method

Restoring, repairing and strengthening of RC columns is necessary for prolonging the structures life, to restore the degraded column, for enhancement in load carrying capacity over the design value, and to ensure the ductility, etc. Researchers and practitioner engineer have developed various RC column strengthening measures (Ref). Some of them are highlighting here under subsiding paragraphs.

3.1 Concrete Jacketing

Amongst several confining techniques, the first method of repairing damaged concrete is concrete jacketing. So many studies focused on strengthening and repairing of RC columns by using concrete jacketing to the original column [37–40]. From the past study, it was concluded that concrete jacketing enhancing the strength and ductility of damaged RC columns. However, high operating costs and the increased size of the repaired columns have definitely restricted its usage.

3.2 Steel Jacketing

One most effective ways of increasing the ductility of damaged concrete columns were the steel. Several researchers investigated the use of steel jackets in strengthening the RC columns [41]. However, the bonding between the steel plates will considerably affect the efficiency of the use of jackets. For members with a right angle of cross-section (square or rectangular) efficiency was considerably low, as does not improve flexural stiffness effectively.

3.3 FRP Jacketing

The wrapping of RC columns by FRP is another technique for strengthening and ductility enhancement. Several researchers have examined the use of FRP jackets to reinforced RC columns [42-52]. As an alternative to above two methodologies, the FRP composites presented it as an attractive option.

4. Techniques of confinement

There are two types of confinement techniques, defined as an active confinement technique and passive confinement technique. Lateral confining pressure acting on the column is the main difference between active and passive confinement [53]. In the active confinement, the confining pressure is applied laterally on the elements prior to the loading. This employs a little lateral stress on the concrete resulting in an increase in load-carrying capacity of the element. Passive confinement method is used to gaining a higher strength and ductility. The passive confinement is generally done by wrapping the FRP to the circular or rectangular column. The passive confinement has been achieved through the application of circumferentially unidirectional fiber epoxy-bonded sheets [54]. Confinement effectiveness of RC columns depends on various factors; e.g. cross-sectional aspect ratio (length to width), the radius of the corner, type of confinement, type of column, etc. Up till now, research has mainly focused on describing the nature of circular and square cross-sectional columns but it's far less understood the efficiency of FRP confinement for the rectangular columns [6]. The efficiency of confinement is stronger in the circular RC columns due to equal nature of stress distribution than the prismatic columns which has limited to just a portion of a prismatic cross-section. From Figure 1 it is clear that axial load acting on FRP confined rectangular concrete columns, the confinement pressure concentrated at the corner of the section. The confining pressure concentrated at the corner of the section is very less or negligible.

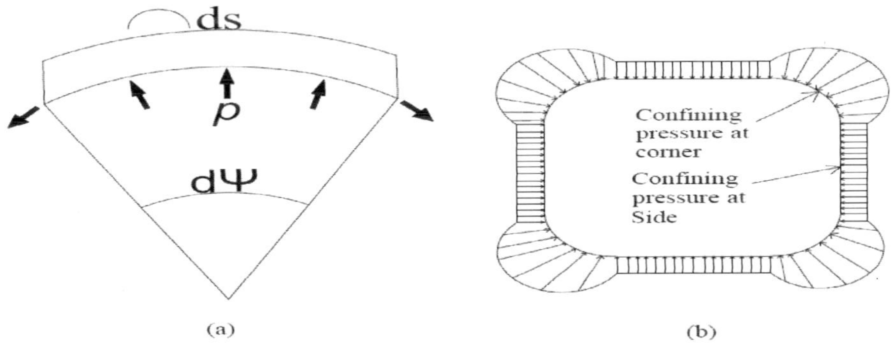

Figure 1. Confine behaviour at the corner of the section (a) Mechanism of the tension force (b) distribution of confining stress (adapted from [24])

5. FRP confinement in rectangular RC Columns

Axial compressive load carrying capacity of rectangular column increases intensely with wrapping the FRP sheets. In non-circulars (square or rectangular) columns FRP wrapping is not as effective as circular column [55]. Due to the equal distribution of fiber on entire section, the effectiveness of FRP confinement in the circular column is very high, whereas in square or rectangular column, the corner confined properly but other part remains unconfined. This efficiency loss is usually shown with corner-defined parabola areas as shown in Figure 2. Also, this efficiency is highly dependent on the cross-section aspect ratio and the radius of the corner [54].

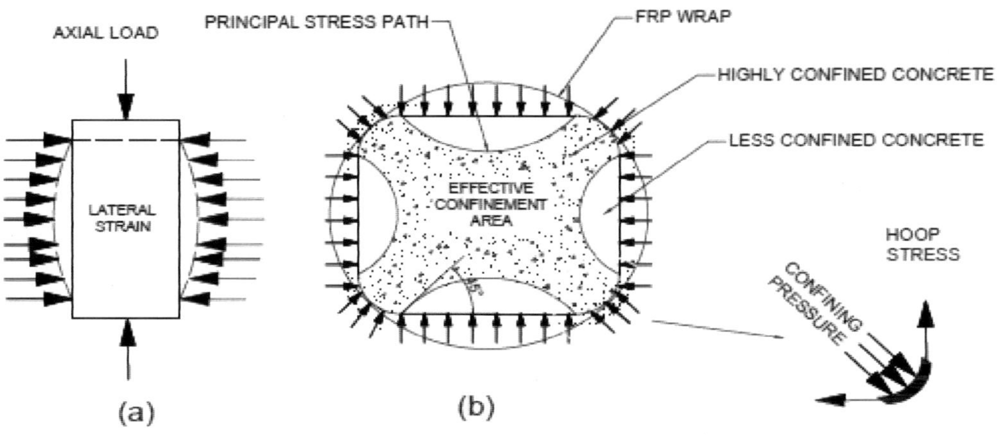

Figure 2. Confinement pressure in a rectangular column (a) axial loading on column (b) confined behaviour of rectangular column (adapted from [30])

5.1 Confining models for rectangular RC columns [30]

The total compressive load taken by an RC column is the combination of the compressive load taken by concrete and tensile load taken by longitudinal steel bars. They take reduction factors as per the codes. The nominal axial compression capacity (Pn) of non-slender, normal-weight concrete member confined with transverse fiber sheets can be calculated as below equation 1.

$$Pn = \alpha[0.85f'_{cc}\left(A_g - A_{st}\right) + f_y A_{st} \quad (1)$$

Where,

Pn = nominal axial compression capacity of column

α = Coefficient to cater for minimum eccentricity

0.85 for non-prestressed members with existing spiral steel reinforcement

0.80 for non-prestressed members with existing steel-tie reinforcement

f'_{cc} = Maximum compressive strength of concrete

A_g = Gross cross-sectional area of column

A_{st} = Total area of reinforcement

f_y = Yield strength of reinforcement

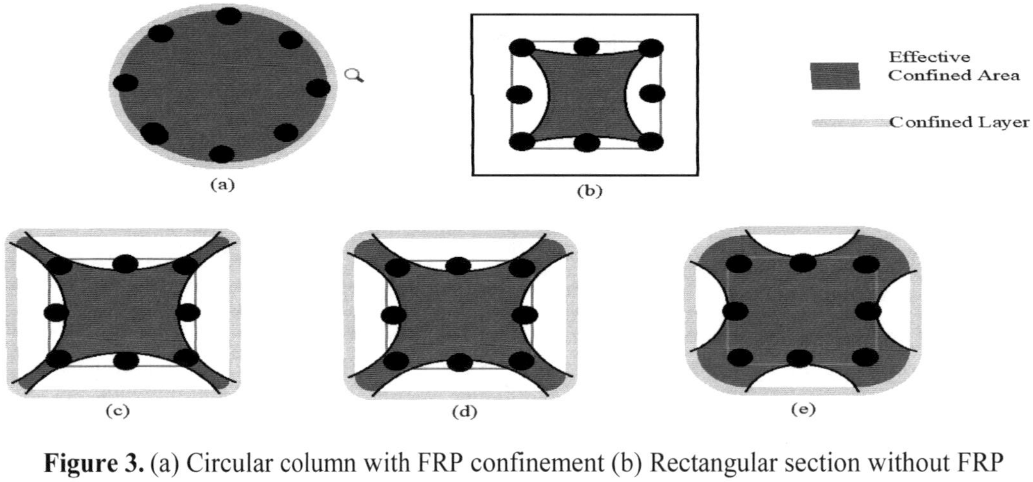

Figure 3. (a) Circular column with FRP confinement (b) Rectangular section without FRP confinement (c) (d) (e) FRP confined Rectangular columns with different corner radius (adapted from [58])

Hoop stress is also depending on the corner edges. It directly affects the behaviour of FRP confined RC column. The stress level at rectangular corners is the main reason for the untimely failure of FRP [56]. At the time of FRP confinement in rectangular RC columns, FRP laminates bent first when wrapped around the column. The effectiveness of FRP laminate reduces due to bending, this action depends on the edges of the corner. The radius of the corner greatly affects the efficiency of confinement. For achieving the better result of confinement the corner radius must be large [57]. Different from the circular cross-section, when the square cross-section is subjected to axial compression the lateral stress is not uniform over the perimeter of the section [58]. Cross-section change from circular to a square, the lateral (circumferential) stress over the section perimeter is not uniform for axial compression. From Figure 3 it should be noted that corner radius plays a very important role in confinement.

5.2 Effect of aspect ratios on the confinement

As discussed earlier, the confinement efficiency is directly related to the shape of a section. Figure 4 shows the loading scheme of a rectangular column in a multi-storey building. The compressive pressure gets distributed unequally if the square and rectangular sections are not uniformly confined with FRP. From the previous research, the shape of a section directly marks the efficiency of the FRP confinement due to the difference of stress-strain relationship [4]. When the cross-sectional aspect ratio increases, the ultimate strength of a column is decreased. Moreover, it can be observed that the strengthening ratio decreases with an increase in the aspect ratio [59]. Figure 5 shows various values of the cross-sectional aspect ratio generally adopted in structural design [59]. For analysing the effect of aspect ratio on RC column, a parametric study was conducted by researcher and it was concluding that increasing in aspect ratio from 1:1 to 1:6 strengthening value decreased up to 10%. [60]. In order to find out the effect of aspect ratio and confined behaviour of RC column an experimental study was conducted by and concluded that increasing in aspect ratio, the ultimate strength of CFRP confined RC columns get reduce [61]. The maximum value of the cross-sectional aspect ratio was limited up to 1.5 [28], and it was revised to the value of 2.0 [30].When the aspect ratio reaches more than 2.0 the strength improvement in RC confined column become insignificant [4].

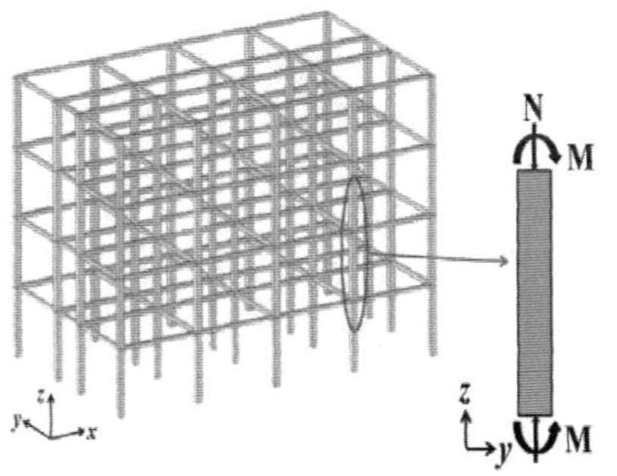

Figure 4. Loading scheme of a rectangular column

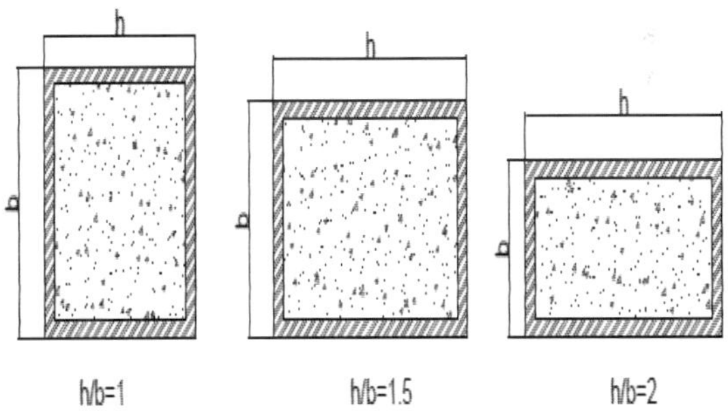

Figure 5. Aspect ratios of column

Wu and Wei 2010 experimental study, RC short column was strengthened with CFRP wrapping, and the axial load was applied afterwards up to the failure of an RC column. Different values of aspect ratio and a various number of CFRP layers were used for the strengthening of short RC columns. They have examined the behaviour of a specimen in the axial and transverse directions. From the test results, it indicates that the strength gaining in confined column relative to the unconfined column, mean compressive strength (fcc/$f$co) decreases as the aspect ratio increases. Table 1 defines the mean value of the compressive strength and the corresponding strength gained in the confined columns.

Table 1. Mean compressive strength and corresponding fcc/fco (adapted from [4])

Aspect ratio h/b	Without Wrapping (MPa)	(1-layer CFRP wrapping) (MPa)	Mean Compressive Strength (fcc/*f*co)	(2-layer CFRP wrapping) (MPa)	Mean Compressive Strength (fcc/*f*co)
1.0	35.30	41.23	1.17	60.37	1.71
1.25	35.30	38.77	1.10	51.36	1.45
1.5	35.30	38.39	1.09	43.87	1.24
1.75	35.30	37.68	1.07	40.51	1.15
2.0	35.30	37.44	1.06	38.97	1.10

5.3 High temperature or fire effect on FRP wrapping a concrete column

In civil engineering area, utilization of FRP materials is a major challenge owing to the lack of standard codes and complex behaviour of FRP materials at various temperatures. The main disadvantage of utilizing FRP in the repair and retrofit of structures is its poor performance under the effect of high temperature or fire exposure [62]. While designing a structural element, it needs to be designed so as to offer a satisfactory level of fire safety and minimized risk from heat and smoke. Fire is characterized as a period of resistance to structural integrity, stability and transmission under fire condition by structural components [63]. With the use of [64, 65] standards, the fire resistance of a structural component can be obtained by conducting full-scale fire tests on a representative sample of a structural member in the furnace. This test is expensive and needs a long time to be performed [66]. In context to the cost-effectiveness, the performance-based method is upright for the calculations and numerical simulations. Variation in temperature produces high thermal stresses in RC structures. Thermal stresses are produced due to the restraining of free expansion, contraction or rotation. When the temperature increases more than100°–150 °C, it directly affects the mechanical and physical properties of FRP [62]. The polymer resins in FRP improves strength when they are placed in the transverse direction. When the temperature exceeds the transition temperature (T_g), the bond between FRP and concrete gets affected.

The polymer resin is a type of epoxy material, which works as a binding agent and transmits the stresses between fibres in FRP. Epoxy can be utilized as a bonding agent between composites and structural elements. Polymer resins of FRP get deteriorated rapidly when their temperature reaches 200-250 °C above the glass transition temperature (T_g), [28]. Under the effect of fire on FRP, polymer resins in it change their state to the flexible/ rubbery state from the stiff/ solid-state [62]. When the temperature reaches the disintegration temperature, resins get deteriorated thermally and the fire spreads along the surface of the material. As the fire spreads extensively, the smoke is generated from the burning of FRP. Burning of this polymer resin directly affects the mechanical and bonding properties of FRP and this would be totally worsening [62].

6. Conclusions

The present study has assessed the various databases related to FRP confinement, the effect of cross-sectional aspect ratio and corner radius on rectangular columns, the effect of high temperature on FRP confined columns etc. and the following conclusions are drawn from the literature.

FRP confinement is more effective as compare to other confinement for strengthening and repairing of RC columns.

The passive protection provided by FRP wrapping can significantly improve the load-carrying capacity and ductility.

IOP Conf. Series: Materials Science and Engineering **814** (2020) 012025 doi:10.1088/1757-899X/814/1/012025

FRP confinement is less effective in rectangular columns due to its corner edges and uneven stress distribution.

Corner roundness is an essential parameter for changing stress concentration in rectangular columns.

FRP wrapping improves the performance of rectangular RC columns under an axial loading but enhancement level decrease with increase in aspect ratio.

Axial and lateral stresses at the corner are significantly higher than in the middle of the sides in rectangular columns.

With an increase in temperature, the load-carrying capacity of FRP confined RC column decrease continuously.

The FRP wrap system was loss of effectiveness at temperatures below the Tg of epoxy adhesive used. The tensile strength of the FRP wrap at such temperatures is thought to decrease.

References

[1] Hideo K, Yoshirou K and Toshikazu T 1988 *ninth world conference on earthquake engineering* (tokyo-kyoto,japan)
[2] Wu G, Lü Z T and Wu Z S 2006 *Constr. Build. Mater.* **20** 134–48
[3] Abdel-Hay A S 2014 *HBRC J.* **10** 279–86
[4] Wu Y and Wei Y 2010 *Eng. Struct.* **32** 32–45
[5] Siddiqui N A 2009 *Lat. Am. J. Solids Struct.* **6** 343–62
[6] Mastrapa J C, Pico O, Engr S Des, Smith W and Rd L 1998 175–85
[7] Peled A 2007 sleeves *J. Compos. Constr.* **11** 514–22
[8] Ilki A, Peker O, Karamuk E, Demir C and Kumbasar N 2008 *J. Mater. Civ. Eng.* **20** 169–88
[9] Zhou Y W, Liu X M, Sui L L, Xing F and Zhou H J 2015 *Mater. Res. Innov.* **19** S6125–31
[10] Panjehpour M, Farzadnia N, Demirboga R and Ali A A A 2016 *J. Civ. Eng. Manag.* **22** 56–64
[11] Gu D S, Wu G, Wu Z S and Wu Y F 2010 *J. Compos. Constr.* **14** 531–40
[12] Iacobucci R D, Sheikh S A and Bayrak O 2003 *ACI Struct. J.* **100** 785–94
[13] Gay D 2014 Composite materials: design and applications (CRC press)
[14] Bakis C E, Bank L C, Brown V L, Cosenza E, Davalos J F, Lesko J J, Machida A, Rizkalla S H and Triantafillou T C 2002 *J. Compos. Constr.* **6** 73–87
[15] Fam A, Mandal S and Rizkalla S 2005 *J. Compos. Constr.* **9** 34–43
[16] Kong W K and Kim H C 2004 *Can. J. Civ. Eng.* **31** 178–88
[17] Pessiki B S, Harries K A, Kestner J T, Sause R and Ricles J M 2001 *Gg* 5 237–45
[18] Carey S A and Harries K A 2005 *ACI Struct. J.* **102** 596–604
[19] Cui C and Sheikh S A 2010 *J. Compos. Constr.* **14** 553–61
[20] Lim J C and Ozbakkaloglu T 2014 *Mag. Concr. Res.* **66** 1020–35
[21] Mimiran A and Shahaway M 1997 B Amir Mirmiran 1 and Mohsen Shahawy / Members, ASCE *ASCE J. Struct. Eng.* **123** 583–90
[22] Monti M R S and G 1985 *Kinet. Catal.* **26** 201–4
[23] Xiao and wu 2000 *J. Mater.* **12** 139–46
[24] Youssef M N, Feng M Q and Mosallam A S 2007 *Compos. Part B Eng.* **38** 614–28
[25] Pellegrino C and Modena C 2010 *J. Compos. Constr.* **14** 693–705
[26] Triantafillou T C, Choutopoulou E, Fotaki E, Skorda M, Stathopoulou M and Karlos K 2016 *Mater. Struct. Constr.* **49** 651–64
[27] Harajli M H, Hantouche E and Soudki K 2006 *ACI Struct. J.* **103** 672–82
[28] ACI 440.2R 2002 Guide for the Design and Construction of Externally Bonded FRP Systems for Strengthening Concrete Structures
[29] GB-50367 2013 Code for Design of Strengthen Concrete Structure
[30] ACI 440.2R 2008 Guide for the Design and Construction of Externally Bonded FRP Systems
[31] Abdel-hafez L M, Abouelezz A E Y and Hassan A M 2014 *Hous. Build. Natl. Res. Cent.*

[32] Wolf R and Miesseler H-J 1989 New Materials for Restressing and Monitoring Heavy Structures
[33] Li P, Sui L, Xing F, Li M, Zhou Y and Wu Y-F 2019 *J. Compos. Constr.* **23** 04019001
[34] Fardis M N and Khalili H 1981 *ACI J.* **440** 6.
[35] ACI 440.2R-17 2017 Guide for the Design and Construction of Externally Bonded FRP Systems for Strengthening Concrete Structures
[36] CAN 2002 Design and Construction of Building Components with Fibre-Reinforced Polymers (CAN/CSA S806-02) *Csa S806-02* 177
[37] Julio E S, Branco F and Silva V D 2003 *Prog. Struct. Eng. Mater.* **5** 29–37
[38] Elbakry H M F and Tarabia A M 2016 *Alexandria Eng. J.* **55** 57–67
[39] Xie J, Fu Q and Yan J B 2019 C *Constr. Build. Mater.* **204** 643–58
[40] Vandoros K G and Dritsos S E 2008 *Constr. Build. Mater.* **22** 264–76
[41] Belal M F, Mohamed H M and Morad S A 2015 *HBRC J.* **11** 201–12
[42] Alotaibi K S and Galal K 2017 *Compos. Part B Eng.* **114** 467–79
[43] Laterza M, D'Amato M, Braga F and Gigliotti R 2017 *Compos. Struct.* **176** 910–22
[44] Al-Nimry H and Neqresh M 2019 *Eng. Struct.* **196**
[45] Zeng J J, Lin G, Teng J G and Li L J 2018 B *Eng. Struct.* **174** 629–45
[46] Turgay T, Polat Z, Koksal H O, Doran B and Karakoç C 2010 *Mater. Des.* **31** 357–64
[47] Eid R and Paultre P 2017 *Eng. Struct.* **132** 518–30
[48] Mazzucco G, Salomoni V A, Majorana C E, Pellegrino C and Ceccato C 2016 *Constr. Build. Mater.* **111** 590–9
[49] Wang Y C and Hsu K 2008 *Compos. Struct.* **82** 132–9
[50] Juntanalikit P, Jirawattanasomkul T and Pimanmas A 2016 *Eng. Struct.* **125** 400–18
[51] Gao P, Wang J, Li Z, Hong L and Wang Z 2018 *Eng. Struct.* **162** 109–20
[52] Sayyar Roudsari S, Hamoush S A, Mohamad Soleimani S and Madandoust R 2019 *Eng. Struct.* **178** 680–93
[53] Rousakis T C and Tourtouras I S 2014 *Compos. Part B Eng.* **58** 573–81
[54] Triantafillou T C, Choutopoulou E, Fotaki E, Skorda M, Stathopoulou M and Karlos K 2016 *Mater. Struct.* 651–64
[55] Mukherjee A, Boothby T E, Bakis C E, Joshi M V and Maitra S R 2004 *J. Compos. Constr.* **8** 97–103
[56] Mostofinejad D, Moshiri N and Mortazavi N 2015 *Mater. Struct. Constr.* **48** 107–22
[57] Wang L M and Wu Y F 2008 *Eng. Struct.* **30** 493–505
[58] Faustino P, Chastre C and Paula R 2014 *Part B Eng.* **57** 187–98
[59] Li X, Lu J, Ding D-D and Wang W 2017 *Mag. Concr. Res.* **69** 1011–26
[60] Maalej M, Tanwongsval S and Paramasivam P 2003 *Cem. Concr. Compos.* **25** 263–76
[61] Cole C and Belarbi A 2001 *Science 80* 823–32
[62] Hawileh R A, Abu-Obeidah A, Abdalla J A and Al-Tamimi A 2015 *Constr. Build. Mater.* **75** 342–8
[63] Feasey R and Buchanan A 2002 *Fire Saf. J.* **37** 83–105
[64] ASTM E119 2016, ASTM International, West Conshohocken, PA *ASTM Int. West Conshohocken, PA* 552
[65] ULC-S101 2004 Scarbrgh. Underwrit. Lab. Canada
[66] Cree D, Chowdhury E U, Green M F, Bisby L A and Bénichou N 2012 *Fire Saf. J.* **54** 86–95

IOP Conf. Series: Materials Science and Engineering **814** (2020) 012026 doi:10.1088/1757-899X/814/1/012026

Buckling analysis of laminated composite plate due to localised in plane loading

Achchhe Lal [1], Anant Parghi [1], Anil Kumar Mahto [1], Rahul Kumar [2]

[1] Assistant Professor, Department of Mechanical Engineering, SVNIT, Surat, India
[2] Research Scholar, Department of Mechanical Engineering, SVNIT, Surat, India

E-mail: lalachchhe@yahoo.co.in

Abstract. Laminated composite plates are inevitable parts of structure due to its superior properties than that of conventional material. Many times, these structural elements are subjected to the partial loadings during various applications which arises buckling in the elements. This necessitates proper analysis of these types of structural member for better safety and stability. In the present paper buckling analysis of laminated composite stiffened plate under in-plane localized edge loading is presented. Buckling load for various cases of partial edge loading are investigated for different fiber orientations and loading extent. Finite element method-based MATLAB program is developed for the present analysis. Validation study is done, which shows well agreement with the results already published.

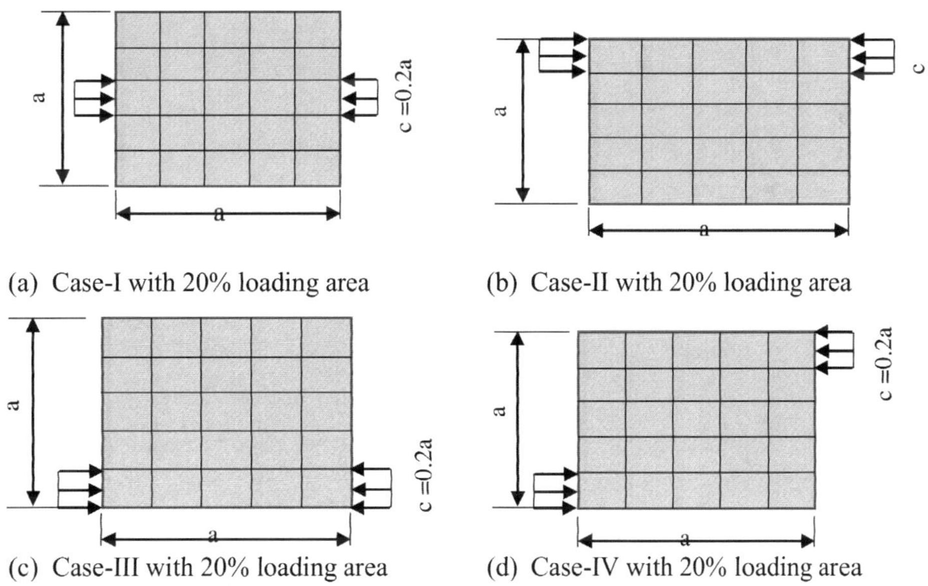

(a) Case-I with 20% loading area (b) Case-II with 20% loading area

(c) Case-III with 20% loading area (d) Case-IV with 20% loading area

Figure 1. Partial edge loading of Laminated composite plate at (a) centre (b) top of vertical sides (c) bottom of vertical sides (d) bottom to the left side and at top of right side.

1. Introduction

Application of composite panels in thin walled structures like automobiles, aerospace vehicles, submarines and others weight sensitive area has increased in last few decades. In the course of their services these structural members are usually subjected to in-plane uniform compressive loading. Therefore, analysis of these panels under localised uniform loading must be carried out for their better design and stability. Large number of research work are available for buckling behaviour of composite plates under uniform loading [1-3], while limited research is available for such plate under partial in-plane compressive loadings. Leissa [2] presented a review study of buckling of composite plate and discussed the post-buckling behaviour and influence of geometrical imperfections. Chia [3] used analytical method for analysis of post-buckling response of laminated elastic plates.

Analytical analysis of buckling of plate subjected to non-uniform in-plane edge loading is done by some researchers [4-5]. Brown [4] investigated elastic stability of thin plate under concentrated load. Kaldas and Dicknson [5] done calculation for stability of plates under in-plane stress using Rayleigh-Ritz integration. Liew and Chen [6] used a numerical method for buckling analysis of Mindlin plates under various locally distributed in-plane loadings and boundary conditions. Daripa and Singha [7] done nonlinear buckling analysis of isotropic and composite plates under localised in-plane compressive loading using four noded quadrilateral element and reveal the influence of location and variation of in-plane edge loading on critical buckling load. Using first order shear deformation theory with finite element buckling analysis of cross-ply laminated plate is done by Cagdas and Adali [8] for various boundary conditions when plate is subjected to varying edge loadings. Investigation is carried out for the optimal layer thickness of plate for maximum buckling load with various boundary restrains. A nonlinear bucking analysis using 3D cubic element of functionally graded plates under various non-uniform in-plane load patterns is done by Shariyat and Asemi [9] and buckling loads are determined. First order shear deformation theory using eight noded isoparametric shell element is used by Patel and Sheikh [10] for buckling response investigation of laminated composite plate under partial in-plane loadings. The influence of parameters like loading pattern, fiber orientation, and number of layers on buckling load is investigated. Kumar et al. [11] done nonlinear buckling analysis of cylindrical composite panels using higher order shear deformation theory (HSDT) under localized in-plane mechanical and thermal loading.

2. Mathematical formulation

A finite element based MATLAB program with HSDT using nine noded isoperimetric quadrilateral element is developed for buckling behaviour analysis of laminated composite plate subjected to partial edge in-plane loading. Material properties of composite plate considered for the present study are tabulated in Table 1.

Table 1. Material properties consider for the present paper.

E_1 (GPa)	E_2 (GPa)	$\nu_{12}=\nu_{21}$	ν_{23}	G_{12} (GPa)
82.64	9.8034	0.3	0.0356	3.8545

Square plate having dimension 1 x 1 x 0.1 is taken for present analysis.

2.1. Displacement field

Displacement field for present paper is based on HSDT and can be given as,

$$\begin{aligned} u_1 &= u + z\theta_x + z^2\phi_x, \\ u_2 &= v + z\theta_y + z^2\phi_y, \\ u_3 &= w, \end{aligned} \tag{1}$$

2.2. Strain-displacement relation

$$\varepsilon_x = \frac{\partial u_1}{\partial x}, \quad \varepsilon_y = \frac{\partial u_2}{\partial y}, \quad \varepsilon_{xy} = \frac{\partial u_1}{\partial y} + \frac{\partial u_2}{\partial x}$$

$$\varepsilon_{yz} = \frac{\partial u_3}{\partial y} + \frac{\partial u_2}{\partial z}, \quad \varepsilon_{zx} = \frac{\partial u_1}{\partial z} + \frac{\partial u_3}{\partial x} \tag{2}$$

2.3. Stress-strain relations

Stress in x-, y- direction and in xy-plane can be evaluated using reduced constitutive matrix and Eq.1 as,

$$\begin{bmatrix} \sigma_x \\ \sigma_y \\ \tau_{xy} \\ \tau_{yz} \\ \tau_{zx} \end{bmatrix} = \left[\bar{Q}\right] \begin{bmatrix} \varepsilon_x \\ \varepsilon_y \\ \gamma_{xy} \\ \gamma_{yz} \\ \gamma_{zx} \end{bmatrix} \tag{3}$$

Where $\left[\bar{Q}\right]$ is reduced constitutive matrix given in appendix (A1).

Boundary condition applied in the present study is given as,

$u = 0 \quad at \quad x = 0$

2.4. Finite element approach

Displacement vector can be given as,

$$U^e = \left[u_1 \ v_1 \ u_2 \ v_2 \dots u_n \ v_n\right]^T \tag{4}$$

$$U_1 = \sum_{i=1}^{N} N_i^e u_i \quad and \quad U_1 = \sum_{i=1}^{N} N_i^e v_i \tag{5}$$

Where, N_i^e is elemental interpolation function.

Strain vector can be given as

$$\{\varepsilon\} = [B]\{U^e\} \tag{6}$$

The elastic stiffness matrix of laminated plate can be given as,

$$[K_1] = \int_{-1}^{+1}\int_{-1}^{+1} [B_c]^T [\bar{Q}][B_c] |J| \, d\xi \, d\eta, \tag{7}$$

Where, $[B_c]$ is kinematic matrix, $|J|$ is determinant of Jacobian matrix and $[\bar{Q}]$ is transformed constitutive matrix.

The geometric stiffness matrix of plate can be written as,

$$[K_{gc}] = \int_{-1}^{+1}\int_{-1}^{+1} [B_{gc}]^T [N_0][B_{gc}] |J| \, d\xi \, d\eta,$$

whre, $[B_{gc}]$ is geomeric kinematic matrix and $[N_0]$ is load vecter. (8)

$$[N_0] = [N_x \quad N_{xy}; \ N_{xy} \quad N_y],$$

Where, $[N_x] = 1/\text{thickness}$, $[N_{xy}] = 0$ and $[N_y] = 0$,

For partial edge loading $[N_x] = C_x$, where, C_x is x-coordinates of nodes of edge on which load is applied.

$$[B_{gc}] = \begin{bmatrix} 0 & 0 & \partial N_i/\partial x & 0 & 0 & 0 & 0 \\ 0 & 0 & \partial N_i/\partial x & 0 & 0 & 0 & 0 \end{bmatrix}, \tag{9}$$

2.5. Governing equation for buckling

The governing equation for buckling response of plate acted up on by in-plane partial loading may be given as,

$$\left[[K]-\lambda_{cr}[K_g]\right]\{\vee\}=0, \tag{10}$$

Non-dimensionalised critical buckling load for partial in-plane loading may be given as,

$$\lambda_{cr}=\frac{\lambda c b^2}{E_2 h^3},$$

Where 'c' is percentage of width of plate on which partial load is acted upon.

3. Results and discussion

Table 2 Shows validation study for non-dimensional buckling load for a laminated plate with simply supported boundary conditions having stacking sequence (0/90/0/90/0) under in-plane partial compressive load for plate thickness ratio a/h=20 and a/h=100. It is revealed from the Table 2, that outcomes are in good agreement with the published result.

Table 2. Validation of non-dimensionalised critical buckling load for a simply supported laminated composite plate with stacking sequence (0/90/0/90/0) subjected to in-plane partial edge loadings.

c/a	a/h = 20		a/h =100	
	Present	Ref. [12]	Present	Ref. [12]
0.2	15.4163	16.287	18.7168	19.563
0.4	17.9857	18.417	21.8363	21.590
0.6	20.5550	21.810	24.9558	25.110
0.8	23.1244	26.487	28.0752	30.205
1.0	30.3919	31.958	36.8987	35.962

Table 3. Influence of loading position on non-dimensional critical buckling load along with loading extent having stacking sequence (0/90/0/90/0) and a/h=100.

Loading Positions	Extent of partial edge loading (in % of with of plate)				
	0.2	0.4	0.6	0.8	1
Case- I	18.7168	21.8363	24.9558	28.0752	36.8987
Case-II	20.4184	23.8214	27.2245	30.6275	36.8987
Case-III	20.4184	23.8214	27.2245	30.6275	36.8987
Case-IV	25.5229	29.7768	34.0306	38.2844	36.8987

Table 3 Shows influence of non-dimensional buckling load with different loadings cases as shown in Figure 1 along with the percentage extent of loadings for laminated composite plate (0/90/0/90/0), plate span to thickness ratio a/h=100 when subjected in-plane partial uniformly distributed edge loadings. from Table 3 it is clear that, buckling load is maximum in Case-IV type of loadings which approx. 36 % more than that in case-I loading. With increase in extent of loading the critical buckling load increases in each case. For case-II and case-III the critical buckling is same due to symmetricity of loadings. For case-I critical buckling load is 97% less, when partial loading is applied on 20% of span than that when loading is on 100 % span of the plate.

Table 4. Influence of fiber orientation on buckling load for laminated plate with a/h=100, having loading condition as Case-I under 20% partial in-plane compressive edge loadings.

Fiber orientation	Non-dimensionalised critical buckling load
(0/90/0/90/0)	18.7168
(60/-60/60/-60/60)	24.3129
(45/-45/45/-45/45)	30.2476
(30/-30/30/-30/30)	26.9531

Table 4 shows influence of fiber orientations of critical buckling load for case-I i.e. when the loading is applied at 20 % of width of plate at mid for a/h=100. It is observed that laminated composite plate is more stable with 45^0 fiber orientation. The plate with cross ply (0/90/0/90/0) shows minimum critical buckling load, because well-known fact that fibers have minimum strength along transverse direction.

4. Conclusions

Buckling responses of plate subjected to partial compressive in-plane loading to different extent (%) of width of plate is analysed in the present paper. Some conclusion based on results can be given as:

- Laminated composite plate with higher span to thickness ratio shows higher non-dimensional critical buckling load.
- With fiber orientation 45^0 laminated composite plate is more stable.
- Laminated composite plate with cross play shows minimum value of non-dimensional critical buckling load.
- When laminated plate is acted upon by symmetrical loading system then buckle easily.

References

[1] Timoshenko S P and Gere J M 1961 *Theory of elastic stability* 2nd ed. *New York: McGraw-Hill.*
[2] Leissa A W 1987 *Appl. Mech. Rev.* **40** 579-591
[3] Chia C Y 1988 *Appl. Mech. Rev.* **41** 439–51
[4] Brown C J 1989 *Comput Struct.* **33** 1325–7
[5] Kaldas M M and Dickinson S M 1981. *J Sound Vibr* **75** 151–62
[6] Deolasi P J and Datta P K 1995 *Comput. Struct.* **54** 73–82
[7] Daripa R and Singha M K 2009 *Thin Walled Structures* **47** 601-605
[8] Cagdas I U and Adali S 2011 *J. Thermoplast.* **26** 193-215
[9] Shariyat M and Asemi K 2014 *Composite Structures* **112** 397-408
[10] Patel S N and Sheikh A H 2016 *Int J Numer Meth Eng* **17** 322-338
[11] Kumar R Ramachandra L S and Banerjee B 2016 *Proc Indian Natn Sci Acad.* **82** 271-288
[12] Dey T and Ramachandra L S 2014 *Int J Nonlin Mech* **64** 46-56

IOP Conf. Series: Materials Science and Engineering **814** (2020) 012027 doi:10.1088/1757-899X/814/1/012027

Application of Light Interference Technique for Locating Groundwater Veins

Vijay V Muthekar[1], Pratik S Sasane[2], Rupam R Phalak[2], Shivani S Gour[2] and Avinash G Kharat[3]

[1]Assistant Professor, Civil Engineering Department, JSPM NTC, Pune, MH, India
[2]UG Students, Civil Engineering Department, JSPM NTC, Pune, MH, India
[3]Director (Academics), JSPM's RSCOE, Tathawade, Pune, MH, India

Email: vvmuthekar@gmail.com

Abstract. Water is one of the most needful resources on Earth. The increasing population, growing urbanization, and enhanced industrialization over the years have led to increased demand for water all over the world, causing a scarcity of resources and thereby made the world dependent on other sources of water than surface water. Among all, a majorly available and reliable source is groundwater. With diminishing groundwater levels, it endures the need for precise groundwater exploration. The groundwater investigation has been a more significant issue during the last few years due to intolerable distraction, depreciating water quality, and emerging pollution threats to potable resources. To overcome limitations of contemporary groundwater exploration methods present study proposes a new technique utilizing the 'Light Interference Technique,' a concept from fundamental Physics. An instrument, NaAvmeter, developed based on Light Interference Technique, was used for groundwater exploration in the current study. A literature survey shows that earth radiations are much dominant over groundwater locations. Empirical investigations were carried out in the Pune city under various site conditions to check the feasibility of application of 'Light Interference Technique' for groundwater exploration in the Deccan trap flood basalt terrain. An attempt is also made in the current study to propose a simplified groundwater exploration method using NaAvmeter; an instrument developed based on the 'Light Interference Technique.'

1. Introduction

There is a growing concern over freshwater all over the world. It is unavoidable that the pressure is and will be on the detection of subsurface water storage [1]. From the literature, it has observed that most groundwater (GW) exploration work that has been carried out from ancient times involves fieldwork and the use of knowledge gained from ancestors, limited to the local area. However, from the starting of the 19th century, significant updating in utilizing the various GW exploration methods has been observed [2]. As the cost of water exploration through modern methods is invariably unaffordable for the average person, traditional knowledge of exploring water sources is still in vogue among many rural and urban communities. In the recent past, science and technology have taken long strides in exploring GW as a result of which many techniques are available now to identify the GW [3]. The main focus of current research is to develop a GW exploration method that utilizes modern

IOP Conf. Series: Materials Science and Engineering **814** (2020) 012027 doi:10.1088/1757-899X/814/1/012027

techniques. The present study attempted to use an instrument designed based on light interference technique for GW exploration in the study area as an alternative to current GW exploration methods.

2. Background of GW exploration

The literature survey shows that many methods/techniques exist for GW exploration. Various techniques, methods available for GW exploration are as shown in Figure-1.

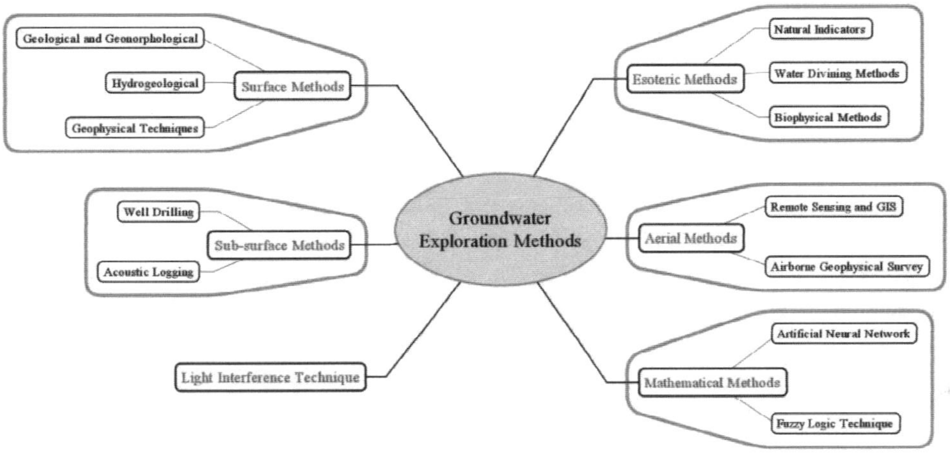

Figure 1. GW Exploration Methods

2.1. Esoteric method

An esoteric method is practiced from ancient times. It considers the human body as the sensor for the detection of earth radiations caused due to movement of GW. It utilizes various instruments such as dowsing rods, Y- twig, coconut, pendulum, etc. The excessive Earth radiations at the surface of the GW location cause deflection to the instruments. Also, it makes use of knowledge of bio-indicators such as flora and fauna for the identification of GW at places.

2.2. Subsurface method

It is destructive testing, which involves borehole logging, drilling of cores, and interpreting the obtained data to explore the GW. Stratification is determined from the acquired data, and accordingly, traces of GW are determined at particular locations. Accuracy of results is high but is uneconomical considering field test setup and data interpretation.

2.3. Surface method

Surface method is a non-destructive type of testing. Unlike subsurface method, stratification is determined by using geophysical techniques such as ERT, SRT etc. without disturbing to the ground surface. Along with the other applications, one of the significant forms of the surface method is GW exploration. It ensures much reliable and accurate results of GW exploration due to which it is widely accepted all over the world. Undoubtedly cost is a significant disadvantage.

2.4. Aerial method

It includes airborne surveys and electromagnetic surveys for identification of GW potential zones. The data from such studies and other mapping details like thermal imaging, topography, drainage pattern, etc., are overlapped together, and a thematic map is generated. Thematic map shows the exact locations of GW potential zones. It covers large area. The method is uneconomical in terms of the requirement of large setup, skilled supervision for data collection, and interpretation.

2.5. Mathematical modeling

Mathematical modeling for GW exploration comprises of artificial neural networking and fuzzy logic modeling. It is data-based coding that works on a system of input and output layers. It performs an input-output mapping via a series of simple processing nodes or neurons. A layer represents a set of parallel processing units (or nodes). Fuzzy-logic modeling gives the output in the range of 0 to 1, where the nearest value to 1 shows the maximum probability of the GW zone and vice versa. This method requires substantial data sets and highly skilled personnel for data interpretation. The accuracy of results highly depends upon the input data.

The research findings related to the various GW exploration methods have bought out certain characteristic deficiencies while adopting them in the field. It appears that the contemporary exploration methods are more efficient in terms of accuracy, speed, field coverage, field data interpretation that gives a realistic location for GW yield. Although traditional methods are offering human-body specific results and have limitations regarding quantification, the success ratio for utilizing these methods in the field is significant.

Sr. No.		GW Exploration method	Cost for field work	Effect of Season	Field Test setup	Cost for data interpretation	Mathematical model	Dependency on other test	Field Result Repeatability	Result Accuracy
1		**Esoteric Methods**								
	a.	Water Divining	Less	Slight	Small	Small	Not Available	High	Medium	Medium
	b.	Biological Indicators	Less	Slight	Small	Small	Not Available	High	Less	Less
	c.	Biophysical Methods	Less	Slight	Small	Small	Not Available	High	Medium	Medium
2		**Surface Method**								
	a.	Geological and Geomorphological	Medium	Medium	Small	Large	Available	Less	Medium	Medium
	b.	Hydrogeological	Medium	High	Small	Large	Available	Less	Medium	Medium
	c.	Geophysical Techniques	Medium	High	Medium	Large	Available	Less	High	High
3		**Subsurface method**								
	a.	Test Drilling	High	Medium	Large	Medium	Available	High	High	High
	b.	Tracer Technique	High	High	Medium	Less	Available	Less	Medium	Medium
	c.	Geophysical Logging	High	High	Large	Large	Available	Less	High	High
4		**Aerial Method**								
	a.	Remote Sensing and GIS	High	Slight	Large	Large	Available	High	High	High
	b.	Airborne Geophysical survey	High	Slight	Large	Large	Available	High	High	High
5		**Mathematical Methods**								
	a.	Artificial Neural Network	Small	Slight	Small	Large	Available	High	High	High
	b.	Fuzzy Logic Technique	Small	Slight	Small	Large	Available	High	High	High

Legends:

Most Preferred **Acceptable** **Not Acceptable**

Figure 2. Comparison of the various GW exploration methods

Observing the essentiality of GW utilization, the development of various simple methodologies, instruments is a societal need. Currently, available different destructive, non-destructive methods, techniques, devices for GW exploration are much lengthy, expensive to operate, time-consuming, and complicated to interpret the results. Also, if there are any special field issues, then the use of various techniques in combination increases the cost, complexities in result interpretation and reliability on other field or laboratory tests. Handling several experimental setups on the ground at a time is also very cumbersome. Various researchers have attempted combinations of Esoteric,

IOP Conf. Series: Materials Science and Engineering **814** (2020) 012027 doi:10.1088/1757-899X/814/1/012027

Surface, Subsurface, Aerial, and Mathematical methods of GW exploration for identifying the zones of GW with certain limitations. Figure-2 shows the comparison of various GW exploration methods.

A primary objective of the proposed study is to explore GW with accuracy and economy, taking into accounts the GW exploitation due to increased demand. While developing this objective, the literature review shows that in current practices, many techniques exist for GW exploration. All exploration techniques have advantages and disadvantages in a particular domain (Figure-2). An attempt is made in the current study, to develop a method for GW exploration using the interaction of light with earth radiation for investigation of GW and which will also overcome limitations for contemporary practices.

3. Light Interference Technique

Many researchers have exercised the combination of various GW exploration techniques successfully. However, site conditions, financial aspects, the requirements of accuracy, and urgency of the result required, knowledge of an expert are the various factors that limit the execution and correctness of the exploration method employed. A fundamental concept of 'Light Interference' from Physics was observed to have the potential to be utilized for the exploration of GW. The idea was further elaborated to develop an instrument, NaAvmeter, for GW investigation [4]. Prima facie it has observed that the limitations of contemporary GW exploration methods have addressed while developing and utilizing the instrument. Figure-3 shows the field setup of the NaAvmeter.

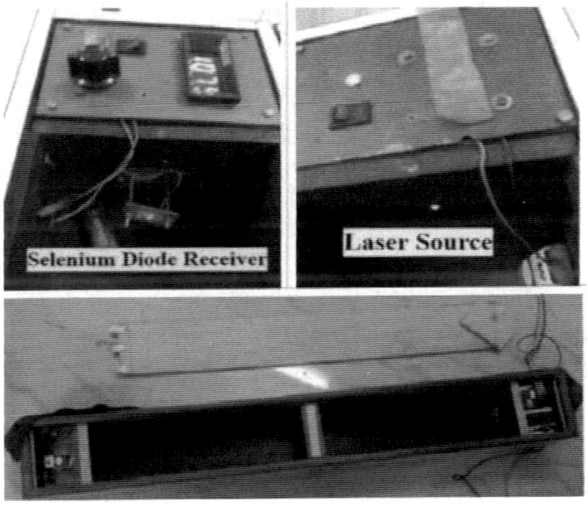

Figure 3. Field setup of the NaAvmeter

3.1. Construction of NaAvmeter

NaAvmeter is a rectangular wooden box having a 1.0m length between the source and the receiver on the sides of the box. Materials that are attenuating the magnetic field are avoided, such as metals or polymers. The gap between source and receiver is for the interaction of the light beam emitted from the source. The source is a monochromatic coherent Laser beam produced by a semiconductor diode which travels and falls on selenium photodiode used as a detector. The selenium diode and current meter are connected to measure the current of the laser beam falling on the selenium diode [5].

3.2. Earth radiations

In the Earth's crust, there exist several natural fields such as Gravitational field, Magnetic field, Electric field, Radioactive field, Seismic field, Geothermal field, Geochemical field [6]. A variation in one or more of these fields may have some detectable effects. Some radiations are excited from the Earth's surface toward the underground, and after reaching a specific impedance interface, they reflect and return to the ground surface. Thus, by identifying these waves, the exact depth of the impedance interface can be calculated. The transmitted radiation pulses are a complex combination. After reflecting from underground to the ground surface, waves reflect the various underground media characteristics [7]. One school of thought had elaborated that there exist such earth radiations at the location where GW is existing [5]. These earth radiations are having electromagnetic nature and are measurable [8, 9]. Scientists have observed that at GW zones, these earth radiations are significant, and at the location where GW is absent, these radiations are not substantial. Due to the various absorption effects of ions contained in the GW, it gives rise to the Earth's radiation. Specific metallic ions or metal particles contained in GW or flowing in streams through the rock mass, sand, soil strata of the natural ground are inducing an electromagnetic field [7].

3.3. Concept of NaAvmeter based on Light Interference Technique

The reflection and refraction of light are well-established from the literature. Laser exhibits the same behavior as that of ordinary light despite having unique properties like coherence and monochromaticity. The Earth's surface has an electromagnetic field in addition to various complex Earth radiations. Literature mentions that light in any form can bend when it passes the neighborhood of an electrically or magnetically charged object [10]. One stream of thought is that, in the presence of the electromagnetic field, the 'Delbruck scattering' effect causes the Laser photons to disintegrates into charged electron and positron for a shorter period. These charged electron and positron recombines into Laser photons after having interaction with the electromagnetic field. There is more probability of producing more photons due to such recombination causing scattering of light and thus causing light splitting into several rays [11]. Another stream of thought is that, at GW locations, the existence of earth radiations is evident, which may cause electromagnetic radiation tending to deflect the laser beam [5]. By this strong motive, an instrument, NaAvmeter, has been developed by combining the concept of Light Interference at the field to identify the locations having perturbations in the geo-electromagnetic field, resulting in GW locations.

NaAvmeter has a gap between the Laser light source and a selenium photodiode detector. Selenium diode converts light into electric current, and the indicator displays the output in terms of intensity of electrical current or intensity of light. In the case of perturbation in geo-electromagnetic fields are existing in this gap, then interaction occurs between photons of Laser light and geo-electromagnetic fields, which changes the output of the detector. If no perturbations in the geo-electromagnetic fields are present in the gap of source and detector, then no interaction occurs; hence, the output of the sensor remains the same on the field [4].

4. Use of NaAvmeter to identify GW locations

The methodology utilized for locating GW zones comprised of water dowsing that has been well established since ancient times. L-rod dowsing is well-known in the GW location study. The well-established electrical resistivity method (ERM) can be utilized to confirm the existence of GW. The presence of GW is evident by data processing of ERM results using master curves. Similarly, NaAvmeter can also be used in the identified locations to explore the existence of GW. The following procedure shall be adopted at the field to locate GW zones using NaAvmeter [9].

- A grid of around 10m×12m shall be plotted around the location under consideration.
- Five (05) profile lines at an interval of 3.0m aligned in North-South direction and each profile line with eleven (11) grid points at a range of 1.0m shall be marked.
- NaAvmeter readings shall be recorded at each grid point.
- Following the mentioned procedure, readings shall be recorded for each grid point of each profile. For one location under observation, five profile readings shall be recorded. In this way, field readings using NaAvmeter shall be filed at all sites under consideration in the field.

- Duration of keeping NaAvmeter at a point while taking readings is one of the significant affecting factors. Experience gained through field observations, saturation in the NaAvmeter readings was observed after 2-3 minutes keeping at the place. Hence, readings shall be recorded after three (03) minutes of keeping the NaAvmeter at the location.
- NaAvmeter readings then shall be graphed using suitable contour plotting software for the readings observed for all rotations on that particular grid point in the study area. The graph plots show distinct variation in NaAvmeter readings based on which GW location can be identified.

5. Application of NaAvmeter in the field

The basic working of NaAvmeter has discussed above. The emission of earth radiations and interacting with the laser beam in the NaAvmeter causing a change in the value of current in microampere on the selenium photodiode detector of the NaAvmeter is, in fact, a very complex phenomenon. It has already established that earth radiations are in existence everywhere on the Earth's surface, however at the locations of GW zones, such earth radiations have measurable perturbations. Authors have used NaAvmeter successfully at various locations in the Pune region, i.e., in the Deccan trap flood basalt region.

5.1. The physical setting of the Study Area in the Pune region

The study area is situated in the valley of Mula–Mutha rivers on typical rugged Deccan volcanic topography. The mountain and hill range in the West, river basins with undulating topography at the center, and in the East highland plateaus with more extensive river basins mark the physical landscape of the region [12]. In the West of the study area, the average height is about 900m above sea level [12]. The entire study region of Pune is mainly formed by two major parts of the Western Ghats, and 2,000m thick layers of Deccan volcanic basalts formed 65 to 70 Mya during the Late Cretaceous period [13]. The plateau after the catastrophic phase of faulting has remained relatively stable and has undergone a succession of cycles of erosion [14]. The region has a basaltic base on which there are alluvial deposits in river valleys on the terraces and old flood plains. Fault lines have scattered throughout the Pune region, and these areas are minor earthquake-prone [12]. Alluvial deposits have observed on both the banks of Mula and Mutha rivers. The alluvium is about 5 to 15m thick. It has a lateral extent of over 2000m on both sides of the Mutha river. This alluvium also occurs in patches in the Paleo depressions [12].

The occurrence of GW in limited quantity within hard rock terrains such as volcanic Deccan Traps in Western India is well known. The main aquifer horizons of the Pune region are vesicular, fractured, and weathered basalts [15]. The inter-trap formations, such as red bole, also form aquifer horizons at places. The study area has the domination of zeolites as cavity fillings and veins in the basalt. Pores and voids of the natural zeolite structure have characterized by water molecules [16]. Seasonal GW level fluctuation analysis between pre-monsoon and the post-monsoon season for the year 2011 reveals that the rise in GW levels has found between 0.20m to 12.35m. In the major part of the Pune region in North-Western and South-Eastern part, GW level fluctuation is varying from 0 to 2m. The central and eastern part of the Pune region has a variation between 2m to 6m. GW level fluctuation of more than 6m has observed in isolated pockets. Falls in the GW level between pre and post-monsoon season of the year 2011 has also seen at various places, and it varies from 0.05m to 2.00 m [17].

5.2. Locations identified for the current study

From the study area, two (02) sites have identified for conducting experiments. Details of the site have given in Table-1. An empirical investigation using NaAvmeter had conducted at the above-identified locations. Result interpretations using 'Surfer,' contour plotting software has been discussed in the following section.

IOP Conf. Series: Materials Science and Engineering **814** (2020) 012027 doi:10.1088/1757-899X/814/1/012027

Table 1. Details of field Locations.

Location No.	Field Location	Satellite Locations		
	Plot Number	Latitude	Longitude	Altitude (m)
1	RH-6 D	18°34'0.9"	73°32'40.9"	627
2	D-7A	18°33'45.1"	73°32'20.5"	736

5.3. Interpretation of field investigation

The NaAvmeter readings show a decrease in current readings over the GW locations and show incremental readings over the non-GW zone. Figure-4 shows the contour plots of the NaAvmeter readings where contour legends are the representative of the same. On the color scale, blue color shows lesser NaAvmeter readings, whereas the red color shows increased readings. Literature indicates that there is a dominant zone of electromagnetic radiation at places of GW locations. It has also been seen in the form of a significant decrease in readings of NaAvmeter. From the field observations regarding geology, geohydrology, biological indicators, etc. confirms the GW potential at the locations. The same can be seen in the contour plots via distinguishing color scale. Blue colored zone is a clear indicator of GW zones at the places, whereas the red color zone indicates the probability of the non-existence of the GW zone. By considering all the parameters in the account, the location of the GW vein and its width can be easily identified.

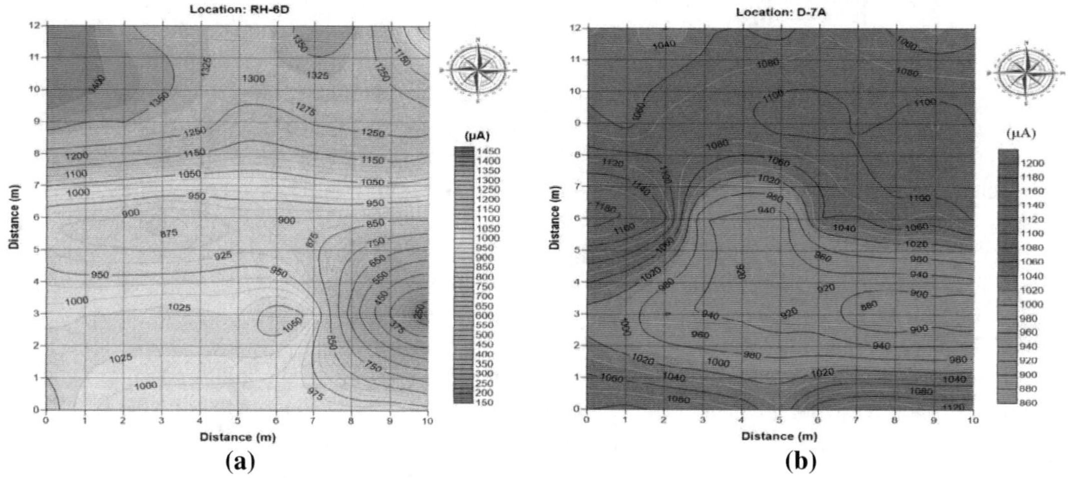

(a) (b)

Figure 4. GW zone identification using NaAvmeter at two locations under consideration

6. Summary

The two locations from the study area have been selected on the basis of availability of data and other geological, geomorphological, and geohydrological details. The preliminary investigation includes the collection of aquifer data, morphological data, biological indicators, availability of existing bore well point and its working condition, rainfall data, and respective GW fluctuation levels, etc. A brief study on the obtained data confirmed the presence of GW at the locality. NaAvmeter readings at both locations were taken by procedure elaborated in the above section. The obtained readings were then tabulated and interpreted using a contour plotting software 'Surfer.' Taking into account, contour plots and acquired field data, the exact location of GW has been identified.

7. Conclusion

Field experiments were conducted in the study area using NaAvmeter. One school of thought is that electromagnetic field perturbations due to the availability of GW may cause scattering of light while passing through such zones, which causes a decrease in current readings. The NaAvmeter readings recorded at the identified GW zone shows a sharp reduction in comparison with non-GW zone over remaining locations in the study area. The present work interprets the results obtained through NaAvmeter to provide a quantifiable result to identify the zones of GW. The contour plots give the quantification of GW veins in terms of its width, which can be used further to provide a borewell point for GW exploration. The current work contributes to the advancement of knowledge regarding the investigation of GW resources. NaAvmeter can be used successfully to identify the GW locations. Use of NaAvmeter avoids the practical limitations of contemporary GW exploration methods/instruments.

References

[1] Joshi R R, Kulkarni U D, Vartak A V and Khan I A. Dakargalam – An Ancient Water Exploration Technique Based on Varahamihira's Brihat Samhita. *Asian Agri-History*. 2013; **17**(2): p. 175-181.
[2] Badrinarayan T S. Groundwater Exploration- *An Introduction Kollidam*; 2009.
[3] Mohapatra P K. Is water divining a science. *Orrisa Review*. 2008 **June**: p. 55-56.
[4] Dharmadhikari N P, PhD thesis on Detection of Geopathic Stress Zone and It's Effects on Human System. *PhD thesis. Pune, Department of Environmental Science*; 2011.
[5] Dharmadhikari N P, Mahajan C S, Hambarde S M, Patil M S and Kharat A G. 2018 *Int. Journal of Computer Engineering and Applications* **12** 1-4.
[6] Chadwick D G and Jensen L. The detection of magnetic fields caused by groundwater. Study report. Utah State: *Utah Water Research Laboratory*; 1971. Report No.: **PRWG78-1**.
[7] Jishan H., 1995 *J. Central South University Technology* **2** 1-6.
[8] Thurnell-Read J. Geopathic Stress and Subtle Energy Cornwall: *Life-Work Potential*; 2006.
[9] Dharmadhikari N P, Muthekar V V, Mahajan C S, Basavaiah N, Kharat A G and Barde S I, 2019 *Journal of Applied Geophysics*. **171** 1-11.
[10] Kim J Y, Lee T. Light bending by non-linear electrodynamics under strong electric and magnetic field. 2011.
[11] Khan. https://www.quora.com. [Online].; **2017** [cited 2018 February. Available from: *https://www.quora.com/Can-you-bend-a-photon-with-a-magnetic-field*.
[12] Bhakare P. Ph.D. thesis on Spatio-temporal changes in population structure of Pune division of Maharashtra. *Kolhapur, Department of Geography*; 2010. Report No.: **10603/9907** (http://shodhganga.inflibnet.ac.in/handle/10603/9907).
[13] Basavaiah N, Satyanarayana K V V, Deenadayalan K and Prasad J N. Does Deccan Volcanic Sequence contain more reversals than the three-Chron N–R–N flow magnetostratigraphy? a palaeomagnetic evidence from the dyke-swarm near Mumbai. *Geophysical Journal International*. 2018 February; **213**(3): p. 1503-1523.
[14] Gabale S M. Impact of urbanization on geomorphic environment of Pune and surrounding. *PhD Thesis. Pune: Department of Geography*, Tilak Maharashtra Vidyapeeth; **2016**.
[15] Pawar N J. 1993 *Journal of the Geological Society of India* **41** 119-131.
[16] Querol X, Moreno N, Uman˜a J C, Alastuey A, Herna'ndez E and Lo'pez-Soler A, 2002 *International Journal of Coal Geology* **50** 413-423.
[17] Mishra S S P. *Groundwater Information of Pune district, Maharashtra*. Nagpur; 2013. Report No.: **1810/DBR/2009**.

IOP Conf. Series: Materials Science and Engineering **814** (2020) 012028 doi:10.1088/1757-899X/814/1/012028

Stabilization on Expansive soil using sea shell powder and Rubber powder

Vinod B R[1], Shobha R[1], A B Raghavendra[2], Rakesh M[2], Pallavi S[2]

[1] Assistant Professor, Department of Civil Engineering, BMS Institute of Technology and Management, Bangalore-560064, India.
[2] UG Students, Department of Civil Engineering, BMS Institute of Technology and Management, Bangalore-560064, India.

E-mail: vinodbr@bmsit.in

Abstract. BCS, type of expansive soil which is responsible for the volumetric changes with the change in the moisture content due to the excessive presence of montmorillonite. The stabilization of soil is the process of enhancing the properties of expensive soil and makes it stable to withstand the load acting on it. The stabilization is done by controlled compaction and addition of different types of stabilizers and admixtures. Hence this stabilization is done by various admixtures like cement, fly ash, bagasse, coir fibre, granite dust, rubber powder, sea shell powder etc. this paper represents the study of rubber powder and sea shell powder as different admixture to refine BCS properties. The unconfined compressive strength of BCS is found by adding sea shell powder and rubber powder as admixture, here this paper presents the results of an attempts that are performed.

1. Introduction

1.1 Black Cotton Soil (BCS)
Expansive soil are also known as black cotton soil. Expansive soil are mainly concentrated in the region of central India and Deccan plateau , cotton is more efficiently Grown in this type of soil . The prime factor for the development of BCS is due to the climatic conditions along with parent rock.

1.2. Soil stabilization
Improvement of the sub base to the load capacity to the support of pavements and foundation is performed by stabilization in engineering properties .The improvement in soil strength and resistance is increased by soil stabilization by water bonding method of the particles together. The basic and normal stabilization forms are drainage and compaction (if water enters out of wet soil, it gets more grounded). The different procedure is by expanding degree of molecule measure and further improvement can be accomplished by adding folios to weak soils. Soil adjustment can be cultivated by few techniques. Each and Every one of these strategies comes into two general classifications in particular

- The Mechanical stabilization.
- The Chemical stabilization.

1.3. Mechanical stabilization

By altering the physical nature of the native soil particles, through incorporating different physical properties like barriers and nailing or by induced compaction and vibration technique, stabilization can be achieved

1.4. Chemical stabilization

The Soil stabilization essentially relies primarily upon synthetic responses accordingly soil minerals (pozzolanic materials) and stabilizer (cementitious material) to accomplish the ideal quality.

1.5. Stabilization by using different types of admixture is as follows.

Lime , cement , fly ash , bituminous , rice husk , thermal , electrical , geo textile , recycled products , fabrics , waste products .

2. Materials and methodology

In the present work some of the tests were carried out by adding Sea Shell Powder (SS) and rubber powder as an admixture by varying its %.

2.1. Materials used

- Black Cotton Soil.
- Sea Shell Powder.
- Rubber Powder.

2.1.1. Black cotton soil (BCS)

Figure 1. Black cotton soil.

They exhibits high swelling of soil under rainy season and shrink under Dry condition .The sample (Figure 1) was collected in the place near koppal district , Bannikoppa village , Karnataka .The following results under various test @ 2m depth .The soil was pulverized to pass through micron sieve of IS -425, air dried and oven dried @ 110 C before testing , general properties are shown table 1.

2.1.2. Sea shell powder

Shell powder Seashells are the exoskeletons of mollusks, for example, snails, shellfishes, clams and numerous others. The shell powder is the buildup powder that is made by powdering of the normally accessible hard shells which is the exoskeleton of Mollusks. Studies uncover that lime stabilization is one of the conventional and monetary procedures and shows a superior improvement. Such shells have various layers and are composed mostly of calcium carbonate with only a small quantity of protein. The sea shells were finely grained and the sea shell powder passed through IS 150 Micron sieve and retained on IS 75 Micron Sieve. The sea shells are cleaned and dried and are heated at 100°c to 250° shown in Figure 2. General important properties and chemical composition of sea shell powder are shown in table 2, 3 respectively.

Table 1. Properties of Black Cotton Soil.

Sr. No	Property		Value
1	Specific gravity (G)		2.68
2	Particle size distribution	Gravel	0.53
		Clay	54.57
		Silt	20.02
		Sand	24.89
		Cu	10
		Cc	1.5
3	Free Swell Index (%)		37
4	Liquid Limit (%)		74.2
5	Plastic Limit (%)		37.57
6	Plasticity Index (%)		36.63
7	Shrinkage Limit (%)		12
8	OMC(%)		27
9	Maximum dry density(kN/m^3)		13.11
10	Unconfined Compressive Strength(qu, kN/m^2)		158.06

Table 2. Properties of Sea Shell Powder

Sr. no	Properties	Value
1	Specific gravity	2.45
2	Liquid Limit (%)	42
3	Plastic Limit	Non-Plastic
4	Optimum Moisture Content	18.28
5	Maximum Dry Density	14.2

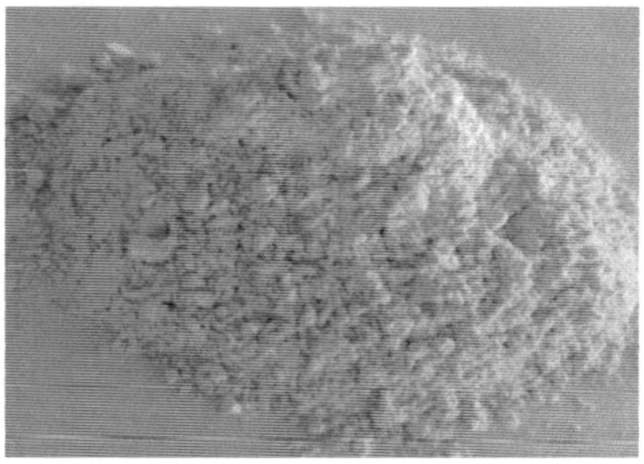

Figure 2. Sea Shell Powder.

2.1.3. Rubber powder

For the improvement of engineering properties of the problem crumb rubber powder , clay was preferred as an additive and its properties are shown in table 4 and table 5. Disintegrate rubber is the term normally preferred to reuse rubber from the Truck scrap tires. During the reusing procedure steel and fluff is evacuate leaving tire rubber with a granular consistency. kept handling with a granular and/or wafer factory perhaps with the guide of mechanical methods, diminish the size of the molecule further Right now rubber is frequently utilized as an added substance in bituminous solid blend. As of now India is confronting a gigantic rubber waste removal. To address the above concerns Crumb Rubber Powder is utilized as an added substance to improve the building properties of dark cotton soil. The Crumb Rubber Powder which is utilized in the examination are of 1.18mm down size

(IS strainer) (Figure 3). In the present examination, Crumb Rubber Powder is utilized as an added substance in the preliminary investigation to get wanted engineering properties.

Table 3. Chemical composition of Sea Shell Powder.

Sr. no	Oxide	%
1	Sio_2	1.60
2	Al_2O_3	0.92
3	Cao	51.56
4	Mgo	1.43
5	Na_2O	0.08
6	K_2O	0.06
7	H_2O	0.31
8	LOI	41.84

Table 4. Properties of rubber powder

Sr. No	Properties	Values
1	Specific gravity	0.92
2	Water absorption	2 %
3	Dry density (kN/m^3)	12.54
4	Moisture content (%)	19.65

Figure 3. Crumb Rubber Powder

Table 5. Chemical Properties of Rubber powder

Sl. No	Material/Element	Mass %
1	Rubber	54
2	Carbon black	29
3	Sulphur	1
4	Textile	2
5	Oxidize zinc	1
6	Additives	13

3. Methodology

The expansive soil was admixture with Sea shell powder of varying % of 6 %, 8, 10, 12, 14 and rubber powder of 2, 3, 4, 5 and 6 % also and admixed with BC soil. The specific gravity was also conducted as per IS 2720-1980, for expansive soil sea shell powder. Grain size analysis was conducted as per IS 2720 (Part 4)-1985 for expansive soil. Atterberg's limits were conducted as per IS 2720(Part 5) 1985 and IS 2720 (Part 6) 1972, for expansive soil independently and by adding admixtures of sea shell powder (6, 8, 10, 12 and 14 %), rubber powder (2, 3, 4, 5 and 6 %).

IOP Conf. Series: Materials Science and Engineering **814** (2020) 012028 doi:10.1088/1757-899X/814/1/012028

3.1. Compaction test

The OMC and MDD of expansive soil alone and admixed with sea shell powder and rubber powder are determined by Mini Compactor test, having an internal diameter of 3.81cm and 10 cm height and 4.61cm external diameter .It has a sample mould assembly with a removable collar of 3.50 Height and a detachable base plate . Height of fall is 10cm, with hammer weight of 1kg with 36 blows/layer. The compaction test were done to determine OMC and MDD with varying % (6 , 8 ,10 , 12 and 14 %) of sea shell powder and (2, 3, 4, 5 and 6 %) of rubber powder.

3.2. Unconfined strength of soil in compression

The Unconfined strength of soil in compression examines are conducted on cylindrical specimens having height of 76mm and 38mm. The test was conducted as per IS 2720 part 10. Soil passing 4.75mm is used for preparation of moulds. Exact amount of soil is taken with respect to MDD of soil obtained through compaction and mixed thoroughly with water which is calculated from OMC and specimen is prepared. The sample is prepared by varying % (6, 8, 10, 12 and 14 %) of sea shell powder separately and the test was carried out of different curing periods. Again the samples are prepared by varying % (2, 3, 4, 5 and 6 %) of rubber powder separately and the test was carried out for different curing periods. The samples with optimum dosage of sea shell powder and rubber powder are prepared by mixing thoroughly desired amount of expansive soil then it is mixed thoroughly to obtain uniform moisture and then specimen is prepared. The required amount of water was calculated and added. After preparation of moulds, are wrapped with air tight bad and kept for curing in desiccators for 0 and 3 days. All the specimens are prepared with OMC and MDD as obtained by mini compaction (Figure 4).

Figure 4. Mini-Compaction

Figure 5. Sample kept for testing

4. Results and discussions

Table 6. Results of tests conducted in the present study

Sr. no	Program	Compaction test		UCS (kN/m^2)	
		γ_d (Kn/m^3)	ω (%)	Un-cured	3 Day
1	Expansive soil	13.9	27	158.06	157.59
2	ES+6 % SS	11.9	19.4	138.24	131.76
3	ES+8 % SS	12.55	17.2	175.68	156.47
4	ES+10 % SS	15.25	20.2	234.37	183.92
5	ES+12% SS	15.79	19.3	258.9	237.10
6	ES+14% SS	15.76	17.4	216.44	182.59
7	ES+2% CRP	11.38	30.82	223.95	226.61
8	ES+3% CRP	12.03	33.7	202.62	199.95
9	ES+4% CRP	12.36	34.76	239.94	221.28
10	ES+5% CRP	13.52	36.89	271.94	250.61
11	ES+6% CRP	11.03	38.62	245.28	189.29
12	ES+12% SS+ 2% CRP	15.01	22.2	183.96	201.43
13	ES+12% SS+ 3% CRP	16.05	21.79	220.23	257.84
14	ES+12% SS+ 4% CRP	17.25	20.37	255.15	279.33
15	ES+12% SS+ 5% CRP	18.32	19.87	279.32	297.61
16	ES+12% SS+ 6% CRP	16.29	22.47	236.35	183.98

4.1. Variation of compaction of soil characteristics on different % of sea shell powder.
The compaction test was conducted on the BCS mixed with Sea shell powder of varying % to obtain the optimum sea shell powder content for the soil which is taken in this study.

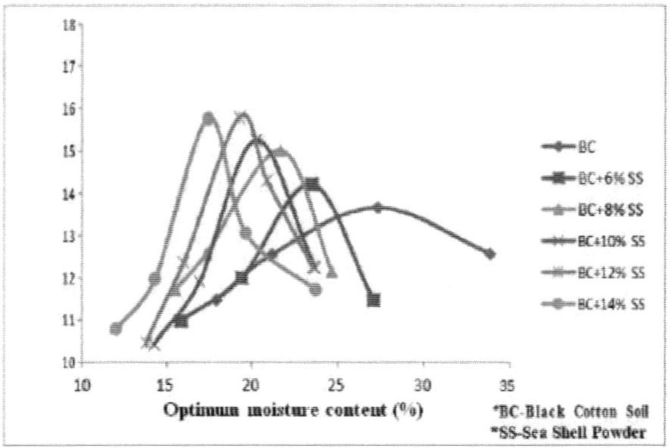

Figure 6. Differential of and MDD for different % of sea shell powder.

4.2 Variation of compaction soil characteristics on various level of rubber powder and ideal sea shell powder.

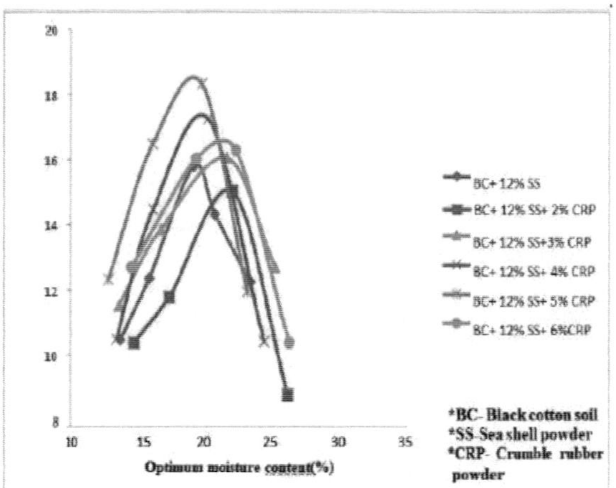

Figure 7. Differential of OMC and MDD for ideal sea shell powder and distinctive % of rubber powder.

Compaction test was directed with various rate (2, 3, 4, 5 and 6 %) of disintegrated rubber powder and ideal sea shell powder. Compaction curve is consequently acquired for expansive soil with various level of rubber powder. Figure 7 Shows that there was an expansion in dry unit weight and abatement in the moisture content with ideal % of sea shell powder and differing level of rubber powder. Impact of expansive soil with rubber powder is controlled by including rubber powder at various rate.

Consequently right now is seen that there is a significant increment in the dry density when the level of added soil to rubber powder expanded from 2 to 5 % in light of the fact that the dry density of soil is expanded up to 5 % and afterward it is diminished. Thus the rubber powder of 5 % is considered as the ideal rubber powder content from the thought of expanded dry density of the expansive soil.

4.3. Impact of Sea shell powder upon unconfined compressive strength of BCS.

Unconfined strength of soil in compression test is led broadly for expansive soil by treating with various level of sea shell powder. In the present examination the test is finished by blending the BCS with 12 % ideal Sea shell powder relieved for a time of Un-Cured and 3 day so as to decide the effect of curing on unconfined compressive soil strength of soil. Stress versus strain chart is plotted to decide the unconfined compressive strength of the BCS.

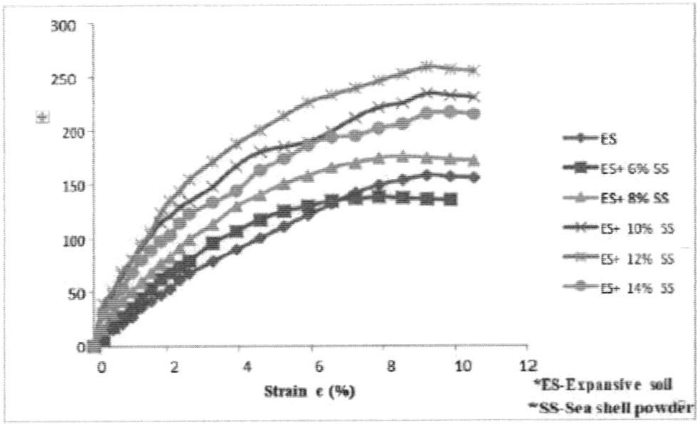

Figure 8. Stress-Strain attributes of BCS for various % of Sea Shell Powder for Un-Cured.

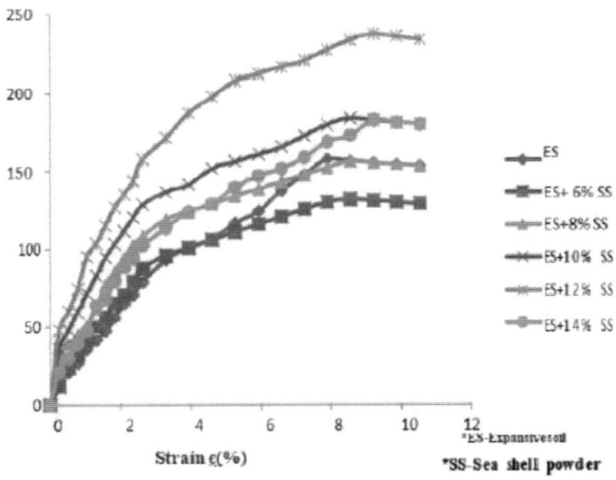

Figure 9. Stress-Strain Attributes of BCS for various % of sea shell powder for 3 Days of curing.

Figure 8 and 9 shows the graphical representation of the unconfined strength of soil in compression of the soil when mixed with sea shell powder (6, 8, 10, 12 and 14 %) for a curing period of Un Cured and 3 days respectively. Optimum Sea shell powder content taken in this study is 12 % which increase the strength of BCS, beyond which the strength of the treated soil decreases at different curing periods. The strength gained for other proportions are less when compared to obtained %. Hence here 12 % of sea shell powder is considered as optimum for BCS.

Figure 10 shows the variation of unconfined strength of soil in compression expansive soil ad mixed with sea shell powder (6, 8, 10, 12 and 14 %) for different curing periods. Compared to Un-Cured and 3 days curing period. Addition of sea shell powder, increases the bonding soil between clay particles there by it increases the cohesive strength of the soil.

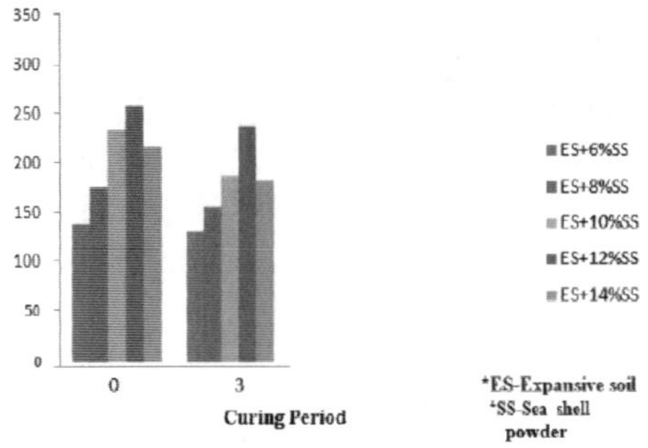

Figure 10. Unconfined strength of soil in compression of expansive soil admixed with Sea shell powder for different curing period.

Table 7. Unconfined compressive strength of BCS treated with sea shell powder for different %

% of Sea shell Powder	Un Cured	3 Days Curing
0	158.06	157.59
6	138.24	131.76
8	175.68	156.47
10	234.37	183.92
12	258.90	237.10
14	216.44	182.59

The above Table 7. Shows the unconfined strength of soil in compression of soil treated with different % of Sea shell powder at different curing period. Here the maximum strength is observed at 12 % of Sea shell powder that is 237.10 kN/m^2.

4.4 Effect of Rubber powder on unconfined strength of soil in compression of BCS.

Unconfined strength of soil in compression test is conducted for expansive soil by treating with 5 % Rubber powder. In the present study, the experiment is conducted by mixing the BCS with varying rubber powder (2, 3, 4 and 5 %) cured for a period of Un-Cured and 3 days in order to determine the effect of curing on unconfined compressive strength of soil. Stress versus strain graph is plotted to determine the unconfined compressive strength of the BCS.

Figure 8, 9, 11 and 12 shows the graphical representation of the unconfined soil compressive strength of the soil when mixed with rubber powder (2, 3, 4, 5 and 6 %) for a curing period of 0, 3, 7 and 14 days respectively. Optimum Rubber powder content taken in this study is 5 % which increase the strength of BCS, beyond which the strength decreases at different curing periods. The strength gained for other proportions are less when compared to obtained %. Hence here 5 % of rubber powder is considered as optimum for BCS.

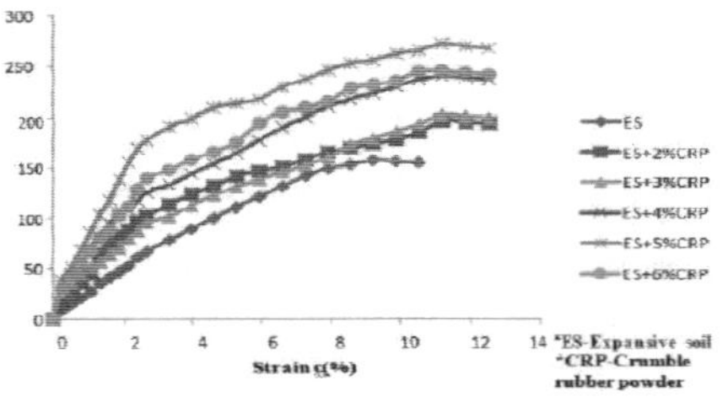

Figure 11. Stress-Strain characteristics of BCS for various % of rubber powder for Un-Cured.

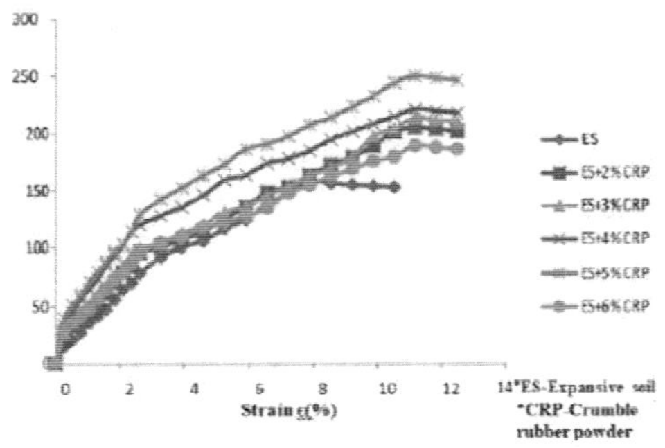

Figure 12. Stress-Strain characteristics of BCS for various % of rubber powder for 3 days of curing.

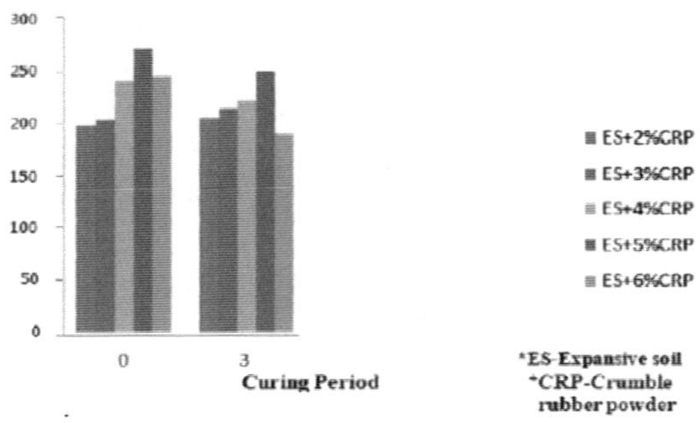

Figure 13. Unconfined strength of soil in compression of expansive soil admixed with rubber powder for different curing period.

Figure 12 shows the variation of unconfined strength of soil in compression expansive soil mixed with rubber powder (2, 3, 4, 5 and 6 %) with different curing periods. Compared to Un-Cured and 3 days curing period, the strength is continuously increases. In addition of rubber powder, increase the bonding between clay particles there by it increases the cohesive strength of soil.

The above Table 8 shows the unconfined strength of soil in compression of soil treated as in with different % of rubber powder at different curing period. Here the maximum strength is observed at 5 % of Crumb rubber powder that is 250.61kN/m^2.

Table 8. Unconfined strength of soil in compression of BCS treated with rubber powder for various %

% of Rubber powder	Un-Cured	3 Day Curing
2	223.95	226.61
3	202.62	199.95
4	239.94	221.28
5	271.94	250.61
6	245.28	189.29

The above table 8 shows the unconfined strength of soil in compression of soil treated as in with different % of rubber powder at different curing period. Here the maximum strength is observed at 5 % of Crumb rubber powder that is 250.61kN/m^2.

5. Conclusion

- BCS treated with shell powder demonstrated higher estimation of MDD and Unconfined Compressive strength for 12 %. From this we presume that the 12 % as ideal rate for shell powder.
- By including 12 % of ideal shell powder content the quality of the strength is expanded by 6 % for a relieving time of 3 days.
- The test outcomes demonstrate that the unconfined compressive strength for shell powder treated with BCS expanded radically with the curing period (up to 12 %) and afterward it diminishes for 14 % of shell powder.
- The compaction test consequences of BCS treated with shell powder and rubber powder yield higher estimation of rubber for the mix of 12 % shell powder and 5 % of rubber powder.

- The blend of 12 % of sea shell powder and 5 % rubber powder demonstrated higher unconfined compressive strength contrasted with other mix.
- The expansion of 12 % of sea shell powder and 5 % of crumb powder expands the compressive strength by 14 % for a curing time of 3 days.
- The test outcomes demonstrate that the unconfined compressive strength of BCS treated with optimum % of sea shell powder and 5 % of rubber powder expanded radically with the curing time.
- The results consistently showed higher values for its unconfined compressive strength tests when compacted at its MDD and OMC for all %s of sea shell powder and rubber powder.
- The treatment of BCS with sea shell powder and rubber powder has a significant effect on its consistency limits.
- The engineering properties of BCS treated with optimum % of sea shell powder are improved owing to the increase in the MDD and unconfined compressive strength.

References

[1] Purushotham G. Sarvade and Prashant R Shet 2012 *International Journal of Industrial Engineering and Management Science* **2** 27-32.

[2] Kavish S. Mehta, Rutvij J. Sonecha, Parth D. Daxini, Parth B. Ratanpara, Miss Kapilani S. Gaikwad 2014 *International Journal of Engineering Research and Applications* **4** 25-32.

[3] Shiva Prasad.A, P.T.Ravichandran, R.Annadurai, P.R.Kannan Rajkumar 2014 *International Journal of Geomatics And Geosciences* **4** 579-584.

[4] Ravikumar Badiger, Maheshwari G.Bisanal 2015 *International Journal of Innovative Research in Science Engineering and Technology* **4** 5875-5882.

[5] MohanChand, V.RameshBabu 2016 *International Journal of Scientific and Engineering Research* **7** 3 2229-5518

[6] B. Sri Vasavi, Dr.D.S.V.Prasad, A.C.S.V. Prasad 2015 *International Journal of Innovative Research In Technology* **2** 2349-6002

[7] Divya Krishnan.K R.P. Koperung kuzhali 2017 *International Journal of Civil Engineering and Technology (IJCIET)* **8** 576–581.

IOP Conf. Series: Materials Science and Engineering **814** (2020) 012029 doi:10.1088/1757-899X/814/1/012029

Life cycle assessment modelling for a developing city's water system

Pooja Shrivastava[1], Dr. M.K.Verma[2], Dr. Meena Murmu[3], Dr. Ishtiyaq Ahmad[3]

[1]Research scholar, National Institute of Technology, Raipur
[2]Professor, Civil Engineering Department National Institute of Technology, Raipur
[3]Assistant professor, Civil Engineering Department, National Institute of Technology, Raipur

Email: pooja.shrivastava04@gmail.com

Abstract. The research objective to analysis the urban water system (UWS) includes water treatment and distribution, wastewater collection, treatment, and disposal for Raipur city of Chhattisgarh state in India. The city has a dry tropical climate and a warm summer. Developing city is being under pressure due to rapid population expansion, urbanization and industrialization that causes a considerable change in water resources and their services Using SIMA-PRO LCA technology with ReCiPe and IPCC methods, the environmental impact assessment for the urban water system in Raipur was calculate by conducting a life cycle assessment (LCA). LCA tool focuses on water treatment and supply, wastewater, storm water, energy consumption, emission of pollutants and heavy elements to the water bodies and to sewage sludge and others spatial considerations such as population, economic and social assessment, water quality along with stakeholder engagement. LCA is a useful tool for identifying alternative environmental impact solutions and urban water operation scenarios for future strategic planning. In assessment, we found that water losses in distribution network and electricity consumption in water treatment have highest impact on environment while reducing this two by 15% and 2% then it reduces impact about 16%. We studied the existing conditions and different scenarios in this research to enhance the system's environmental quality and to find solutions to improve sustainability.

1. Introduction

In developing countries like India population growth and urbanization has been, pressurize the urban water infrastructure of the cities. Nowadays half of the population in the world is shifting to the urban vicinities. This proportion will increase day by day and with an increase in the demand of natural water resources and pollutants loads to the environment. Water scarcity is therefore a major problem worldwide and there is no safe drinking water for about 1.5 billion worldwide. With urbanization, industrialization and population expansion, this condition is getting worse. According to UNESCO, sustainability is the solution for the urban water system. It is necessary to provide sustainable urban water system in the cities should protect human health, resources and ecosystem. Along with the water consumption, energy, chemicals, nutrients and pollutants impact should consider to analyses the environment sustainability. It should identify these harmful effects for environmental sustainability as an indicator.

Urban water systems in the metropolitan's cities are going through a phase of generous changes. To minimize their dependency on withdrawal from natural water resources, many cities are finding options for water resources supply. Wastewater collection and treatment system have loads of high levels of nutrients removal and disposal of solids that increases the system complexity. Integrated urban water management (IUWM) is therefore mandate to make available adequate water for existing and future generations. It needs to examine the situation at the local level and find solutions for an IUWM, in order to manage such systems, it is necessary to map water demand, leakage, water consumption, wastewater treatment.

In this research, we had done the environmental impact assessment (EIA) of the urban water system of a developing city Raipur of Chhattisgarh, India regarding to use the water resources and environmental loads due to pollutants. The work only deals with the functional aspects of an urban water system and not including the infrastructure destruction and improvement of the water system. This will required the precise investigation for environmental impact assessment by using Life Cycle Assessment (LCA) method. The method is often helpful in decision-making framework with an combination of manifold life cycle phases and impact categories. It also allows the tradeoffs without involving the results of weighing and a framework provided by ISO standards to make sure for consistency and avoid their mishandling or misinterpretation. There results have been used to goal particular design and for different functional movements to encourage the environmental policies and distinguish future needs for research. From last many years, Life cycle assessment (LCA) is a tool to quantify environment impacts coupled with urban water system i.e. water supply, water distribution, wastewater storm-water and integration of the urban water system. This work described the influence of the Life Cycle Assessment (LCA) method for developing an enhanced perceptive of integrated urban water systems and to implicate the expected changes in how water supplied to urban population, three scenarios considered for the water treatment, distribution and wastewater services to urban population in Raipur.

2. Materials and methods

2.1. Literature review

For small-scale sewage treatment, LCA tool used for environmental impact assessment for water utility providing a valuable approach in life cycle impact assessment. The following important areas identified through LCA i.e. Ensure the energy attainment in aeration systems at an optimum level; defined the waste products can be reuse and recycle; and planned the company policy on an environment that sound should purchase and contracting [5]. Environment sustainability indicators (ESI) also revealed that they have to an increased understanding and scrutinizing to support the necessities and challenges of the urban water system (UWS) [8]. Including a case study-based real-time functional information taking into account water supply, water abstraction, water treatment and distribution, wastewater treatment, desalination and reclaim by using a holistic approach [1].

This tool also operated to generate information about the environmental impacts causes by water treatment process. Tool can analyse and compare different process or systems through their contribution to worldwide environmental impacts. The important in this is to using LCA tool that define a functional unit for each system to allow a clear comparison of different stages or systems in LCA [2]. As an environmental tool, LCA have unique capabilities to quantitative the life cycle can deal with complexity of systems, to balance environmental impacts [3]. One of the research highlights that appropriate analysis of positive and negative environmental impacts needs a life cycle impact assessment and weighing of costs and benefits for competing environment [6]. LCA applications used in the City of Scottsdale, Arizona, where comprehensive environmental impacts compared to water supply, recycling and desalination suggest that desalination of seawater has the highest impact, while recycling has a lower environmental impact [7]. LCA is a very helpful method for considering alternative future planning scenarios. Systems include analytical decisions in life cycle analysis of significant water and wastewater, which are testing through several phases of the system [11].

IOP Conf. Series: Materials Science and Engineering **814** (2020) 012029 doi:10.1088/1757-899X/814/1/012029

2.2. Study Area

LCA's study area is Raipur City, is one of Chhattisgarh's largest city and became capital on November 1, 2000, before it, Chhattisgarh was the former south-eastern part of undivided Madhya Pradesh. After division Chhattisgarh, state touches the boundary in an east to Orissa, west to Madhya Pradesh and Maharashtra, north to Uttar Pradesh, Jharkhand, and south to Andhra Pradesh. City have fertile plains located between 22 ° 33' N and 21 ° 14'N Latitude and 82 ° 6 'to 81 ° 38'E Longitude. It surrounded by Raigarh and part of the state of Orissa in the East and Durg in the West, Bilaspur in the North, Bastar and part of the state of Orissa in the South. The southeastern part of the upper Mahanadi river and south and west border region occupied by the city.

City located at an attitude of 298.15 m and area under Raipur as per master plan scheme is 226 km2. According to 2011 census, the population of Raipur city reaches up to 10, 10,087. The city's climate is tropical, with temperatures ranging from 5 ° C to 48 ° C and humidity ranging from 30 to 85%. The average annual rainfall receives of 1300 mm in June to October. There are 13 tehsils and 15 revenue blocks which administrated by district. The location map of the study area shown in Figure.1

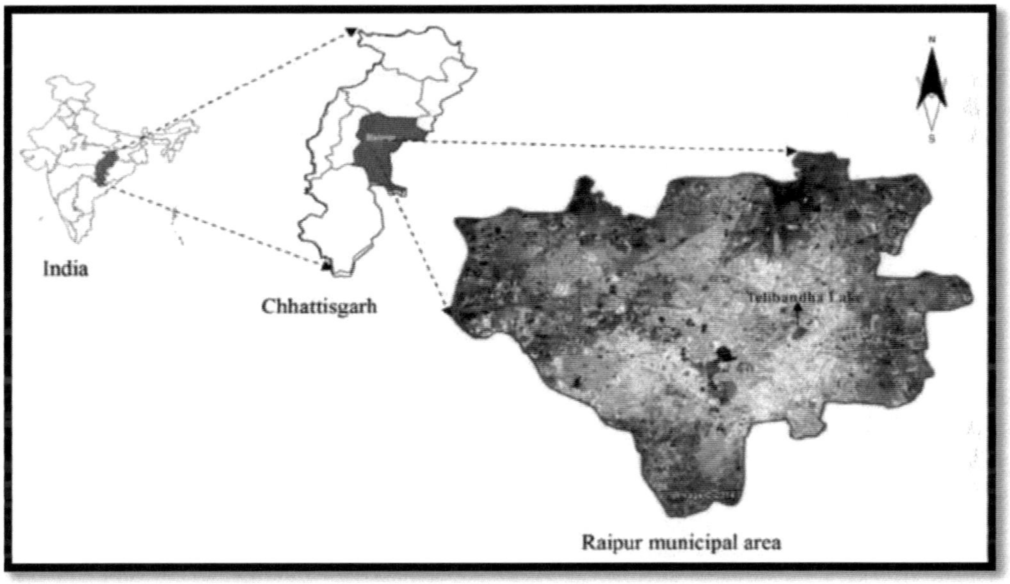

Figure 1. Location map of the study Area (Source: Google)

2.3. Methodology

Life cycle assessment defines term sustainability of a product or services by the realistic analysis of its whole life cycle. This research helps us to assess the environmental impact of products or services over their lifespan. There are three fundamental issues that have to focus in LCA are to protected the human health, conservation of natural resources and preservation of natural environment.

This includes Life Cycle Impact Assessment (LCIA) in which device inventories in many areas into potential impact measures. Because of emission of a particular substance, from a product or from services, which follows a complex series of stages and phases that may cause either unexpected damage or degradation occurs in climate change, environment and human health. The measures of these damages are defining as endpoint indicators as they focus on concern impact.

LCA is a standard tool of EIA that gives its consistency and simplicity. The International Organization for Standardization (ISO) describes guidelines for this method in ISO 14040 and 14044, defining the four main stages of a LCA:

1. Goal and scope
2. Inventory analysis
3. Impact assessment
4. Interpretation

1. Goal and Scope definitions: Goal defines the LCA will perform as a model for product life cycle, service or system. LCA model for the simplification of a complex reality and its simplification will not distort reality in some other way. So this is important for the model developer, LCA simplifications and interpretations should not distort the findings. This will address by defining the objectives and scopes of the LCA study. And it describes the reason for executing the LCA, and it's precise the life cycle of products or services and system boundaries.
2. Inventory Analysis: In the second step analysis for LCA inventory is performing with including all environmental inputs and outputs of concern products or services such as raw material, fuels, energy, emission of pollutants, energy emission.
3. Life cycle Impact Assessment (LCIA): At this stage, the conclusion will remark for better business decisions and translate them into global warming or human health and environmental policies.. It gives the important for choices to a level of integration of results.
4. Interpretation: During the phase of interpretation stage, conclusions need to check it will sustain or not according to ISO 14040 and 14044 standard.

3. Results and discussion

Life cycle assessment (LCA) used for the environmental impacts couple with existing infrastructure for urban water systems in Raipur and identify key opportunities for reducing their impacts. We consider three situations for carrying out LCA for the urban water system. From the survey of the existing state of the urban water system in Raipur, three different scenarios are performing, i.e. reduction of leakage of water in the distribution network up to 15 percent, reduction of electricity in treatment up to 2 percent, and last one combination of this two scenarios. For modelling of LCA, define a functional unit and system boundary for study area is as follows;

3.1. Functional unit

The functional unit identified for LCA Study is the supply of one-meter cube of water at the city's point of intake. The maximum drinking water demand in the Raipur Municipal Corporation (RMC) is 126MLD.

3.2. System boundary and description:

There are following system boundaries are considering in the LCA for Raipur:

1. Water treatment
2. Water distribution and sewage collection
3. Wastewater treatment and disposal
4. Receiving water bodies

3.2.1. Water treatment

There is only one treatment plant in Raipur located near Ravanbhata at the south of the city. This plant serves the population of about 8 lakhs. Design capacity of the treatment plant is 47.67MLD and its augmentation is in progress to increase the capacity by 80 MLD. In plant, conventional treatment is practice for treatment of water with following process includes coagulation, flocculation, filtration and chlorination. It stores the clean water in a sump of the treatment plant. Chlorine gas is use for chlorination of treated water after filtration. When chlorine gas cylinders are not available, bleaching powder slurry prepared in buckets is use for chlorination. Raipur Municipal Corporation is responsible for the water treatment and water supply in the city. The Superintending Engineers are

IOP Conf. Series: Materials Science and Engineering **814** (2020) 012029 doi:10.1088/1757-899X/814/1/012029

overall in charge of water supply, assistant Engineers are in charge of Head Works and the executive Engineers are responsible for distribution, remaining infrastructure and administration.

3.2.2. Water distribution:
There are 12 service reservoirs in the different area for supplying treated water in the city. The present rate of water supply is 70 lpcd for 1 hour in the morning and evening. RMC meets the total demand of water supply through the water from the treatment plant along with groundwater. The groundwater and surface water are mixed in service reservoirs before distribution. In city about there are 3319 public stand posts and 26,536 private connections. During the field survey, it has been finding that the public stand posts and the distribution system based on the bore wells do not have taps. City divides into seven zones for water supply and these zones are to be looking after by sub-engineer. No organized leak detection programme in Raipur and these leakages are repair based on public complaints and as notice by water supply department. It has been a report by personnel of RMC that about 10% of water is Unaccounted for Water (UFW) because of leakages, wastage, etc. This reported percentage is low and need be a test.

3.2.3. Wastewater and Sewage treatment plant:
There is no well-organized sewerage treatment system in Raipur city. There is an open drainage system in most parts of the city. In some, develop residents to give aesthetic look they have covered drains. There are three oxidation ponds at Dalalseoni, Birgaon and Changorabhata. Though the sewerage system constructed and commissioned in 1984, the connections from sewerage line to individual houses did not charge and domestic waste discharged in open drains, which goes to the nearby lake. After the Raipur became the capital, the Corporation applies charges for individual sewer line i.e., the drain connection from private premises up to public sewerage line compulsory and work for such connections is in progress. Heavy algal bloom along with water hyacinth observed in these lakes because of bathing, washing of cloths and other human activities.. However, these lakes are not the parts of organized water supply system of RMC.

3.2.4. Disposal:
Number of nullahs that discharge industrial effluents and domestic sewage without treatment into the Kharun river. The river has been the lifeline of Raipur city for over 40 years. In the last one-decade, the quality has deteriorated from untreated sewage and industrial effluents finding its way into the river. Now, the water quality has declined so much that it is not even fit for bathing, let alone consumption.

Sima-Pro 8.5.2.0 used for Life cycle assessment of the urban water system. The result of LCA states that the energy consumed for scenario 1 and scenario 2 are maximum while energy required for the scenario 3 is the minimum. This will be achieved by eliminating water losses in the distribution network and reducing the supply of power in water treatment throughout the phases. LCA shows the detail results of the ReCiPe endpoint approach in figure 2. Present urban water system have 1499.361points, while for scenario 1, scenario 2 and scenario 3 holds 1471.185, 1288.44 and 1260.264 impact points. As compared to present condition of water system, scenario1 and 2 having high impacts, while in scenario 3 impacts are decreasing by 15.94% under the category of resources (16.58%), human health (15.53%) and ecosystem quality (15.11%). The results of the different scenario regarding present condition as shown in figure 3. Another assessment made by IPCC GWP 500a V1.01 kg Co2 eq. for all four conditions that also shows that scenario 3 has a minimum emission of kg Co2 eq. (figure 4).

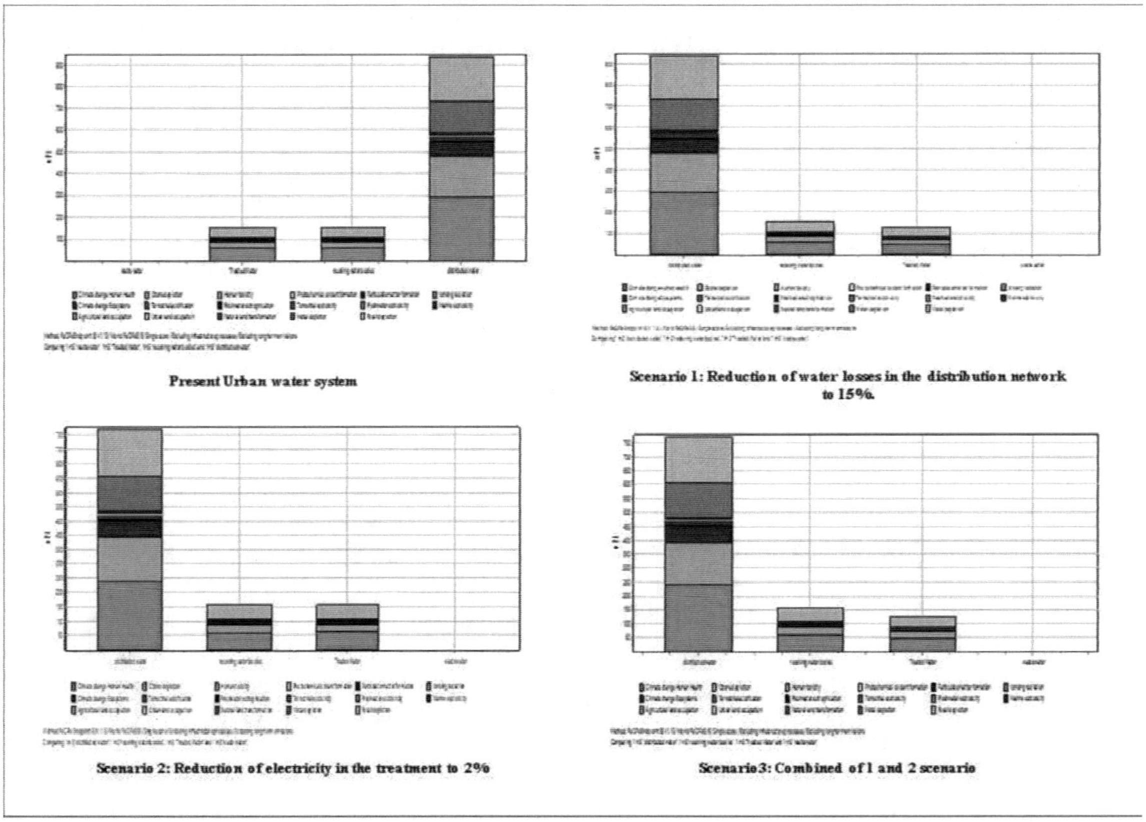

Figure 2. Detailed ReCIPe endpoint (E) V.1.13/World ReCIPe E/E

Figure 2. Detailed ReCIPe endpoint (E) V.1.13/World ReCIPe E/E

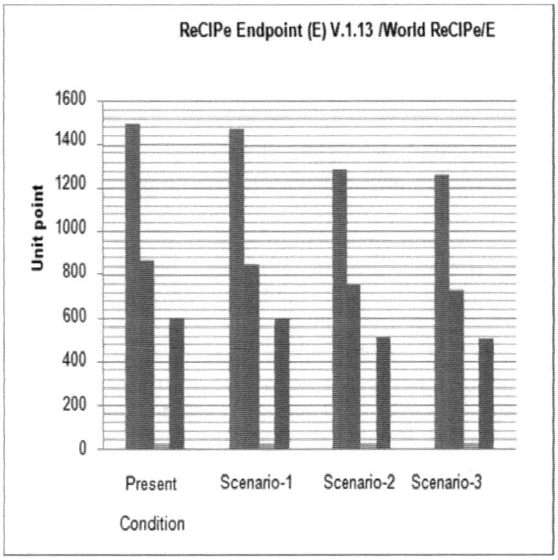

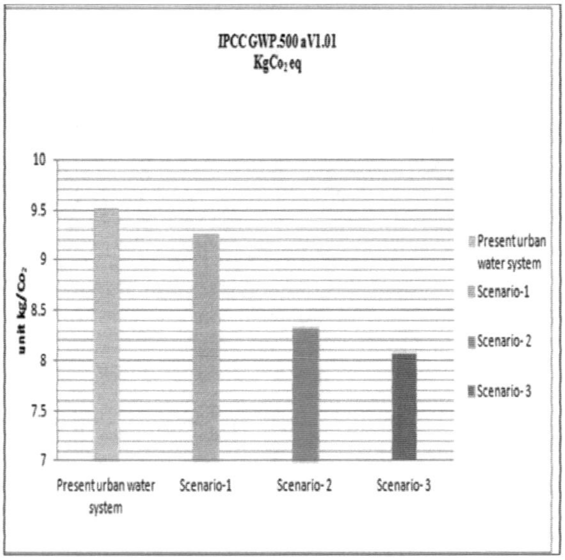

Figure 3. ReCIPe endpoint (E) V. 1. 13/World ReCIPe E/E for present water system and three different scenarios

Figure 4. IPCC GWP 500a V1.01 kg Co_2 eq.

Figure 3. Detailed ReCIPe endpoint (E) V.1.13/World ReCIPe E/E for present water system and three different scenarios

4. Conclusion

The environmental impacts of the urban water system for the Raipur municipal corporation (RMC) analyzed for year 2018-2019. Along with scenario analysis in addition are improvements need to put into practice at the same time such as more reduction in water losses, wastewater treatment plant set up near receiving water bodies. The proposed improvements based on decision-making process regarding future investments towards sustainability of the urban water system environment. This study provides the life cycle assessment of the urban water system for Raipur city in India. The environmental impacts of the existing urban water system described in the different categories.
When considering the data from the observation that used for the LCA, it must be that much information estimated from field survey and literature review. From the results, we can conclude that;

1. Water leakage and electricity consumption is the major source of impacts.
2. Key finding an impact in this category is because of distribution of water.
3. Study shows that the by reducing electricity by 2% and water leakage reduce by 15% will diminish the environmental impact of the urban water system of Raipur by almost 16%.

The city has already starting to set up a waste treatment plant near the Kharun River. The study based on the existing water supply conditions of the Raipur city.

References

[1] Amores, M.J., Meneses, M., Pasqualino, J., Antón, A. and Castells, F 2013 *Journal of Cleaner Production* 43 84-92
[2] Bonton, A., Bouchard, C., Barbeau, B., & Jedrzejak, S 2012 *Comparative life cycle assessment of water treatment plants Desalination* **284** 42–54
[3] Buckley, C., Friedrich, E., & Von Blottnitz, H 2011 *Life-cycle assessments in the South African water sector: A review and future challenges Water SA* **37(5)**
[4] CPHEEO Manual
[5] Emmerson, R. H. C., Morse, G. K., Lester, J. N., and Edge, D. R 1995 *Water and Environment Journal* **9** 3 317–325
[6] Foley, J., de Haas, D., Hartley, K., & Lant, P *2010* Water *Research* **44** (5) 1654–1666
[7] Lyons, E., Zhang, P., Benn, T., Sharif, F., Li, K., Crittenden, J., Costanza, M., Chen, Y.S 2009 *Water Supp* **9** (4) 439-448
[8] Lundéhn C., Morrison G.M 2007 An assessment framework for urban water systems – a new approach combining environmental systems with service supply and consumer perspectives. In:
[9] Morrison G.M., Rauch S. (eds) *Highway and Urban Environment. Alliance For Global Sustainability* Bookseries vol 12. Springer, Dordrecht
[10] Punmia, B. C., Jain, A. K., Jain, A. K. 1995 *Water supply Engineering: Environmental Engineering.*
[11] Parliament of India 1993: IS 1172: *Code of Basic Requirements for Water Supply, Drainage and Sanitation (Fourth Revision)*: BIS 1172:1993
[12] Lundie, S., Peters, G. M., & Beavis, P. C 2004 *Environmental Science & Technology* **38** (13) 3465–3473
[13] V. Bhakar, N. Sihag, R. Gieschen, S. Andrew, C. Herrmann, and K. S. Sangwan 2015 *Procedia* (CIRP) **29** 468–473

IOP Conf. Series: Materials Science and Engineering **814** (2020) 012030 doi:10.1088/1757-899X/814/1/012030

Anaerobic Biological Treatment of Distillery Wastewater – Study on Continuous Stirred Tank Reactor

D. D. Mohite[1], Dr. S. S. Salimath[1]

[1]VTU Regional Research Centre, Belgavi, Karnataka 590018, India

E-mail: ddmohite@rediffmail.com

Abstract. Various studies have verified that anaerobic treatment with the recovery of biogas appears to be the most promising technology for the treatment of distillery wastewater. The technologies currently used by distilleries for treatment of wastewater are bio-methanation followed by two-stage biological treatment and disposal in watercourses or for utilization on land for irrigation and composting with or without bio-methanation. These technologies treat the wastewater up to a certain extent. However, there are limitations posed by these technologies for full compliance with prescribed pollution control standards. For the better understand the performance of continuous stirred tank reactor (CSTR), anaerobic treatment of distillery effluent having very high chemical oxygen demand (COD) (110000–140000 mg/L) and biochemical oxygen demand (BOD) (55000–65000mg/L) was studied on CSTR. Under various organic loading rates (OLR), optimum conditions for maximum COD removal and biogas generation was found and are observed to be 0.09 kg COD/d to 0.12 kg COD/d OLR hydraulic retention time (HRT) of 14d and volatile fatty acids (VFA) to alkalinity ratio of around 0.2. Maximum COD removal efficiency was found to be around 72%. These performance figures are significant when operating the anaerobic bio digesters for the treatment of distillery effluent. Anaerobic CSTR can effectively be employed for the treatment of distillery effluent, but post-bio-methanation effluent still contains high organic concentration and needs to be treated further to meet the safe and acceptable pollution control limits for disposal into surface water or on land.

1. Introduction

The generation of wastewater from distillery is in the order of 12-15 times of the generation of alcohol [1]. The distillery effluent has very high values of biochemical oxygen demand (BOD) (40000-50000 mg per L) and chemical oxygen demand (COD) (100000-125000 mg per L) [2]. The distillery effluent is highly acidic (pH 4.0–4.3) with high suspended solids (2.0–2.5 kg per m^3) [3]. This effluent is toxic to surrounding environment [4]. Besides high organic content, distillery wastewater also contains nutrients in the form of nitrogen (1660–4200 mg per L), phosphorus (225–3038 mg per L) and potassium (9600–17475 mg per L) [5].

The distillery industry is one of the heavily polluting industries identified by Ministry of Environment and Forests, Govt. of India [6]. There are around 400 distilleries are operating at present in different parts of the country, average generation of distillery effluent is up to 15L per litre of alcohol produced, depending on continuous or batch process and quality of molasses used, etc [7].

The secondary and tertiary treatments on their own are not technically and economically viable options for mitigating the problems associated with treatment and disposal of high strength distillery effluent.

IOP Conf. Series: Materials Science and Engineering **814** (2020) 012030 doi:10.1088/1757-899X/814/1/012030

[8]. Because of the high BOD and COD values, distillery effluent is having great potential to produce energy in the form of biogas. From the available technologies, anaerobic digestion of distillery effluent for energy recovery is reported to be the most effective way [9]. Anaerobic treatment of distillery effluent with biogas recovery is being adopted nowadays in India by many alcohol industries [10]. Various types of anaerobic bio digesters have been tried at both pilot scales as well as in full scale operations [11]. For distillery effluent in particular, anaerobic digestion is suitable [12].
The main focus of the present study was to assess the influence of organic loading rates on the performance of CSTR with respect to process parameters and biogas production in order to investigate the feasibility of Continuous Stirred Reactor (CSTR) for anaerobic bio digestion of distillery wastewater. Various factors affect the performance of CSTR digestion process. Parameters of importance relating activity of microorganisms, substrate utilization, and biogas formation on which operations of the reactor is based include Alkalinity, VFA, COD exertion, and TSS [13].

2. Material and Method

2.1 Substrate:
For the present study, effluent and seed culture were collected from the Lokmangal Sugar and distillery Industry located at Solapur, Maharashtra, India. Table 1 shows the characteristics of distillery effluent.

Table 1. Key Properties of Distillery Effluent

Parameters	Values of Raw spent wash
pH	3.0–4.5
BOD5	55000-65000 mg/L
COD	110000-140000 mg/L
Total Solids (TS)	90000-140000 mg/L
Total Volatile Solids (TVS)	80000-100000 mg/L
Total Suspended Solids (TSS)	11000-14000 mg/L
Total Dissolved Solids (TDS)	80000-130000 mg/L
Chlorides	7000-8000 mg/L
Phosphate	2300-3000 mg/L
Total Nitrogen	6000-8000 mg/L

2.2 Startup of CSTR:
In the present investigation, the plastic tank of 20 litres capacity was used as a reactor with 12 litres for liquid volume and 8 litres for gas volume, as shown in Figure 1. The reactor has provision for inlet, outlet, and overflow of effluent. The outlet of biogas was connected to the kerosene lamp, which was partially filled with water and used as a flare stack to burn the biogas produced. Mechanical stirrer with motor was mounted at the top of the bottle. The entire setup was checked for liquid as well as gas leakage, if any. Activated seed culture from full-scale digester was used as a seed material and fed into the reactor with dilution factor 3 to initiate the process. The key characteristics of activated seed were as shown in Table 2.

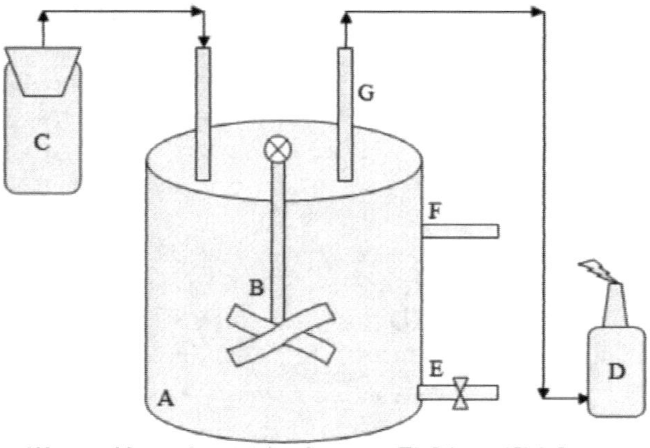

(A)anaerobic continuous stirred rractor; (B) Stirrer; (C) influent; (D) flair stack; (E) outlet; (F) Overflow effluent; (G) biogas outlet.

Figure 1. Schematic diagram of anaerobic continuous stirred bioreactor.

Table 2 Characteristics of activated seed

Parameter	Value
pH	7.8
Temperature	35 °C.
VFA	1300 mg/L
Alkalinity	31000 mg/L
TDS	57000 mg/L
TSS	34000 mg/L
COD	36000 mg/L

Continuous feeding of wastewater to the reactor was started at a very low rate of 0.01 kg COD per day with continuous stirring of reactor liquid. The steady-state condition refers to very little variation in COD of overflow effluent, Alkalinity, or VFA of digester sample [14]. Constant VFA to alkalinity also considered as the steady-state condition. It can also be examined by stable gas production [15]. The pH was tried maintained within the optimum range of 6.5 to 7.5 to enhance the growth and activity of anaerobic bacteria throughout the study. The performance of CSTR for different organic loading rates (OLRs) was assessed in terms of COD removal efficiency and corresponding biogas generation.

3. Results and Discussions

Characteristics of raw spent wash (RSW), Reactor effluent, and overflow effluent were analyzed as per the standard methods for the examination of water and wastewater [16]. The effluent was highly acidic, with a high concentration of TDS and TSS. The COD and BOD values were observed to be 126000 mg/L and 57000 mg/L, respectively, resulting in COD to BOD ratio 2.2, which indicated the effluent is highly suitable for biological digestion.

3.1 Organic and Hydraulic loading rates:

The anaerobic digestion of distillery spent wash was carried out with different organic loading rates from 0.01 kg COD per day to 0.16 kg COD per day with an increment of 0.01 kg COD. As the COD of influent wastewater was not content, hydraulic loading rate (HLR) (L/d) was varying as per the required organic loading rate (OLR) (kg/d). The hydraulic loading rate and corresponding hydraulic retention time (HRT) (Days) for CSTR model were shown in table 3.

Table 3. Influent COD and OLR details

Influent COD (kg/L)	Hydraulic Loading Rate (L/d)	Organic Loading Rate COD (kg/d)	Hydraulic Retention Time (d)
0.122	0.08	0.01	146.4
0.122	0.16	0.02	73.2
0.122	0.25	0.03	48.8
0.122	0.33	0.04	36.6
0.130	0.38	0.05	31.2
0.130	0.46	0.06	26.0
0.130	0.54	0.07	22.3
0.130	0.62	0.08	19.5
0.130	0.69	0.09	17.3
0.128	0.78	0.1	15.4
0.128	0.86	0.11	14.0
0.128	0.94	0.12	12.8
0.128	1.02	0.13	11.8
0.119	1.18	0.14	10.2
0.119	1.26	0.15	9.5
0.119	1.34	0.16	8.9

3.2 Effect of OLRs on CSTR parameters:

From the commencement of the process, the temperature of the reactor was observed as the stability and efficiency of the anaerobic treatment process is greatly influenced by temperature [17]. Temperature is one of the essential parameters in anaerobic digestion and, in most cases, correlates the biogas formation [18]. Reaction rate, biological reactions and activities are some of the areas known to be affected by temperature [19]. Hence, paying attention to the reactor temperature is important, since minor temperature changes can considerably influence the CSTR performance and the biogas yield [20].

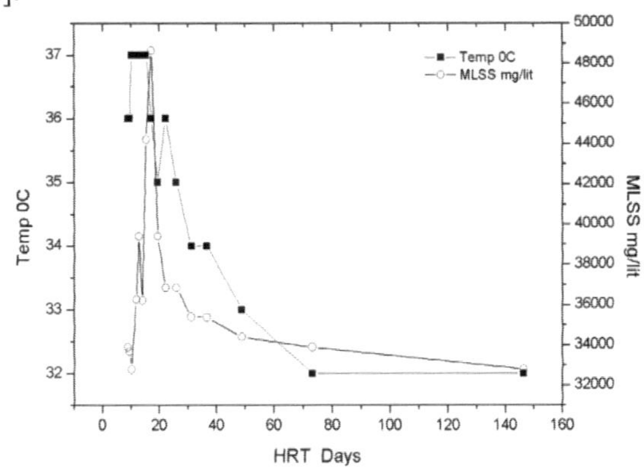

Figure 2. Effect of HRT on MLSS and Temperature.

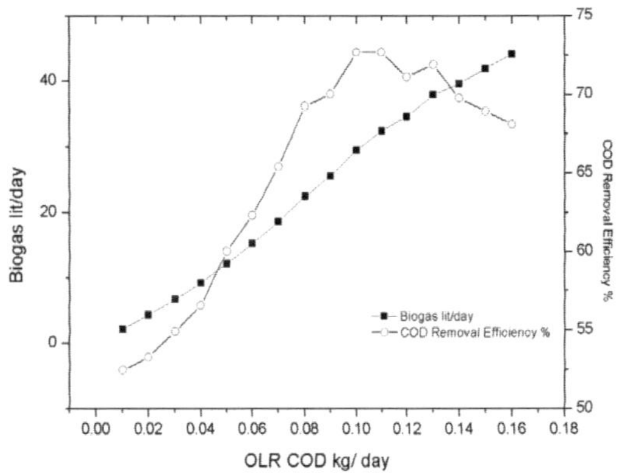

Figure 3. Effect of OLR on Biogas formation and COD removal efficiency.

The growth and decay rates are different at different temperatures; hence, mixed liquor suspended solids (MLSS) concentration was correlated to temperature variations (Figure2.). The steady growth of MLSS was observed for constant reactor temperatures, whereas higher growth in MLSS was recorded for the rise in CSTR temperature. Maximum COD removal efficiency (72%) was recorded at temperature 37 ± 1°C when MLSS was around 36000-44000 mg/L. This efficiency is slightly at lower side as comprised of membrane anaerobic bioreactor which results 76% COD removal efficiency at 37°C [17]. Figure 2 summarizes the variation of MLSS and Temperature at steady-state conditions. Reactor temperature increased up to 37 °C. Further, a decrease in HRT (up to 8 days), results in high COD in overflow effluent, this could be the combine effect of high substrate availability and low net biomass growth rate.

The optimum organic loading for maximum COD removal from the distillery effluent has been examined and shown in Figure 3. It was observed that COD removal reached a maximum of 72.66% when OLR is 0.11 kg COD per day, with 9.166 kg COD/m^3/d. Perez-Garcia et al. (2005) have reported hydraulic shock loading conditions would result with rapid drop in COD reduction activity in case of UASBR with fixed film [21]. In our study, the comparable performance was observed. CSTR performance was on the slightly lower side in terms of the percent removal of COD with respect to a similar range of OLRs. This may be because of a better reaction rate shown by fixed film reactors than the continuous stirring of matter. At higher OLRs more than 0.11 kg COD/d, it was observed and predicted that there is a gradual drop in COD reduction, unlike the fixed-film reactors, which shows a sharp drop in COD removal efficiency. Biogas production is the prime advantage in the anaerobic treatment of highly organic distillery wastewater. An increasing volume of biogas was observed during the treatment process, which indicated the presence of a growing number of methanogenic bacteria. Characterization of seed present in the digester at different stages of bio digestion is required to be done in order to understand the performance of CSTR towards biogas generation better. Generally, biogas production is due to the degradation of organic matter present in the activated sludge in digester and volume of biogas can be calculated as;

$$\text{Volume of biogas} = \alpha\, Q\, (S_{in} - S_{out})$$

Where Q, the feed-flow rate in m^3/d, S_{in}, and S_{out} are the influent and effluent substrate concentration (kg/m^3), and α is the conversion coefficient of the substrate in biogas. For biogas produced by the degradation of COD as a substrate a conversion coefficient α = 0.45 applies [22]. In our study, from the recorded biogas quantities from full-scale CSTR, the conversion coefficient was calculated, and it was found to be 0.405, and the same is used to calculate the biogas generation to get relevant results. It was observed that optimum biogas produced was in the range of 29L to 32L with maximum COD exertion of 72% to 73%. The anaerobic process is very sensitive to temperature as mentioned earlier; it was observed that temperature increases from 32°C to 37°C from the commencement of process study in order to produce maximum COD exertion of 72% for HRT of 14 days.

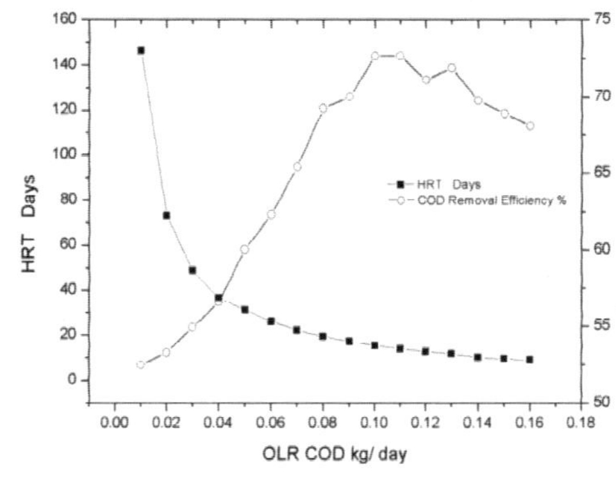

Figure 4. Effect of OLR on HRT and COD removal efficiency.

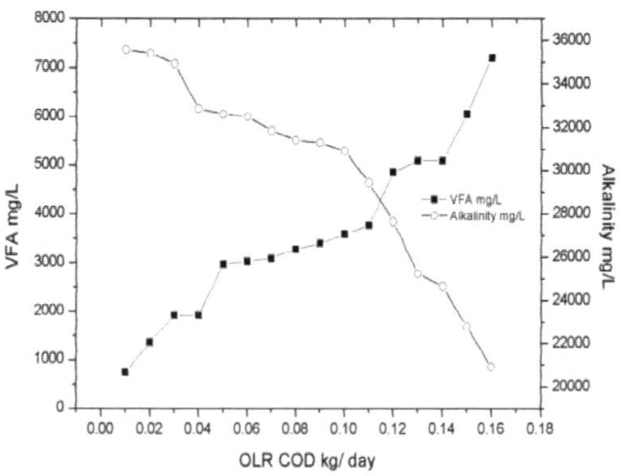

Figure 5. Effect of OLR on VFA and Alkalinity.

As indicated in Table 3, influent COD values are not constant, and variation in COD values of distillery effluent maybe because of the variation in the fermentation and distillation processes in the distillery. During the present study, COD variations observed between 119000 mg/l to 130000 mg/l. Figure 4 shows variation in HRT with OLRs and the corresponding removal of COD. It was observed that the maximum efficiency of COD removal (72.6%) occured when OLR increased from 0.10 kg/d to 0.11 kg/d with the decrease in HRT from 15 d to 14 d. these observations are similar to the study conducted by T. Benabdallah El-Hadj et al. (2007) and were in accordance with the fundamentals of anaerobic mesophilic treatment. Further, it is observed that a decrease in HRT from 14 d to 8 d by increasing OLRs from 0.11 kg/d to 0.16 kg/d, COD removal efficiency decreased from 72.6% to 68%. Effluent shows higher value VFA, and VFA to alkalinity ratio at this stage was recorded as 0.34. For various organic loads, Volatile Fatty Acids and Alkalinity of CSTR samples were observed, and variations in the values are shown in Figure 5. The variation in pH of the reactor is very less during the study and it was well maintained in the favorable range. This indicates that the efficiency of the reactor decreased due to sulfide inhibition rather than VFA inhibition. Similar findings were reported by Gupta S.K. et al (2007) with the anaerobic hybrid reactor. This indicates biogas production is strongly correlated with the OLR [23]. VFA has been recognized as one of the important intermediates during anaerobic digestion and is considered a central parameter for anaerobic treatment. The impact of VFA accumulation was reflected in the decrease in COD removal.

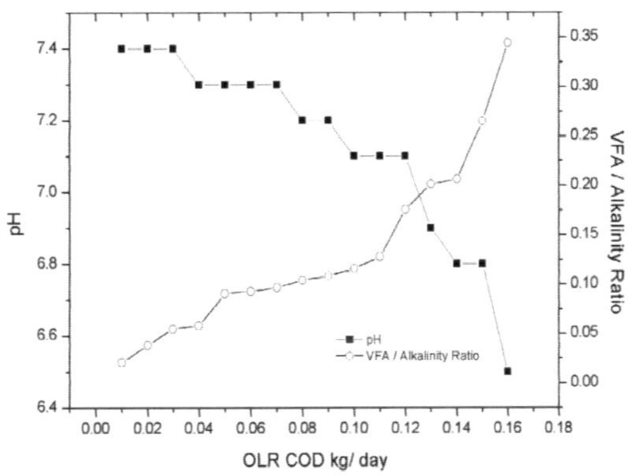

Figure 6. Effect of OLR on pH and VFA/Alkalinity ratio.

Reactor operations were carried out by observing the pH between 6.5 and 7.5. During the operations, variation in VFA to alkalinity ratio was observed to be from 0.021 to 0.344 at all OLRs. VFA has been identified as one of the vital characteristics during anaerobic digestion and is considered a central parameter for anaerobic treatment [24]. The study shows that methano-genesis appears to be an alkalizing process, as it consumes hydrogen and $H3O^-$ ions [25]. Figure 6 shows the variation of pH and VFA to alkalinity ratio for all ranges of OLRs. It is extremely difficult to maintain the pH of the reactor constant as the reaction rate is variable in this phase. VFA production was found to increase with an increase in organic loading due to the high metabolic activity of acid-forming bacteria, and the Alkalinity of the digester is considerably affected by the organic loading rate. The decrease in alkalinity with the increase in OLR can be attributed to an increase in VFA concentration in the reactor effluent. Further, better results could be achieved by increasing the buffering capacity of the reactor.

4. Conclusions

CSTR can effectively be adopted for the treatment of distillery wastewater. The maximum COD removal efficiency of the CSTR was observed to be 72 to 73% when operated in the favorable pH and temperature ranges. Optimum conditions for COD removal and biogas generation were found to be for OLR 0.10 kg/d to 0.11 kg/d, 15d to 14d HRT, and for VFA to Alkalinity ration around 0.12. Optimum biogas generation with a conversion coefficient of 0.405 was observed to be around 30 L/d for steady-state conditions. Post-methanation effluent still contains high COD and needs to be treated further. The performance towards COD reduction and Biogas formation can further be enhanced by increasing the buffering capacity of the reactor. Characterization of seed and or response of seed to micronutrients can also be required to focus in order to enhance the performance of CSTR.

Acknowledgements
We convey our deep-felt gratitude to Mr. Sagar Thombare and Lokmangal Sugar Industry, Solapur, Maharashtra, India, for extending their support to conduct the present study.

References

[1] Shankaran K., Divakar s, Nagarajan S.M. "Analysis on biodegradation and colour reduction of distillery effluent spentwash." *Asian journal of science and technology*, 2011: **2(3)** 015-021.

[2] Chidanand P., P.B.Kalburgi, Mugdha Ghorpade. "Performance and Evaluation of Sugar & Distillery Effluent Treatment Plant." *International Research Journal of Engineering and Technology*, 2015: **2(3)** 1445-1457.

[3] Tapas N., Sunita Shastry,S.N.Kaul. "Wastewater management in a cane molasses distillery involving bioresource recovery." *Journal of Environmental Management*, 2002: **65(1)** 25-38.

[4] Pankaj C., Nawaz Khan and Ram Naresh Bharagava. "Distillery Wastewater: it's Impact on Environment and Remedies." *Environmental Analysis & Ecology Studies*, 2018: **1(2).**

[5] Satyawali Y., M. Balakrishnan. "Wastewater treatment in molasses-based alcohol distilleries for COD and color removal: A review." *Journal of Environmental Management*, 2008: **86(3)** 481–497.

[6] Dikshit D, Mrityunjay Singh Chauhan and Anil Kumar. "Indian Distillery Industry: Problems and Prospects of Decolourisation of Spentwash." *International Conference on Future Environment and Energy*. IPCBEE. 2012, (**28**).

[7] CPCB. Draft Guidelines For Co-Processing Of Distillery Spent Wash Concentrate In Cement Industry. (Ministry of Environment & Forests, Govt. of India), 2017.

[8] Sowmeyan R., G. Swaminathan . "Effluent treatment process in molasses-based distillery industries: A review." *Journal of Hazardous Materials*, 2008: **152(2)** 453–462.

[9] Rajeshwari K., M. Balakrishnan, A. Kansal, Kusum Lata. "State-of-the-art of anaerobic digestion technology for industrial wastewater treatment." *Renewable and Sustainable Energy Reviews*, 2000: **2(4),** 135-156.

[10] CPCB, India. "Guidelines on Techno–Economic Feasibility of implementation of zero liquid discharge for water polluting industries". http://www.indiaenvironmentportal.org.in /files /file/Final-ZL, 2015.

[11] Ravikumar R., R.Saravanan, N.S.Vasanthi, J.Swetha, N.Akshaya, M.Rajthilak and K.P. Kannan. "Biodegradation and decolourization of biomethanated distillery spentwash." *Indian Journal of Science and Technology*, 2007: **1(2)** 1-6.

[12] Moletta R. "Winery and distillery wastewater treatment by anaerobic digestion." *Water Science and Technology*, 2005: **51(1)** 137–144.

[13] Bhavik K., Sarayu Madamwar, Mohana Datta. "Anaerobic treatment of distillery spentwash – A study on upflow anaerobic fixed film bioreactor." *Bioresource Technology*, 2008: **99(11)** 4621–4626.

[14] Prakash N., V. Sockan. V.S.Raju. "An aerobic digestion of distillery spent wash." *Journal of science and technology*, 2014: **4(3)**

[15] Patel H., Madamwar, D. "Effects of temperatures and organic loading rates on biomethanation of acidic petrochemical wastewater using an anaerobic up-flow fixed-film reactor." *Bioresource Technology*, 2002: **82(1)** 65–71.

[16] APHA. Standard Methods for the Examination of Water and Wastewater. Washington, DC: APHA, 2005.

[17] Gao W., K.T. Leungc, W.S. Qin, B.Q. Liao. "Effects of temperature and temperature shock on the performance and microbial community structure of a submerged anaerobic membrane bioreactor." *Bioresource Technology*, 2011: **102(19)** 8733–8740.

[18] Chapleur O., Mazie L, Gordon JJ, Bouchez T. "Asymmetrical response of anaerobic digestion." *Applied Environmental Microbiology*, 2016: **100(3)** 1445–1457.

[19] Appels L., Baeyens J, Degréve J, Dewil R. "Principles and potential of the anaerobic digestion of waste-activated sludge." *Progress in Energy and Combustion Science*, 2008: **34(6)** 755-781.

[20] Muzaffar A., Athar Hussain and Chanchal Verma. "Design considerations and operational performance of anaerobic digester: A review." www.cogentoa.com. 2016: **(3)** https:// www.cogentoa.com/article/ 10.1080/ 23311916.2016.1181696.pdf.

[21] Perez-Garcia M., Romero-Garcia, I.L., Rodriguez-Cano, R., Sales- Marquez, D. "Effect of the pH influent condition in fixed-film reactors for anaerobic thermopillic treatment of wine-distillery wastewater." *Water Sciences and Technology*, 2005: **51(1)** 183–189.

[22] David B., Paolo Pavanb, Paolo Battistonic, Franco Cecchia,. "Mesophelic anaerobic digestion of waste activated sludge influence of solid retention time in the waste water treatment process." *Process Biochemistry*, 2005: **(40)** 1453- 1460.

[23] Gupta S., Gurdeep Singha. "Biodegradation of distillery spent wash in anaerobic hybrid reactor." *Water Research,* 2007: **41(4)**, 721–730.

[24] Rajesh J., S. Kaliappan, M. Rajkumar and Dieter Beck. "Treatment of spentwash in anaerobic mesophilic suspended growth reactor (AMSGR)." *Journal of Environmental Biology*, 2006: **27(1)**, 111-117.

[25] Patel H. "Biomethanation of low pH petrochemical wastewater using up-flow fixed film anaerobic bioreactors." *World Journal of Microbiology and Biotechnology*, 2000: **16(1),** 69–75.

IOP Conf. Series: Materials Science and Engineering **814** (2020) 012031 doi:10.1088/1757-899X/814/1/012031

Review on Passive Heat Enhancement Techniques in Pool Boiling Heat Transfer

Anil Katarkar [1], Biswajit Majumder [1], Swapan Bhaumik [2]

[1]Ph.D Scholar, Mechanical Engineering Department, National Institute of Technology Agartala, 799046, India.
[2]Associate Professor, Mechanical Engineering Department, National Institute of Technology Agartala, 799046, India.

E-mail: anilkatarkar@gmail.com

Abstract. Improvement of pool boiling heat transfer is deeply connected to heat enhancement methods for several years. Nevertheless, the developing progress and ideas in this field make it more intensified with the recently applied methods for enhancing the CHF and HTC. This paper gives a critical analysis and thorough investigative study in the literature discussing the passive heat enhancement techniques in the pool boiling heat transfer. Heat improvement consequence within the pool boiling mainly depends on thermo-physical features of material & the working fluid, modification of surface techniques (pin-fins treated, roughness, extended, artificial cavities pattern, micro-channels), thin-film (nano-scale, micro-scale, macro-scale surfaces, and hybrid scale), coating method, working fluid (water or refrigerants) and additives for fluids. The current status of these passive heat enhancement methods for pool boiling heat transfer with various kinds, improvement potentials, future challenges, surface modification techniques, and their potential industrial and technical applications are also examined in this work.

1. Introduction

Through the years, excessive utilization of conventional energy resources has given rise to several issues like climatic variations, global warming, and depletion of energy resources. Hence, the enhancement of energy efficiency and protection of the natural environment has become an essential factor in the world. By enhancing the performance efficiency of applications of various heat exchangers, evaporators in the refrigerators, boilers in power generation plants, the impacts of global warming can be mitigated. Notably, several heat exchangers work on the boiling heat transfer technique [1]. NPBT method has been categorized into passive, active and compound methods which are given in Table 1. The passive methods need no other external power supply; another hand, the active methods need an external power input to conduct the improvement. However, on the other hand compound methods are the fusion of both passive and active methods. Passive methods have greater fidelity when contrasted to active methods, because of the lack of moving parts, it offers energy and also materials reductions supplemental energy [2].

Table 1. Different type of enhancement methods [3]

Passive techniques	Compound Enhancement	Active Techniques
Swirl flow devices,		Mechanical aids
Rough shapes and surfaces,		Magnetic or
Extended and Treated surfaces		Electrostatic fields
Surface tension equipments		Surface vibration
Displaced enhancement devices	Two or More active and	Suction or injection
Additives for fluids and gases	passive techniques that are	Fluid vibration
Enhanced surfaces, Finned surface	employed together.	Jet impingement
surface tension devices, Coiled tubes		liquid and/or Surface
Displaced enhancement devices		rotation
Micro channel structure		suction or injection

2. Boiling Heat Transfer

Nucleate boiling is a competent thermal process for the heat transfer mechanism. This technique is mostly employed in technical and industrial appliances due to its capability to carry considerable heat with a very less variation of temperatures. Boiling is considered to be a liquid-to-vapour phase-shifting method.

Besides this, it is essential to understand the distinction between flow boiling and pool boiling as shown in Figure 1. Pool boiling occurs in the stationary fluid. For the pool boiling, water in jar placed over heater is example of pool boiling. Pool boiling more classified forced and natural boiling. Where flow boiling takes place in flow proceed. External source is used to lift/suck water over the inside channel or heated channel. Flow boiling classified in diabatic and adiabatic boiling.

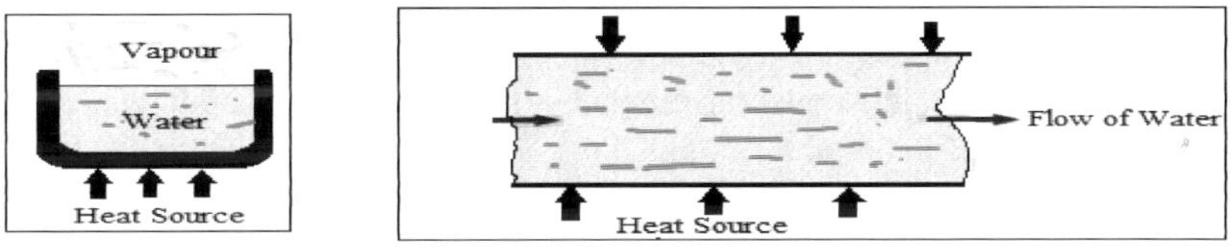

a. Pool boiling b. Flow boiling

Figure 1. Boiling Process [4]

The benefit of the pool boiling technique is, a high heat flux can be expelled passively by keeping less superheat contrasted with the natural or applied convection without the phase change. Nevertheless, heat evacuation capability is restricted by higher limits of cooling, i.e., the CHF, where the HTC reduces significantly as the boiling management has been varied from nucleate to the film boiling. Hence, the improvement of CHF is of high importance to the researchers and technicians. Investigations on the improvement of CHF by modifying the surface in a saturated pool boiling usually considered as (1) modulated or uniform porous coatings and (2) structures of different sizes established on the heated surface. In common, CHF improvement in the saturated pool boiling is a consequence of the impacts of (a) expansive surface area (i.e., metallic foams, micro fins, and porous coatings), (b) nucleation locality density (i.e. nanostructures, microcavities, and sandblasting), (c) capillary wicking, (d) wettability (i.e., hydrophobic and hydrophilic coatings), and (e) wavelength reduction on the basis of the revised hydrodynamic stability design.

3. Surface Coating Techniques

In the present scenario, numerous technologies are being employed for decreasing the size of the material into micro and nano-thicknesses. By reducing the material to nano-size, it has led to the development of novel and unusual performances of those materials in the areas of applications like optoelectronics, electro-optics, dielectrics, and so on. Micro and nano surface coatings methods are employed as inventive methods to enhance the performance of heat transfer to hold periodic dry out. Thin films perform a major function in the advancement and investigation of the materials with unique and novel characteristics. A thin layer of nano- micron material on the surface is termed as a thin film, where the material thickness is in the range of a few nanometers to the number of micrometers. By these thin films, a layer-by-layer (LbL) method is one of the promising techniques of fabricating multi-layer coatings, composition, accurately regulated design architecture, and nanometer thickness level. The thin films have a prominent influence on recent technology. These films can be manufactured by adopting physical and chemical depositions.

3.1. Chemical Vapor Deposition (CVD)

CVD is a method in which the substrate is exhibited to volatile antecedents, which decay and/or react on the surface of the substrate in order to form a deposit of thin films. This process is a chemical process which is employed to deliver efficient, good-performance, better-quality materials. Some of the examples are Chemical bath deposition, Electroplating technique, Sol-gel technique, etc.

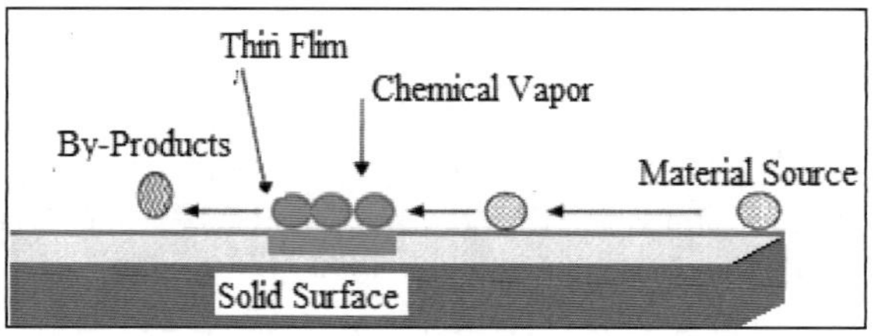

Figure 2. Diagram CVD Processes

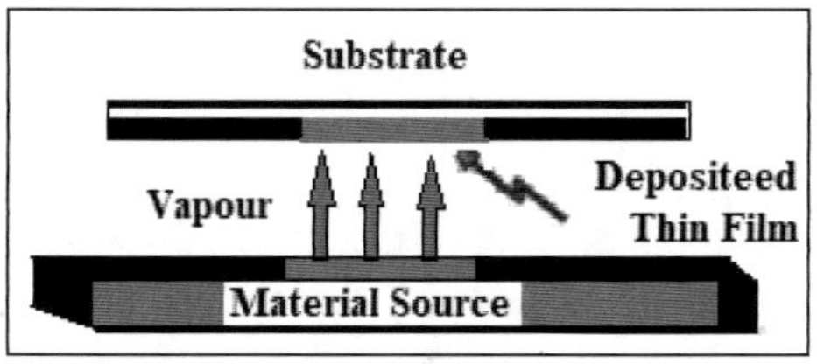

Figure 3. Diagram of PVD Processes

3.2. Physical Vapor Deposition (PVD)

PVD is defined as a kind of vacuum ousting methods that are employed to create coatings and thin films. In PVD, the transition of material takes place from a condensed to vapour state and later reverse to the condensed state. Some of the examples are Evaporation techniques like Ion plating, Molecular beam, Vacuum thermal, Electron beam, Laser beam, and Sputtering techniques like RF and DC sputtering.

4. Heat Enhancement Techniques

In general, improving the performance of heat transfer is accomplished by just modifying the characteristics of the surfaces like modifying the coating, roughness, wettability and contact angle (CA) or by enhancing the area of the surface for heat exchanging like enlarged surfaces, coiled tubes, tube inserts, metallic foams, micro fins and porous coatings. Other improvement methods summon for the inclusion of supplements in working fluid like nanofluids and surfactants. The above-suggested methods come under passive heat transfer enhancement methods [5].
Many published reports are present in the literature, which discusses the enhancement of boiling heat transfer employing the modification of surfaces technique. All these reports either discuss the improvement of pool boiling, accompanying with additional heat transfer methods are applied solely to the enhancement of pool boiling techniques.

4.1. Macro-finned surfaces

Manufacturing various quadrilateral or squared fins on the heating surface is a prevailing way to increase the heat transfer. Whereas the apparent advantage of this method is increasing the area of heat transfer, the main complication in this is reducing the size and spacing of the fins in the view of the best performance of cooling.

Rainey and You [6] summarized that the performance of NBHT for FC-72 with the surface thickness of 1-mm wide and 1-mm spacing fins was tremendously improved with the growing height of the fins up to approximately 5 mm, over which the tip of the fins temperature has dropped to a very low value to sustain the boiling, which decreased the 'efficient' surface area of boiling with no extra heat transfer improvement.

4.2. Porous mesh and Porous foam attachments

4.2.1. Porous mesh

Earlier investigations concerning to employment of porous mesh over the heat-diffusing surface point to various NBHT advantages, involving plainness, an enhanced quantity of effective nucleation localities, and the breakup of huge vapour bubbles, as well as detachment of paths of liquid and vapour for the betterment of fitter surface liquid replenishment.

They also Melendez and Reyes [7] discovered that the porous metal coating of the surface has enhanced the boiling HTC for the binary ethanol-water compounds.

4.2.2 Porous foam

In recent days, a high porosity metallic foam has captivated important consideration due to its mixed advantages of heat transfer improvement and lightweight. Yang et al. [8] discovered that welding of the copper foam over surface heating has decreased the initial superheat for H_2O. They gained the greatest improvement with a PPI of sixty and further showed that the performance efficiency of boiling was affected by PPI as well as the foam cover thickness.

Yang et al. [9] observed that with the copper foam, the orientation of varying surfaces from parallel upward-facing to the perpendicular facing has only a trivial influence on the boiling performance.

4.3. Microscale surface modification

4.3.1. Surface roughening

The growing coarseness of the heat-dispelling surface, accomplished by sandblasting, developing various small artificial cavities or by mechanical roughing, are successful methods for enhancing the performance efficiency of nucleate boiling. Figure 4 show the cross section of rough surface.

Marto and Rohsenow [10] observed that for sodium, the roughness of the surface is highly influenced by the nucleate boiling HTC as well as the boiling incipience. Jones et al. [11] observed that the improvement of the roughness of the surface was mainly reliant on fluid.

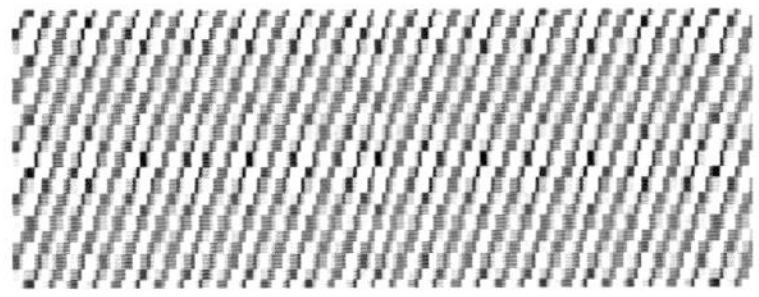

Figure 4. Cross section of rough surface

Increasing the roughness of the heat-dissipating surface, gained by sandblasting, by developing various small synthetic holes, mechanical hardening, or by chemical etching is a traditional technique for improving the nucleate boiling performance [12].

4.3.2. Micro-fin surfaces

Modeling micro-fins over the surface of heating enhances the area of heat transfer and additionally takes benefit of effected capillary discharge plus aids in segregating the ways of escaping replenishment fluid and vapor. Those supplemented advantages need a precise configuration of micro-fin design and also reducing the width of the fin, spacing, and length [13]. The surfaces of the micro-fin are available in different types of molds, like parallel inclined fins that are aimed to catch vapor embryos, rectangular cross-sectional parallel vertical fins which are likewise known as open micro-channels, cylindrical fins as given in Figure 5.

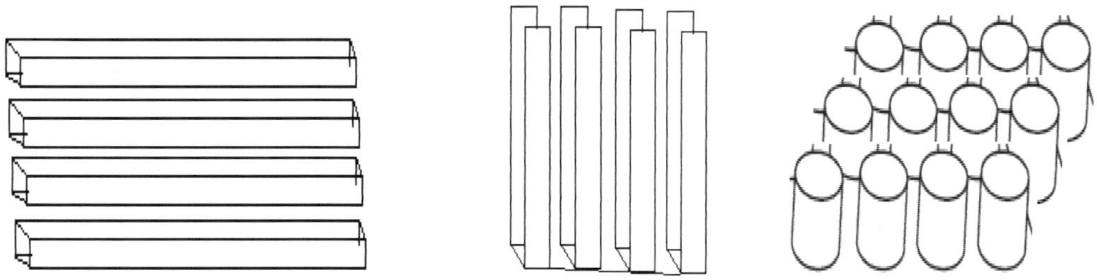

Figure 5. Models of the patterns of micro-fins employed for boiling improvement.

Mitrovic and Hartmann [14] increased a unique composition of micro-fin composing of circular fins with orbicular points using an electro-coating process. By R-141b, they demonstrated that the superheat from the wall continued fairly constant with increasing heat flux.

4.3.3. Tunnel surfaces and reentrant holes

Ramaswamy et al. [15] analysed the boiling method over the surface of microporous material for FC-72 with a design lightly similarly to [16]. Nevertheless, the surface design prepared by wafer dicing and wet etching, driving in an interrelated tiny micro-channel system.

4.4. Nano Enhancement (Nanotubes, Nanowires, Nanofibers)

4.4.1. Nanotubes

With the advantage of good mechanical and high thermal conductivity properties, carbon nanotubes (CNTs) are utilized for boiling enhancement from a decade. CNTs are very thin carbon graphite tubes with lengths ranging from 1 to 50 lm and diameters from 1 to 100 nm. Also, CNTs are said to increase the boiling performance by generating TiO2 nanotubes on substrate. In [17], authors have examined a copper surface with CNTs coating and included a layer of diamond between the copper surface and CNTs for adhesion improvement. They also obtained initiation of boiling when compared to copper surface that is bare; this is attributed to increased roughness and high hydrophobicity.

4.4.2. Nanowires

Nanoscale rod with some diameter of few nanometers, a large ratio of diameter to length is called a nanowire. Lu et al. [18] studied wafer-scale electroless etching in which saturated water pool boiling on plain surfaces of silicon coated with few thicker arrays of a silicon nanowire. An increase in CHF has been observed, resulting in enhanced liquid spread ability on the surface of the nanowire. Assuming that the vapor jet's critical Helmholtz wavelength emitting normal to the surface persisted unchanged compared to that of a bare surface, but by changing ratio of surface area covered by jets to the surface area in total, they gave the good agreement of their CHF data with hydrodynamic instability model of Liang and Mudawar [19].

4.4.3. Nanofibers

Nanofiber is one kind of fiber created of polymers having diameters smaller than 100 nm. Jun et al. [20] examined electrospun copper nanofibers boiling performance. They estimated BHTC for ethanol and water 3-8 times those from these bare exterior but witnessed no improvement in CHF. They recommended that heat transfer enhancement during nucleate boiling was the outcome of this technique's ability to improve the average temperature of surrounding liquid bubbles, through fostering more efficient bubble growth.

4.5. Nano coating

4.5.1. Nano film coating

Many studies have suggested that the inclusion of a coating of heat transfer surfaces with few nanoscale-thickness layers to improve boiling performance by enhancing surface wettability. In [21], authors manufactured a polymer nanoparticle membrane coating on a nickel substrate employing a layer by layer technique. They obtained excellent surface adhesion along with adsorption of the following layers over the surface. Their experiments revealed more than 100% improvements in both CHF for water and NBHTC, which were associated with enhanced nucleation and improved wettability.

4.5.2 Nanoporous coating

Tang et al. [22] developed nanoporous copper surfaces possessing 50–200 nm pores upon a copper substrate, by which they managed to achieve a 63.3% decline in incipience superheat, including a 172.7% rise in nucleate BHTC for water. But a change in porosity and the three-dimensional structure were observed.

Jones et al. [23] used molecular dynamics simulations to introduce one mechanism for maintaining superheated liquid inside hydrophilic pores to water with giving a new pathway during wetting by vapor condensation inside surface textures. This idea remained to be efficient for minute spacing, notably in the nanoscale.

4.6. Nanofluid

Nanofluids hold suspended nanoparticles that have particles with volumes tinier than 100 nm. Nanofluids are stable when compared to the milli- and micro fluids, because they experience less gravitational and erosion ousting over more extended periods. The boiling of these nanofluids became an exciting topic over the last decade. Several researchers examined the deposition of nanoparticles on the heated surface. This increases wettability and surface roughness. CHF is also observed to be improved significantly. Nonetheless, questionable results were achieved for HTC.

Narayan et al. [24] described boiling of nanofluids over the aspired surface can change out to become a novel coating technique. For that, one must be able to handle the thickness of the deposition.

4.7 Structure of Micro Channel

Micro channel surface produced by chemical etching is a traditional technique for improving the nucleate boiling performance as shows in Figure 6.

Figure 6. Different structure of micro channel surface.

Hao et al. [25] applied titanium nickel which is a part of shape memory alloy to manufacture a distorted compounded structure of micro-channel, the form of which can be changed while the process of boiling; this method produced the best performance of nucleate boiling in ethanol compared to the closed as well as open micro channel-structures produced the best performance of nucleate boiling in ethanol compared to the closed as well as open micro channel-structures. Many innovative designs have been created on the channel walls to improve the thermal efficiency of mini- and micro-channels [26].

5. Conclusion

This paper reviewed thoroughly the issues reported and discussed by other researchers about the passive heat enhancement techniques in PBHT. This paper captures the glimpses of the involved techniques used for coating, structured surfaces, nano-scale, micro-scale and macro-scale surfaces, the structure of micro-channel, and nanofluid methods. Besides this, the study also points out difficulties associated with the execution of experimentation techniques and its probable solutions. Important conclusions from this study are compiled as given below:

1. Improvement of Macroscale methods involves the application of expansive surfaces, porous foam and mesh. They render different levels of improvement of CHF as well as NBHTC, simultaneously with the exclusion of the superheat.
2. Microscale improvement methods comprise roughening of the surface and application of the open micro-channels, micro-fins, sintered and graphite particle coats, brazed coats, tunnels with re-entrant holes and structures of dendritic microporous. The main advantage of these methods is the improved density of nucleation site, that improves the NBHTC, however their influence on CHF changes.
3. Nano-scale improvement methods include surface coating with nanofibers, nanowires, nanotubes, layers of nanofilms and nanoparticle as well as nanoporous deposition by nanofluid boiling. The essential advantage of surface improvement at this range is boiling improvement by the capillary wicking inside the nano-structures.
4. In spite of the important advances in the performance of NPBHT essentially through recently formed micro-nano coating materials and methods, the technical utilization is not yet feasible. New fabrication methods are required to be scrutinized to guarantee the less cost of surfaces and nano-coatings.
5. The smooth surfaces explicate more limited heat improvement contrasted to micro-structured surfaces. Particularly, utilizing the heavily-wetted structure surface, the occasional dry out is increased, i.e. the quality of the critical vapour increases.

Nomenclature

A	surface area (m^2)
h	heat transfer coefficient (W/m^2 K)
q	heat flux(W/m^2)

Abbreviation

CHF	Critical Heat Flux
HTC	Heat transfer coefficient
LbL	layer-by-layer
NBHTC	Nucleate boiling heat transfer coefficient
R	Refrigerant

References

[1] Park, Ki-Jung, Dongsoo J and Sang E 2009 Nucleate boiling heat transfer in aqueous solutions with carbon nanotubes up to critical heat fluxes *Int. J, of Multiphase Flow* **35** 525–532.

[2] Paul G, Chopkar M, Manna I, and Das P K **2010** Techniques for measuring the thermal conductivity of nanofluids: A review *Renew. Sustain. Energy Rev*, ***14*** 1913–1924.

[3] Manglick R M 2003 *Heat transfer Enhancement* John Wiley & Sons Hoboken NJ USA 1029–1130.

[4] Yunus A, Cengel and Afshin J G 2011 *Heat and Mass Transfer* McGraw-Hill Higher Education 4th ed 515-560.

[5] Wasekar V M and Manglik R M 2002 *Int. J. of Heat and Mass Transfer* **45** 483-493.
[6] Rainey K N and You S M 2000 Pool boiling heat transfer from plain and microporous, square pin-finned surfaces in saturated FC-72, J. *Heat Transfer* **122** 509–516.
[7] Melendez E and Reyes R 2006 The pool boiling heat transfer enhancement from experiments with binary mixtures and porous heating covers Exp. *Therm. Fluid Sci.* **30** 185–192.
[8] Yang Y, Ji X and Xu J 2010 Pool boiling heat transfer on copper foam covers with water as working fluid *Int. J. Therm. Sci.* **49** 1227–1237.
[9] Yang Y, Ji X and Xu J 2010 Effect of inclination angle on the pool boiling heat transfer of ultra-light copper foams *Heat Mass Transfer* **46** 695–706.
[10] Marto P J and Rohsenow W M 1966 Effects of surface conditions on nucleate pool boiling of sodium *J. Heat Transfer* **88** 196–203.
[11] Jones B J, McHale J P and Garimella S V 2009 The influence of surface roughness on nucleate pool boiling heat transfer *J. Heat Transfer* **131** 121009.
[12] Wen C D and Mudawar I 2004 Emissivity characteristics of roughened aluminum alloy surfaces and assessment of multispectral radiation thermometry (MRT) emissivity models *Int. J. Heat Mass Transfer* **47** 3591–3605.
[13] Mudawar I and Anderson T M 1990 Parametric investigation into the effects of pressure, subcooling, surface augmentation and choice of coolant on pool boiling in the design of cooling systems for high-power-density electronic chips, *J. Electron* **112** 375–382.
[14] Mitrovic J Hartmann J 2004 A new microstructure for pool boiling *Super latt. Microstructure* **35** 617–628.
[15] Ramaswamy C, Joshi Y, Nakayama Y and W.B. Johnson, 2002 High-speed visualization of boiling from an enhanced structure, Int. *J. Heat Mass Transfer* **45** 4761–4771.
[16] Nakayama W, Daikoku T, Nakajima T 1982 Effects of pore diameters and system pressure on saturated pool nucleate boiling heat transfer from porous surfaces, *J. Heat Transfer* **104** 286–291.
[17] Dharmendra M, Suresh S, Kumar C S and Yang Q 2016 Pool boiling heat transfer enhancement using vertically aligned carbon nanotube coatings on a copper substrate, *Appl. Therm.Eng.* **99** 61–71.
[18] Lu M C, and Chen, Srinivasan V, Carey V P, and. Majumdar A 2011 Critical heat flux of pool boiling on Si nanowire array-coated surfaces *Int. J. Heat Mass Transfer* **54** 5359–5367.
[19] Gangtao Liang and Issam Mudawar 2019 Review of pool boiling enhancement by surface modification, *Int. J. of Heat and Mass Transfer* **128** 892–933.
[20] Jun S, Sinha-Ray S and Yarin A.L. 2013 Pool boiling on nano-textured surfaces, *Int. J. Heat Mass Transfer* **62** 99–111.
[21] Hu H, Xu C, Zhao Y, Ziegler A L, and Chung J N 2017 Boiling and quenching heat transfer advancement by nanoscale surface modification *Sci. Rep.* **7** 6117.
[22] Tang Y, Tang B, Qing J, Li Q and Lu L, 2012 *Appl. Surf. Sci.* **258** 8747–8751.
[23] Jones P R, Elliott A R and Patankar N A 2016 *Langmuir* **32** 12947–12953.
[24] Narayan, Anoop K B, Das S K, 2007 *J. Appl. Physics* **10** 074317.
[25] Hao W, Wang T, Jiang Y Y and Guo C 2017 *Int. J. Heat Mass Transfer* **112** 236–247. Heat Source
[26] Naqiuddin N H, Saw L H, Yew M C Yus M C Ng T C and Yew M K 2018 *Renew. Sust. Energ. Rev.,* **82** 901-914.

IOP Conf. Series: Materials Science and Engineering **814** (2020) 012032 doi:10.1088/1757-899X/814/1/012032

A Systematic Ergonomics Approach of Maintenance Workstation

N S Mujmule[1], S D Katekar[2], A P Gaikwad[2], S C Mali[2], S R Koli[1]

[1]Student, M. Tech. Mechanical Design Engineering, SKN Sinhgad College of Engineering, Pandharpur, 413304.
[2]Assistant Professor, Mechanical Department, SKN Sinhgad College of Engineering, Pandharpur, 413304.

E-mail: nitinmujmule101@gmail.com

Abstract. The transportation systems contribute very vital role in growth of country, either it may private, public or own transportation system, and to keep the automobiles or machine in good condition without any interruption consistent corrective or preventive maintenance has to be done, and this maintenance activities should take the least time and cost of maintenance along with good quality, and to achieve this it is preferred to design a good maintenance workstation with a systematic ergonomic consideration which has a great influence on time, service, comfort, quality, productivity. For the design of a maintenance workstation, systematic ergonomic procedures are considered these procedures should concern for good ergonomically design of workstation. In an actual design of workstation ergonomically approach is needed concern of the worker anthropometric data with the various parameter of the workstation like adequate posture, work table or chair height, position of the operating nobs and levers comfortable sufficient working areas that help to reduce MSD (Musculoskeletal disorder) and prevent injuries of worker along with enhancing productivity. The main objective of this paper is to propose a systematic ergonomically designed maintenance workstation. Existing workstation of Maintenance in MSRTC (Maharashtra State Road Transport Corporation) Solapur workshop, is not designed with consideration of Ergonomic principle, so the ergonomic analysis has been made for different posture of maintenance activity, and modifications in workstation has been suggested. For analysis CATIA software is used, Manikin (simulated posture of worker) has been developed based on the anthropometric data collected from MSRTC for RULA analysis under quasi-static loading and dynamic loading conditions.

1. Introduction

Ergonomics is designing a work suitable to the worker, not worker suitable for a given work so that the work is safer as well as more efficient. A systematic ergonomic approach can make employees more comfortable along with increased productivity. Ergonomics is vital because when you're performing a work or an activity and your body is strained by a clumsy posture, repeated movements; effect on your musculoskeletal. Symptoms like tiredness, awkwardness, and pain may begin to hold by your body, which may be the primary marks of a musculoskeletal disorder. Each industry has its aim to enhance the productivity of Worker, particularly with higher frequency industrial jobs, and it is challenge for the profitmaking executives.

To enhance the efficiency and reduce the injuries and MSDs of the worker, there is need of ergonomic assessment and optimization of Industrial workplace and tools. Ergonomics improves performance along with the productivity, safety and health, and of the workers.

2. Anthropometric Measurements

The anthropometric measurement were presented in the Table 1 having different variables with 98% accuracy relating to five workers of Maharashtra, (MSRTC, Solapur) Indian, used for building a manikin in CATIA V5R19 (software).

Table 1. Anthropometric Data Analysis Sheet

Sr. No.	Parameters (in cm except wt. and year)	Subject No. W1	W2	W3	W4	W5	Mean
1	Weight (Kg)	64	52	65	70	60	62.2
2	Stature	165	163	168	165	153	162.8
3	Eye Height	153	154	157	154	143	152.2
4	Acrominal Height	139	140	142	141	135	139.4
5	Axilla Height	122	121	125	120	128	123.2
6	Chest Height	119	118	120	115	124	119.2
7	Tenth Rib Height	108	107	110	102	110	107.4
8	Iliocristale Height	98	97	101	99	99	98.8
9	Waist Height	101	102	105	104	97	101.8
10	Olecranon Height	103	105	107	108	98	104.2
11	Elbow height	105	107	108	110	102	106.4
12	Crotch Height	77	72	78	80	75	76.4
13	Knee Height	47	45	50	46	47	47
14	Chest Breadth	25.8	24.6	25.7	27.8	26.3	26.04
15	Hip Breadth	28.5	27.8	28.2	27.9	26.2	27.72
16	Hand Length	17.5	18.4	19.2	20.3	19.5	18.98
17	Palm Length	10.4	10.8	10.2	11.2	10.3	10.58
18	Hand Breadth (At Metacarpal-III)	7.5	8.2	7.4	7.6	8.2	7.78
19	Grip Diameter(Inside)	4.5	4.2	4.1	4.5	4.2	4.3
20	Grip Diameter (Outside)	5.9	5.6	4.9	5.2	5.4	5.4
21	Foot Length	23.5	22.4	22.2	24.2	23.3	23.12
22	Foot Breadth	9.2	9.4	9.2	9.5	8.9	9.24
23	Age (years)	33	37	23	31	27	30.2

IOP Conf. Series: Materials Science and Engineering **814** (2020) 012032 doi:10.1088/1757-899X/814/1/012032

3. Rapid Upper Limb Assessment (RULA)

The RULA technique assesses individuals' work postures, muscle activities and forces which shows repetitive strain injuries. Its gives a resulted in the form of score form one (1) to seven (7), with colour zones as presented in table 2. Zone colour varies from green to red with respect to the score wherever higher scores point out greater levels of risk, low score of result doesn't assurance of workplace is free from ergonomic risks, and a high score doesn't promise that a severe problem present. It was established to have rated supplementary attention towards noticed postures of work or risk factors.

Table 2. Explanation of RULA score in basic mode

Sr. No.	Score	Colour	Meaning
1	1 and 2	Green	If the posture is not retained or Repeated for longer period then its acceptable
2	3 and 4	Yellow	Additional examination is essential and modifications may also be necessary.
3	5 and 6	Orange	Examination and modifications are desirable soon.
4	7	Red	Examination and modifications are necessary instantly.

4. Modelling and Ergonomic Analysis of Existing Maintenance Workstation

Results of RULA analysis are obtained using CATIA V5R19 software. From the video or photographs of maintenance workstation, posture is designed using the ergonomics module in CATIA V5R19 software. In MSRTC, Solapur workshop using Digital Human Modelling Technique in CATIA V5R19 software, RULA analysis of job-related disorders is performed.

This study was conducted at MSRTC, Solapur workshop Solapur District. The chosen workstation for the present study is related to maintenance department. A simple rating was given to the subject to calculate the postural awkwardness experienced. Also, the subject was asked to do his working series routinely and the process was noted through photography. limited postures of the maintenance person from his routine work are selected and simulated as a manikin in the CATIA V5R19 software. Further, the RULA analysis performed on the manikin with exact duplication to calculate the workers posture level of awkwardness. The maintenance practice involving some of the complicated postures of the worker was identified and designed using CATIA V5R19 software for ergonomic analysis. With the help anthropometric data worker, the manikin was modelled.

To select a critical posture among the activities carried out in the process of maintenance, RULA analysis has been carried out for the four postures as shown in figure 1. The results are presented in table 3, which tells us that the posture (d) is critical posture among the other posture as the score of posture is 7. (Note: score is less posture is ergonomically good, score is more posture is ergonomically not good).

Figure 1. picutre of postures while maintenance process in going on

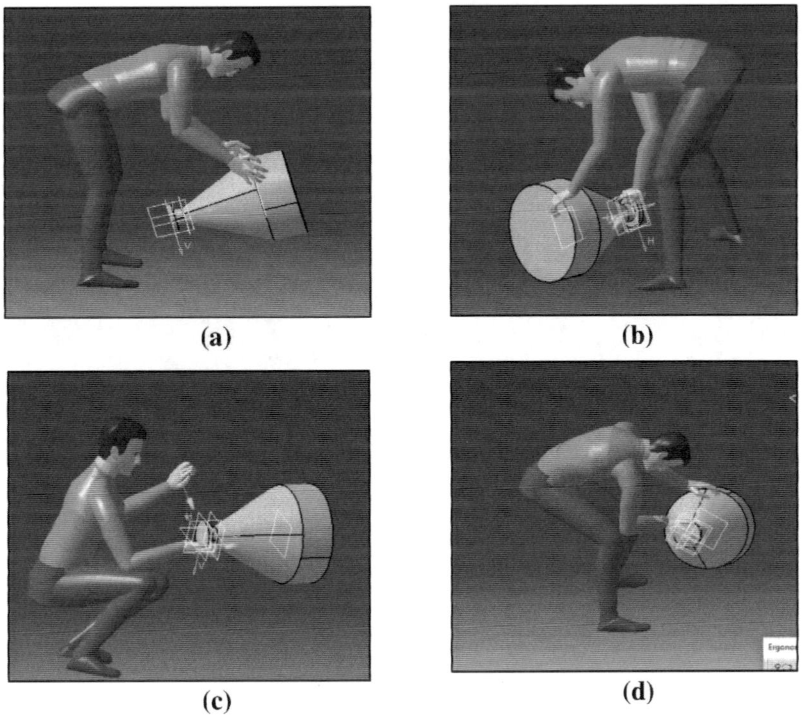

Figure 2. Mainkin modeling with job for posture a, b, c and d.

Table 3. RULA analysis result: posture a, b, c and d.

Posture	RULA analysis Score					
	Type of Loadings					
	Static		Intermitted		Repeated	
	Left hand	Right hand	Left hand	Right hand	Left hand	Right hand
a	7	6	6	5	7	6
b	7	7	7	6	7	7
c	7	7	7	6	7	7
d	7	7	7	7	7	7

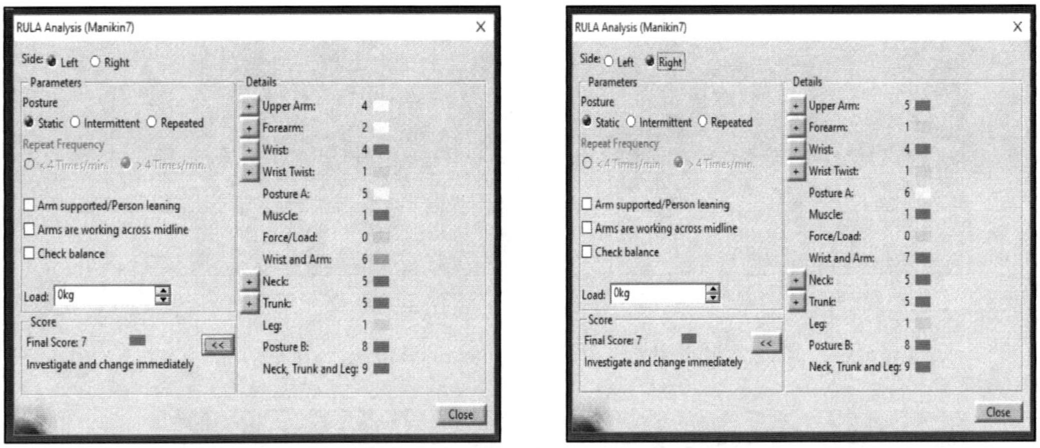

Figure 3. RULA analysis result of Posture (d) for static loading

Figure 3 shows static loading of both hands of manikin and corresponding posture score 7 and colour is red. This means posture should Examination and modifications are desirable instantly. Trunk, neck and leg are the stressful parts.

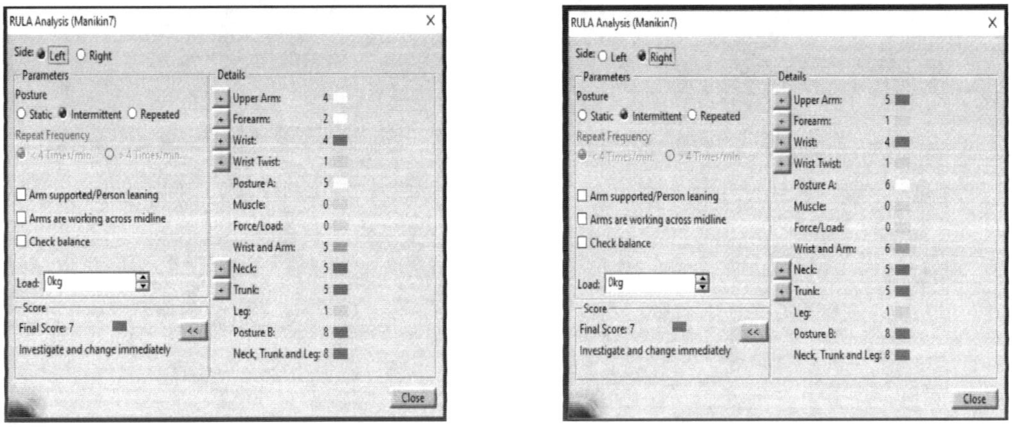

Figure 4. RULA analysis result of Posture (d) for Intermittent loading.

Figure 4 shows intermittent loading of both hands of manikin and corresponding posture score 7 and colour is red. This means posture should Examination and modifications are desirable instantly. Trunk, neck, and leg are stressful parts.

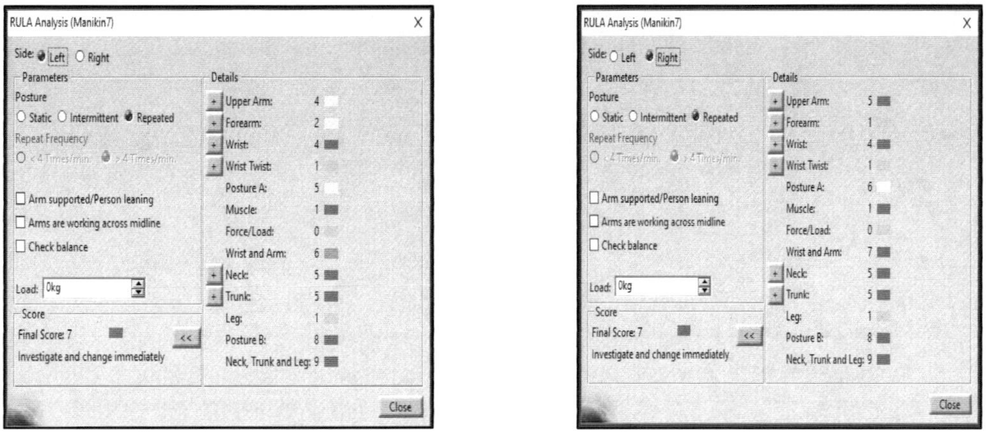

Figure 5. RULA analysis result of Posture (d) for Repeated loading

Figure 5shows repeated loading of both hands of manikin and corresponding posture score 7 and color is red. This means posture should Examination and alterations are desirable immediately. Trunk, neck and leg are the stressful parts.

From above analysis for different condition gives average score of existing workstation as a 7 which means Additional examination is essential and alterations may be necessary. As there is no proper maintenance workstation is available for the maintenance person, so it is advised that maintenance person should have a maintenance workstation designed with the consideration of agronomical principals, so we tried to the same and presented in this study as follows.

5. Modelling and Ergonomic Analysis of Modified Maintenance Workstation

A number of references permissible to safeguard workers ease and safety are possible. Repetitive work may result on the workers' wellbeing even if the load is minor. The workplace restoration may ensure a worthy and easy work atmosphere through modifying the systematic ergonomically designed maintenance workstation. It is recommended that modifying as shown in fig. 6 can be used to make workers postures safe. This construction of the platform allows the worker to do the work with minimum expenditure of energy. Excessive stress on neck, trunk, forearm, leg, etc. is reduced considerably, which allow efficient working of worker for large duration. Further health hazards of workers can be avoided. This will automatically increase the productivity, and reduce MSD and prevent injuries of worker along with enhancing productivity.

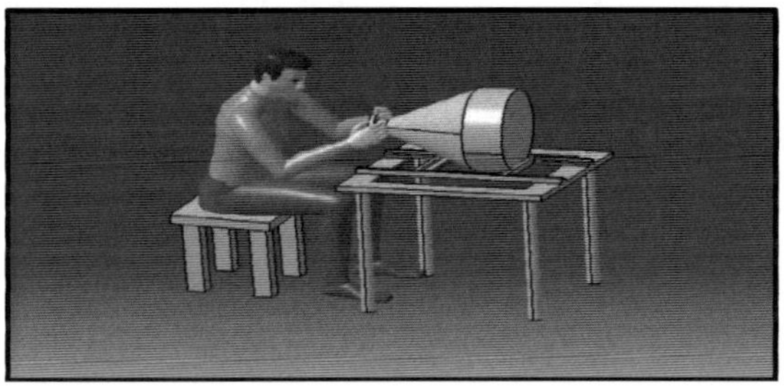

Figure 6. Modified Posture.

Figure 7 shows Static loading on both hands of manikin and corresponding posture score 3 and colour is Yellow. Additional examination is essential and alterations may be essential. The Trunk, neck and leg parts are in Yellow colour it menaces stresses are optimum.

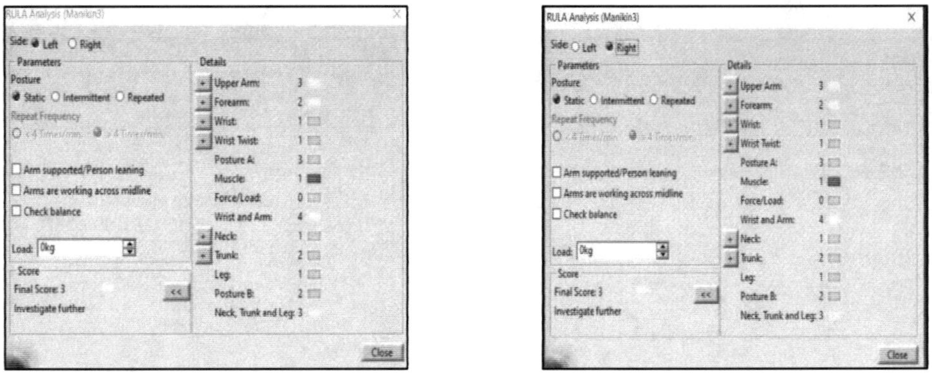

Figure 7. RULA Analysis results of Modified posture of static loading.

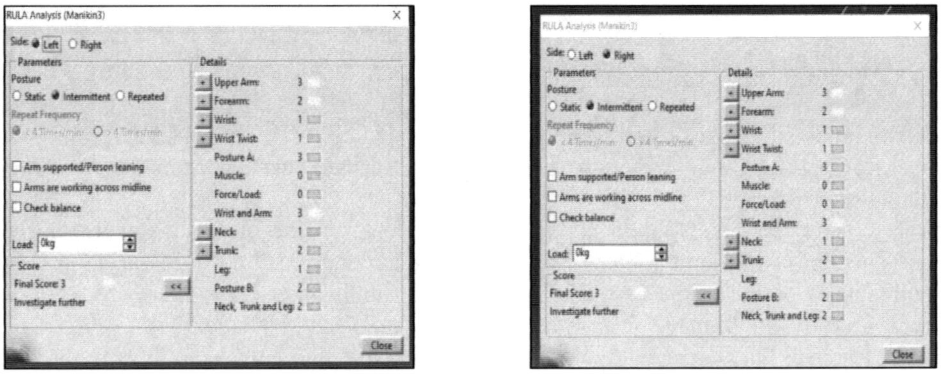

Figure 8. RULA analysis result of Modified Posture for Intermitted loading

Figure 8 shows intermitted loading on both hands of manikin and corresponding posture score 3 and colour is Yellow. Additional examination is essential and alterations may also be essential. The Trunk, neck and leg parts are in Yellow colour it manes stresses are optimum

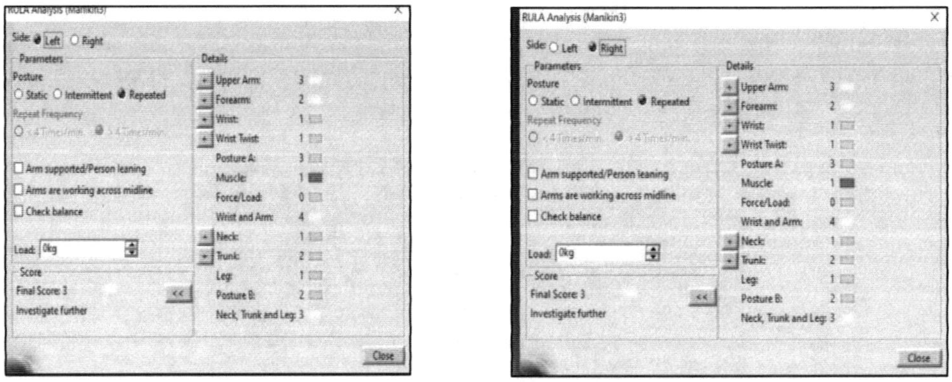

Figure 9. RULA analysis result of Modified Posture for repeated loading.

Figure 9sShows repeated loading on Both hands of manikin and corresponding posture score 3 and colour is Yellow. Additional examination and changes are essential.

According to RULA analysis the percentage score reduction comparing existing workstation with redesigned workstation is given below.

IOP Conf. Series: Materials Science and Engineering **814** (2020) 012032 doi:10.1088/1757-899X/814/1/012032

$$\% \text{ score reduction} = \frac{\text{Avg. score of Existing WS} - \text{Avg. score of Modified WS}}{\text{Avg. score of Existing WS}} \times 100$$

$$\% \text{ score reduction} = \frac{7-3}{7} \times 100$$

$$\% \text{ score reduction} = 57.14$$

6. Conclusions

The conclusions drawn from the study are as follows.

- Performance of worker at a workstation for finishing task has a significant effect on productivity. Hence, a systematic ergonomically modified workstation plays an important role in any maintenance workstation.
- RULA analysis for existing maintenance workstation shows the overall score 7, and hence there is need of immediate Examination and modifications.
- RULA analysis for redesigning workstation shows the overall Avg. Score 3 i.e. near about 57.14 % of reduction in RULA score was observed as compared to existing workstation.
- Modified workstation reveals more comfort than the existing workstation. This increases efficiency of worker along with productivity.
- Thus to reduce MSD as well as potential accidents and health problems and to enhance the performance at workplace ergonomic approach is very imperative.
- With reference to the analysis carried out, probably it is possible to increase comfort, efficiency of workers and productivity of the maintenance industry with the use of a systematic ergonomically designed workstation.

Acknowledgement

We would like to sincerely thankful to MSRTC, Solapur for imparting the provision to convey our research work and be inclined to thank Mr. Rajendra Koli (Maintenance Manager, MSRTC, Solapur).

References

[1] A. Kumar, D. Mohan, R. Patel, M. Varghese 2002 *Applied Ergonomics* **33** 503–508.
[2] Mali S. C., and Vyavahare R. T. 2015 *Int. J. of Current Engineering and Tehnol.* **5 (3)** 1820.
[3] Rahman, C. M. 2014 Int. J. of Engg. **5 (3)** 8269.
[4] Vyavahare R. T. and Kallurkar S. P. 2016 *Int. J. of Industrial Ergonomics* **53** 80-85.
[5] Vaclav S., Senderska, K., Mares A. *article written for grant project VEGA* **1** 0130.
[6] Das B., and Sengupta A. K. 1996 *Applied Ergonomics* **27 (3)** 157-163.
[7] Prakash R. A., Arun, B. S., Narayanan, N. and Yeshwanth V. 2013 *Int. J. R. in Engg. and Tech,* **2(1)** 349-355.
[8] Vyavahare R. T., Kallurkar S. P. 2012 *Agricultural Engg. Inter. CIGRJ* **14 (4)** 102-114.
[9] Kale P. N., Vyavahare R. T. 2016 *Int.. J.of Current Engg. and Tech.* **6 (4)** 1271-80.
[10] Ansari N. A., Shende P. N., Sheikh M. J., Vaidya R. D. 2013 *Inter. J. of Engg and Adv. Tech (IJEAT)* **2** 3.
[11] Chandra A., Chandna P., Deswal S. 2011 *Int.. J. of Engg,* **5 (3)** 242-256.
[12] Stefan V., Katarina S., Albert M. 2008 *Article written for grant project.* **1** 0130.
[13] Ansari N. A., Sheikh M. J. 2014 *IOSR J. of Mech. and Civil Engg.* **11 (4)** 18-23.
[14] Shikdar A., Garbie I., Khadem M. 2011 *Int. Conf. on Indu. Engg. and Operations Management (IEOM)*, Kula Lumpur, Malaysia **5 (3)** 233-240.
[15] Mohamad D., Deros B. M., Ismail A. R., Daruis and Sukadarin, E.H. 2013 *Advanced Engineering Forum* **10** 9-15
[16] Yadhu G, Raghunathan Rajesh, 2014 *Int.. J. of Innov. R. in Science Engg. and Tech.* **3 (9)** 15894- 15902.

IOP Conf. Series: Materials Science and Engineering **814** (2020) 012033 doi:10.1088/1757-899X/814/1/012033

The nonlinear deflection response of CNT/nanoclay reinforced polymer hybrid composite plate under different loading conditions

Achchhe Lal[1] and Nand Jee Kanu[2]

[1]Assistant Professor, Department of Mechanical Engineering, SVNIT, Surat-395007, Gujarat, India
[2]Ph. D. Research Scholar, Department of Mechanical Engineering, SVNIT, Surat-395007, Gujarat, India

E-mail: nandssm@gmail.com, lalachchhe@yahoo.co.in

Abstract. The carbon nanotubes (CNTs) are used as significant reinforcement materials in the advanced composites due to their improved mechanical properties. The carbon nanotubes reinforced nanocomposites have improved mechanical strength, and stiffness properties in addition to their reduced weight. Only a few percentages of CNTs (2–5% by weight) could be added to these nanocomposites as more volume fraction will deteriorate their mechanical properties. A computational approach is used in the present investigation to find the mechanical properties of the hybrid nanocomposite which is modeled by adding nanoclay particles into the conventional CNTs reinforced epoxy system. Using Halpin-Tsai approach (which is based on the micromechanical modeling) the mechanical properties of hybrid nanocomposites are evaluated. The effect of the CNTs and the nanoclay particles on the nonlinear transverse central deflection response of nanocomposite plate under different loading conditions is studied in detail. The simulation results of present investigation have been validated with the data sets of the deflection responses available in the references.

1. Introduction

The investigations about the material properties of CNTs are making way to variation of the elasticity modulus within *200 GPa* to *5.6 TPa* due to the number of the CNT walls, length, diameter, etc. Now-a-days, the carbon nanotubes (CNTs) are in limelight due to their improved physical and chemical features [1]. The increase in CNTs in the matrix could enhance the required mechanical properties of the smart CNTs reinforced composites. It is important to note that increase in CNT volume fractions will lead to improvement in stiffness of CNTs reinforced polymer composites. On the contrary, the experimental data however, advocates about the reduction in stiffness after addition of CNTs volume fractions beyond a certain limit in the matrix [2]. It is due to the agglomeration of CNTs within the matrix material. In the present investigation, we aim to investigate the elastic nonlinear deflection response of the hybrid nanocomposite plate under different loading conditions using the shear deformation theory based on the secant function (SFSDT) [1, 2].

2. Modeling of the hybrid nanocomposite plate

The modeling of the CNTs/nanoclay based nanocomposite plate is proposed in two stages such as micromechanical modelling and numerical modelling using finite element method (FEM) [1]. The elastic properties of effective CNT fibers and Clay particles including CNTs, and Clay interphases are computed using modified Halpin–Tsai model as shown in the set of equations (1).

$$\left.\begin{aligned} &\frac{P}{P_m}=\frac{1+\left(\varepsilon_1\eta_{CP}V_{CP}+\varepsilon_2\eta_{CF}V_{CF}\right)}{1-\left(\eta_{CP}V_{CP}+\eta_{CF}V_{CF}\right)};\\ &\eta_{CP}=\frac{\left(P_{CP}/P_m\right)-1}{\left(P_{CP}/P_m\right)+\varepsilon_1};\\ &\eta_{CF}=\frac{\left(P_{CF}/P_m\right)-1}{\left(P_{CF}/P_m\right)+\varepsilon_2};\\ &V_m+V_{CP}+V_{CF}=1 \end{aligned}\right\} \tag{1}$$

where, $\varepsilon_1=\left(2l/d_p\right)$, $\varepsilon_2=\left(2l/d\right)$ and $\varepsilon_1=\sqrt{3}\log\left(l/d_p\right)$, $\varepsilon_2=1$ are used for the longitudinal and transverse moduli of elasticity, respectively. The symbols *"l"* and *"d"* are used to represent the length and diameter of the CNTs. In the above equation (1), V_m, V_{CP}, and V_{CF} are the volume fractions of matrix, Clay, and CNT, respectively, and ε_1 and ε_2 are parameters used for the Clay and CNT, respectively. Further, thicknesses of the effective Clay particle and CNT can be calculated as $d_p=d_c+2d_1$, and $d=2r$, respectively [1].

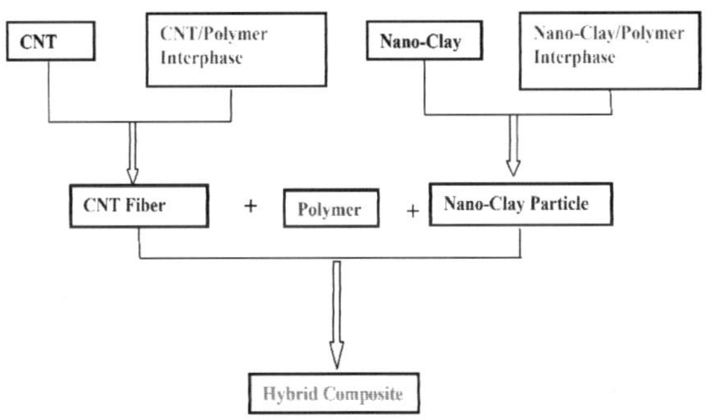

Figure 1. The layout for evaluating elastic properties of the SWCNTs reinforced hybrid nanocomposite plate [1] (Under CC Attribution 3.0 License. Copyright 2016 IOP Publishing Ltd)

Figure 2. The schematic representation of unit volume element of the CNTs/nanoclay reinforced polymer [1] (Under CC Attribution 3.0 License. Copyright 2016 IOP Publishing Ltd)

In Figure 1 the layout for evaluating the elastic properties of single walled carbon nanotubes (SWCNTs)/nanoclay reinforced hybrid nanocomposite plate using Halpin–Tsai approach is illustrated. In Figure 2 the schematic representation of unit volume element of the CNTs/nanoclay reinforced polymer nanocomposite plate is demonstrated.

2.1. Governing equations

The displacement field is expressed using the shear deformation theory (SFSDT) based on the secant function [3, 4], which involves the nonlinear shear stress distribution. The displacement fields for laminated Single walled carbon nanotube (SWCNT) plate are shown in equations (2), and (3).

$$\left.\begin{aligned}\bar{u}(x,y,z)&=u(x,y)-z\frac{\partial w}{\partial x}+\left(g(z)+z\Omega\right)\theta_x(x,y);\\ \bar{v}(x,y,z)&=v(x,y)-z\frac{\partial w}{\partial y}+\left(g(z)+z\Omega\right)\theta_y(x,y);\\ \bar{w}(x,y,z)&=w(x,y)\end{aligned}\right\} \tag{2}$$

where,

$$g(z)=z\sec\left(\frac{rz}{h}\right);\ \Omega=\frac{\sec\left(\frac{r}{2}\right)}{\left(1+\frac{r}{2}\tan\left(\frac{r}{2}\right)\right)} \tag{3}$$

In the above equation (3), the symbol $'r'$ is used for the transverse shear stress and its value is ascertained by the inverse method in post processing step, the symbol $'\Omega'$ is used for a constant and its value is evaluated by including transverse shear stress boundary conditions to nullify values of transverse shear stresses at the boundary, and the symbol $'g(z)'$ is used for the stress strain functions. The complexity and difficulty involved while making a choice of C^1 continuity are well known, and hence a C^0 continuous element would be sufficient for the finite element analysis. Using independent variables such as φ_x and φ_y, C^1 continuity is modified to C^0 continuity as shown in equation (4).

$$\varphi_x=\frac{\partial w}{\partial x};\ \varphi_y=\frac{\partial w}{\partial y} \tag{4}$$

The displacement field vector $\{\Lambda\}$ can be expressed as shown using equation (5).

$$\{\Lambda\}=\left[u\quad v\quad w\quad \theta_y\quad \theta_x\quad \varphi_y\quad \varphi_x\right]^T \tag{5}$$

2.1.1 Strain-displacement relationships. The final strain is shown using equation (6).

$$\varepsilon=\varepsilon_l+\varepsilon_{nl} \tag{6}$$

The Von-Karman theory is used to express the relationships between strains and displacements as shown in equation (7).

$$\left.\begin{aligned}\varepsilon_x&=\frac{\partial u}{\partial x}+\frac{1}{2}\left(\frac{\partial w}{\partial x}\right)^2;\\ \varepsilon_y&=\frac{\partial v}{\partial y}+\frac{1}{2}\left(\frac{\partial w}{\partial y}\right)^2;\\ \varepsilon_{xy}&=\frac{\partial u}{\partial x}+\frac{\partial v}{\partial y}+\left(\frac{\partial w}{\partial x}.\frac{\partial w}{\partial y}\right);\\ \varepsilon_{xz}&=\frac{\partial u}{\partial z}+\frac{\partial w}{\partial x};\\ \varepsilon_{yz}&=\frac{\partial v}{\partial z}+\frac{\partial w}{\partial y}\end{aligned}\right\} \tag{7}$$

The linear strain vectors are shown using equations (8), and (9).

$$\{\varepsilon_l\}=\left\{\varepsilon_1^0\quad \varepsilon_2^0\quad \varepsilon_6^0\quad k_1^0\quad k_2^0\quad k_6^0\quad k_1^2\quad k_2^2\quad k_6^2\quad \varepsilon_4^0\quad \varepsilon_5^0\quad k_4^2\quad k_5^2\right\}^T \tag{8}$$

where,

$$\left.\begin{aligned}&\varepsilon_1=\varepsilon_1^{\,o}+z\left(k_1^o+z^2k_1^2\right);\\&\varepsilon_2=\varepsilon_2^{\,o}+z\left(k_2^o+z^2k_2^2\right);\\&\varepsilon_3=0;\\&\varepsilon_4=\varepsilon_4^{\,o}+z^2k_4^2;\\&\varepsilon_5=\varepsilon_5^{\,o}+z^2k_5^2;\\&\varepsilon_6=\varepsilon_6^{\,o}+z\left(k_6^o+z^2k_6^2\right)\end{aligned}\right\} \tag{9}$$

The nonlinear strain vector ε_{nl} corresponding to displacement fields are shown using equation (10).

$$A=\frac{1}{2}\begin{bmatrix}\frac{\partial w}{\partial x} & 0\\ 0 & \frac{\partial w}{\partial y}\\ \frac{\partial w}{\partial x} & \frac{\partial w}{\partial y}\\ 0 & 0\\ 0 & 0\end{bmatrix};\theta=\begin{Bmatrix}\frac{\partial w}{\partial x}\\ \frac{\partial w}{\partial y}\end{Bmatrix} \tag{10}$$

2.1.2 Constitutive relation. Stress–stain relationship can be expressed as shown using equation (11).

$$\begin{Bmatrix}\sigma_x\\ \sigma_y\\ \tau_{xy}\\ \tau_{yz}\\ \tau_{xz}\end{Bmatrix}=\begin{Bmatrix}\sigma_1\\ \sigma_2\\ \sigma_6\\ \sigma_4\\ \sigma_5\end{Bmatrix}=\begin{bmatrix}\bar{Q}_{11} & \bar{Q}_{12} & \bar{Q}_{16} & 0 & 0\\ \bar{Q}_{12} & \bar{Q}_{22} & \bar{Q}_{26} & 0 & 0\\ \bar{Q}_{16} & \bar{Q}_{26} & \bar{Q}_{66} & 0 & 0\\ 0 & 0 & 0 & \bar{Q}_{44} & \bar{Q}_{45}\\ 0 & 0 & 0 & \bar{Q}_{45} & \bar{Q}_{55}\end{bmatrix}\begin{Bmatrix}\varepsilon_x\\ \varepsilon_y\\ \gamma_{xy}\\ \gamma_{yz}\\ \gamma_{xz}\end{Bmatrix} \tag{11}$$

where, the symbols 'Q_{ij}' and 'θ_k' are used for the elastic constants and fiber orientation of the lamina.

3. Results and discussions

Using MATLAB code, the nonlinear transverse central deflection of single walled carbon nanotubes reinforced composite plate, is evaluated. In Table 1 [1], the material properties of all phases in 5-phase composite are listed and the results have been validated in Table 2 for different CNTs and Clays conditions.

Table 1. Proposed material properties [1]

Material Specification	Modulus of Elasticity (GPa)	Poisson's Ratio (v)	Details
Epoxy	2.026	0.4	$25\times 25mm$
CNT	1054	0.25	$r_i=0.315mm; r_o=0.650mm$
Clay	178	0.28	$t=4mm; d_c=1nm$
Interphase (CNT/Polymer)	16.10	0.4	$r_{\text{int}}=1.404nm$
Interphase (Clay/Polymer)	16.10	0.4	$d_i=3nm$

In Table 2, the effects of nanoclay particles and CNTs in nanocomposite are discussed through the variations in Clays and CNTs. The results show that with increase in Clay and CNT particles, the material properties will be improved as shown in Table 2.

The convergence of the transverse central deflection was observed at (4×4) mesh size as shown in Table 3. Therefore, a minimum of (4×4) mesh size is considered in the present investigation to reduce the computational effort.

Table 2. Validation of the material elastic property with variation in CNT and Clay particle

Particle variation	E_1		E_2		G_{23}		ν_{12}		ν_{21}	
	Present	[1]	Present	[1]	Present	[1]	Present	[1]	Present	[1]
$1CNT + 1Clay$	5.941	5.3245	2.563	2.420	1.836	0.7868	0.397	0.3938	0.384	0.5382
$1CNT + 4Clay$	7.667	10.2846	2.948	3.0908	2.529	0.9929	0.382	0.3917	0.312	0.5564
$1Clay + 4CNT$	10.039	10.3201	2.820	2.6521	2.556	0.8381	0.396	0.3935	0.381	0.5820

Table 3. Convergence study using different elements for square plate

Mesh density	Present result (W_o)	Published result [5]
	$a/h = 10$	
3×3	0.3177	0.3000
4×4	0.3168	
5×5	0.3168	

The boundary conditions are employed for nonlinear responses in the nanocomposites: (a) for clamped four edges (CCCC) ($u_0 = v_0 = w_0 = \theta_x = \theta_y = \varphi_x = \varphi_y = 0$; at $x = 0,\ a$, and $y = 0,\ b$), (b) for simply supported four edges (SSSS) ($u_0 = w_0 = \theta_y = \psi_y = 0$; at $x = 0,\ a$; and $u_0 = w_0 = \theta_x = \psi_x = 0$; at $y = 0,\ b$).

In Figures 3 and 4, the values of the transverse central deflection using different load parameters for various practical variations, and loading conditions, respectively as per boundary condition such as $CSCS; a/h = 10$; and stacking sequence as $(0^o / 45^o / -45^o / 90^o)$, are shown. The different combinations of CNTs and Clays under uniformly distributed loading conditions have been tested in Figure 3 to investigate the maximum transverse central deflection in case of 1Clay layout. It is observed about the plate (having 1Clay) that it is less stiff than other combinations considered for analyses under the plot (Figure 3) so far.

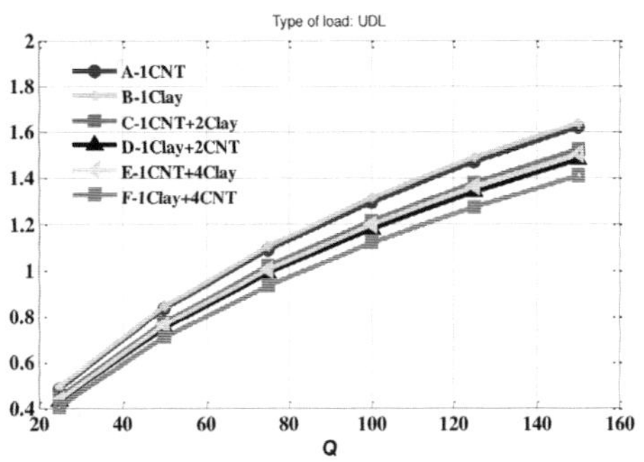

Figure 3. Values of transverse central deflection using different load parameters for various practical variations as per boundary condition

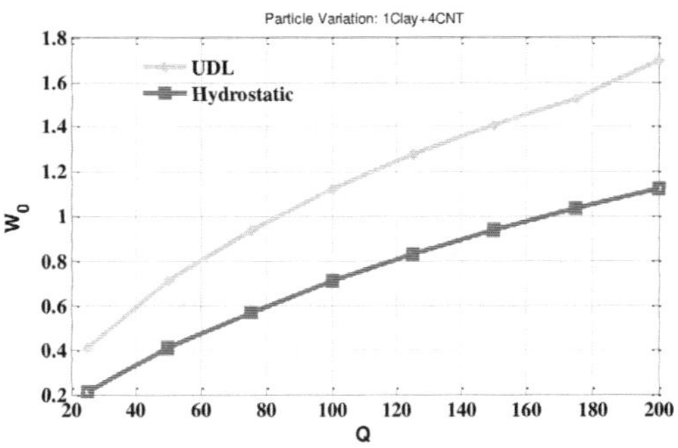

Figure 4. Values of transverse central deflection using different load parameters for different loading conditions as per boundary condition such

such as $CSCS; a/h = 10;$ and stacking sequence as $(0^o/45^o/-45^o/90^o)$

as $CSCS;$ $a/h = 10;$ and stacking sequence as $(0^o/45^o/-45^o/90^o)$

In Figure 5, variation in the transverse central deflection using different load parameters for different boundary conditions while $a/h = 20;$ and stacking sequence as $(90^o/0^o/0^o/90^o)$, is shown. The transverse central deflection response is however maximum when edges of the nanocomposite plate are simply-supported. In that way the nonlinear transverse central deflection response is minimum when edges of the nanocomposite plate are clamped.

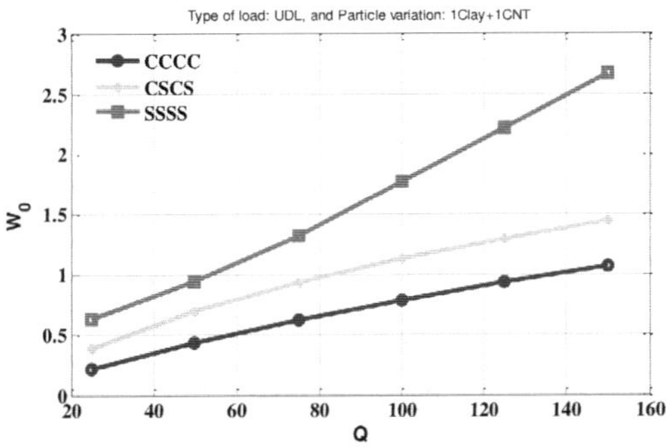

Figure 5. Variation in transverse central deflection using different load parameters for different boundary conditions while $a/h = 20;$ and stacking sequence as $(90^o/0^o/0^o/90^o)$

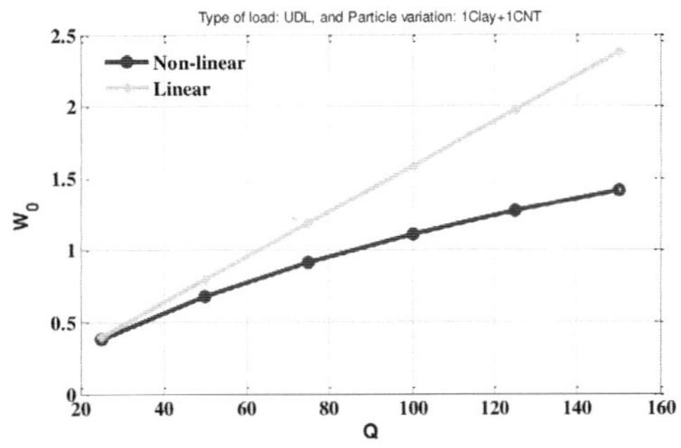

Figure 6. Variation in transverse central deflection using different load parameters for linear and nonlinear analyses as per boundary condition such as $CSCS;$ $a/h = 20;$ and stacking sequence as $(-45^o/90^o/90^o/-45^o)$

In Figure 6, variation in the transverse central deflection using different load parameters for linear and nonlinear analyses as per boundary condition such as $CSCS;$ $and\ a/h = 20;$ and stacking sequence as $(-45^o/90^o/90^o/-45^o)$, is shown.

4. Conclusions

In the present investigation, the nonlinear deflection response of CNTs reinforced composite plate (modeled using Halpin–Tsai Model in MATLAB environment based on the present mathematical model) is successfully carried out so far. The transverse central deflection is more in case of the uniformly distributed loading conditions; however, it is less in case of hydrostatic loading conditions under different load parameters (plotted along X-axis). It is observed that transverse central deflection is reduced while increasing number of CNTs to four. The response of nonlinear transverse central deflection is less in case of CCCC boundary condition among the other support conditions considered in the present investigation. The plate, whose edges are simply supported, has shown highest transverse central deflection during analyses.

References

[1] Arvind Kumar Thakur et al. 2016 *IOP conf. ser., Mater. sci.* **115** 012007
[2] Kulmani Mehar and Subrata Kumar Panda 2017 *Adv Polymer Technol.* **37(6)** 1–15
[3] Achchhe Lal et al. 2016 *Comp Mater Sci.* **5(4)** 1650020
[4] Grover Neeraj et al. 2014 *Latin Am. J. Solids Struct.* **11(7)** 1275–1297
[5] N. S. Putcha and J. N. Reddy 1986 *Compos. Struct.* **22(4)** 529-538

IOP Conf. Series: Materials Science and Engineering **814** (2020) 012034 doi:10.1088/1757-899X/814/1/012034

Effect of structured surface on contact angle using Sessile Droplet method

Biswajit Majumder[1], Anil Katarkar[2], Swapan Bhaumik[3]

[1] Ph.D. Scholar, National Institute of Technology, Agartala, India and Assistant Professor, Tripura Institute of Technology Narsingarh, Tripura.
[2] Ph.D. Scholar, National Institute of Technology, Agartala.
[3] Associate Professor, Mechanical Engineering Department, National Institute of Technology, Agartala, 799046.

E-mail: bmajumdertit@gmail.com

Abstract. At the age of energy-crisis, enhancement of boiling heat transfer is the most crucial and advanced investigation field. Effects of surface characteristics, specially, surface wettability on pool boiling is yet to understand. Wetting, the ability of solid surface to reduce surface tension of a liquid in contact, can specify by measuring contact angle. Sessile drop method is a very sensitive process to measure contact angle, mainly used for evaluating, comparing and monitoring test surfaces. In this experimentation, a set of copper surfaces are roughened by emery grit size of 600, 1000, 1500 and 2000. Data on surface roughness (generated by AFM and FEG-SEM) and on dynamic contact angle (by macroscopic contact angle (CA) meter) are used to find relation between them. De-ionized water is used as working fluid to protect submerged secondary heater. In this experimentation, it is observed that advancing contact angle grows for surface roughened with higher emery grit-size.

1. Introduction

In our surroundings, everywhere in the nature, it can be observed that liquid encircles solid surface, and is most common natural phenomenon. This liquid encircling above the solid surface is known as Wetting. It is a kind of surface ability to trim its surface tension down. This can be noticed for a liquid in contact with it. In this era of innovation, it is very possible to detect test surface properties, especially by measuring contact angle. In literatures available in this field from the ages of Young to recent months, researchers are struggling to formulate the model to describe the inter-relation between surface wettability and contact angle [1, 2, 3, 4-8]. Papers are there amplifying contact angle from the end of surface roughness [9- 12]; some papers even discuss the possibilities of changing surface properties by forming thin film [13, 14] and in some papers effort to narrate surface roughness by wettability[15-18] is also visible. As a consequence, an opportunity, to expand a mathematical model forforecastingthe test plane roughness, is growing.

Again,due to the above study, it is very evident that few effort was made to develop coorelation amongliquid drop geometries, wettability and surface irregularity. Here, it is tried to scrutinize the dependency of contact angle on surface roughenedby different grits of emery paper and other geometric parameters. These parameters are associated with the drop which forms contact angle on surfaces roughened by different grit of emery. Intension of this work is to anticipate surface roughness and surface free energies when only data available is contact angle.

2. Substrate Characterization

2.1 Test surface preparation

Copper (C10800) is mainly used as tubing in refrigeration and air conditioning industries. Thus it is taken as experimental substrate due to its higher thermal and electrical conductivity, low cost.In figure1, shape of the substrate is illustrated; it is cylindrical in shaped with flat top and bottom. It has two step-wise sections; upper section is smaller in diameter than lower section.Dimension of the upper section is 09 mmϕ × 01 mm and that of the lower section is 11mmϕ × 02 mm. For testing, end surface of upper section is used after surface modifications. Total fournumbers of surfaces are used for this experiment; one each are polished by emery grit size of 600, 1000, 1500 and 2000 respectively. Then all surfaces are cleaned by acetone bath and rapped by tissue paper and kept with silica gel.

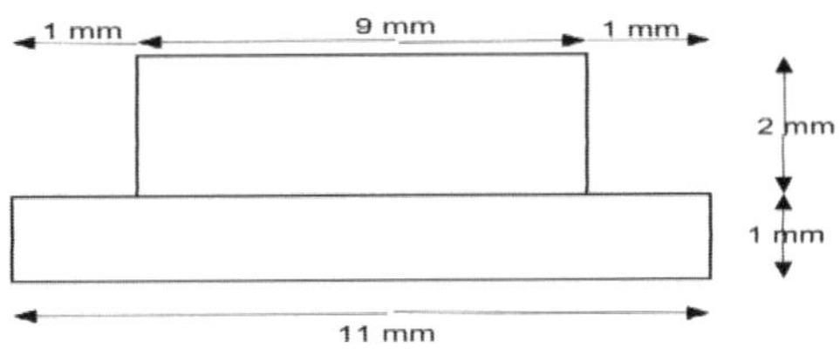

Figure 1. Dimension of Substrate

2.2 Surface Characterization

Two ways of characterizations, i.e., structural characterization and wettability characterization are applied for generation of data set.

2.2.1. Structural Characterization.

For measuring surface roughness for structured surface, AFM was conducted at Physics Department, NIT Agartala and related 1st set of AFM images were presented in a tabular form below:

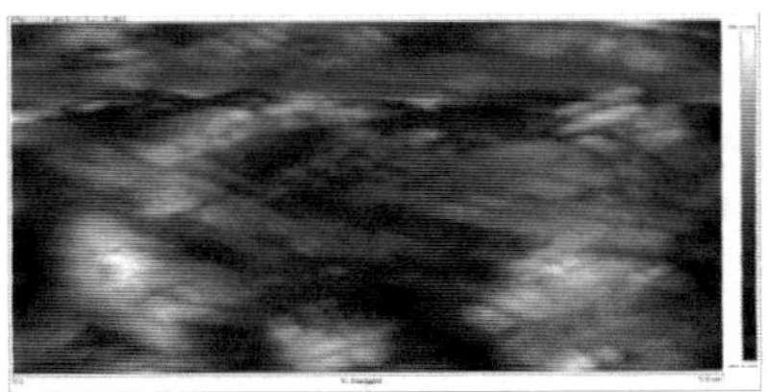

For 600 grit: R_a = 19.6 nm

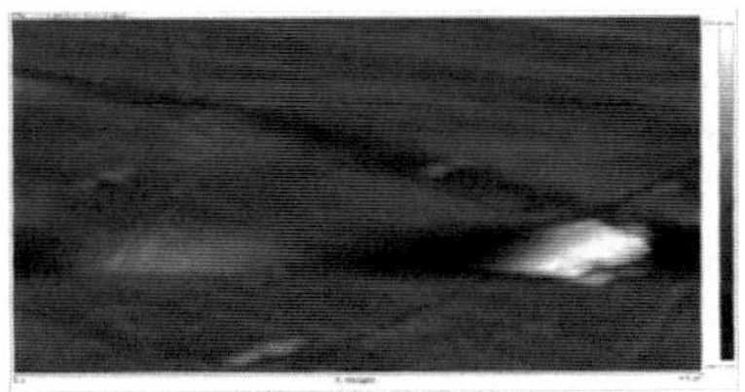

For 1000 grit: R_a = 20.9 nm

For 1500 grit: R_a = 38.7 nm

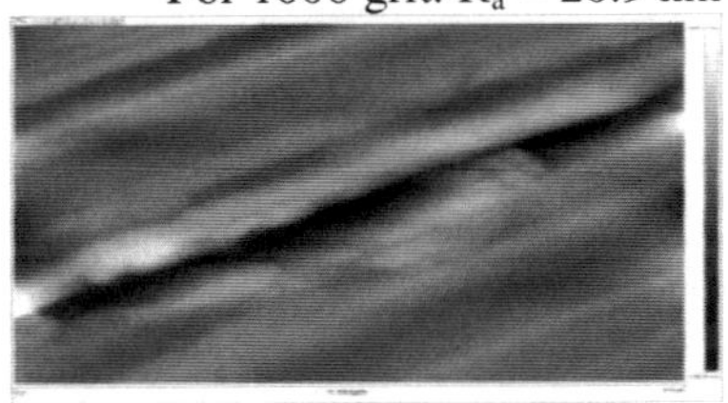

For 2000 grit: R_a = 51.0 nm

Figure 2. 2D AFM Images for roughened surface by different emery paper

2.2.2. Wettability Characterization.
Water has been used as test fluid under sub-cooled condition at 20°C and atmospheric pressure with 45% RH. For measuring contact angle, Sessile drop method has been used. All trials are conducted on Microscopic Contact Angle Meter at IIT Patna. To generate contact angle profiles interface software FAMAS for Contact Angle measurement Format Version: 3.4are used.

2.3. Computation
2.3.1. Anticipating Apparent Contact Angle (CA) and Surface Roughness.
It is known that for plain and ideal surface, surface free energy (SFE) operated in liquid -solid interface and angle formed by liquid with solid surface, has one relation as proposed by Young [19], which is,

$$\Upsilon_{sv} = \Upsilon_{sl} + \Upsilon_{lv}.\cos\theta \quad (1)$$

or,

$$\Upsilon_{s} = \Upsilon_{sl} + \Upsilon_{l}.\ \cos\theta \quad (2)$$

2.3.2. Anticipating wettability of the surface from data of Dynamic CA.
It has been postulated that the investigational advancing contact angle can be the good estimate of the Young's contact angle [20].If result is H < 50°, test surface is homogeneous in nature; and, for H > 50°, test surface is of heterogeneous nature.
When in eq^n. (2), θ is replaced by θ_a, it will provide,

$$\Upsilon_{s} = \Upsilon_{sl} + \Upsilon_{l}.\ \cos\theta_a \quad (3)$$

Then with the help of Chibowski [1, 2] correlation SFE of the solid surface can be determined, considering the value of surface tension of water at 20°C is $\Upsilon_{lv} = \Upsilon_l = 72.8$ mN/m [21]
Using this value of Υ_l in the eq^n. 3, the value of Υ_{sl} can find out. Smaller values of Υ_{sl} indicate weaker SFE which indicates increasing wetting tendency.
In this present experiment, flat and solid surface is used. So, the wettability can be predicted by the Young–Dupre equation for the condition that neither Υ_s nor Υ_{sl} can be larger than the sum of other two surface energies. It imposes restriction due to which forecasting of complete wetting (when $\Upsilon_s > \Upsilon_{sl} + \Upsilon_l$) or zero wetting (when $\Upsilon_{sl} > \Upsilon_s + \Upsilon_l$) becomes possible. This wetting phenomenon can be measured both experimentally and numerically. 'Affinity of the liquid for the solid' is always opposed by the 'friction between liquid molecules and solid molecules on the substrate'. This concept is used for measuring wetting phenomena. It is recognized as spreading coefficient and expressed as,

$$S = \Upsilon_s - (\Upsilon_{sl} + \Upsilon_l) \quad (4)$$

When S greater than 0, liquid spreads all over, complete wetting occurs; whereas, when S is less than 0, incomplete wetting occurs and well- defined contact angle formed.
As per the method described above, characteristics of test surfaces in trialling are anticipated and results tabled in Table no.1.

2.3.3. Determining wettability of the surface from Apparent CA.
As an equilibrium CA, apparent CA can be calculated macroscopically on solid surface of rough or heterogeneous nature.This parameter along with geometrical structure of substrates helps to calculate

the wetting of the substrate. Equilibrium condition can be reached with the spread of drop, displacing all air with liquid. According to Tadmor – Chibowski model [22],

$$\theta_{ap} = \arccos\left(\frac{r_a cos\theta_a + r_r cos\theta_r}{r_a + r_r}\right) \quad (4a)$$

$$\text{Where, } r_a = \left(\frac{sin^3\theta_a}{2-3cos\theta_a+cos^3\theta_a}\right)^{\frac{1}{3}} \text{ and, } r_r = \left(\frac{sin^3\theta_a}{2-3cos\theta_r+ cos^3\theta_r}\right)^{\frac{1}{3}}$$

On real (rough or contaminated) surface, contact angle hysteresis is observed. So, local equilibrium contact angle may differ from point to point on surface [23].

2.3.4. Anticipating Surface Roughness Ratio (r).

With the augment of surface roughness, effective surface area increases [24].When Wenzel state is considered,

$$\cos(\theta_{ap}) = r \cos(\theta_y) \text{ [valid for } r > 1] \quad (5)$$

where, θ_{ap} is apparent contact angle; r =[real rough surface area/ projected perfectly smooth surface]; and (θ_y) is the Young's contact angle. But at r = 1,above relation becomes $\cos(\theta_{ap}) = \cos(\theta_y)$.

At the valley of the rough surface at the solid-liquid interface [25],

$$cos\theta_{ap} = f_{LS}\left[\cos\theta_y + 1\right] - 1 \quad (6)$$

For real heterogeneous rough surface, it becomes necessary to combine eq^n. (7) and eq^n. (9)

$$cos\theta_{ap} = r f_{LS}\left[\cos\theta_y + 1\right] - 1 \quad (7)$$

Now, using the correlation presented by Whyman et al. [22], and putting values of θ_{ap}, $\theta_y$and $\emptyset_{LS}$, surface roughness ratio (r_p) can be anticipated.

But, using values obtained from AFM data, Surface roughness ratio (r_{AFM}) can be defined and as shown in Table 1.

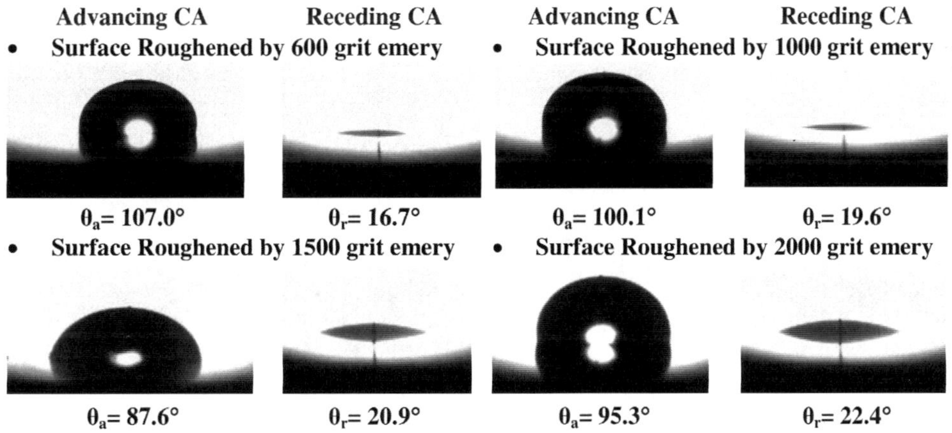

Figure 3: Advancing and receding angle for respective surfaces 'roughened by different grit of emery'. (Images are from CA meter)

2.4. Uncertainty while measuring Dynamic Contact Angle

For Contact angle Meter, uncertainty of the dynamic contact angle is approximated as ±1°. For emery roughened surface, percent uncertainty of surfaces is smaller (less than 1%). As uncertainty are calculated 3.72% of the experimental values, within the tolerable range of ± 5%, making the data set acceptable [26].

3. Result and Discussion

Presently surface roughness ration ® is calculated using Young – Laplace Model on assumption that surfaces are homogeneous perfectly flat and smooth; and the liquid used is pure. But in experiments, cases are not real. There are heterogeneity in the surfaces and liquid used are not 100% pure.

Table 1: Calculatingroughness of test surface roughened by emery paper.

Copper Surface	Treated surface by emery grit size			
	600	1000	1500	2000
Θ_a (°)	107.00	100.10	87.60	95.30
Θ_r (°)	16.70	19.60	20.90	22.40
Hysteresis (H)	90.30	80.50	66.70	72.90
Anticipated SFE on solid surface Υ_s [mN/m]	12.30	14.32	14.73	12.50
Anticipated SFE on solid-liquid interface Υ_{sl} [mN/m]	25.70	15.72	9.40	24.77
Anticipated Apparent CA ($\theta_{ap\ predicted}$)	49.27	49.01	51.05	57.41
Mean Apparent CA ($\theta_{ap\ mean}$)	61.85	59.85	54.25	58.85
Anticipatedr_p	0.83	0.85	0.84	0.79
$r_m$derived from mean apparent CA	0.80	0.75	0.83	0.77
$r_p$by AFM data	0.97	0.98	0.98	0.97

Relations among different parameters for each surfaces acquired from Contact Angle Meter are plotted below in a tabular form.

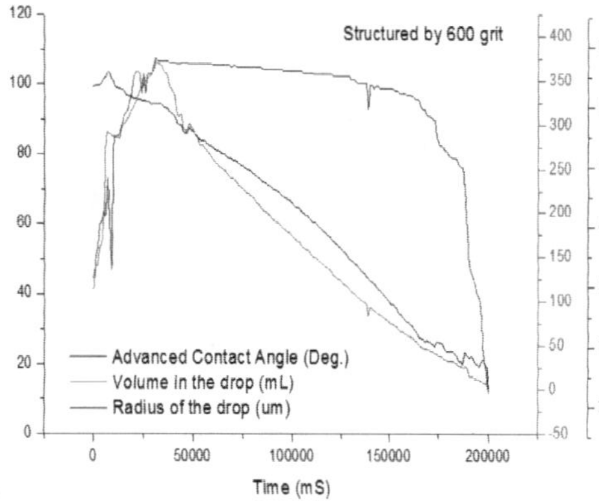

(a)

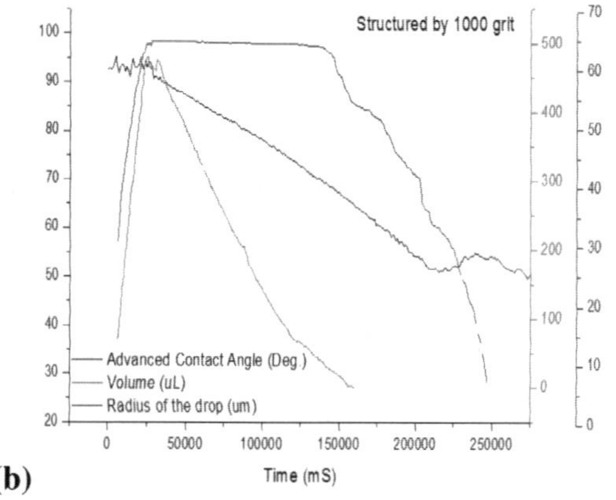

(b)

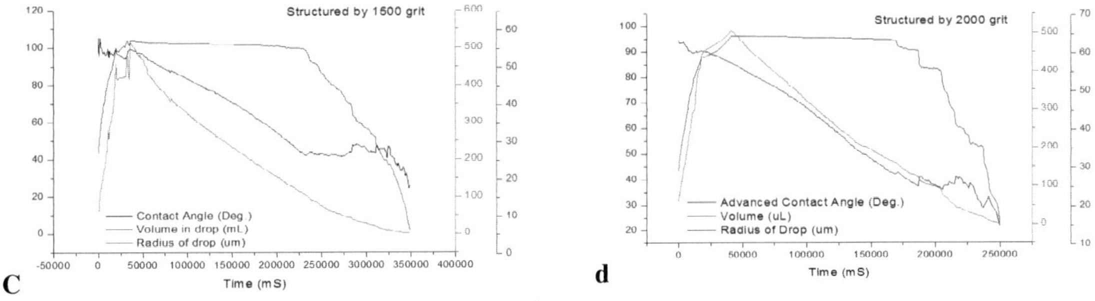

Figure 4. Relationship among Contact angle, volume of water in the drop and Radius of drop among test surface roughened by different emery.

In figure 4, for test surface roughened by emery grit 600, 1000, 1500 and 2000, has been depicted separately by figure 4a to figure 4d respectively. Graphs are analyzed and presented below in a tabular form:

	Emery grit size			
	600	1000	1500	2000
CA	At 0 sec., CA 100°	At 0 sec., CA 93°	At 0 sec., CA 107°	At 0sec., CA 95°
	After 20 sec., CA 0° Linearly decreasing.	At 20sec., CA 50° Linearly decreasing.	At 23sec., CA 40° Linearly decreasing.	At 25sec., CA 0° Linearly decreasing.
Volume (same vol^{m}. In each drop)	Max. vol. at 3 sec. Linearly diminishes to 0mL at 20 sec.	Max. vol. at 2.5 sec. Linearly diminishes to 0mL at 16 sec.	Max. vol. at 3.5 sec. Linearly diminishes to 0mL at 35 sec.	Max. vol. at 3.5 sec. Linearly diminishes to 0mL at 25 sec.
Radius	Max radius at 3 sec.	Max radius at 2.5 sec.	Max radius at 2.5 sec.	Max radius at 4.5 sec.
	Retain the radius upto 16 sec.	Retain the radius upto 15 sec.	Retain the radius upto 25 sec.	Retain the radius upto 17.5 sec.
	Sharp drop to 0 at 20 sec.	Sharp drop to 0 at 25 sec.	gradually drop to 0 at 35 sec.	step-wise dropped to 0 at 25 sec.
Remarks	It is wettable and suitable for heat transfer applications.	Though other two parameters reach to zero level, but CA never touches to zero. Surfaces showing hydrophobic nature. Not suitable for heat transfer applications.		It is wettable and best than other surface under test as it provides 5sec extra time to carry heat.

In simple line, these trends are also supported by the table 1. The observed values of calculated surface roughness shown in table 1 are different from the surface roughness measured directly by the Atomic Force Microscopy for the test surface with cross section of 5µm X 5µm. Here, it is also noticed that SFE in solid surface is lowest for surface roughened by the emery 2000.

4. Conclusion

According to the result and discussion, it can be pointed that liquid-volume present in the drop, drop-radius, contact angle and time are inter-related. Surface roughened by emery grit 2000 is most wettable than other roughened surface under present experimental condition. Surface roughness shows growing trend with rising emery grit-size. Further, it is also noticed that with the rise of surface roughness, contact angle also augmented. Similar trends for all parameters observed in graphs for surfaces roughened by 600, 1000, 1500 and 2000 emery.

References

[1] Chibowski, E., Ontiveros-Ortega, A., Perea-Carpio R., 2002, *J. Adhesion Science and Technology*, **16** 1367-1404
[2] Chibowski, E., 2003, Advances in Colloid and Interface *Science,* **103** 149-172.
[3] Rishi Raj, Enright, R., Zhu, Y., Adera, S., and Wang, E. N., 2012 'Unified Model for Contact Angle Hysteresis on Heterogeneous and Superhydrophobic Surfaces', *J. Langmuir* **28** 15777 – 15788.
[4] Gao, L., and McCarthy, T.J., How Wenzel and Cassie were wrong, 2007, *Langmuir* **23** 3762–3765.
[5] Munshi, A. M., Singh, V. N., Kumar, M, and Singha, J. P., 2008 Effect of nanoparticle size on sessile droplet contact angle, *J. of Applied Physics* **084315** 103-107.
[6] Cansoy, CE, Erbil, HY, Akar, O, and Akin, T, 2011 Effect of pattern size and geometry on the use of Cassie–Baxter equation for superhydrophobic surfaces, *Colloids Surf. A - Physicochem. Eng. Aspects* **386** 116-124.
[7] Yi, Z D, Wei, Q, and Deng, W L., 2011 Hydrophobic mechanism and criterion of lotus-leaf-like micro-convex-concave surface, *Chinese Science Bulletin* **56** 1623–1628.
[8] Bottiglione, F., and Carbone, G., 2015 An effective medium approach to predict the apparent contact angle of drops on super-hydrophobic randomly rough surfaces, *J. of Physics: Condensed Matter* **27** 015009 – 015018.
[9] Hong, K. T., Imadojemu, H., and Webb, R. L., 1994 Effects of Oxidation and Surface Roughness on Contact Angle, *Experimental Thermal and Fluid Science* **8** 279-285.
[10] Erbil, H.Y., McHale, G., Rowan, S.M., and Newton, M.I., 1999 Determination of the receding contact angle of sessile drops on polymer surfaces by evaporation, *Langmuir* **15** 7378-7385.
[11] Extrand, C.W., 2003 Contact angles and hysteresis on surfaces with chemically heterogeneous islands, *Langmuir*, **19** 3793–3896.
[12] Zhou, C. X., Yuan, X.G., Fan, L. T., Zeng, A. W., Yu, K.T., Kalbassi, M., and Porter, K. E., 2010 Experimental study on contact angle of ethanol and N-propanol aqueous solutions on metal surfaces, *Distillation Absorption* 359-364.
[13] Deshmukh, S.G., Kheraj, V., Karande, K.J., Panchal, A.K., and Deshmukh, R. S., 2019, Hierarchical flower-like Cu_3BiS_3 thin film synthesis with non-vacuum chemical bath deposition technique, *Materials Research Express* **6** 084013.
[14] Deshmukh, S.G., Kheraj, V., and Panchal, A.K., 2017, Development of Cu_3BiS_3 thin films by chemical bath deposition route, *J. of material science: Materials in Electronics*, **28** 11926-11933.
[15] Wenzel, R.N, 1936 Resistance of solid surfaces to wetting by water, *Industrial and engineering chemistry* **28** 988.
[16] Quere, D., 2002 Rough ideas on wetting, *Physica A* **313** 32 – 46.
[17] Quere, D., 2008 Wetting and Roughness, *The Annual Review of Materials Research* **38** 71–99.

[18] Kubiak, K.J., Wilson, M.C.T., Mathia T.G., and Carval, 2011, Wettability versus roughness of engineering surfaces, *Wear* **271** 523–528.
[19] Young, T., An essay on the cohesion of fluids, *Philosophical Transactions of the Royal Society of London*, **95** 65-87.
[20] Kwok, Y., 1998 J. of Colloid and Interface Science, **206** 44-51.
[21] Carey, V P, 1992 Liquid – Vapour phase change phenomena, Hemisphere Pub. Corp., Ed. 1st, Chapter - **3** 57-76
[22] Whyman, G., Bormashenko, E., and Stein, T., 2008 The rigorous derivation of Young, Cassie–Baxter and Wenzel equations and the analysis of the contact angle hysteresis phenomenon, *Chemical Physics Letters (Elsevier)* **4503** 55–59.
[23] Nishino, T., Meguro, M., Nakamae, K., Matsushita, M., Ueda, Y., 1999 *Langmuir* **15**, 4321-4323.
[24] Wenzel, R.N., 1936 Resistance of solid surfaces to wetting by water, *Ind. Eng. Chem* **28** 988-994.
[25] Cassie, A.B.D., and Baxter, S., 1944 Wettability of porous surfaces, *Trans. Faraday Soc.* **40** 546-551.
[26] Deshmukh, S.G., Patel, S.J., Patel, K.K., Panchal, A.K., and Kheraj, V., 2017, Effect of Annealing Temperature on Flowerlike Cu_3BiS_3 Thin Films grown by Chemical Bath Deposition, *Journal of Electronic Materials*, **46** (10) 5582-5588.

IOP Conf. Series: Materials Science and Engineering **814** (2020) 012035 doi:10.1088/1757-899X/814/1/012035

Carbon Dioxide Sequestration by Mineral Carbonation Using Alkaline Rich Material

Janhavi Desai[1] K S Sreekeshava[2] Manish S Dharek[3] Prashanth Sunagar[4] C R Ganesh[5]

[1]Jyothy Institute of Technology Bengaluru affiliated to Visvesvaraya Technology University Belagavi.
[2]Jyothy Institute of Technology Bengaluru affiliated to Visvesvaraya Technology University Belagavi, Karnataka, India.
[3]BMS Institute of Technology Bengaluru affiliated to Visvesvaraya Technology University Belagavi, Karnataka, India.
[4]Ramaiah Institute of Technology Bengaluru affiliated to Visvesvaraya Technology University Belagavi, Karnataka, India.
[5]Jyothy Institute of Technology Bengaluru affiliated to Visvesvaraya Technology University Belagavi, Karnataka, India.

E- mail: janhavi.desai@jyothyit.ac.in, sreekeshava.ks@jyothyit.ac.in

Abstract. Mineral carbonation is the carbonation of alkaline silicate Ca/Mg minerals, which is replica of the regular weathering processes. Mineral carbonation comprises of the reaction between the minerals and carbon dioxide to form a safe and stable mineral carbonate product. The present investigations deal with the usage of industrial alkaline rich waste (IAW) containing considerable amount of metal oxides as the possible raw material for the carbonation process. The suitability of applying the aqueous scheme of mineral carbonation by using steel slag is studied. The material is economical and is available easily near the large point sources of carbon dioxide emissions, and due to the chemical instability tend to react with carbon dioxide. Various process variables were considered to check the extent of carbonation like the temperature, pressure, liquid to solid ratio, reaction time. Chemical analysis was conducted to check the extent of carbonation. The experiments were carried out in artificial carbonation reactor. The particle size of the samples considered was less than 90 microns and distilled water was used as the medium to carry out carbonation. A maximum of 44% of carbonation was achieved for steel slag at a temperature of 100 °C, carbon dioxide pressure of 10 Kg/cm^2, liquid to solid ratio of 6 and reaction time of 2 hours.

1. Introduction

Carbon dioxide emissions primarily emitted from large area sources like the power plants and other industrial processes involving burning, incineration and combustion processes. The smaller and distributed sources of carbon dioxide include automobile engines emissions, furnaces used in residential and commercial purposes. When carbon dioxide is released in the atmosphere it remains in the atmosphere until it is absorbed in some form. Unlike other Green-house gases carbon dioxide does not absorb the light waves from the sunlight instead when the light waves strike the earth surface it is converted into heat waves which have longer wavelength and carbon dioxide readily absorbs this, thereby preventing the heat waves from escaping the earth's atmosphere [4, 10].

The increasing carbon dioxide emissions due to various anthropogenic activities have potentially devastating consequences such as global warming and climate change. The main carbon dioxide emissions are due to the burning of fossil fuels and industrial activities. In order to prevent this increasing emission it requires that the net anthropogenic emissions of CO2 be reduced to a small fraction of their current value. To prevent this increasing carbon dioxide emission there arises necessity for a technology that can contribute in capturing these emissions and storing it for a long term in a stable form [4, 14].

The technology of capturing carbon dioxide and its long term storage is called as carbon dioxide sequestration. This technology involves the transportation of carbon dioxide from large point sources to the storage sites where either the emissions are forced deep into the geological formations from which the emissions do not escape into the atmosphere [3]. The carbon dioxide sequestration includes many options in which it can be carried out. The different ways in which the carbon dioxide can be captured and stored include burying the emissions into geological formations, forcing the emissions into the ocean beds, using it in enhanced oil recovery, mineral carbonation.

In mineral storage carbon dioxide is made to react exothermically with available metallic oxides, to produce the end products in the form of stable metal carbonates. Limestone formation is an example for natural mineral carbonation, however in artificial carbonation at high temperatures and pressures carbon dioxide is reacted with the metal oxides to form stable metal carbonates.

The basic principle behind the mineral carbonation is to mimic the natural weathering process in which the magnesium and calcium oxides are converted into calcium and magnesium carbonates [9, 15].

$$(Ca, Mg)O(s) + CO2\ (g) \longrightarrow (Ca, Mg)\ CO3\ (s) \qquad (1)$$

It is based on the principle that the carbon dioxide reacts with the metallic oxides to form their respective carbonates, calcium and magnesium metals are being the common ones [13]

As mineral carbonation is similar to the natural weathering process which proceeds in a slow way, it cannot be relied upon for the sequestration. Hence the mineral carbonation process needs to be accelerated for the capture of carbon. In order to accelerate the process of mineral carbonation aqueous carbonation route is one of the solutions. In this process the carbon dioxide is passed in to the slurry of Ca/Mg rich raw materials. [1, 2]

Industrial waste such as the steel slag is rich in Ca/Mg oxides and is alkaline in nature. The characteristics of the waste make it a potentially possible feedstock for the mineral carbonation process. Major advantage of using steel slag over the ores are it is present near the large point sources of CO_2,by making use of these wastes for CO_2 [8] sequestration it creates an opportunity to reclaim some of the CO_2 emitted during the manufacturing process.

2. Experimental Sequences

2.1 Chemical analysis of steel slag

Steel slag is the industrial waste generated in the iron and steel industry. It is produced in the industry during the separation of molten steel from the impurities. Initially it will be in the molten liquid state which upon solidification on cooling results in complex mix of silicates and oxides. The steel usually consists of 30 to 40% of CaO and 5 to 10% of MgO in its chemical composition. The chemical composition steel slag is alkaline waste with the pH ranging between 8 to10. Due to this chemical composition and natural alkalinity steel slag proves to be a possible feedstock for the mineral carbon sequestration.The steel slag used in this work was procured from the FOSROC pvt limited. The particle size of the procured steel slag obtained was 4.75 mm passing. This slag was tested for its CaO and MgO content in accordance with IS-4032:1985 [5].

IOP Conf. Series: Materials Science and Engineering **814** (2020) 012035 doi:10.1088/1757-899X/814/1/012035

2.2 Mineral Carbonation of Steel Slag

The carbonation experiments were carried out in the reactor by varying different parameters. One parameter was varied at a time keeping the remaining parameters constant.10g of 90microns passing, moisture free sample was taken and mixed distilled water according to the prior chosen liquid to solid ratio stirred to form the slurry and then poured into the reactor. The reactor was then sealed with lid and made sure it was air tight. The reactor was then heated to the desired temperature once the desired temperature was reached the carbon dioxide was bubbled through the slurry for the desired reaction time. Care was taken to maintain the desired temperature and pressure during the reaction time using the appropriate regulating valves. Figure 1 shows the reactor used for carbonation.

Figure 1. Reactor for mineral carbonation

After the reaction time elapsed the reactor was depressurized by opening the exhaust valve. Then the reactor was cooled to 30° C by passing cold water through the inlet of the cooling system. The reactor lid was opened and the slurry was filtered on a 42 number Whattmann paper. The carbonated sample retained on the filter paper was dried in the hot air oven at 50° C overnight. The dried carbonated sample was stored in air tight containers until tested to know the extent of carbonation.

2.3 Analysis of extent of carbonation

Thermal gravimetric analysis (TGA) was carried out on the carbonated sample to know the extent of carbonation. According to the TGA method the change in mass at temperatures of 500° C and 1000°C indicates the percentage of carbonates in the sample. TGA results give the amount of change in weight as the function of temperature. Carbonate decomposition occurs at these temperatures. Hence the change in weight or the loss in weight gives the extent of carbonation.

The carbonated samples were weighed in the silica crucible and heated to temperatures of 500° C for 15min and the loss in weight was taken as the organic fraction of CO2 which indicates the presence of carbonates of Magnesium. The same sample was heated for 15min at 1000° C and the loss in weight was noted. This fraction of weight loss at 1000° C indicates the inorganic CO2 that is the presence of carbonates of calcium.

A clean silica crucible was weighed empty and then 1g of carbonated sample was placed in it and the sample was heated for 15min at temperature of 500° C and 1000° C respectively. The loss in weight at each of the temperatures indicates the extent of carbonation [6].

The steel slag containing CaO and Mgo were used for the sequestration of carbon dioxide. The parameters varied and checked for their influence on carbonation are reaction time, pressure, temperature and liquid to solid ratio. The extent of conversion of Ca and Mg oxides into the

IOP Conf. Series: Materials Science and Engineering **814** (2020) 012035 doi:10.1088/1757-899X/814/1/012035

carbonates was tested using thermo gravimetric analysis. The influence of various process variables on extent of carbonation are as discussed below.

2.3.1 Effect of reaction time
The time duration for which carbon dioxide remains in contact with the slurry in the reactor at the pre-determined temperature and pressure is called as reaction time. It is the time taken by the gas to interact with the slurry. The reaction time was varied between 1 hour to 4 hours, and the behavior of carbonation is as tabulated below in Table 1.

Table 1. Effect of reaction time on carbonation

Reaction time in hours	Empty wt	Empty wt + 1g sample	Wt at 500°C	Wt at 1000°C	Loss in wt at 500°C	Loss in wt at 1000°C	% carbonation of MgO	% carbonation of CaO
1	13.29	14.29	14.28	13.92	0.01	0.372	1.08	37.2
2	48.52	49.52	49.52	49.10	0.0045	0.4171	0.45	41.71
3	48.52	49.52	49.51	49.10	0.0212	0.426	2.12	42.6
4	48.52	49.52	49.51	49.10	0.0183	0.4322	1.83	43.22

2.3.2 Effect of temperature
Temperature is another important process variable. The slurry is heated to the desired temperature and when it reaches the desired temperature the carbon dioxide gas is passed to the slurry. Until the decided reaction time elapses the temperature is maintained by operating the exhaust gas valve and cooling systems. The variation in the extent of carbonation at various temperatures is as tabulated below in Table 2.

Table 2. Effect of temperature on carbonation

Temperature in °C	Empty wt	Empty wt + 1g sample	Wt at 500°C	Wt at 1000°C	Loss in wt at 500°C	Loss in wt at 1000°C	% carbonation of MgO	% carbonation of CaO
50	48.517	49.517	49.513	49.102	0.004	0.417	0.4	41.7
100	48.541	49.541	49.523	49.101	0.018	0.442	1.8	44.2
150	48.541	49.541	49.539	49.222	0.001	0.320	0.1	32
200	48.541	49.541	49.524	49.321	0.017	0.221	1.7	22.1

2.3.3 Effect of liquid to solid ratio
The liquid to solid ratio is the ratio of water to slag. In aqueous path of mineral carbonation water acts as a medium for calcium oxide and magnesium oxide to dissolve in water for reaction with the carbon dioxide. At lower liquid solid ratio good mixing cannot be achieved hence the sequestration of carbon dioxide is not effective. The liquid to solid ratio was varied between 2 to 10 and the results are as tabulated below in Table 3.

IOP Conf. Series: Materials Science and Engineering **814** (2020) 012035 doi:10.1088/1757-899X/814/1/012035

Table 3. Effect of liquid to solid ratio on carbonation

Liquid to solid ratio	Empty wt	Empty wt + 1g sample	Wt at 500°C	Wt at 1000°C	Loss in wt at 500°C	Loss in wt at 1000°C	% carbonation of MgO	% carbonation of CaO
2	48.5344	49.5344	49.5279	49.1079	0.0065	0.4265	0.65	42.65
4	48.5344	49.5344	49.5102	49.1094	0.0242	0.425	2.42	42.5
6	48.5344	49.5344	49.5211	49.0928	0.0133	0.4416	1.33	44.16
8	48.5418	49.5418	49.5233	49.1013	0.0185	0.4405	1.85	44.05
10	48.5344	49.5344	49.5278	49.1088	0.0066	0.4256	0.66	42.56

2.3.4 Pressure

Pressure refers to the pressure of the carbon dioxide inside the reactor. The reactor pressure is maintained by operating the inlet and exhaust gas valves accordingly. Throughout the reaction time the pressure in the reactor should be maintained at the desired constant pressure by operating the necessary valve. At lower temperatures the pressure of the carbon dioxide is of not of much significance however at higher temperature when the dissolution of the carbon dioxide in water becomes difficult higher pressure of the carbon dioxide gas is desired. Usually higher pressure is desirable for the mineral carbonation process. Carbon dioxide was varied between 4 Kg/cm2 to 10 Kg/cm2 and its influence on the extent of carbonation was observed. The variation in carbonation at different pressures of carbon dioxide is as shown below in Table 4.

Table 4. Effect of pressure on carbonation

Pressure	Empty wt	Empty wt + 1g sample	Wt at 500°C	Wt at 1000°C	Loss in wt at 500°C	Loss in wt at 1000°C	% carbonation of MgO	% carbonation of CaO
4	48.5315	49.5315	49.5297	49.5012	0.0018	0.0303	0.18	3.03
6	48.5312	49.5312	49.531	49.3959	0.0002	0.1353	0.02	13.53
8	48.5314	49.5314	49.5302	49.3002	0.0012	0.2312	0.12	23.12
10	48.5315	49.5315	49.5311	49.0885	0.0004	0.443	0.04	44.3

3. Summary and Conclusions

The work demonstrates the suitability of using industrial steel slag for carbon dioxide .The experiments were carried out in a prefabricated artificial carbonation reactor. The samples were exposed to the continuous supply of carbon dioxide at the pre decided temperature, pressure and reaction time. The particle size of the samples was 90 microns passing and was kept constant throughout the work. The process variables temperature, pressure, reaction time and the liquid to solid ratio were studied. The results of this work indicate that steel slag could be an attractive and potential method for carbon dioxide sequestration. The results show that the maximum carbonation of steel slag was achieved of the order 44.3% at temperature of 100 °C pressure of 10 Kg/cm2 for a liquid to solid ratio of 6 and for a reaction time of 2 hours.

It was seen the extent of carbonation increased with high pressure and temperature. However when temperature was increased beyond 100 °C the carbonation decreased this was due to the fact that carbon dioxide dissolution in the water decreases at higher temperature. The carbonation process also needs a good amount of reaction time for the reaction between the slurry and carbon dioxide. The chemical composition of the material showed significant amount of metallic oxides which showed the potential use of these materials. The carbonation was successfully carried out and higher pressures could not be applied because of the inefficiency of the reactor. More efficiencies of carbonation could be achieved with higher order of pressures of the inlet carbon dioxide pressure.

This approach can result in a breakthrough involving utilization of alkaline rich waste materials like the industrial steel slag waste; the use of such waste also reduces the cost of procurement of raw materials for mineral carbonation. The sites where these material is generated is also the sites where there is large emissions of carbon dioxide hence by using the waste at sites it is possible to reclaim some of the carbon dioxide emitted during their manufacturing process.

References

[1] Dahlin, D.C., O'Connor, W.K., Nilsen, D.N., Rush, G.E., Walters, R.P., And Turner, P.C, 2000 *A Method for Permanent CO2 Sequestration: Supercritical CO2 Mineral Carbonation. Proceedings of the 17th Annual International Pittsburgh Coal Conference Pittsburgh,* PA September 11-15, pp.14.

[2] Hamdallah Béarat, Michael J. McKelvy, Andrew V. G. Chizmeshya, Deirdre Gormley, Ryan Nunez , R. W. Carpenter, Kyle Squires, and George H. Wolf. 2006 *Environmental science and Technology* **40** 4802-4808.

[3] Hyomin Lee, Jinhyun Lee, Jinyeon Hwang, Younghoon Choi, Byeongseo Son. 2015 *World Academy of Science, Engineering and Technology Geological and Environmental Engg.* **2**.

[4] IPCC Fourth Assessment Report, Summary for policymakers, *climate change April 2007: Impacts, adaptation and Vulnerability (WGII).*

[5] IS 4032:1985 for determining the *chemical composition of steel slag.*

[6] IS 4032:1985 for procedure related to determining *carbonates in the samples.*

[7] J.C.M.Pires, F.G.Martins, M.C.M.Alvim-Ferraz, M.Simões. 2011 *Chemical Engineering Research and Design Science Direct.*

[8] Johan Sipilä, Sebastian Teir and Ron Zevenhoven. 2008 *Carbon dioxide sequestration by mineral carbonation* literature review update 2005-2007.

[9] Jong Soo Cho, Soon Mi Kim, Hee Dong Chun, Gun Woo Han, Chang Hoon Lee. 2011 *International Journal of Chemical Engineering and Applications 2.*

[10] Jos G.J. Olivier (PBL), Greet Janssens-Maenhout (IES-JRC), Marilena Muntean (IES-JRC), Jeroen A.H.W. Peters (PBL). *Trends in global CO2 emissions: 2013 Report*, PBL Netherlands Environmental Assessment Agency.

[11] Joshuah K. Stolaroff, Gregory V. Lowry, David W. Keith. 2005 *Energy Conversion & Management* **46** 687-699.

[12] Sanni Eloneva, Sebastian Teir, Justin Salminen, Carl-Johan Fogelholm, Ron Zevenhoven. 2008 *science direct*, *Elsevier Energy* **33** 1461-1467.

[13] W.K. Connor, D.C. Dahlin, D.N Nilsen, G.E. Rush, R.P. Walters and P.C. Turner. 2004 *Carbon Dioxide Sequestration by Direct Mineral Carbonation: Results from Recent Studies and Current Status*, Albany Research Centre Office of Fossil Energy.

[14] W.J.J. Huijgen, and R.N.J. Comans, 2010 *Energy Resource Center of the Netherlands ECNC* **3** 016.

[15] Wouter J.J. Huijgen, Geert Jan Witkamp and Rob N.J. Comans. 2005 *Environmental Science and Technology* **39** 9676-9682.

Optical and Gas sensing properties of $Ba_xSr_{1-x}TiO_3$ (with x=0.5 and x=0.7) thin films grown using Pulse Laser Deposition

Nipa M. Shastri[1,*], Adhish V. Raval[2,*], V.G. Joshi[3]

[1,3] Department of physics, Veer Narmad South Gujarat University, Surat, Gujarat-395007, India.
[2]Department of Applied Physics, S. V. National Institute of Technology, Surat, Gujarat-395007, India

Email. shastrineepa2004@yahoo.com, adhishraval0777@gmail.com

Abstract. Thin films of Barium Strontium Titanate (BST) have been fabricated using Pulse Laser Deposition (PLD) technique. Two stoichiometries of $Ba_xSr_{1-x}TiO_3$ with x=0.5 and x=0.7 are deposited on well cleaned quarts substrates in high vacuum chamber with oxygen pressure of 0.13mbar at 700°C. XRD pattern shows perovskite structure of the films. Energy Dispersive X-ray (EDX) is used to determine the element concentration of both the films. Transmittance spectra of UV characterization shows decrement in transparency of the film as Barium (Ba) concentration increase and sharp cut-offs are observed at the band edge. Refractive Index (n), absorption coefficient (α), extinction coefficient (k), optical conductivity (σ) and optical band gap (Eg) are derived from UV-Vis spectroscopy. The calculated parameters for $Ba_{0.5}Sr_{0.5}TiO_3$ and $Ba_{0.7}Sr_{0.3}TiO_3$ are presented in this work. The present work also includes study of BST thin films as gas sensors.

1. Introduction

Barium Strontium Titanate (BST) is the material which shows very useful properties for opto-electronic and tuneable microwave devices, IR sensors and chemical sensors for sensing humidity and gases [1]. Because of high dielectric constant, high optical nonlinearity [2] and Curie temperature depending on composition [3], much attention has been paid to $Ba_xSr_{1-x}TiO_3$ ferroelectric materials. These BST materials are suitable for applications in phase conjugation, two beam coupling, dynamic holography and optical signal processing [4]. As BST has prominent properties like large electro-optic coefficient [5] and relatively low optical propagation loss [6], BST thin films show much more interest in opto-electronic device applications. These BST thin films can give great contribution in optical memory, Optical storage devices and low voltage optical switching.

The optical spectroscopy technique is used to investigate optical properties of Barium Strontium Titanate (BST) thin films. From these basic spectrometric measurements, related optical parameters like refractive index, absorption coefficient etc. are deduced by the method of Swanepoel [7] and reported using the measured transmittance response in the range of 200nm to 900nm.

Currently there is a great interest in the synthesis and investigation of material response to the variations in the gaseous environment. Sensors can detect the pollutants from environment [8, 9],

IOP Conf. Series: Materials Science and Engineering **814** (2020) 012036 doi:10.1088/1757-899X/814/1/012036

which are very useful to environment monitoring system. Semiconducting oxides based gas sensitive resistors are very strong and simple devices to fabricate. The perovskite oxides have a good steadiness in chemical and thermal atmospheres, so they have huge area of interest as gas sensors. Over a last decade the perovskite oxide materials like $BaTiO_3$ [10, 11], and (Ba, Sr) TiO_3 [12-16] have created interest in chemical sensors.

In present work we have calculated the optical parameters of $Ba_xSr_{1-x}TiO_3$ (x=0.5, 0.7) thin films based on the UV spectroscopy data. The H_2S gas sensitivity of the samples has also been experimentally studied.

2. Experimental Details

$Ba_xSr_{1-x}TiO_3$ (BST) solid solutions with two desire molar formula $Ba_{0.5}Sr_{0.5}TiO_3$ (BST0.5) and $Ba_{0.7}Sr_{0.3}TiO_3$ (BST0.7) were prepared by Mixed Oxide Conventional route. Starting materials like $BaCl_2.2H_2O$, $SrCl_2.6H_2O$ and TiO_2 of 99% purity were weighted according to molecular wt%, mixed and grinded well using mortal pestle to prepare BST0.5 and BST0.7 compounds. Acetone is selected as medium for mixing-milling process. Both the mixtures were annealed in an electrical furnace at 1020˚C for 12hrs, again milled well, mixed with Polyvinyl Alcohol (PVA) and pressed to yield pellet of about 5mm thickness and 15mm in diameter. These two pellets (BST0.5and BST0.7), sintered at 1020˚C for 12hrs were used as PLD target.

For the film deposition, multi step process was followed to clean the quartz substrates. The first step was to clean all substrate with deionised water properly. Second step was acetone soak for 10 mins followed by rinsing in deionised water. Third step was ultrasonically cleaning in the mixture of acetone and isopropyl alcohol for 15 mins at 70˚C followed by rinsing in deionised water. Then all substrates were dried in a process oven with air atmosphere at 80˚C.

The Pulse Laser Deposition process has been carried out using high power Nd: YAG (EKSPLA, Lithuania) laser having wavelength of 355nm and the laser pulse energy density of $200mJ/cm^2$. Inside PLD chamber the rotating target and quartz substrate were kept parallel to each other and both were spaced at few centimetres apart. The laser was focused on the rotating target resulting into ablation. This process results in deposition of quartz substrate. The frequent laser shots can damage the target material. In order to avoid it the target is rastered and rotated continuously in specified manner during the deposition process. Initially the PLD chamber was evacuated in the order of 10^{-6} mbar using a diffusion pump supported by rotary pump and pressurized with 0.13mbar oxygen (O_2) gas. The laser was pulsed at 10Hz for 35k pulses and for deposition of both the films 700˚C substrate temperature was maintained.

3. Result and Discussion

3.1. X-Ray Diffraction Analysis:

Figure 1 shows the X-Ray diffraction pattern of $Ba_{0.5}Sr_{0.5}TiO_3$ (BST0.5) and $Ba_{0.7}Sr_{0.3}TiO_3$ (BST 0.7) thin films grown on quartz substrate using PLD at 700˚C. The 2θ range is from 20˚ to 90˚ with step size $\Delta\theta = 0.02^\circ$. The X- ray diffractograms of films show polycrystalline structure which is in well agreement with standard peaks and found to be perovskite cubic structure (JCPDS card 39-1395) [17] with Pm3m space group for $Ba_{0.5}Sr_{0.5}TiO_3$ and teragonal strucutre (JCPDS card 44-0093) [18] with P4mm space group for $Ba_{0.7}Sr_{0.3}TiO_3$. Due to amorous nature of quartz substrate the broad spectrum was obsereved from 20˚ to 30˚.

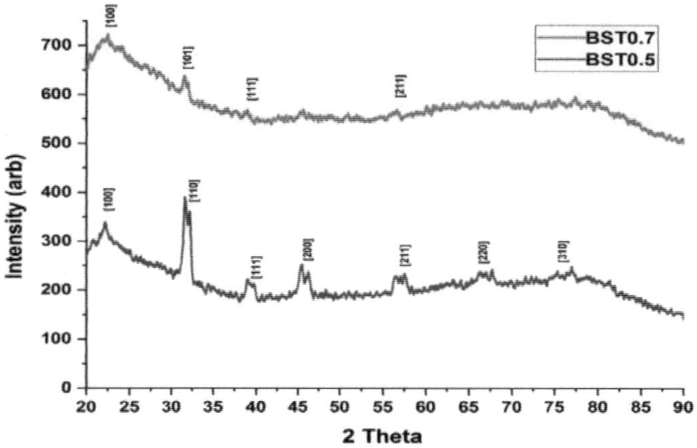

Figure 1. XRD pattern of $Ba_xSr_{1-x}TiO_3$ (x=0.5, x=0.7) thin films

3.2. Energy Dispersive X-ray (EDX) Analysis:

Energy Dispersive X-ray analysis of both the thin films $Ba_xSr_{1-x}TiO_3$ (x=0.5 and x=0.7) are shown in the figure 2. Figure 2(a) and 2(b) shows the element's concentration for both the films $Ba_{0.5}Sr_{0.5}TiO_3$ and $Ba_{0.7}Sr_{0.3}TiO_3$ respectively.

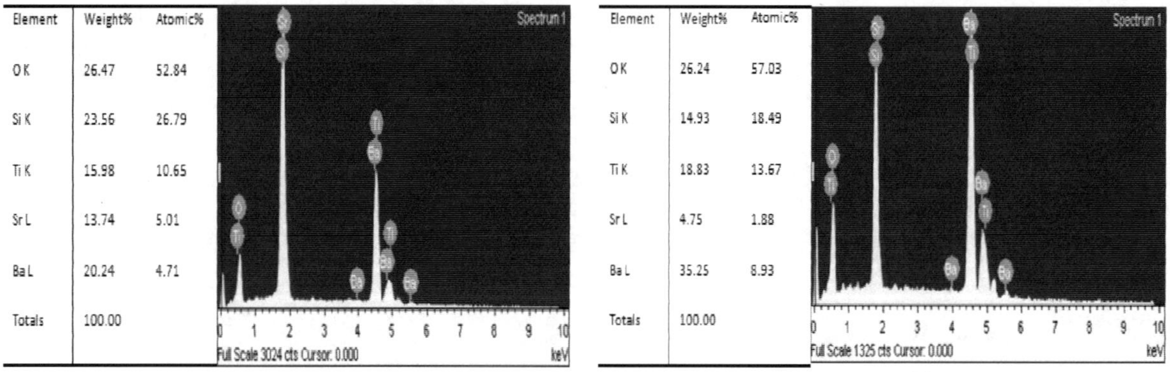

Element	Weight%	Atomic%
O K	26.47	52.84
Si K	23.56	26.79
Ti K	15.98	10.65
Sr L	13.74	5.01
Ba L	20.24	4.71
Totals	100.00	

(a)

Element	Weight%	Atomic%
O K	26.24	57.03
Si K	14.93	18.49
Ti K	18.83	13.67
Sr L	4.75	1.88
Ba L	35.25	8.93
Totals	100.00	

(b)

Figure 2. Energy Dispersive X-ray of (a) $Ba_{0.5}Sr_{0.5}TiO_3$ and (b) $Ba_{0.7}Sr_{0.3}TiO_3$

3.3. Optical Properties:

The profile of optical transmittance spectra for the thin films BST0.5 and BST0.7 is shown in figure 3. BST0.5 has good transmittance behavior as compare to BST0.7 film. It is evident that transmittance suddenly decrease in the UV region of the spectrum band (320nm – 220nm).

Refractive index has been calculated from optical transmittance spectra using envelop method which is proposed by Swanepoel [7]. The Index of Refraction (n) for both the films BST0.5 and BST0.7 has been obtained from the equation as follows;

$$n = \left[M + (M^2 - n_s^2)^{\frac{1}{2}} \right]^{\frac{1}{2}} \qquad \text{..........(1)}$$

Where,

$$M = \frac{2n_s}{T_m} - \frac{(n_s^2 + 1)}{2}$$; for all regions which are optically transparent. ....(2)

and

$$M = \frac{2\,n_s\,(T_M - T_m)}{T_M T_m} + \frac{(n_s^2 + 1)}{2}$$; for all regions which are medium and weak absorbents ....(3)

n_s is the refractive index of the quartz substrate (1.45). T_M and T_m are the values of maximum and minimum transmission respectively at the definite wavelength.

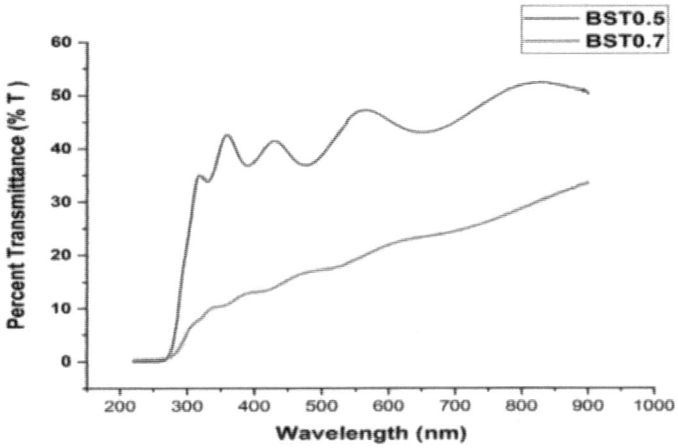

Figure 3. Percent Transmittance vs wavelength of $Ba_xSr_{1-x}TiO_3$(x=0.5, 0.7) thin films

The index of refraction can be determine by extrapolating envelopes corresponding to maximum transmission (T_M) and minimum transmission (T_m), using Swanepoel's method [7]. The calculated average refractive index is 2.19 for $Ba_{0.5}Sr_{0.5}TiO_3$ thin film and 1.94 for $Ba_{0.7}Sr_{0.3}TiO_3$ thin film for the wavelength 800-900nm.
The extinction coefficient (k) can be derived from the following equation [19],

$$k = \frac{\alpha * \lambda}{4\pi}$$ ....(4)

Where λ = wavelength, α = absorption coefficient and can be derived from the following equation [19],

$$\alpha = \left(\frac{1}{d}\right)\log\left(\frac{1}{T}\right)$$ ....(5)

Where d = film thickness and T = transmittance for definite wavelength.

From definition Extinction coefficient (k) is the imaginary part of complex index of refraction and it defines the several measures of absorption of light in a medium. The variation in extinction coefficient (k) with photon energy is shown in the figure 4(a). It is observed that for BST0.5 and BST0.7 thin films, extinction coefficient increase with photon energy up to 5eV and 4.62eV respectively and above these values it starts decreasing.

IOP Conf. Series: Materials Science and Engineering **814** (2020) 012036 doi:10.1088/1757-899X/814/1/012036

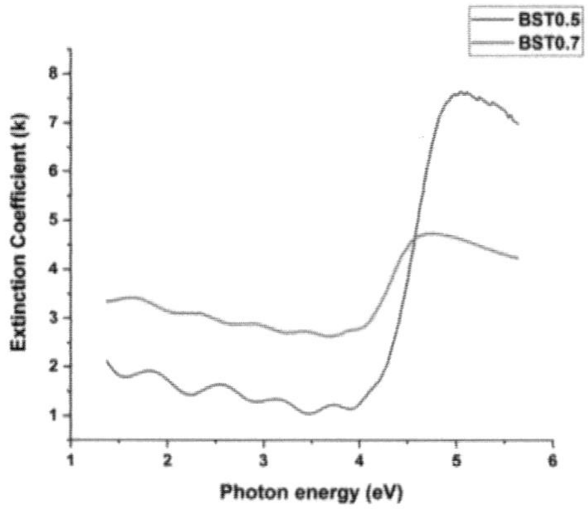

(a)

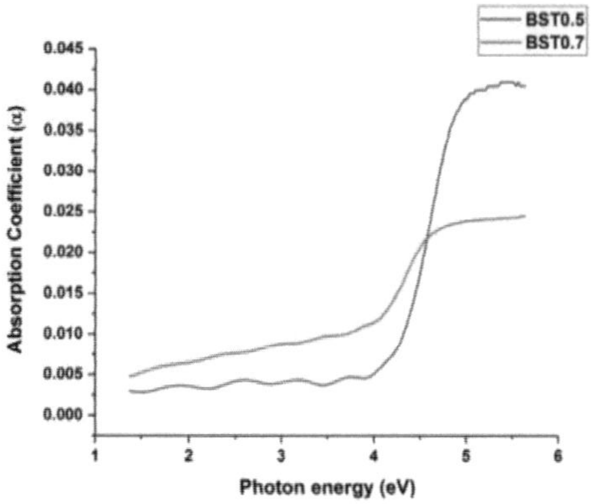

(b)

Figure 4. For $Ba_xSr_{1-x}TiO_3$(x=0.5,0.7) films (a) Extinction co-efficient(k) vs. Photon energy (eV) (b) Absorption coefficient (α) vs. photon energy (eV)

The absorption coefficient is dependent on material type and wavelength of the absorbed light. Figure 4(b) shows the behavior of absorption coefficient with respect to photon energy. The optical band gap can be derived from the variation of absorption coefficient (α) with respect to photon energy. From graph, the energy band gap is 4.25ev for BST0.5 and 3.7ev for BST0.7 thin films. These values are close in agreement with the literature value 4.5eV for BST0.5 [2] and 4.3eV for BST0.7 [20] thin films.

The optical conductivity (σ) is determined by following relation [19],

$$\sigma = \frac{\alpha n c}{4\pi} \qquad \text{..........(6)}$$

where α=absorption coefficient, n=refractive index, c= velocity of light.

Figure 5 shows behaviour of optical conductivity (σ) with respect to photon energy. Optical conductivity (σ) is depends on absorption coefficient and refractive index. Figure 5 shows that optical conductivity increase sharply above photon energy 4.2eV and 4eV for BST0.5 and BST0.7 respectively.

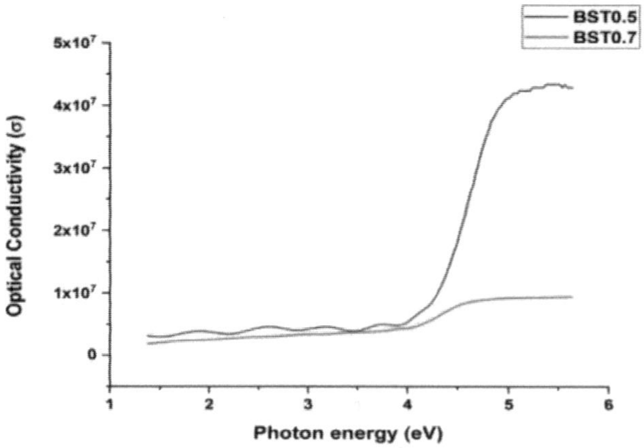

Figure 5. Optical conductivity (σ) vs. Photon energy (eV) for $Ba_xSr_{1-x}TiO_3$(x=0.5, 0.7) thin films.

IOP Conf. Series: Materials Science and Engineering **814** (2020) 012036 doi:10.1088/1757-899X/814/1/012036

3.4. Gas Sensing Property:

In present work, sensitivity (Response) of the BST thin films, as gas sensor has been studied. Response of any sensor mainly depends on the interaction between gas and surface of sensing element. As resistance of sensor material changes, the change in Sensitivity or Response of the sensor can be determine. It is define as ratio of change in the resistance of sensor in the presence of gas to the resistance of sensor in ambient air [21], which is denoted as,

$$\text{Response} = \left(\frac{Rg - Ra}{Ra}\right) \times 100\ \%$$

Where, Rg = Electrical resistance of sensor in presence of gas and Ra = Electrical resistance of sensor in ambient air (in the absence of test gas).

We have used $Ba_{0.5}Sr_{0.5}TiO_3$ and $Ba_{0.7}Sr_{0.3}TiO_3$ thin films as gas sensor and studied the H_2S gas (800 ppm) response against temperature. Figure 6 depicts behaviour of H_2S gas Response against temperature. Figure 6(a) and 6(b) shows that response increases with temperature and reach maxima at 330 °C, and then start decreasing.

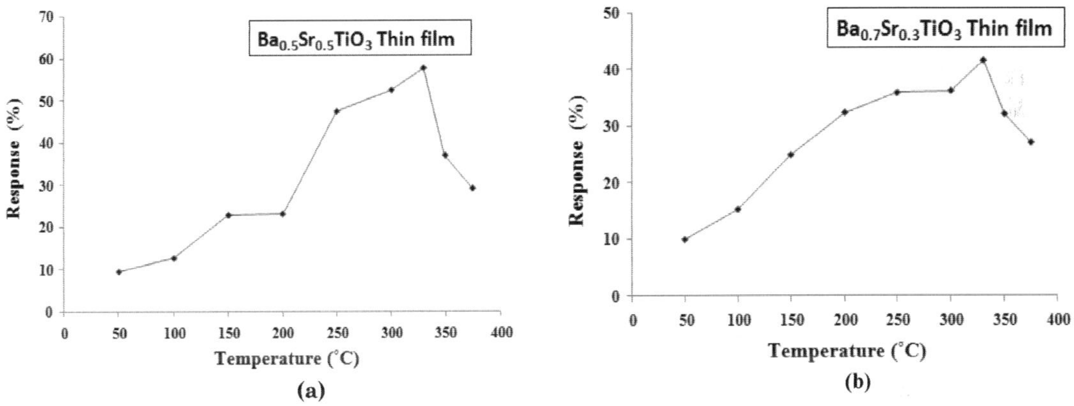

Figure 6. H_2S (800 ppm) gas response against temperature for (a) $Ba_{0.5}Sr_{0.5}TiO_3$ thin film and (b) $Ba_{0.7}Sr_{0.3}TiO_3$ thin film.

Figure 6 shows, in operating temperature range, the Gas response decrease as Barium (Ba) concentration increase in BST thin film. $Ba_{0.5}Sr_{0.5}TiO_3$ shows 57.57 % while $Ba_{0.7}Sr_{0.3}TiO_3$ shows 41.61 % gas response to H_2S gas (800ppm) at 330 °C.

4. Conclusion

$Ba_{0.5}Sr_{0.5}TiO_3$ (BST0.5) and $Ba_{0.7}Sr_{0.3}TiO_3$ (BST0.7) thin films are fabricated by Pulse Laser Deposition method on quartz substrate. XRD shows perovskite structure of both the films. From UV-Vis spectroscopy, optical parameters have been calculated for both the samples. Calculated average refractive index is 2.19 for $Ba_{0.5}Sr_{0.5}TiO_3$ thin film and 1.94 for $Ba_{0.7}Sr_{0.3}TiO_3$ thin film for 800-900nm wavelength.

Absorption coefficient (α) and so the Extinction coefficient (k) is large of BST0.7 than BST0.5. Optical band gap found for BST0.5 is 4.25ev while for BST0.7 is 3.7ev. Gas sensing test revealed that sensor based on BST0.5 sample exhibit high sensitivity for H_2S gas (800 ppm) than BST0.7 sample.

IOP Conf. Series: Materials Science and Engineering **814** (2020) 012036 doi:10.1088/1757-899X/814/1/012036

References

[1] S. J. Patil, A. V. Patil, C. G. Dighavkar, K. S. Thakar, R. Y. Borase, S. J. Nande, N. G. Deshpande and R. Ahire 2015 *Front. Mater. Sci.* **9** 14-37.

[2] S. Behera, A. Khare 2019 *AIP Conf. Proc.* **2082** 040006.

[3] M. Kocanda, I. Abdel-Motaleb 2009 *J. App. Phy.* **106** 123916.

[4] F. Tcheliebou, H. S. Ryu, C. K. Hong, W. S. Park, S. Baik 1997 *Thin Solid Films* **305** 30.

[5] M. Gaidi, M. Chaker, P. F. Ndione, R. Morandotti, B. Bessais 2007 *J. App. Phy.* **101** 063107.

[6] Yu-Fu Kuo, Tseung-Yuen Tseng 1999 *Materials Chemistry and Physics* **61** 244-250.

[7] R. Swanepoel 1983 *J. Phys. E: Sci, Instrum.* **16**.

[8] S. Korgaokar, M. Moradiya, Om prajapati, P. Thakar, J. Pala, C. Savaliya, S. Parikh, J. H. Markna 2017 *AIP Conf. Proc.* **1837** 040050

[9] G. H. Jain, V. B. Gaikwad, D. D. Kajale, D. Kajale, R. Chaudhari, R. Patil, N. Pawar, M. Deore, S. Shinde and L. Patil 2008 *Sensors and Transducers journal* **90** 160-173.

[10] D. N. Suryawanshi, I. G, Pathan, A. R. Bari, L. A. Patil 2017 *AIP Conf. Proc.* **1953** 030041.

[11] G. H. Jain, S. B. Nahire, D. D. Kajale, G. E. Patil, S. D. Shinde, D. N. Chavan and V. B. Gaikwad 2011 *New developments and Appl. In Sen. Tech. LNEE* **83** 157-167.

[12] G. H. Jain, L. A. Patil, P. Patil, U. Mulik and K. Patil 2007 *Bull. Mater. Sci.* **30** 9-17.

[13] G. H. Jain, L. A. Patil 2006 Bull.*Mater. Sci.* **29** 403-411.

[14] C. Huck, A. Poghossian, M.Backer, S. Reisert, J. Schubert, W. Zander, V. Begoyan, V. Buniatyan, M. Schoning 2014 *Procedia Engineering* **87** 28-31.

[15] S. Sharma, A. Sharma, M. Toamar, N. Puri and V. Gupta 2014 *Procedia Engineering* **87** 1067.

[16] Ade Kurniawan, Dani Yosman, Ardian Arif, Jajang Juansah, Irzaman 2015 *Procedia Environmental Sciences* **24** 335-339.

[17] Anuradha Kumari, Barnali Dasgupta Ghosh 2018 *Advances In Applied Ceramics* DOI : 10.1080/17436753.2018.1491166.

[18] M. Enhessari, A. Parviz, K. Ozaee, H. Habibi Abyaneh 2011 *Int. J. Nano Dim.* **2** 85-103.

[19] Martin Kocanda, Syed Farhan Mohiudin, Ibrahim Abdel-Motaleb 2012 *Crystal Structure Theory and Applications* **1** 17-20.

[20] Zhimou Xu, Yuichiro Tanushi, Masato Suzuki, Keita Wakushima, Shin Yokoyama 2006 *Thin Solid Films* **515** 2326-2331.

[21] Jeffrey W. Fergus 2007 *Sensors and Actuators* B **123** 1169-1179.

IOP Conf. Series: Materials Science and Engineering **814** (2020) 012037 doi:10.1088/1757-899X/814/1/012037

Theoretical Analysis of LC Filter to Three Level Diode Clamped Inverters in Low/Medium Voltage Applications

Santoshkumar M. Hunachal[1], Bhagyraj Chodankar[1], Rahul Salgonkar[1] Durvesh Nakhwa[1]

[1]Department of Electrical Engineering, VPMs Maharshi Parshuram College of Engineering, Velneshwar, Guhagr, Maharshtra, India.

E-mail: santoshkumar.hunachal@gmail.com

Abstract. The employing of power electronic devices for converters plays a vital role in the conversion of AC to DC vice versa. Applications of these converters more demand in renewable energy, hybrid vehicles, Drives, HVDC, etc. In renewable energy, especially a solar-based power generation stations require inverter stations. These inverters stations have the multilevel inverters. Multilevel inverters are emerging day by day in the sector of the renewable energy sector. The increasing level of multilevel inverters can be effectively utilized in low and medium voltage applications to get the desired voltage and frequency. But output voltage is not in sinusoidal. However, most of the loads require a pure sinusoidal form of voltage. To get sinusoidal voltage across loads a high switching frequency is required to switch the MOSFETs or IGBTs. Along with a suitable value of LC, filters make the sinusoidal nature of waveform with less total harmonic distortion (THD) levels. To get a sinusoidal wave, a proper theoretical analysis of filters is required. In this paper theoretical analysis has been done for LC filters and checked with simulation results for various loads. The performance of theoretical LC filters values checked using diode clamped multi-level inverters have been examined through MATLAB/SIMULINK by POD PWM by POD PWM.

1. Introduction

Remarkable advancement in semiconductor technology leads to the using of applications of power electronics technology is widely in the areas of smart grid, solar, wind energy conversions, marine engineering, biomedical applications, etc. In the converter technology power electronics plays a vital role during conversions of dc to ac and vice versa. Also, the applications of power electronics are finding in shaping, conditioning and transforming frequency, voltage, power factor, etc. The major advantages of these devices are is possible to get a quick response using digital controllers in a closed-loop system for the dispensation of power from renewable energy sources is remarkable.

Naik Raghuram, et.al.[2] has discussed a review of grid-connected WECS for the three-level diode clamped voltage source inverter with different modulation techniques. The paper covered, SPWM, Multicarrier method, SPWM using multiple references frame, carrier Phase Variation method, SVPWM, Hysteresis control, soft-switching algorithm has been discussed concerning its advantages and disadvantages of these techniques. With the use of SPWM the use of DC bus utilization is less, in Multicarrier method suffers in shortcoming such as narrow pulse width, capacitor voltage imbalance, in SPWM using multiple reference frame suffers in the switching the frequency will be doubled and losses hence it is preferable and recommended. For higher voltage levels circuit complexity is more in hysteresis control complexity of the circuit, in the soft-switching algorithm, requires supplementary components. But the utilization of DC bus is high in the SVPWM algorithm and said problems are less among compared to said techniques and are also suitable for three-level neutral point clamped inverter in wind energy conversation systems (WECS).

IOP Conf. Series: Materials Science and Engineering **814** (2020) 012037 doi:10.1088/1757-899X/814/1/012037

Bharti Katre, et.al.[5] presented the better utilization of renewable resources such as solar, wind, etc. has led to the establishment of the MLI. An inverter is the power device that converts DC to AC direct as per application requirements. The multilevel inverter is the power electronic devices that are capable of achieving alternating current at the desirable stage at the output by using multiple dc sources at the input. In diode clamp multilevel inverter topology for three levels, there are four switches and two clamped diodes are required. The clamped diodes help in reducing the stress of dc sources from the switching devices. As the levels of inverter increases, the output voltage becomes smoother but with many levels, the inverter becomes complex with the increase in switching devices as well as complexity in controllers function. The pulse width modulation techniques are created, by comparing the combinations of the fundamental frequency and the triangular carrier wave where the intersection point determines the switching points of devices hence establishes a switching frequency. Azah Mohamed *et*.al [6] implemented the dSPACE DS1104 is one of the controllers by using which multilevel inverter is implemented. The simulation is carried out through MATLAB/Simulink with the help of controllers which gives real-time operation. The three-phase inverter gives stable and near to the sinusoidal voltage of 415Vrms with a low THD of 2.26%.

Gautam ghosh, *et*.al [7], an inverter is used in various applications in as example adjustable speed drive, UPS and ac operation appliances with a low THD in output voltage, high efficiency and good power factor from batteries or solar-based PV system. The main disadvantages are that the dc-bus capacitors suffer from voltage balancing across the series-connected. This tends to either overcharge of the DC capacitors or entirely discharges. In such conditions, balancing of capacitor charge by reverts passion to a three-level inverter unless an explicit control technique is devised. Lakshman Rao S.P, *et*.al.[8] presented to meet the load requirement the output voltage of inverter can be fed to the filter to get sinusoidal waveforms which can be again fed to the transformer for theoretical output. S. M. Cheratiet. *et*.al, [9] developed the close-loop current mode PI controller has been implemented in MATLAB/SIMULINK using Sisotool. A Continuous second-order state-space model for resistive load has done. Then a multiple feedback control loop tested with PWM gain is considered the inner loop parameter as an inductor current of the filter and the outer loop parameter as output voltage. K. Deepa et.al, [10] comparative analysis has been done for different PWM techniques 7 level inverter.

In this paper concerted on analysis of the substance of effect of LC filters to get output sinusoidal waveforms for Neutral Point Clamped inverter. The main motto of this paper studies the design of LC filters for multilevel inverters for low voltage applications with less no of switching devices with minimum THD levels.

2. Inverter Topologies

The highly developed multilevel inverter topologies are

- Diode Clamped or Neutral Point Clamped Technique (DC-MLI or NC-MLI)
- Flying Capacitor (FC-MLI)
- Cascaded Full Bridge (CMLI)

In 1981, Nabae, Takahashi, and Akagi proposed first the neutral point three-level diode clamped inverter. The most important idea of this inverter is to reduce the voltage stress of the power devices with the help of diodes. The voltage over each capacitor and each switch is V_{dc}. In this topology, it consists of four switches and two diodes have consisted of a three-level diode clamped inverter. Here the pair of switches work in complimentary mode and access of mid-point voltage balanced by the diodes. By connecting the two DC capacitors (C1 and C2) in series the bus voltage is dividing into three voltage levels. The two diodes D_1 and D_2 are used as clamping diodes to diminish the voltage stress across each switching device is partial to V_{dc}. It is supposed to maintain the total dc-link voltage is V_{dc} and mid-point is synchronized at half of the dc-link voltage, the voltage across each capacitor is $V_{dc}/2$ ($V_{c1}=V_{c2}=V_{dc}/2$). In a three-level diode clamped inverter there are three different feasible switching states which apply the staircase voltage on output voltage relating to DC link capacitor voltage rate. To get three-level output voltage simultaneously two switches should be on and so on.

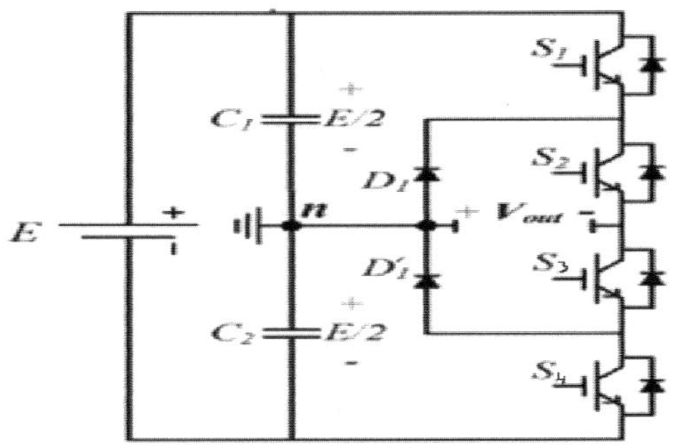

Figure 1. Circuit diagram of one leg diode clamped inverter

Table 1. Switching state of diode clamped three level inverter

Switching Status	State	Voltage
S1=High,S2= High	+ve Voltage	$V_{out}= V_{dc}/2$
S2=High,S3= High	0	$V_{out}= 0$
S3=High,S4= High	-ve Voltage	$V_{out}= -V_{dc}/2$

The classification of Modulation strategies based on fundamental frequency switching,
- Pulse Width Modulation(PWM)
- Sinusoidal Pulse Width Modulation (SPWM)
- Space Vector Pulse Width Modulation (SVPWM)
- Carrier-Based Space Vector Pulse Width Modulation (CB- SVPWM)

3. Theoretical calculations of the LC filter

All loads with a pure sinusoidal waveform from the inverter, which can be done by using a properly designed LC filter. During designing filters the criterion of cost, size, losses, etc are to be considered. Along with an additional criterion is required i.e. minimum reactive power, which helps indirectly, reducing the cost, size, and losses. The harmonics of the LC filter are given in the Fourier series form the closed-form expressions of the inductance and capacitances of the LC filter were not obtained. The reactive power of the LC filter can be calculated as [4].

$$Pr = WrLr(Is^2 + I^2av) + WrCf(Vo^2 + Vo^2av) \tag{1}$$

Where, Is and Vo is the fundamental component of the RMS value of inductor current and load voltage, respectively.
The RMS value of the inductor current can be calculated as

$$Is = [Ior^2 + (Ioi - WrCfVo)^2]^{\frac{1}{2}} \tag{2}$$

Where, I_{or} and I_{0i} are the RMS values of the real and imaginary components of the load current.

Where, C_f is the filter capacitance of $Cf = K\frac{Ed}{Lf^2Voav}$ (3)

and

$$K = \left[\frac{k^2 - \frac{15}{4}k^4 + \frac{64}{5\pi}k^5 - \frac{5}{4}k^6}{1440}\right]^{\frac{1}{2}} \tag{4}$$

Where,

$$factor \qquad k = \sqrt{2} * \frac{V_0}{E_d} \tag{5}$$

V_o= RMS value of Output voltage
E_d=DC Voltage
From equation 3 in 1, the optimum value of the inductance of the filter can be calculated by using equation;

$$L_f = \frac{V_0}{I_0 f_s}\left\{K\frac{E_d}{V_{0,av}}\left[1 + 4\pi^2\left(\frac{f_r}{f_s}\right)^2 K\frac{E_d}{V_{0,av}}\right]\right\}^{\frac{1}{2}} \tag{6}$$

The RMS value of output voltage harmonic over one period of the fundamental output voltage can be obtained as

$$Vo,av = \left[\frac{1}{2\pi}\int_0^{2\pi} Vo^2 d(2\pi f_r)dt\right]^{\frac{1}{2}} \tag{7}$$

Where,
V_0=RMS value of output voltage of inverter

$$Vo = 2\times\frac{Ed}{\sqrt{2}\times\pi} \tag{8}$$

The practical value of load resistance and inductance is considered for calculation of output voltage, current, and THD level for single-phase inverter for R and RL load. The load impedance is taken as, Z=R+jXL= 403.3+j1043.3 Ω.

Table 2. Calculated Inductance and Capacitance values

$E_d(V)$	$f_s(H_z)$	$f_r(H_z)$	$L_f(mH)$	$C_f(\mu F)$	$R_L(ohm)$	$L_l(H)$
48	3000	50	**4.707**	**283.3**	403.33	3.320

4. Simulation results and discussions

The simulation has been carried out in MATLAB/SIMULINK R2016a by using phase opposition pulse width modulation techniques. In POD-PWM techniques two carrier signals (one is for positive half cycle and other is for negative half cycle) have been used and compared with one reference signal which is a sinusoidal wave. Here, the modulation index has been taken 0.9 to get the required RMS voltage. The theoretical and simulated values are compared with filter and without the filter. The required load voltage 220V gets it by the step-up transformer, which is required for R and RL load in low voltage applications. Fig.2. Shows the Simulink model of the three-level neutral clamped inverter. Table 2 shows the calculated and simulated output voltages and current for a real load impedance of R_L=403.33Ω and X_L=3.320H. Table.3. indicates that THD level and output RMS voltage for different modulation index.

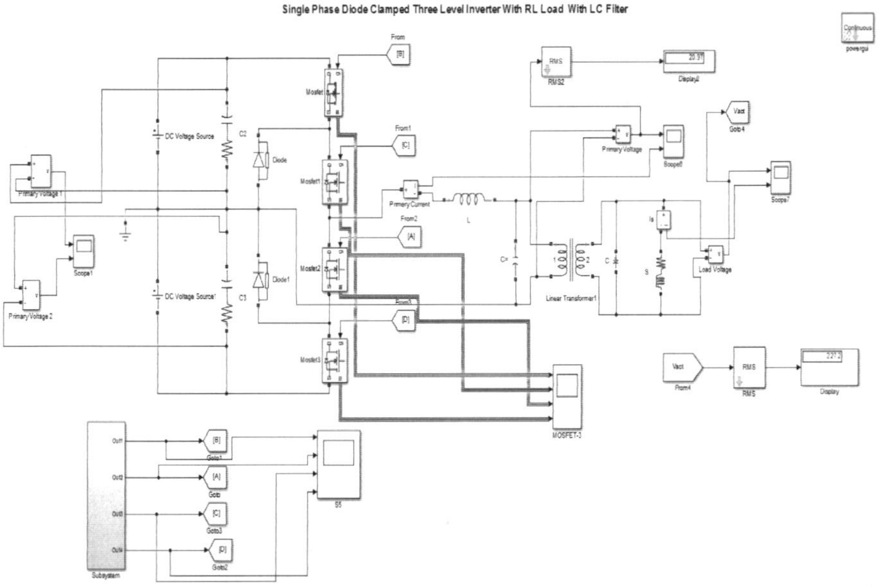

Figure 2. 3 level diode clamped simulink model with transformer

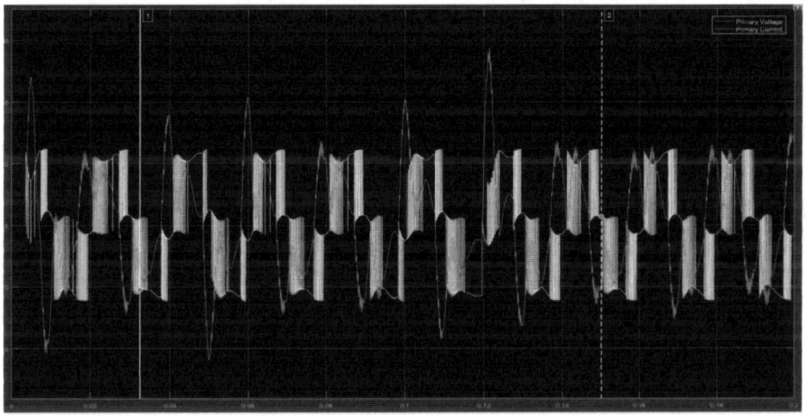

Figure 3. Without filters output voltage and current waveform of inverter

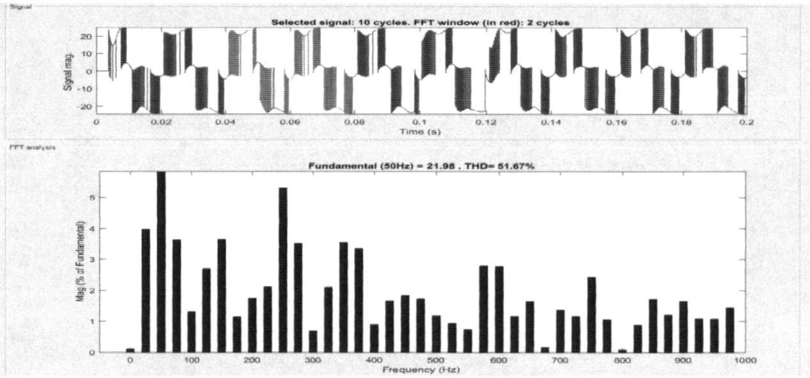

Figure 4. THD level of three level diode clamped inverter without filters

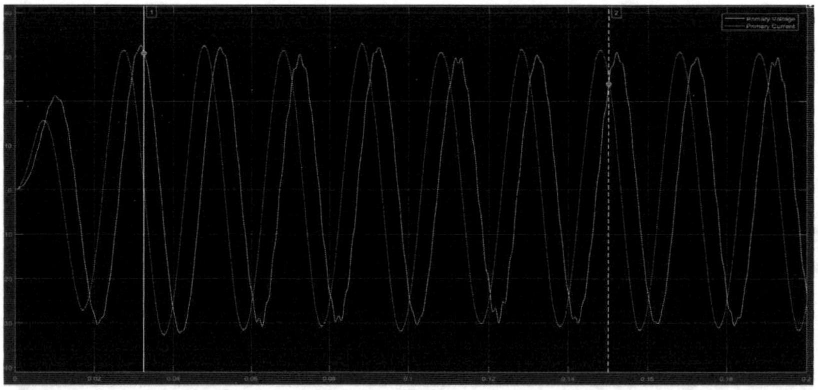

Figure 5. With filters Voltage and current output waveform of inverter (primary side of transformer)

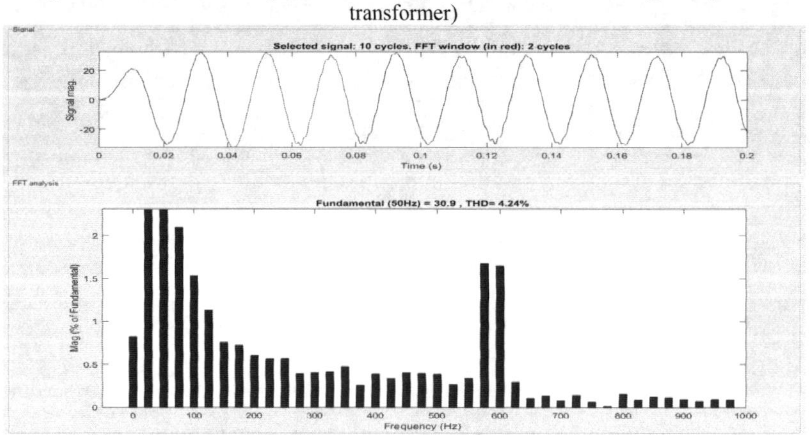

Figure 6. THD level of three level diode clamped inverter with filter

IOP Conf. Series: Materials Science and Engineering **814** (2020) 012037 doi:10.1088/1757-899X/814/1/012037

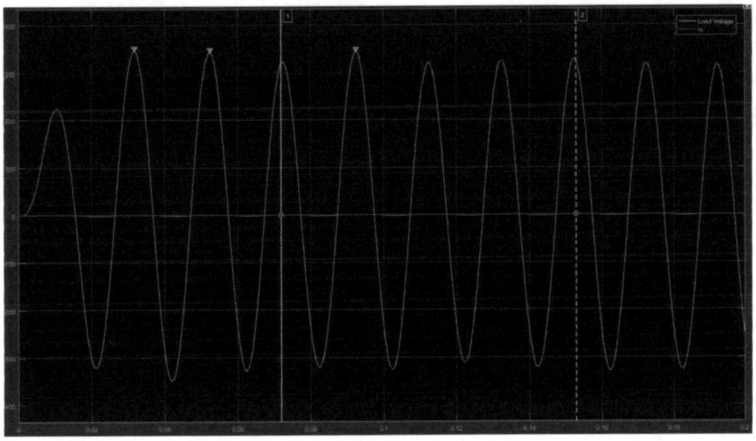

Figure 7. With filters Voltage and current waveform of inverter at load side (secondary side of transformer)

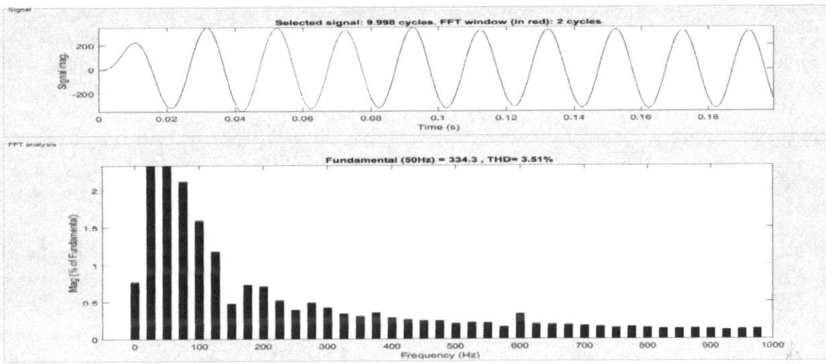

Figure 8. THD level in RL Load with filter

Table 3. A Comparison of theoretical and simulated values for R=403.33Ω and L=3.323H.

Type of Load		Output Voltage (V)	Output Current (A)	THD
For R Load Without Filter	Theoretical Values	21.70	0.0545	48.35%
	Simulated Values	16.90	0.059	55.16%
For R Load With Filter	Theoretical Values	21.61	0.054	48.35%
	Simulated Values	19.90	0.065	44.56%
For RL Load Without Filter	Theoretical Values	21.607	0.054	48.53%
	Simulated Values	17.01	0.600	51.6%
For RL Load With Filter	Theoretical Values	21.607	0.055	48.35%
	Simulated Values	**20.97**	0.053	**3.51%**

Table 4. THD level for RL load at different modulation index

Output Voltage	Output Current (A)	THD
0.5	11.93	1.20%
0.6	13.96	3.74%
0.7	15.48	3.32%
0.8	17.84	2.04%
0.85	19.10	**1.68%**
0.9	**20.97**	3.51%

5. Conclusion

The theoretically calculated values of LC filters are tested on three-level diode clamped inverter by considering R and RL loads. Simulation has been carried out with and without filters for R and RL load using POD-PWM techniques and secondly checked for RL load with different modulation index. The output results show that simulated values are closer to the calculated values. From the theoretical calculated LC, filters are able to bring the THD level within 5% proven. In the second case, the load fixed and varied the modulation index from 0.5 to 0.9, at 0.85 gives very less THD level and output RMS voltage near to the calculated value. At 0.9 THD level is slightly increased and output RMS voltage closest to the calculated value. With proper analysis of LC values, we can get a voltage in the sinusoidal form for the loads. For low voltage applications using less, no switches can accomplish the THD level within the limits instead of high large no of switches.

Reference

[1] Jose Rodriguez, Jih-Sheng Lai, Fang Zheng Peng, 2002 *Multilevel Inverters: A Survey of Topologies Controls, and Applications,* (IEEE Transactions on industrial Electronics), Vol. 49, No. 4

[2] Naik Raghuram. L, Jangamshetti Suresh. H, 2011 1st International Conference on Electrical Energy Systems *A Critical Study of Modulation Techniques for Three Level Diode Clamped Voltage Source Inverter for Grid Connections of WECS*

[3] Subhashree Choudhury,Samikhya Nayak, Tara Prasanna Dash, P K Rout, 2018, IEEE International Conference on Technologies for Smart-City Energy Security and Power *A Comparative Analysis of Five Level Diode Clamped and Cascaded H-Bridge Multilevel Inverter for Harmonics Reduction,* (Bhubaneswar, India)

[4] Pekik A. Dahono, Agus Purwadi, and Qamaruzzaman, 1995 *An LC Filter Design Method for Single-phase PWM Inverters,* IEEE

[5] Bharti Katre, Apurva Dabare, and Santosh Patil, 2017 *Analysis of Single Phase Diode Clamped Three Level Inverter,* International Journal of Trend in Research and Development, vol. **4**(1).

[6] Azah Mohamed, 2009, IEEE Student Conference on Research and development, *Development of Three Phase Photovoltaic Inverter using dSPACE DS1104 Board,* (UPM Serdang, Malaysia) p 116-18

[7] Gautam ghosh,Suman Sarkar, shamik mukherjee, 2017 *A Comparative study of different multilevel inverter,* IEEE

[8] Lakshman rao S.P, Dr ciji pearl kurian, 2014 *Design and Simulation of Grid Connected Hybrid Solar –WECS using SIMULINK/MATLAB,* IEEE

[9] S. M. Cherati, N. A. Azli, S. M. Ayob, A. Mortezaei, 2011, IEEE Applied Power Electronics Colloquium(*IAPEC*) *Design of a Current mode PI Controller for a single-phase PWM Inverter*

[10] K.Deepa, P.Ajay Kumar, V.Sai Krishna, P.N.Koteswara Rao, Mounika.A, D.Medhini, 2017 *A Study of Comparative Analysis of Different PWM Techniques,* IEEE

[11] *Power Electronics* by Dr. P.S. Bimbhra, Khanna Publishers, Fifth Edition 2012.

IOP Conf. Series: Materials Science and Engineering **814** (2020) 012038 doi:10.1088/1757-899X/814/1/012038

Numerical investigations of stress-deformation responses in fractured paediatric bones with prosthetic bone plates

Nand Jee Kanu[1], Devendra Patwardhan[2], Eva Gupta[3], Umesh Kumar Vates[4], and Gyanendra Kumar Singh[5]

[1]Ph. D. Research Scholar, Department of Mechanical Engineering, SVNIT, Surat, Gujarat-395007, India
[2]BE Research Scholar, Department of Mechanical Engineering, JSPM NTC, Pune, Maharashtra-411041, India
[3]Ph. D. Research Scholar, Department of Electrical Engineering, Amity University, Noida, Uttar Pradesh-201313, India
[4]Associate Professor, Department of Mechanical Engineering, Amity University, Noida, Uttar Pradesh-201313, India
[5]Faculty, Department of Manufacturing Engineering, Adama Science and Technology University, Adama-1888, Ethiopia

E-mail: nandssm@gmail.com

Abstract. Nearly *25%* of paediatric injuries happen due to fractures of femur bones, either under accidental or inflicted conditions. In the present investigation, static analyses are performed for fractured paediatric femur bones of children joined with bone plates using finite element method (FEM) approach to find stresses and deformations in the bone and plate assembly under axial compressive and torsional loadings. The investigation aims to suggest the suitable biomaterial for fracture fixation bone plates modeled using two recently-developed biomaterials such as titanium and magnesium alloys, after studying the stresses and deformations at the fracture spots of the paediatric femur bones of children aged between *1-3 years*. The stress-deformation responses are further analyzed in relation to ages of the children.

1. Introduction

The report suggests that children aged between *1-3 years* are relatively at higher risk of fatal paediatric injuries due to their inability to save themselves [1]. In the present investigations it has been thus realized the need of addressing the stress-deformation responses in fractured paediatric bones with prosthetic bone plates for children aged between *1-3 years*. In most of the cases the suitable fracture fixation bone plates are used to join fractured paediatric femur bones of children [2, 3]. The aim of study is to investigate the suitable biomaterial for fracture fixation bone plates modeled using two recently-developed biomaterials such as titanium and magnesium alloys [4], after studying the stresses and deformations at the fracture spots of the paediatric femur bones of children aged between *1-3 years* under axial compressive and torsional loadings [5-7].

IOP Conf. Series: Materials Science and Engineering **814** (2020) 012038 doi:10.1088/1757-899X/814/1/012038

2. Modeling and analysis

2.1. Generation of models using finite element

The computer tomography (CT) scanned data of asymmetric femur bones are used in medical imaging system (MIMICS) in digital formats to assign material properties in cortical bones [2]. The CT scanned data of fractured paediatric femur bones (having *0.5 mm* slice thicknesses) of children aged between *1-3 years* are imported into the PTC Creo software to assemble these with fracture fixation bone plates and screws. In the present investigation the fracture fixation bone plates are modeled using the material properties of two recently-developed biomaterials such as titanium and magnesium alloys as shown in Table 1 for the static analyses of cortical bones under axial compressive and torsional loadings in ANSYS 15.0 software.

Table 1. The mechanical properties of paediatric femur bones and recently-developed fracture fixation bone plates

Properties of Bones [1]				Properties of Bone Plate [4]		
Age	Poisson's Ratio	Bone Density *(kg/m^3)*	Modulus of Elasticity of Bone *(GPa)*	Material	Modulus *(GPa)*	Density *(g/cm^3)*
48 weeks (Male)	*0.342*	*346.6*	*17.25*	Mg alloys Ti alloys	*35-45* *110-120*	*1.79-2* *4.2-4.5*
96 weeks (Female)	*0.342*	*346.6*	*18.48*	Mg alloys Ti alloys	*35-45* *110-120*	*1.79-2* *4.2-4.5*
144 weeks (Female)	*0.342*	*480*	*19.09*	Mg alloys Ti alloys	*35-45* *110-120*	*1.79-2* *4.2-4.5*

2.2. Meshing and boundary conditions

The assemblies of fractured paediatric femur bones of children aged between *1-3 years* joined with bone plates are discretized using the *solid 186* layered solid elements. These 20-noded higher order elements are having three degrees of freedoms for quadratic displacement behaviors. The assemblies are meshed with such 8631 elements, and 15608 nodes for accurate static responses in terms of the stresses and deformations at the fracture spots of the paediatric femur bones of infants and toddlers. The convergence test is carried out for the suitable optimum mesh to reduce the computational efforts.

2.3. Validation of model

The models (assemblies of fractured paediatric femur bones of children aged between *1-3 years* joined with bone plates) are validated using the results available in the references [2, 3]. The mechanical properties of paediatric femur bones joined with prosthetic bone plates used in validation of our models are shown in Table 2.

The eccentric load of *750 Pa* is applied at the head of the femur bone which is fixed at its opposite end as shown in Figure 1. The maximum directional deflection (Figure 1 (b)) and equivalent stress (Figure 1 (d)) are obtained under the compressive loading as *0.000001942 m* and *76346 Pa,* respectively. Thus, the results obtained are in good agreement with the corresponding data sets available in Ref. [3] (Figures 1 (a)-(c)). The maximum deflection under torsional loading *(50 N m)* is *0.013288 m* (Figure 1 (f)) whereas under compressive loading *(700 N)* is *0.00004302 m* (Figure 1 (h)) and thus, the efficacy of present model are validated with the corresponding data sets available in Ref. [2] (Figures 1 (e)-(g)).

Table 2. The mechanical properties of paediatric femur bones [3] and fracture fixation bone [2]

Properties of Bone [3]			Properties of Bone Plate [2]			
Poisson's Ratio	Bone Density *(g/cm³)*	Modulus of Elasticity of bone *(GPa)*	Material	Modulus *(GPa)*	Density *(kg/m³)*	Poisson's Ratio
0.3	*2000*	*2.13*	Titanium	*120*	*4500*	*0.3*

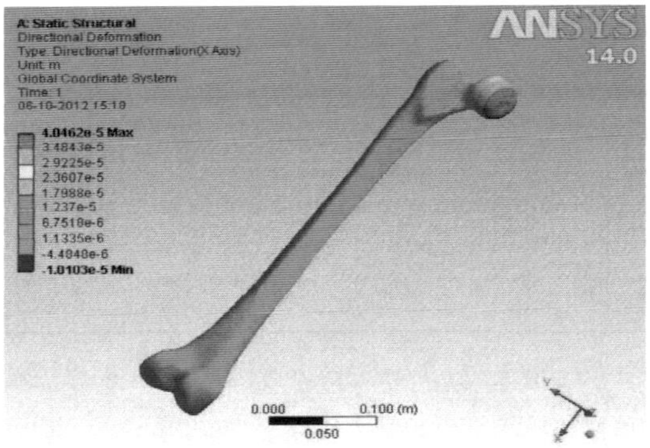

(a)

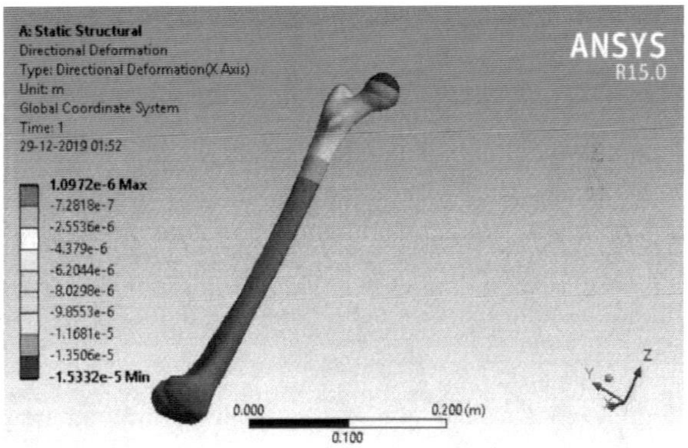

(b)

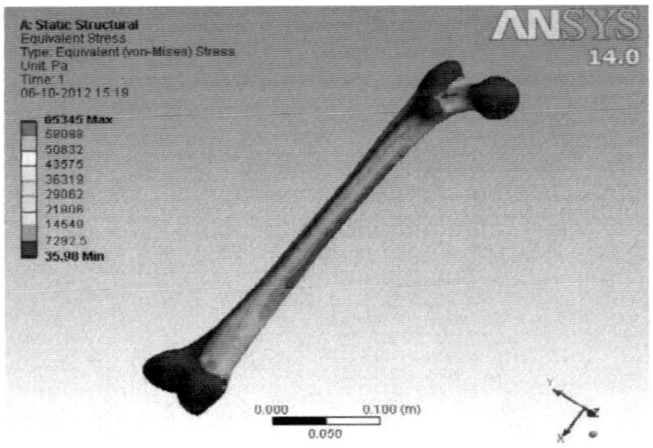

(c)

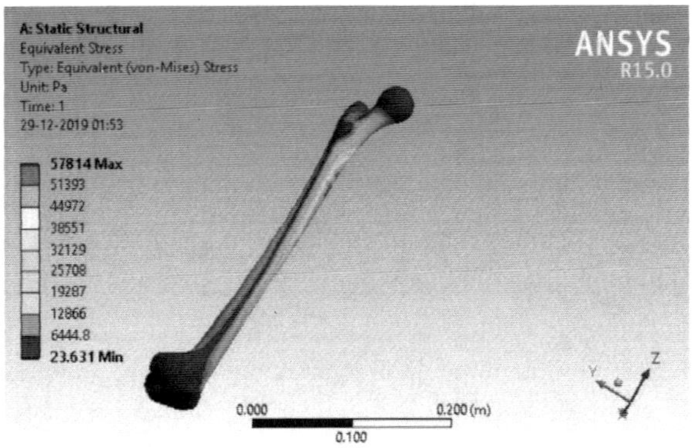

(d)

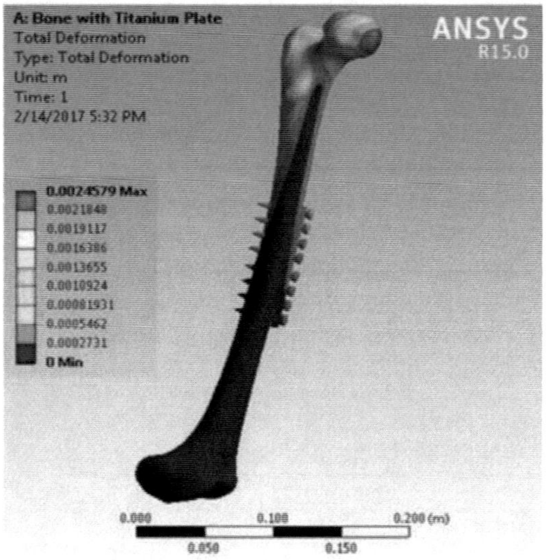

(e)

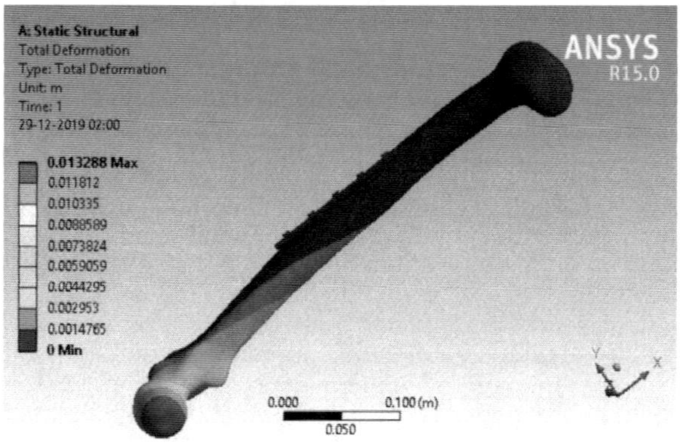

(f)

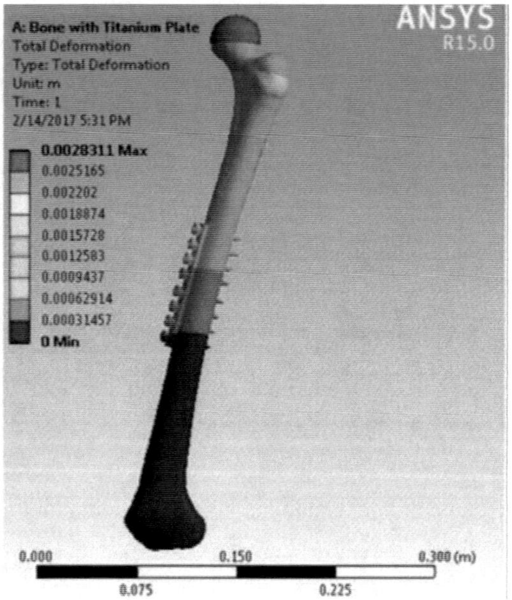

(g)

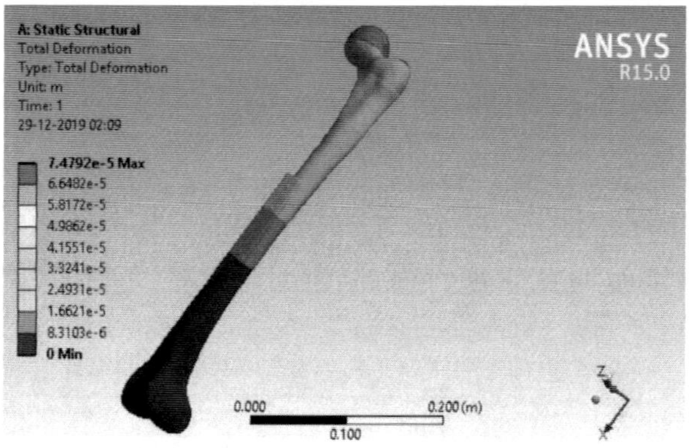

(h)

Figure 1. The stress-deformation responses of cortical bones and bone plate system. The directional deformation in Ref. [3] (a) is used for validation in present investagation (b). The equavalent stress in Ref. [3] (c) is used for validation in present investagation (d). The torsional total directional deformation in Ref. [2] (e) is used for validation in present investagation (f). The compressive total directional deformation in Ref. [2] (g) is used for validation in present investagation (h).

3. Results and discussions

The results are presented for suitable biomaterial for fracture fixation bone plates modeled using two recently-developed biomaterials such as titanium and magnesium alloys, after studying the stresses and deformations at the fracture spots of the paediatric femur bones of children aged between *1-3 years*. The stress-deformation responses are further analyzed in relation to ages of the children as shown in Figures 2 (a)-(d). It is found that titanium alloy is the suitable biomaterial for the bone plates as under axial compressive and torsional loadings, the effect of equivalent stresses at the fracture spots of the paediatric femur bones is significantly reduced.

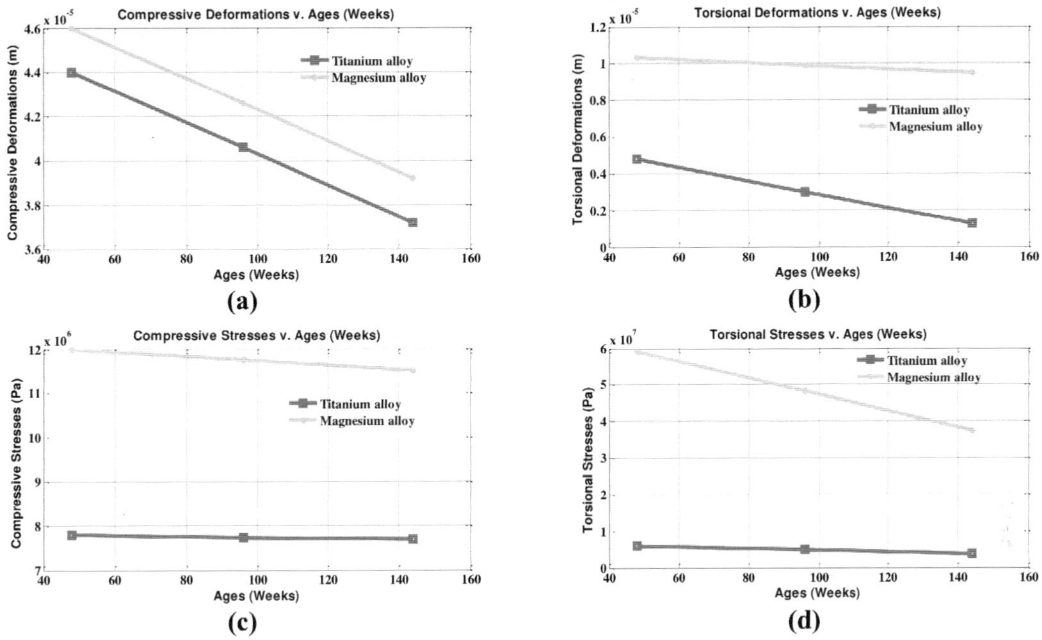

Figure 2. The stress-deformation responses in relation to ages of the children. (a) Compressive deformations *(m)* versus ages *(weeks)*. (b) Torsional deformations *(m)* versus ages *(weeks)*. (c) Compressive stresses *(Pa)* versus ages *(weeks)*. (d) Torsional stresses *(Pa)* versus ages *(weeks)*.

4. Conclusions

In the present investigations, the fracture fixation bone plates are modeled using two recently-developed biomaterials such as titanium and magnesium alloys. After studying the stresses and deformations at the fracture spots of the paediatric femur bones of children aged between *1-3 years* under axial compressive and torsional loadings, it is found that titanium alloy is the suitable biomaterial for the bone plates (used to join fractured paediatric bones of children). Under axial compressive and torsional loadings, the effect of equivalent stresses at the fracture spots of the paediatric femur bones is reduced significantly after modeling the prosthetic bone plates with titanium alloy instead of magnesium alloy. Thus, the present investigation favours the future applications of recently developed titanium alloy as the suitable biomaterial for modeling the prosthetic bone plates instead of commercially available prosthetic bone plates made up of magnesium alloys.

References

[1] Zainab Altai et al. 2018 *Biomechanics and Modeling in Mechanobiology* **17** 1001-1009
[2] Pravat Kumar Satapathy et al. 2018 *IOP Conf Ser.: Mater. Sci. Eng.* **330** 012027
[3] P.S.R. Senthil Maharaj et al. 2013 *Procedia Engineering* **64** 1242-1251
[4] Youwen Yang et al. 2020 *Materials & Design* **185** 108259
[5] Daniel Hedequist et al. 2008 *J Pediatr Orthop.* **28(1)** 6-9
[6] B. de Billy et al. 2014 *Orthopaedics & Traumatology: Surgery & Research* **100 (1)** S139–S148
[7] Sandeep Das and Saroj Kumar Sarangi 2014 *IJBAB* **1(1)** 1-5

IOP Conf. Series: Materials Science and Engineering **814** (2020) 012039 doi:10.1088/1757-899X/814/1/012039

Study on Effect of Abrasive Water Jet Machining Process Parameter on Taper Angle during Machining of Epoxy Resin Glass Fibre

Meghna K Gawade[1], Vijaykumar S Jatti[2], Vaibhav V Phule[3] and Pranita V Sawant[4]

[1,4]PG Student, Department of Mechanical Engineering, D.Y. P. COE, Pune,India
[2]Asso. Professor, Department of Mechanical Engineering, D.Y. P. COE, Pune,India
[3]Ass. Professor, Department of Mechanical Engineering, D.Y. P. COE, Pune,India

E-mail: meghnagawade.mg@gmail.com

Abstract. Abrasive Water Jet Machining (AWJM) is one of the most advanced machining processes, which has the ability to machine various materials. In this paper, experiment was performed on cutting of epoxy fibre glass by AWJM. Epoxy fibre glass is a composite which consists of fibre glass cloth and epoxy resin as binder, achieved under specific heat and pressure. There are various process parameters such as water pressure, standoff distance, abrasive flow rate, traverse rate, abrasive etc. The water pressure and standoff distance was varied while performing the experiment and rest all parameters were kept constant. From this experiment taper angle was measured and it was observed that taper angle increases as the standoff distance increases, this is because the shape of the jet diverges with larger standoff distance. Minimum taper angle was observed as 0.34° at 32 ksi pressure, having 2 mm standoff distance. It was also observed that width of cut has significant effect on taper angle. From further calculations, the deviation of the cut slot for top width was observed as +1% to 5% and for bottom width it was -1% to 2%.

1. Introduction

Abrasive Water Jet Machining (AWJM) is the most advanced and extensively used machining processes. AWJM is convenient in carrying operations on various materials like glass, composites, metals, etc. Momber and Kovacevic [1] has explained various principles on abrasive water jet machining wherein due to plastic deformation, gradual wear and fracture mechanism by high velocity abrasive water jet the material removal takes place in AWJM. When the operation takes place there are no effects of thermal distortion, high machine versatility, high flexibility and also there is absence of heat which makes it cold machining process and thus due to this there is no any damage to the material, which makes it more advantageous than the other machining process used for cutting, drilling etc. [2].

Composites have become one of the most widely used materials due to their high mechanical strength, stiffness and low density, etc. due to their combined material characteristics. Operations carried out by AWJM on composites have various advantages such as any complex shapes can be manufactured withprecision, also good surface finish of the composite material is achieved. Fibre reinforced material is widely used and it needs accurate cutting of processes with efficiency. It cannot

IOP Conf. Series: Materials Science and Engineering **814** (2020) 012039 doi:10.1088/1757-899X/814/1/012039

be achieved using non-conventional cutting processes like drilling, electric discharge machining, ultrasonic machining, etc. But among all of these non-conventional techniques AWJM is most feasible method for cutting fibre reinforced materials [3].

Various parameters are associated with AWJM such as water pressure, abrasive flow rate, traverse speed, standoff distance, etc. which affects the surface quality of the machined material. The purpose of this paper is to study cutting of epoxy fibre glass by AWJM by varying processparameters.

2. ExperimentalWork

2.1 Materials and Methods

In this study, the material used is hybrid composite, manufactured commercial epoxy fibre glass sheet with dimension 550 mm × 200 mm × 5 mm. Density is 1.9 g/cm to 2 g/cm approximately. Glass content is 65 to 70 % and epoxy resin content 30 to 35 % by mass approximately. Properties of the material is taken as follows

Table 1. Properties of epoxy fibre glass

Sr. No.	Properties	Specification	
		Lengthwise	Crosswise
1	Flexural Strength at 20°C	450 N/sq.mm	350 N/sq.mm
2	Impact Strength at 20°C	300 N/sq.mm	190 N/sq.mm
3	Tensile Strength	450 N/ sq.mm	400 N/ sq.mm
4	Modulus of elasticity at 20°C	25 kN/ sq.mm	20 kN/ sq.mm
5	Compressive strength at 20°C	390 N/ sq.mm	500 sq.mm

2.2 Equipment

The experiment was conducted on MAXIEM® 1530 Jet Machining® equipped with CNC, gravity feed type abrasive hopper and work piece table dimensions as 3708 mm × 1727mm. The process parameters associated with the machine were as follows

Table 2. Machining Parameters

Sr. No.	Parameters	Specification
1	Water Pressure	20-43 ksi
2	Diameter of orifice	0.4064 mm
3	Diameter of nozzle	1.0668 mm
4	Flow rate of abrasive	315 g/min
5	Abrasive size	85 mesh
6	Traverse rate	50-200 mm/min
7	Abrasive index	0.85
8	Stand-off distance	1-5 mm
9	Abrasive diameter	0.025 mm
10	Velocity of jet	150-300 s

2.3 Experimental Design

In this study, through cut square slots of 15 × 15 mm were made on the epoxy fibre glass using AWJM. This operation was carried for 30 numbers of slots on the work piece. The abrasive used to cut the slot was garnet. To conduct the experiment, parameters of machine were considered as control variables as mentioned in Table 2. Wherein pressure was varied from 30 ksi to 40 ksi by varying the standoff distance at an interval of one i.e. from 1 mm to 5 mm for each hole and rest all parameters were kept constant.

Figure 1. Experimental setup.

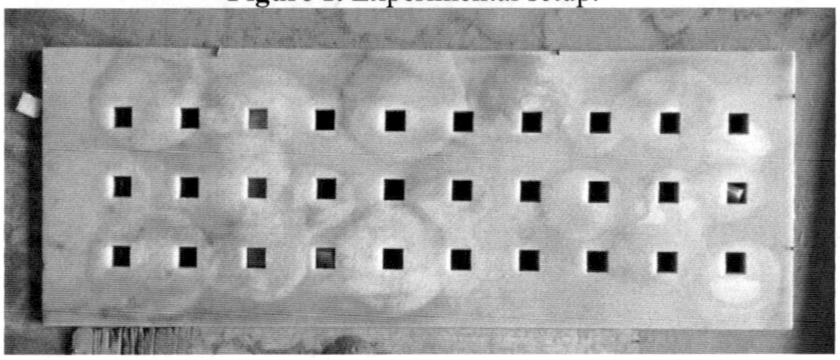

Figure 2. Work piece machined by AWJM

After machining, width of cut at top and width of cut at bottom of each slot was measured using digital vernier caliper. This data was used to calculate the taper angle in both the widths i.e. at top and bottom width of the slot. The taper angle formula is as follows [10]

$$\text{Taper angle } (\theta) = \tan^{-1} \left(\frac{W_t - W_b}{2 \times t}\right) \text{degree}$$

Where

W_t – top width inmm,

W_b – bottom width in mm, t – thickness inmm.

3. Results anddiscussions

3.1 Effect of abrasive particles on width ofcut

Table 3. Details of experiment

Input Parameters				Output Parameters			
Sr No.	Pressure (ksi)	Standoff Distance (mm)	Top Width	Bottom Width	Deviation of top width (%)	Deviation of bottom width (%)	Taper Angle (θ)
1		1.00	15.20	14.88	1.33	-0.80	1.83
2		2.00	15.42	14.94	2.80	-0.40	2.74
3	30	3.00	15.30	14.78	2.00	-1.47	2.97
4		4.02	15.36	14.80	2.40	-1.33	3.20
5		5.08	15.40	14.80	2.67	-1.33	3.40
6		1.00	15.40	15.01	2.67	0.07	2.23
7		2.02	15.38	15.32	2.53	2.13	0.34
8	32	3.00	15.30	15.00	2.00	0.00	1.71
9		4.00	15.42	15.00	2.80	0.00	2.40
10		5.05	15.60	15.00	4.00	0.00	3.43
11		1.02	15.30	15.10	2.00	0.67	1.14
12		2.03	15.34	15.00	2.27	0.00	1.94
13	34	3.00	15.18	14.82	1.20	-1.20	2.06
14		4.06	15.34	14.82	2.27	-1.20	2.97
15		5.05	15.38	14.84	2.53	-1.07	3.09
16		1.02	15.40	15.00	2.67	0.00	2.29
17		1.52	15.40	14.92	2.67	-0.53	2.74
18	36	3.05	15.38	14.88	2.53	-0.80	2.86
19		4.06	15.44	14.84	2.93	-1.07	3.43
20		5.08	15.42	14.80	2.80	-1.33	3.54
21		1.02	15.30	14.92	2.00	-0.53	2.17
22		2.03	15.38	14.92	2.53	-0.53	2.74
23	38	3.05	15.42	14.92	2.80	-0.53	2.86
24		4.06	15.52	14.98	3.47	-0.13	3.09
25		5.08	15.74	15.00	4.93	0.00	4.23
26		1.02	15.34	15.00	2.27	0.00	1.94
27		2.03	15.42	15.00	2.80	0.00	2.40
28	40	3.05	15.30	15.08	2.00	0.53	1.26
29		4.06	15.70	15.10	4.67	0.67	3.43
30		5.08	15.64	15.10	4.27	0.67	3.09

From the above calculations it was noted that the top width is larger than the bottom width of each slot. The deviation of the cut slot for top width was observed as +1% to 5% and for bottom width it was -1% to 2%. Wang and Wong [4] has explained that initially at the time of cut the kinetic energy of the abrasive particles is very high but as it moves downwards the abrasive particles looses, K.E. and relative strength of the jet i.e. it gets narrower as it moves downwards, thus the bottom width is smaller than the top width. We notice that the taper of cut is varying from 0.2° to 4.5°, Khan [5] has explained that when garnet is used as abrasive the taper of cut is comparatively higher. Because the sharpness of the garnet abrasive is reduced and hence the bottom width is also smaller than the top width.

IOP Conf. Series: Materials Science and Engineering **814** (2020) 012039 doi:10.1088/1757-899X/814/1/012039

3.2 Effect of standoff distance on taper angle at differentpressures

Figure 3(a) and 3(b) shows the effect of standoff distance on angle of taper with varying pressures. From below graphs, it is observed that as the standoff distance increases the angle of taper also increases. Khan [5] explained that the nature of jet is dependent on the standoff distance applied. The increase in the taper of angle is due to the reason that as standoff distances increases the jet focus area diverges and thus there is increase in the width of cut.

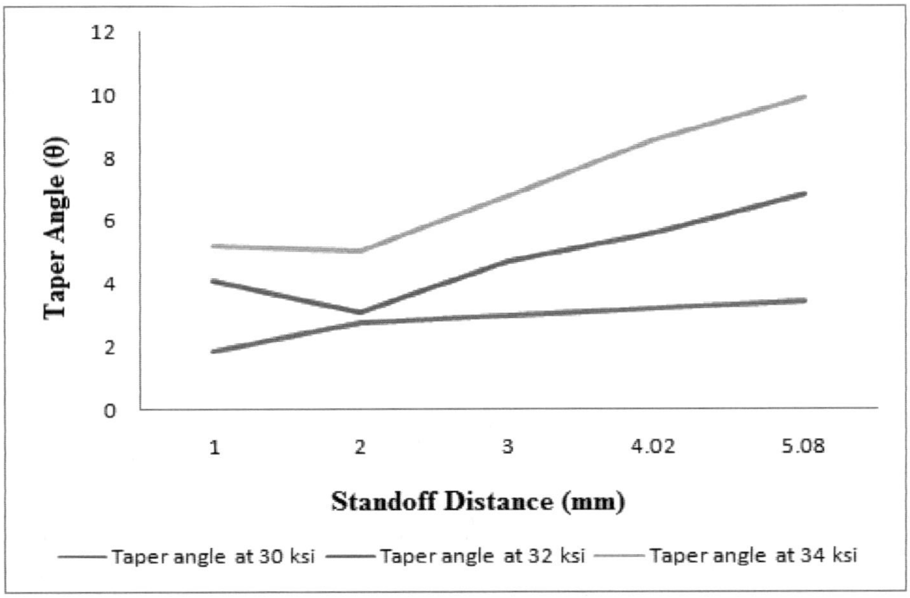

Figure 3(a). Effect of standoff distance on taper angle at 30 ksi, 32 ksi and 34ksi.

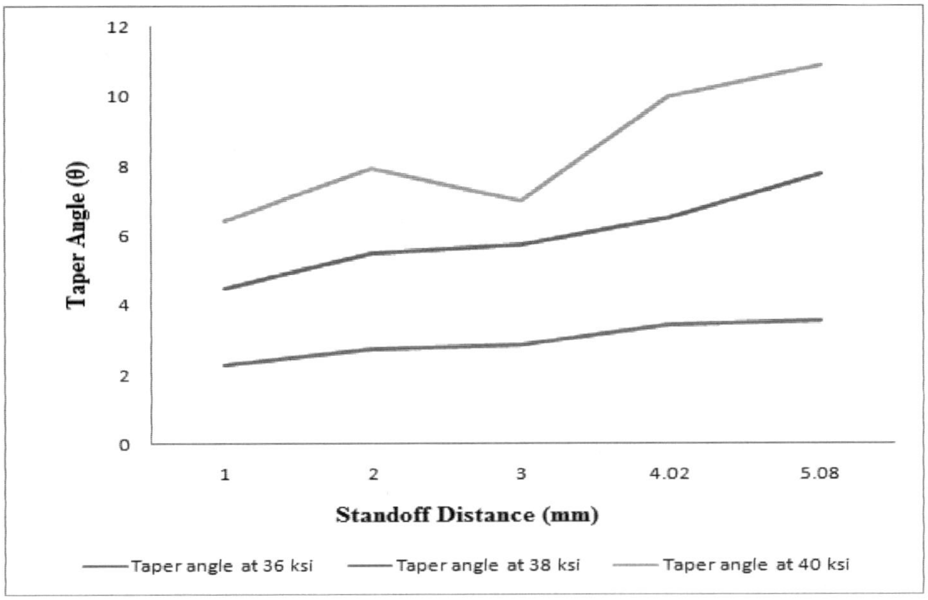

Figure 3(b). Effect of standoff distance on taper angle at 36 ksi, 38 ksi and 40ksi.

4. Conclusion

From the above observations, it can be concluded, as the standoff distance increases then the angle of taper also increases, this happens because as the standoff distance increases, the jet focus area diverges. Minimum taper angle was observed as 0.34° at 32 ksi pressure, having 2 mm standoff distance. It was also observed that the bottom width of cut is smaller than width of cut at top, which is because at impingement of the jet, the kinetic energy is higher and when the jet moves downwards, the abrasive particles loses their kinetic energy. Hence there is deviation in the width of cut. The deviation of the cut slot for top width was observed as +1% to 5% and for bottom width it was -1% to2%.

References

[1] Momber A W and Kovacevic R 1992 *Springer Verlag Limited (London)*
[2] Aich U, Banerjee S, Bandyopadhyay A and Das P K 2014 3rd *Int. Conf. on Materials Processing and Characterisation* **7** 75-85
[3] Komanduri R, Zhang Bi and Visa C M 1991 *ASME Process on Manufacturing Composite Material* **4** 91-36
[4] Wang J and Wong WCK 1999 *Int. J. Mach. Tool Manu.* **398** 55-70
[5] Khan A A and Hague M M 2007 *J. Mater. Process. Technol.*404-07
[6] Azmir M A and Ahsan A K 2008 *J. Mater. Process. Technol.* **198** 122-28
[7] Hocheng and Chang 1994 *J. Mater. Process. Technol.* 287-304
[8] Alberdi A, Suarez A, Artaza T, Escobar-Palafox G A and Ridgway K 2013 *Int. Conf. MESIC Procedia Engg.* 421-29
[9] Azmir M A and Ahsan A K 2009 *J. Mater. Process. Technol.* **209** 6168-73
[10] Hascalik A, Aydas U C and Guru H 2007 *Mater. Des.* **5** 1953-57

IOP Conf. Series: Materials Science and Engineering **814** (2020) 012040 doi:10.1088/1757-899X/814/1/012040

Stress strain characteristics of reinforced hollow concrete block masonry melded with mesh reinforcement

Manish S Dharek[1*], S Raghunath[2], Prashant C Sunagar[3], Aravind H Bhashyam[4], Sreekeshava K S[5]

[1] Civil Engg Dept, BMS Institute of Technology and Management (affiliated to VTU, Belagavi), Bengaluru, India
[2] Civil Engg Dept, BMS College of Engineering (affiliated to VTU, Belagavi), Bengaluru, India
[3] Civil Engg Dept, Ramaiah Institute of Technology, Bengaluru, India
[4] Civil Engg Dept, Christ University, Bengaluru, India
[5]Civil Engg Dept, Jyothy of Technology (affiliated to VTU, Belagavi), Bengaluru, India

*Corresponding Author: manish.shashikant@gmail.com

Abstract. Plain Masonry similar to unreinforced concrete, is resilient in compression and weak in tension. Masonry gains strength with age similar to concrete. Inspite of these resemblances, there exist numerous differences between masonry and concrete. The major difference is the regular pattern of horizontal joints(known as bed joints) at specific intervals along the height of walls introduce due to the method of construction of masonry. These bed joints make masonry a direction dependent material possessing orthotropic properties, unlike concrete which is usually regarded as isotropic atleast in the elastic range. Mechanical properties such as compressive strength, tensile strength, flexural strength are a pre-requisite as part of the design of masonry walls. The present study deals with the experimental study to evaluate the mechanical properties of hollow concrete block masonry specimens for varying cement mortar proportions melded with mesh reinforcement at bed joints. Parameters such as compressive strength, modulus of elasticity, failure pattern have been studied and compared for reinforced and unreinforced hollow concrete block prisms. The study showed higher compressive strength and improved elastic modulus for specimens with higher grade of mortar

1. Introduction

Masonry may be defined as an assembly of masonry units along with mortar laid in a predetermined orientation. The building units may be of stones, bricks or precast blocks of concrete. It is the simplest of all building techniques in which mason stacks the pieces of material over the top of one another with mud or mortar used as a binding material. Masonry, similar to concrete, is strong in compression and weak in tension, gains strength with age and also properties of masonry are affected by site practices such as handling, placing and curing as well as weather condition similar to concrete. In spite of these similarities, there exist several differences between masonry and concrete; the major one being the relative insensitivity of masonry compressive strength to the water-cement ratio of the mortar – indeed, minimal attention is paid to the water-cement ratio of mortar commonly produced on-site. The second major difference is the regular pattern of horizontal joints (known as bed joints) at specific intervals along the height of walls introduced due to the method of construction of masonry.

These bed joints make masonry a direction dependent material possessing orthotropic properties, unlike concrete which is usually regarded as isotropic, at least in the elastic range. While common concrete construction is largely dependent on formwork, the requirement of no formwork for masonry construction is regarded as a significant advantage, because formwork is expensive and construction of formwork always lies along the critical path of most multi-storied building construction scheduling.

Masonry structure are efficient than RC framed structure however, construction practices are limited because the use of huge number of masonry units, requirement of skilled labour, its performance against lateral loading etc. To improve the efficiency of masonry structure, reinforcements are strategically introduced into it. The introduction of reinforcement not only increases its lateral load carrying capacity also improvises its compressive load carrying capacity. In European countries, reinforcement is introduced in bed joint for controlling crack width but its capacity of increasing the load bearing property is neglected. This style of introducing bed reinforcement in masonry is very common in Germany. Placing of vertical reinforcement in case of concrete blocks can be done by making use of blocks which have cores/openings in it to accommodate the reinforcement. This cores can be later plugged with free flowing grout or concrete. This practice of reinforced masonry construction is very popular in the seismically active regions in latin America. This methodology also has proved to be comparably stronger and ductile when compared with R C Framed structures when subjected to earthquakes. The present work deals with comparative study of strength and elastic properties of solid block masonry prisms with mesh reinforcement and hollow block masonry prisms by varying the cement mortar mix proportions.

2. Literature Review

Matthias Ernst et al [1] studied the effect of vertical and horizontal reinforcement of perforated clay block under compression and shear loading. The increase in flexural strength, shear and compression strength as well as an increased ductility of brittle material was observed during experimentations The application of reinforcement in load bearing wall to increase the compression strength is low but reinforcements were widely used to control the crack in the structure. The use of horizontal reinforcement has increased 15% of compression strength and vertical reinforcement is 40% were observed when compared to unreinforced masonry. The ultimate strain for unreinforced masonry was given by approximately 15% (diameter of 6 mm) and vertically (2 number of 12 mm diameter) reinforced specimens by 25%. Mohamad et al [2] carried out experimental studies on brick masonry prisms subjected to axial compressive loading. The variation in elastic modulus of brick units and mortar governed the failure mechanism wherein the nonlinear behavior of masonry was predominantly governed by mortar. Gumaste et al [3] studied the mechanical properties of brick masonry prisms made of table moulded bricks and wire cut bricks of Indian origin for varying grade of mortar. It was concluded that the masonry prisms made with wire cut bricks showed a better correlation. Bryan and Mervyn [4] studied the stress strain characteristics of unconfined and confined clay brick masonry. The presence of confinement plates placed within the bed joints controlled the lateral expansion of joint and reduced the differential expansion between the brick and the mortar joint. About 40% improvement in ultimate strength was observed. Jagadish et al [5] studied the effect of containment reinforcement in masonry buildings to arrest the post cracking deflections. Since the failure pattern of masonry buildings is brittle which can lead to ultimate failure of buildings, introduction of containment reinforcement played vital role in improving the ductility of masonry. Oliveira et al [7] studied the performance of brick masonry prisms under cyclic loading and concluded that the stress strain characteristics was bilinear in nature.

3. Engineering Properties of block and mortar

The materials used as part of current experimentations comprised of hollow concrete block and solid concrete blocks of 150mm thickness acquired locally from Bengaluru. The properties of solid and hollow blocks adopted for casting masonry specimens have been summarized and presented in Table 1 and Table 2 respectively conforming to IS 2185-2003- part 1 and part 2 [6]. The mortar used in the present study was cement sand mortar(CSM) in three different proportions i.e. CSM 1:3, CSM 1:6,

CSM 1:8. The cement used comprised of OPC-53 grade conforming to IS 12269-2013[9] and sand conforming to IS 2116-1980[10]. The properties of mortar are presented in Table 3.

Table 1. Properties of Solid concrete blocks

Property	No. of specimens tested	Solid concrete blocks	COV (%)
Dimensionality-(mm)	20	402.9 x 151.5x201.5	----
Water absorption-(%)	03	6.17	9.88
Initial rate of absorption (kg/m^2/min)	03	0.83	27.37
Block Density-(g/cc)	03	2.036	4.2
Compressive Strength-(MPa)	08	3.46	15.36(%)
Modulus of Elasticity-(MPa)	03	6278	7356

Table 2. Properties of Hollow concrete blocks

Property	No. of specimens tested	Hollow concrete blocks	COV (%)
Dimensionality-(mm)	20	401.5x150.6 x 199.7	----
Water absorption-(%)	03	4.85	3.70
Initial rate of absorption (kg/m^2/min)	03	0.34	23.76
Block Density-(kg/m^3)	03	1.26	4.2
Compressive Strength-(MPa)	08	9.03	15.36
Flexural Strength-(MPa)	03	2.01	7.90

Table 3. Properties of Mortar

Property	**Cement sand mortar (CSM) proportion**			**COV (%)**		
	1:3	**1:6**	**1:8**	**1:3**	**1:6**	**1:8**
Mean Compressive strength (MPa)	13.63	8.86	3.23	12.05	10.58	13.26
Mean Tensile strength (MPa)	1.25	0.98	0.56	13.35	11.64	12.32
Mean Flexural strength (MPa)	5.35	3.52	1.35	9.05	10.36	14.59

4. Experimental programme

In order to determine the compressive strength of masonry, IS 1905-1987 [8] recommends casting and testing of masonry prisms. Masonry prism is an assemblage of individual masonry units and mortar that has been casted as a test specimen in order to evaluate its compressive strength. In the present study, masonry prisms were casted in stack bonded pattern as shown in figure 1(a) for testing of compressive strength in accordance with provisions given in IS 1905-1987 [8]. Suitable precaution was taken to ensure that the verticality was maintained truly plumb to avoid eccentricity if any. The blocks used were pre wet before casting of prisms. In case of the reinforced solid block masonry prisms, horizontal reinforcement comprising of welded wire mesh of 1.25-inch gauge and 2.4mm thickness. Vertical reinforcement of 6mm diameter an 8mm diameter were used in case of prisms casted with hollow blocks. These were filled with free flowing pre mixed cementitious grout obtained from FOSROC chemicals "Conbextra GP1". Slump tests were conducted by varying the water/powder ratio to obtain a slump greater than 300mm to eliminate compaction. A slump value of 340mm was obtained for a water/powder ratio of 1.8. The h/t ratio in case of these prisms were equal to 4.2 within the range of 2 to 5 and a correction factor of 1.314 was applied to obtain the actual prism test. In total, three masonry prisms which were cured for 28 days were tested in a Universal testing machine by placing nominal sheets of 4mm plywood at top and bottom till the specimen underwent failure for various combinations shown in Table 4. The failure load of the specimen was noted down. Displacement values required to compute the strain values were noted down using digital demountable mechanical gauge. The test set up of the same is shown in figure 1 (b). The breakup of the specimens is shown in Table 4.

Table 4. Details of Prisms casted and their nomenclature

Designation of prism	Blocks	Mortar	Detail of reinforcement
UA	Solid	1:3(H1 type)	Unreinforced
UB	Solid	1:6(M2 type)	Unreinforced
UC	Solid	1:8(L1 type)	Unreinforced
HB	Hollow	1:6	Unreinforced
RA	Solid	1:3	Horizontally reinforced 1.25" gauge 2.4mm
RB	Solid	1:6	Horizontally reinforced 1.25" gauge 2.4mm
RC	Solid	1:8	Horizontally reinforced 1.25" gauge 2.4mm
RHB6	Hollow	1:6.	Vertically reinforced 6 mm diameter
RHB8	Hollow	1:6	Vertically reinforced 8 mm diameter

(a) (b)

Figure 1. (a) Masonry prisms stacked after casting, (b) Test Set up of prisms

The corrected prism strength (Mean) which is a product of Prism strength and correction factor for various combination of specimens have been tabulated and indicated in Table 5 above. Typical stress strain curve obtained for Reinforced hollow block masonry prism with vertical reinforcement-6mm is shown in Figure 3.

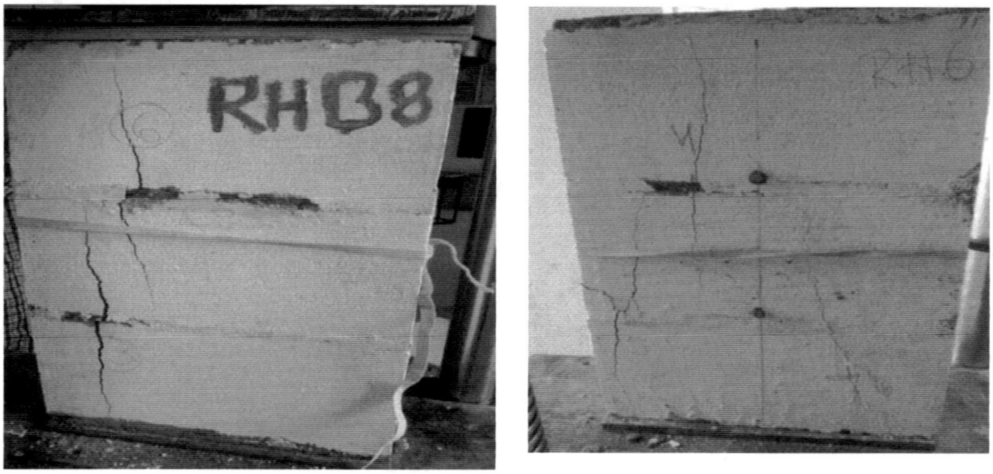

(a) (b)

Figure 2. (a) Vertical cracks originating from top, (b) Splitting failure in prisms

Table 5. Details of Masonry Prism strength tested

Designation of prism	Mean Prism strength N/mm^2	Mean Corrected prism. strength (Correction factor = 1.314) N/mm^2	COV(%)	Elastic Modulus (MPa)
UA	3.22	4.23	12.35	4800
UB	2.88	3.78	10.68	3800
UC	2.25	2.95	13.32	2765
HB	6.10	8.01	8.15	6850
RA	4.10	5.39	11.38	5897
RB	3.53	4.64	12.67	4767
RC	3.50	4.6	13.66	3124
RHB6	6.56	8.62	14.28	10,822
RHB8	7.80	10.25	13.04	13,465

5. Results and Discussions

The Elastic modulus obtained from stress strain curve for the different specimens is tabulated in Table 5.The values clearly indicate that the reinforced masonry prisms have shown higher elastic modulus. In general, the prisms casted using reinforced hollow blocks have shown better prism strength and elastic modulus. These specimens also showed enough resistance before undergoing failure clearly indicating the role played by the vertical reinforcement. The solid block masonry prism with horizontal mesh reinforcement showed an improvement in prism strength and elastic modulus when compared with unreinforced solid block masonry prisms. This clearly indicates the effectiveness of horizontal mesh reinforcement also.

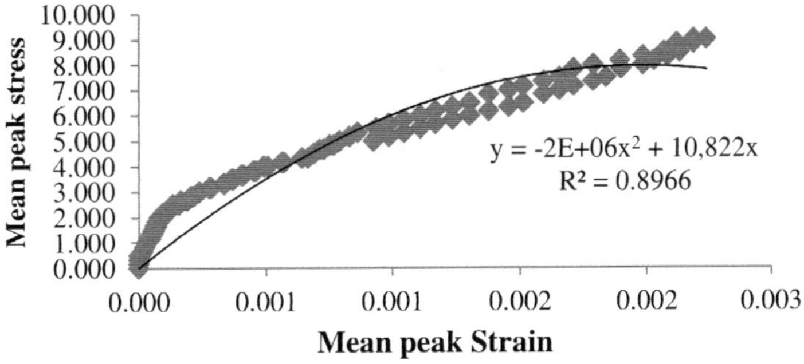

Figure 3. Stress strain curve of Reinforced Hollow Concrete block prisms with 6mm vertical reinforcement

6. Conclusions

The following broad conclusions are drawn from the experimental investigations

- The compressive load carrying capacity of solid block masonry prism with mesh reinforcement was higher with H1 type of mortar. This was about 11.90% higher than the prism of M2 mortar and around 43.33 % higher in case of prism made of L1 mortar
- The compressive load carrying capacity of hollow block masonry prism with 8mm reinforcement was 18.9% higher than the prism with 6mm reinforcement. This was about 27.9% higher than unreinforced hollow block masonry prism.
- Typical compression failure was observed in prisms during testing. It was observed that the vertical cracks started in the middle of the prism, near the masonry unit – mortar interface and propagated toward the ends of the prism. In later stages of loading, spalling of brick or block was observed on the vertical sides of the prisms.
- The effectiveness of mesh reinforcement and vertical reinforcement in improving the load carrying capacities can be clearly seen from the experimental investigations

REFERENCES

[1] Matthias Ernst, Gert Konig 2006 *Shear Strength and Conpressive Strength of Reinforced Perforated Clay Block Masonry Institut fur Massivbau* 221-242.

[2] Mohamad, Lourenço PB, Roman HR 2005 Mechanical behavior assessment of concrete block masonry prisms under compression *International Conference on Concrete for Structures* (INCOS 05), Coimbra, p. 261.

[3] Gumaste K S, Nanjunda Rao K S, Venkatarama Reddy B V 2007 *Mater. Struct.* **40** 241.

[4] Bryan D E, Mervyn J K 2004 *J. Struct. Eng. @ ASCE* 650.

[5] Jagadish K S, Raghunath S, Nanjunda Rao 2002 *J. Struct. Engg* **29** 9-17.

[6] *Specification for concrete masonry units Part II hollow and solid lightweight concrete block,* IS-2185-1983-Part II,Bureau of Indian Standards, New Delhi

[7] F. Oliveira, J. De Hanai. Sao Carlos N *2008 Ibracon structures and materials Jr.* **1** 58-170.

[8] *Code of Practice for structural use of Unreinforced masonry, IS-1905-1987, Bureau of Indian Standards*, New Delhi

[9] *Specifications for 53 grade ordinary Portland cement*, IS 12269-2013, Bureau of Indian Standards, New Delhi

[10] *Sand for masonry mortars – Specifications*, IS 2116-1980, Bureau of Indian Standards, New Delhi

IOP Conf. Series: Materials Science and Engineering **814** (2020) 012041 doi:10.1088/1757-899X/814/1/012041

Blast resistance of steel plate shear walls designed for seismic loading

Prashant Sunagar[1*], Aravind Bhashyam[2], Manish S Dharek[3], Sreekeshava K S [4], Ramegowda R S[5], Lakshmi H S[6]

[1,5,6] Civil Engg Dept, Ramaiah Institute of Technology and Management (affiliated to VTU, Belagavi), Bengaluru, India
[2] Civil Engg Dept, Christ University,Bengaluru, India
[3] Civil Engg Dept, BMS Institute of Technology and Management, Bengaluru, India
[4] Civil Engg Dept, Jyothy of Technology (affiliated to VTU, Belagavi), Bengaluru, India

Email: prashant.sjce@gmail.com

Abstract. Since a blast loading or explosion can create nonlinear wave action and impact pressure on a structure, it necessary to construct a structure to resist blast loading as like other loads. In this study the nonlinear behaviour of a blast loading is simulated by calculating the pressure diagram with respect to time under the guidance of IS 4991 – 1968, code for "Criteria for Blast Resistant Design of Structures for Explosions above Ground". The study carried out for different charge weight (100kg TNT, 200kg TNT and 400kg TNT) and standoff distances of 20metre. Nonlinear behaviour of a Blast loading to steel structures with shear plates of thickness 6 mm, 8 mm and 10 mm are modelled in ETABS and the analysis is carried out to obtain base shear, story displacement, story deformation pattern, column forces, etc.

1. Introduction
The effect of blast load to the structure on different faces (front, rear and side faces) are different. Since the pressure created due to blast load will gradually decreases with time, this can be applied to model by defining time history function in ETABS. Resistance of a structure to these effects will depend on the properties of the structure. It is achieved by providing more stability for any structure and which can be done in different ways.

2. Methodology
Analysis of steel structure modelled in ETABS is done by considering different charge weights and standoff distance of 20m. Performance is measured in terms of lateral displacements and stresses in the members of the structure. Further, with different blast intensities, the analysis is repeated on the same structure and its performance with different blast intensities is studied and compared.

3. Modelling and Inputs
A typical industrial steel structure of three levels (3 storied) is considered for the study, having three bays in X direction with equal bay spacing of 5m and three bays in Y direction with equal bay spacing of 5m. The Geometrical and material specification of the structure is tabulated. The structural steel is of grade FE350. The steel profiles or sections adopted are of Indian standard.

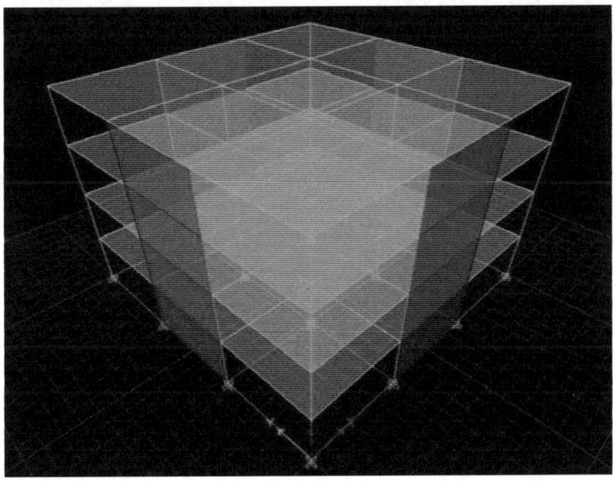

Figure 1. 3D View of Structure modeled in Etabs

Table 1. Dimensions and Properties of structure

Sl No	Item	Specification
1	Material	Structural Steel
2	No. of stories	3
3	No. of bays in	3
4	No. of bays in	3
5	Bay spacing in	5m
6	Bay spacing in	5m
7	Modulus of	$2X10^5$ N/mm^2
8	Density	7850kg/m^3
9	Poisson's ratio	0.3

The 3D structural system is analysed for all other conventional loads such as dead, live wind and earthquake before analysing for blast load to ensure the stability of the structure. Effect of blast loading or explosion on the response of the steel structural system is studied by considering different blast intensities

4. Blast Load Design

Design blast load for any structure is the probable intensity of pressure that is going to act on any face. Since the effect is different for different faces, the pressure diagram is developed by using IS 4991 – 1968. The pressure will maximum for front face which is oriented with the face normal to the direction of propagation of the shock front. The pressure determined on different faces of the building depending upon the clearance time, drag coefficient, geometrical dimensions of the building, Duration of the equivalent triangular pulse, etc.

Table 2. Blast Parameters from Ground Burst of 1 T Explosive

Distance, m (x)	Peak side on over pressure ratio p_{so}/p_a	Match no, M	Positive phase duration, in milli-seconds	Peak elected over pressure ratio p_{ro}/p_a	Duration of equivalent triangular pulse, t_d in milli-seconds
27	1.8	1.6	20.92	5.81	Structural Steel

Table 3. Drag Coefficient, Cd

Shape Of Element	Drag Coefficient, Cd
Front face sloping	0.4
Roof, rear and side faces for	-0.4

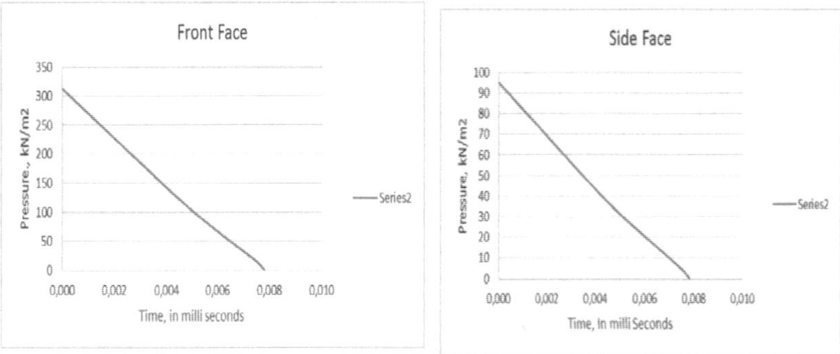

Figure 2. Pressure Diagram for 100kg charge Weight

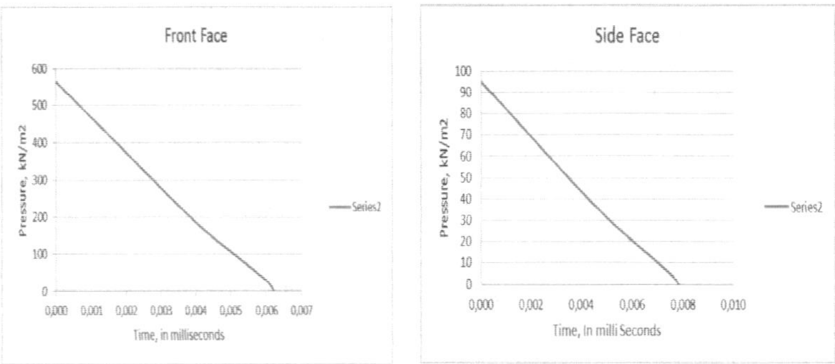

Figure 3. Pressure Diagram for 200kg charge Weight

IOP Conf. Series: Materials Science and Engineering **814** (2020) 012041 doi:10.1088/1757-899X/814/1/012041

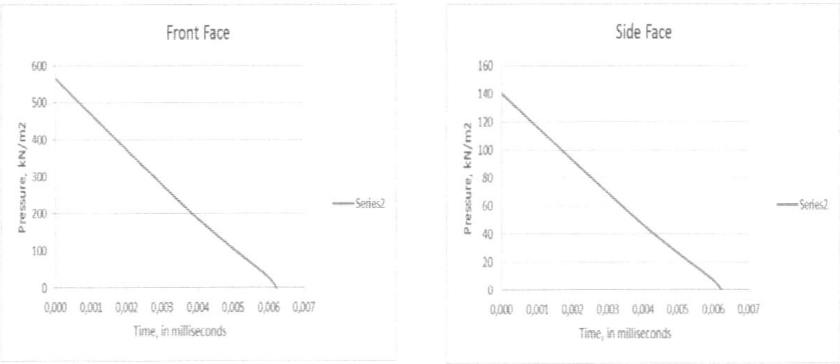

Figure 4. Pressure Diagram for 400kg charge weight

5. Blast Analysis

The analysis of structure is carried even for the dead load, live load, super dead load and earthquake load. The live load and super dead load are taken as 1.5 KN/m^2. Every members of the structure are modelled by using ISMB500 section from IS code for steel and shear plates of 6 mm, 8 mm and 10 mm thick are used at each middle span sides. A 3D steel structure is modelled using one dimensional frame elements for columns and beams and two-dimensional shell elements for slabs. One dimensional frame element having six degrees of freedom at either end or two-dimensional shell elements having six degrees of freedom at each node.

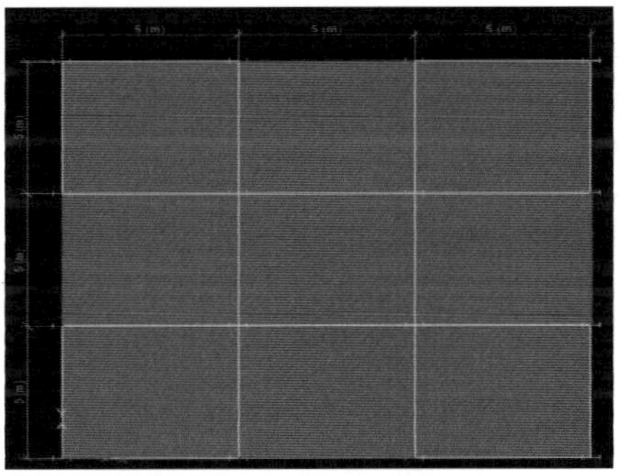

Figure 5. Floor plan of model.

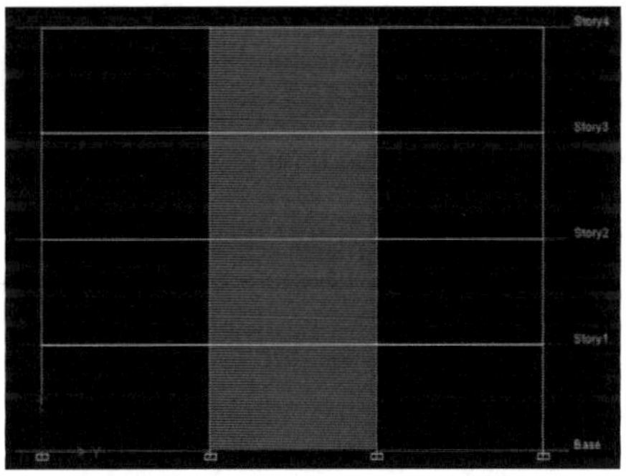

Figure 6. Elevation of Structure

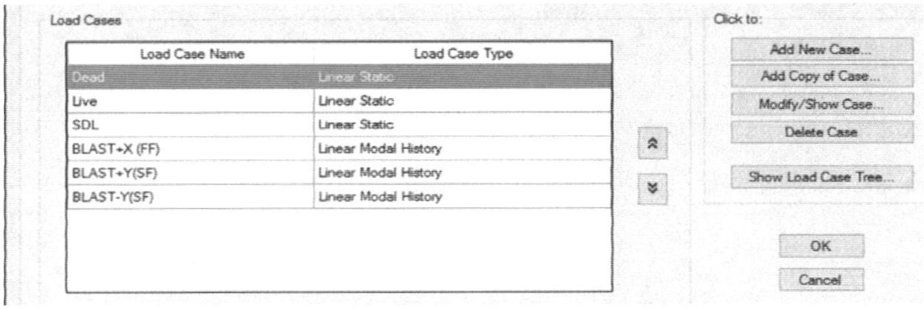

Figure 7. Load Cases Considered for analysis

IOP Conf. Series: Materials Science and Engineering **814** (2020) 012041 doi:10.1088/1757-899X/814/1/012041

6. Results and Comparisons

After analysis of modelled steel structure, the results obtained from ETABS are as follows.
6.1 Base shear

Table 4. Max Base Shear in X and Y-Direction

Model	Base Shear in (KN)					
	X Direction			Y Direction		
	6 mm	8 mm	10 mm	6 mm	8 mm	10 mm
0.1 MTonne Charge weight	9085	10886	12502	2713	3428	4053
0.2 MTonne Charge weight	14470	17293	20121	3695	4625	5452
0.4 MTonne Charge weight	22102	26320	30036	4610	5749	6754

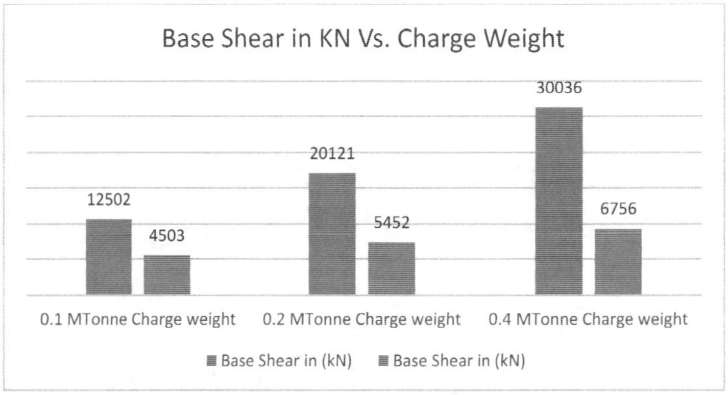

Figure 8. Max Base Shear vs. Charge Weight

In the above table no 4 it is observed that the base shear is gradually increased in both the directions when the charge weight is changed from .1 tonnes to .2 tonnes and then .4 tonnes.in each of the charge weight conditions for any charge weight the base shear is maximum in the direction normal to propagation of blast loads, i.e. in X direction.

6.2 Max Displacements
In the below table no 5 it is observed that the displacement is gradually increased in both the directions when the charge weight is increased from 0.1 tonnes to 0.2 tonnes and then 0.4 tonnes. In each of the charge weight conditions for any charge weight the displacement is maximum in the direction normal to propagation of blast loads, i.e. in X direction

Table 5. Max Displacements in X and Y Direction.

Model	Maximum Displacements (mm)					
	X Direction			Y Direction		
	6 mm	8 mm	10mm	6 mm	8 mm	10mm
0.1 MTonne Charge weight	5.2	5.78	6.22	2.43	2.8	2.96
0.2 MTonne Charge weight	9.2	9.52	10.1	3.3	3.57	3.86
0.4 MTonne Charge weight	14.15	14.38	14.77	4.08	3.76	3.91

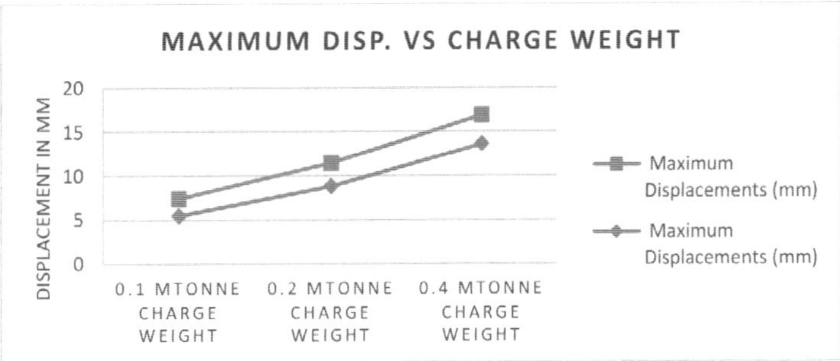

Figure 9. Max Displacements Vs. charge weight

6.3 Max Column Forces

Table 6. Max Column Forces

Forces	0.1MTonne Charge weight			0.2MTonne Charge weight			0.4MTonne Charge weight		
	6 mm	8 mm	10 mm	6 mm	8 mm	10 mm	6 mm	8 mm	10 mm
P (kN)	52.31	60.2	66.5	84.5	97.37	110.2	131.4	151.	167.2
M2 (kN-m)	1.8	1.2	1.4	2.4	2.3	2.3	3	2.9	2.8
M3(kN-m)	99	92	86	157.	145.6	131.5	239	221	204.3
P(kN)	45.9	48.7	51.1	74	79.1	88.6	116.4	124	130.4
M2(kN-m)	1.5	1.5	1.4	2.1	2	1.9	2.6	2.4	2.3
M3(kN-m)	83.7	79	71	131	125.6	102.3	201	189	178
P(kN)	45.9	48.7	51.1	74	79.1	88.6	116.4	124	130.4
M2(kN-m)	1.5	1.5	1.4	2.1	2	1.9	2.6	2.4	2.3
M3(kN-m)	83.7	79	71	131	125.6	102.3	201	189	130.4

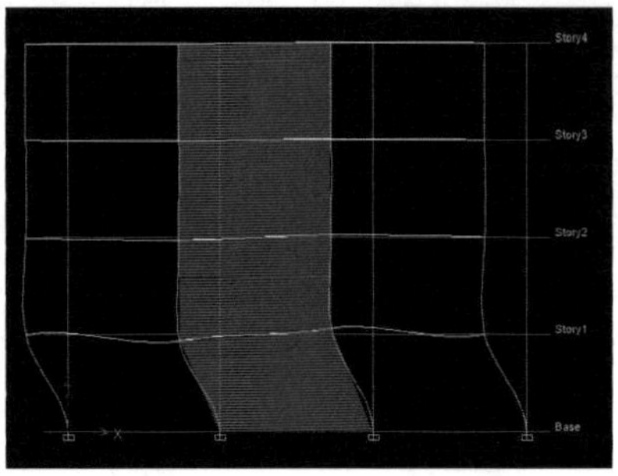

Figure 10. Deformed Shape

The column forces are sorted and tabulated above are, maximum axial (P), minor axis moment and (M2) in ground Story columns Major axis moment (M3). The following graphs. Figure 10,11 and Figure 12 show the variation of design forces in the columns of ground story, as the blast load intensity increased, the axil and major axis bending moment increased significantly whereas the minor axis bending moment showed a drop after the initial increase with the blast load intensity.

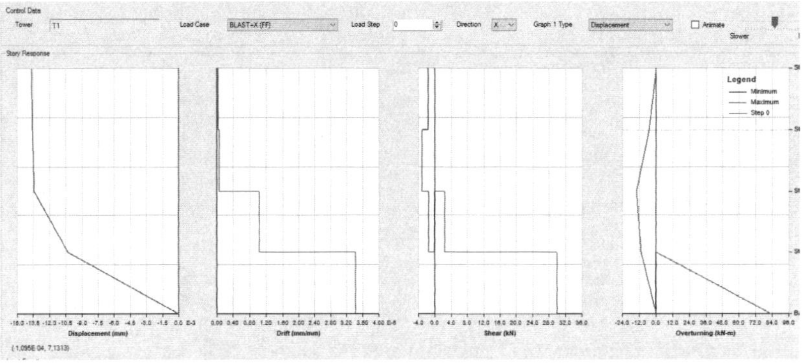

Figure 11. Max combined story response plots for step 1

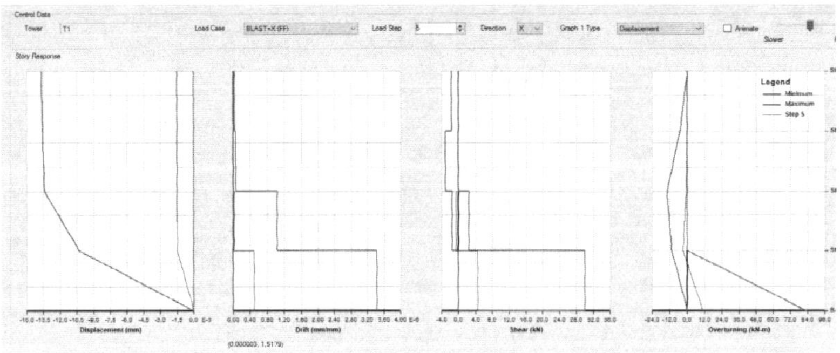

Figure 12. Max combined story response plots for step 2.

IOP Conf. Series: Materials Science and Engineering **814** (2020) 012041 doi:10.1088/1757-899X/814/1/012041

7. Conclusions

- Detailed literature survey has been done to know what are the studies are conducted on the blast loading in different conditions. Pressure time history has been plotted in accordance with the IS 4991 – 1968. A steel structure has been modelled in ETABS to analyse and get the results.
- From the analysis it was observed that the maximum story displacement is exerted in X direction, it was in the fourth story for the case of 0.4 MTonnes charge weight.
- The story displacement has increased for every increase in thickness of shear plate from 6 mm to 8 mm and then to 10 mm.
- Story displacement is also increased for of charge weight from 0.1 Mtonnes to 0.2 Mtonnes and then to 0.4 Mtonnes.
- Max Base shear is observed in case of 0.4 Mtonnes charge weight with 10 mm plate thickness.
- Column forces are more in case of 0.4 Mtonnes charge weight with 10 mm plate thickness, it was in the column number 12.
- Combined story response plots for step 1 and 2 are also shown.

References

[1] Hassan Ostadhossein, Saeid Lotfi 2018 Journal of Solids and Structures **15** 2
[2] T. Ngo, P. Mendis, A. Gupta, J. Ramsay 2007EJSE Special Issue: Loading on structures Hindawi Publishing Corporation Advances in Civil Engineering, Volume 2016, Article ID 2604232, 14 pages, http://dx.doi.org/10.1155/2016/2604232
[3] M. Larcher, M. Arrigoni, C. Bedon, et al. 2016 *Advances in Civil Engineering Article ID 2604232* doi:10.1155/2016/2604232
[4] L. Figuli, D. Papan, 2014 *Applied Mechanics and Materials* **617** 92-95
[5] D. Papán, Z. Papánová 2015 *Journal of Engineering and Technology* **2** 1407
[6] J. M. Biggs, Introduction to Structural Dynamics, McGraw-Hill book company, 1964, ISBN 07-005255-7
[7] Prashant Sunagar, 2013 *International Journal of Innovative Research in Science Engineering and Technology* **4** 7156-7165.

IOP Conf. Series: Materials Science and Engineering **814** (2020) 012042 doi:10.1088/1757-899X/814/1/012042

Spatial variation of climate change issues using remote sensing technique

Kumar R Rao[1], Sreekeshava K S[2*], Manish S Dharek[3], Prashant C Sunagar[4], C R Ganesh[5]

[1, 2*, 5] Civil Engg Dept, Jyothy Institute of Technology(affiliated to VTU, Belagavi), Bengaluru, India
[3] Civil Engg Dept, BMS Institute of Technology and Management, (affiliated to VTU, Belagavi), Bengaluru, India
[4] Civil Engg Dept, Ramaiah Institute of Technology, Bengaluru, India

E-mail: sreekeshava.ks@jyothyit.ac.in

Abstract. The assessment of climate change can be carried out by analysing rainfall variability, temporal climatic changes, spatial interpolation of climatic factors and climatic risking. The climatic changes have consequences on precipitation pattern changes, temperature variation, and crop growing season alteration, increase of sea ice and grounded ice sheets. This has implications on ecosystem, human health, settlement and agriculture. Some of the repercussions attributed to temperature and precipitation increase is an increase of Cyclonic activities with increase of cyclonic wind speeds, Increase in Extreme Precipitation events like floods and droughts, agricultural yield is projected to decrease and negative impact on subsistence agriculture. The Tropical Cyclones intensifying in eastern Indian Ocean and crossing the coasts of Indian subcontinent year after year is a major cause of concern. The Latent Heat Release (LHR) of cyclones of Bay of Bengal carried out using advanced sensors like SSM/I (Special Sensor Microwave/Imager) on-board DMSP satellites (Defence Meteorological Satellite Program of US) for assessing spatial distribution of precipitation since LHR is responsible for maintenance and intensification of tropical cyclones and LHR can also enable measurement of brightness temperatures of atmospheric/ocean surfaces. The images of the clouds in visible and infrared region provide information about cloud aerial extent, type, and growth/decay. The raining clouds have high visible band reflectance and are tall due to upward motion indicating deep convection. These issues pertaining to climate changes through Remote sensing techniques is discussed in the paper.

1. Introduction

The consequence of climatic changes are manifold such as precipitation pattern changes, temperature variation, crop growing season alteration, increase of sea ice, grounded ice sheets etc. The climatic changes can also result in regional disturbances such as agriculture impact, water balance changes, forestry biomass degradation, disturbance to fish apart from effect on human health and settlements [2]. Environmental impacts accrued due to human induced climatic changes are now larger, faster, more widespread predominantly affecting tropics, high latitudes and elevated landscapes, major deltas, coastlines etc. Climatic change study should identify consequences that are vulnerable to populations and societies apart from evaluation of issues which are ecologically and climatologically sensitive [4].

IOP Conf. Series: Materials Science and Engineering **814** (2020) 012042 doi:10.1088/1757-899X/814/1/012042

This can ensure planning for mitigation measures rather than follow the adaptation for climatic changes based upon geographic location.

Hyper spectral remote sensing has become a potent source due to its potential in identification of ocean related parameters. The features can be identified more accurately in comparison to broad band multispectral data. Large volume of data in hyper spectral sensing possesses a challenge for traditional processing techniques such as removal of redundancy and provision of continuous relevant information pertaining to climatic studies. High spectral resolution in region of 0.4 to 2.5 micrometre can be used for identifying large range of ocean cover features that cannot be identified with broad band, low spectral resolution imaging system.

Latent Heat Release (LHR) of cyclones of Bay of Bengal has been done using advanced sensors like SSM/I (Special Sensor Microwave/Imager) on board DMSP satellites (Defence Meteorological Satellite Program of US) for assessing spatial distribution of precipitation as LHR is responsible for maintenance and intensification of tropical cyclones. It also measures brightness temperatures of atmospheric/ocean surfaces. Cloud images (Visible/IR) provide cloud aerial extent, type, and growth/decay [8]. The raining clouds have high visible band reflectance and are tall due to upward motion indicating deep convection.

Satellite measurement of wind speed using SSM/I is useful for meteorological and oceanographic studies like moisture transport, upwelling phenomena, sea state and air-sea interactions [3]. SEASAT active and passive sensors contains radar altimeter, scatter meter and scanning multichannel microwave radar (SMMR) can give wind speed over ocean surface.

2. Remote Sensing methodology for climate studies

Figure-1 gives the methodology for generating outputs using remote sensing data for climatic studies. The methodology include synergistic studies using imaging and non-imaging sensors in visible, infrared and microwave bands for integration as per process models to obtain derived outputs required in climate change analysis.

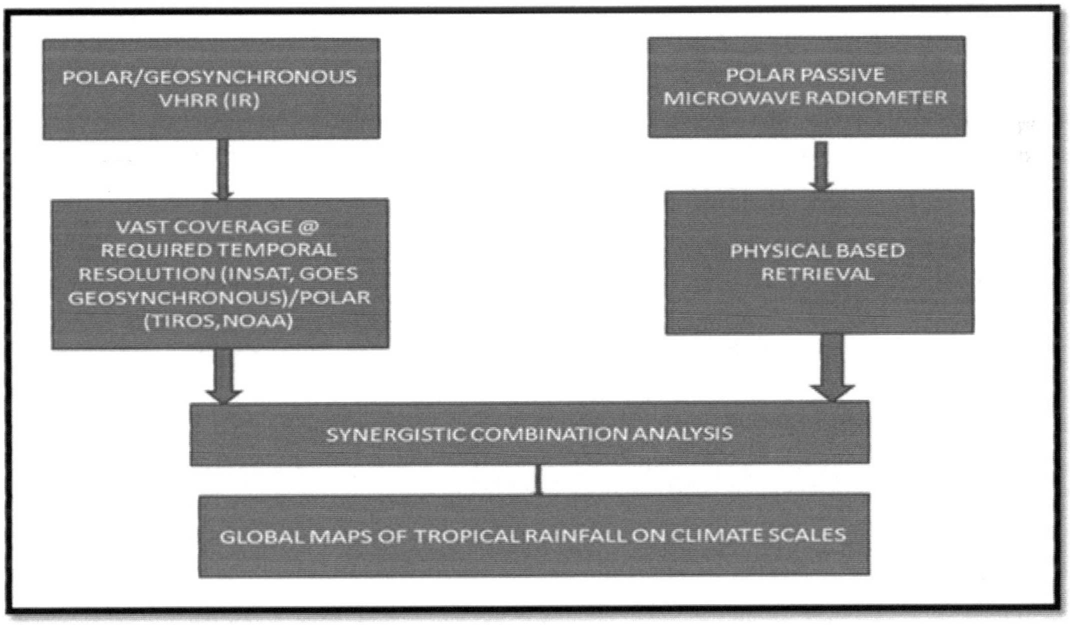

Figure 1. Methodology for climate change studies

The outputs create awareness amongst the people about climate change issues and capacity to cope with climate variability apart from helping local authorities and communities to develop and access a range of adaptive measures along with enabling related policies. The action plans for mitigating the effect of climatic changes involve reforestation, planting agricultural crops and trees more suited to warmer temperatures and drier soil conditions, investing in infrastructure (e.g. drainage systems and conservation schemes), and establishing disaster early warning systems to tackle chronic conditions such as higher temperatures, more or less rainfall, less predictable seasons, sea level rise and saline intrusions. The remote sensing data can be utilized for mapping the decrease in snow covered area attributed to climatic changes [1]. The snow melt runoff due to melting of glaciers and consequent reduction in its area is considered as a negative impact of climatic change. The snow covered area once mapped can be integrated with altitude and other meteorological data into various models for quantifying the snow melt runoff arriving into the river or oceans [5]. The methodology for estimating the snow melt runoff using remote sensing data is given in figure 2.

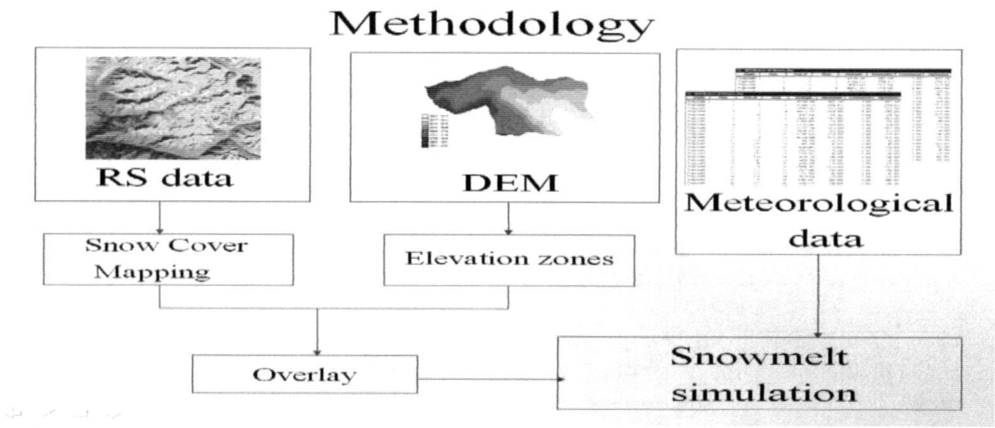

Figure 2. Methodology for snow melts runoff estimation.

The snow melt runoff can be estimated well before the arrival of actual quantity of water into the river site or ocean site. This will enable planning in advance for any eventuality.

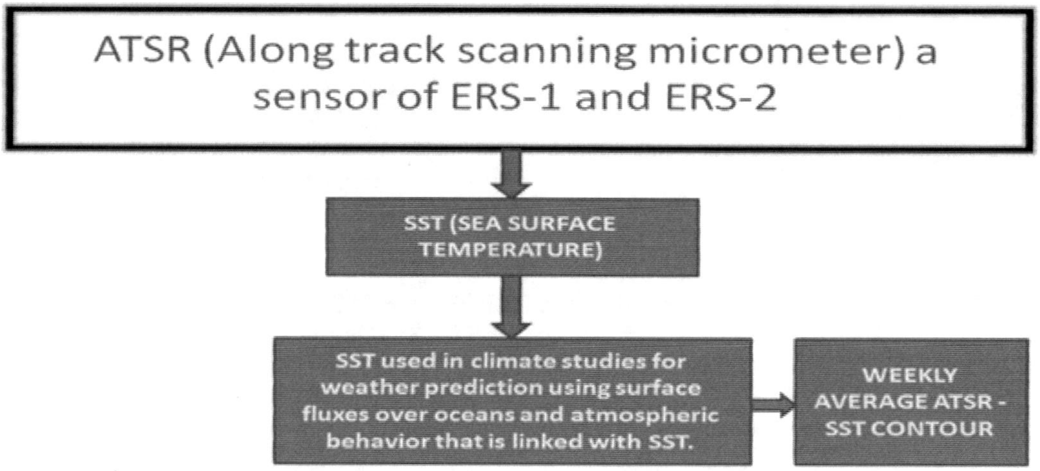

Figure 3. Methodology for determining sea surface temperature

The atmospheric behaviour and the climatic studies for weather prediction using surface fluxes over the oceans are linked to estimation of sea surface temperature (SST) [6]. The along track scanning micrometre (ATSR) sensor on board ERS-1 and ERS-2 is used to prepare SST contour maps on a weekly basis. Figure-3 shows the methodology for arriving at SST contour map using ATSR sensor. The SST determination from satellites helps in monitoring of global climate and related studies and for useful climate application [7]. The accuracy standards of 0.3 degree kelvin have been specified by International Tropical Ocean Global Atmosphere (TOGA). The nutrient concentration such as nitrates, ammonia, phosphorous, dissolved inorganic carbon etc, is linked with phytoplankton concentration and is inversely related to temperature. The relation is dependent upon seasons and geographical locations.

3. Results and Discussions

Himalayas are one of the largest glacier concentrations outside polar region and Temporal Remote sensing observation over extended periods using IRS, TM, and MSS image data can give an assessment regarding Accumulation Area Ratio (AAR), Wherein the AAR is given by

AAR = Snow accumulation area/Total glacier area.

It has been observed that one degree centigrade global warming in the glacial region can melt the glaciers by almost 50% with appreciable drop in specific mass balance values as mass of ice is highly sensitive to minor variation of atmospheric temperature.

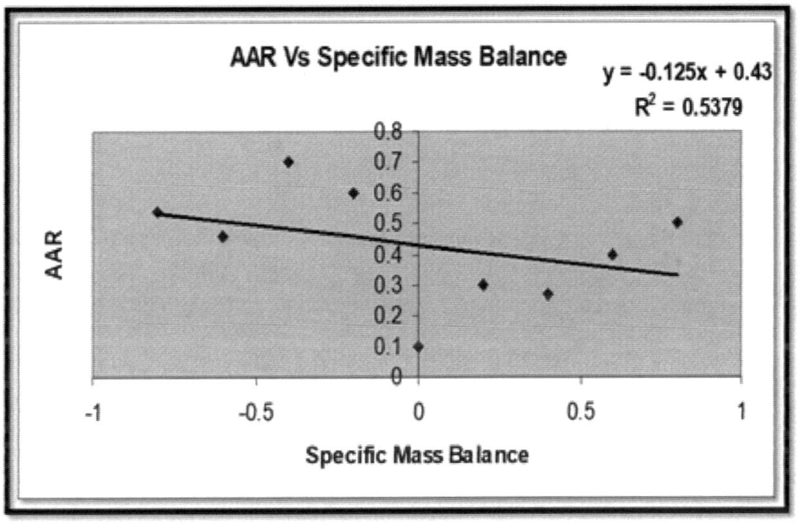

Figure 4. AAR versus Specific Mass Balance Plot

Figure-4 illustrates the graphical representation between AAR and specific mass balance. It is observed that the glacial area with less than 0.42 AAR are losing more ice mass than accumulation of snow and with AAR more than 0.42 glacial accumulation occurs with change in thickness. The data from Coastal Zone Colour Scanner (CZCS) satellite ocean colour information (visible and thermal) can give information pertaining to concentration of phytoplankton pigment at less than one meter depth in high turbid area and up to 25 to 30 meter depth in low turbid area. (This means the pigment concentration < 0.3 mg/cubic meter). In this context figure-5 gives a graphical plot of phytoplankton concentration versus the temperature of water at 50 m depth. The temperature at 50 meters depth is obtained from climatologic atlas since CZCS measures only the surface temperature

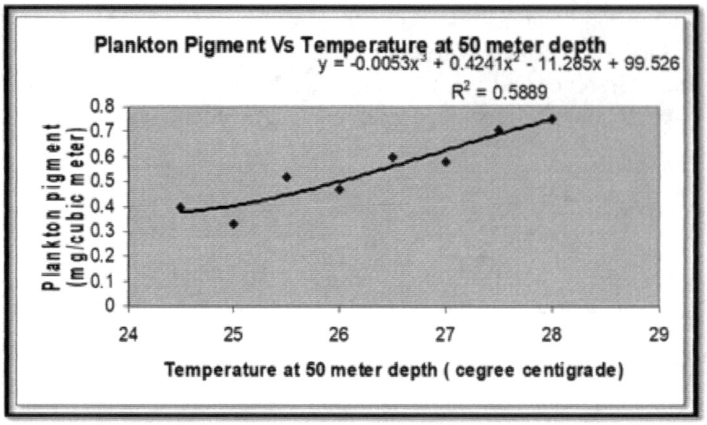

Figure 5. Plankton Pigment versus Sea Temperature plot

In Arabian Sea the nutrient concentration such as nitrates, ammonia, phosphorous, dissolved inorganic carbon etc is linked with phytoplankton concentration due to sufficient light throughout the year and nutrient concentration is related to temperature. The relation depends upon seasonal variation and geographical location. Figure-6 gives a plot of Geo-satellite of US Navy Ku band radar altimeter obtained wind speed (using models) versus the wind speed obtained from SSM/I microwave satellite.

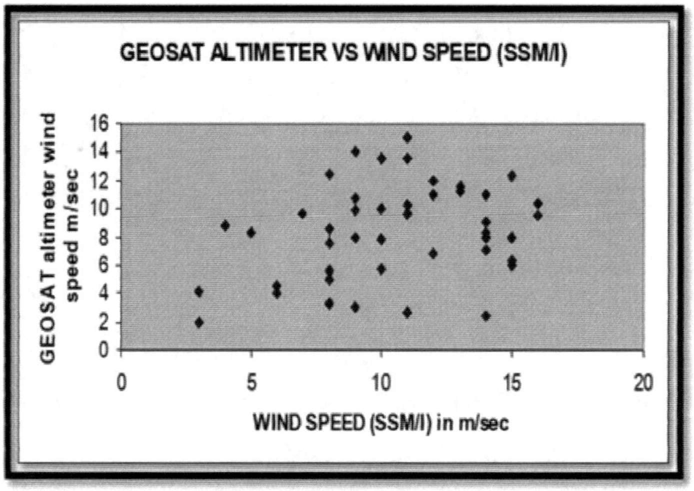

Figure 6. GEOSAT Wind speed SSM/I Wind speed

It is observed that the standard deviation is 4.08 m/sec for difference of two wind speeds for the data obtained in Indian Ocean tropics period varying from Oct to Dec (10-25 deg. N and 40-90 degree E).

4. Conclusion

SSM/I measurement of wind speed obtained at less than 2hrs interval and 0.15 degree spatial resolution based on surface emissions is governed by wave structures and has a RMS error of +/- 0.5m/sec. This can help evaluation of moisture transport, upwelling phenomena, and sea state and air-sea interactions. Upwelling radiation expressed as brightness temperature can be different for raining and non-raining clouds and can be inferred by microwave radiometer against the backdrop of oceans. This brightness temperature is reliable as the oceans have cool and uniform emissions. Large raindrops ($>\lambda$) and ice particles reduces brightness temperature because of scattering of upwelling radiation away from the direction of sensor

IOP Conf. Series: Materials Science and Engineering **814** (2020) 012042 doi:10.1088/1757-899X/814/1/012042

REFERENCES

[1] Ghent D., Kaduk, J., Remedios, J. and Balzter 2011 H. Int. J. Remote Sens. **32** 617–632
[2] Karl T. R 2010 *Procedia Environ. Sci.* **1** 192–205
[3] Karnauskas K. B., Seager R., Kaplan A., Kushnir Y. and Cane M. A 2010 J. Clim. **22** 4316–4321
[4] Large W. G. and Yeager S. G 2012 (1984–2006*). J. Clim.* **25** 6123–6135
[5] Overpeck J. T., Meehl G. A., Bony S., Easterling D. R. 2011 *Science* **331** 700
[6] Pachauri R. K. and Reisinger A. (eds) *IPCC Climate Change 2007: Synthesis Report IPCC, 2007.*
[7] Thies B. and Bendix 2011 *J. Satellite based remote sensing of weather and climate: Recent achievements and future perspectives. Meteorol. Appl.* **18** 262–295
[8] Zhang H 2009 *Bull. Am. Meteorol. Soc.* **90** 31–38
[9] Deser C., Phillips A. S. and Alexander, M. A 2010 *Geophys. Res. Lett* **37** L10701

IOP Conf. Series: Materials Science and Engineering **814** (2020) 012043 doi:10.1088/1757-899X/814/1/012043

Prediction of mechanical properties of high strength steel fiber reinforced concrete with multiple regression technique

Shriganesh S Kadam[1], Vilas V Karjinni[2]

[1]Assistant Professor, SKN Sinhgad College of Engineering, Korti, Pandharpur, Maharashtra, India
[2]Director, Kolhapur Institute of Technology's College of Engineering, (Autonomous), Kolhapur, Maharashtra, India

E-mail: shriganeshkadam@gmail.com

Abstract. High strength steel fiber reinforced concrete which has cube compressive strength greater than 60 N/mm2 was prepared with OPC 53 grade cement confirming IS 269. Sand passing through IS sieve 4.75mm and retaining on IS sieve 150 micron was used in the investigation. Steel fiber with aspect ratio 80 reinforced randomly and at one third position from top of the concrete surface. The percentage of steel fiber varies from 0.5%, 1.0% and 1.5% by volume of concrete specimen. The percentage of silica fume varies from 5%, 10% and 15% by weight of cement used in concrete. Mechanical properties of concrete such as compressive strength, flexural strength were investigated for both the position of steel fiber in concrete. A multiple regression model was employed to predict the mechanical properties of HSSFRC. The regression analysis was performed based on the relationships between the independent variables and the dependent variables. Scatter plot was used to determine the relationship between predictor and criterion variables. The relationship between dependant and independent variable is explained by the magnitude (or the strength) and the direction of the relationship. The magnitude of relationship was calculated using "CORRELATION" formula, which gives the strength of the relationship. The actual shape of the curve determines the direction of the relationship.

1. Introduction

For most of the concrete construction work to save time and money, prediction of concrete strength in advance is required. Multiple regression technique can be effectively utilized to predict concrete strength in early days. M.F.M. Zain et al. [1] suggested the different formulae to predict split tensile strength obtained from compressive. Split tensile strength also related with water binder ratio, concrete age and compressive strength. Mahmoud Sayed-Ahmed [2] studied different matrix mixtures to find effect of matrix mixture on compressive strength. Prepared the statistical tool to forecast the compressive strength of various matrix mixtures at permanentage or at various ages. Julien Michels et.al.[3]shows ductile failure of steel fiber reinforced concrete as compared with concrete without fiber. A O Baarimah and S M Syed Mohsin [5] performed four point bending test on steel fiber reinforced slabs to investigate structural behavior with respect to slab thickness and percentage of steel fiber. The percentage of steel fiber is 0%, 1% and 2% in concrete.

Abdel latif Selmi [6] demonstrated major development in the behavior of concrete due to addition of steel fiber as reinforcing material. Kiambigi, Maina, Gwaya[8] developed empirical correlations that have been used to determine concrete strength estimation.

IOP Conf. Series: Materials Science and Engineering **814** (2020) 012043 doi:10.1088/1757-899X/814/1/012043

2. Materials

Coarse aggregate of aggregate greater than 4.75 mm is considered as coarse aggregate. For present study coarse aggregates passing through 16.5 mm and retained on 10 mm were used. Locally available fine aggregate which is free from organic impurities was used for present work. Sand passing through IS sieve 4.75mm and retaining on IS sieve 150 micron was used in the investigation. Ordinary Portland Cement 53 grade conforming IS 269 were used in the study. Dramix steel fibre used for steel fibre reinforced concrete having aspect ratio 80. It is circular in cross section hooked end type.

3. Preparation of mix

For the present investigation total 9 concrete mixes were prepared by varying percentage of steel fiber and silica fume. For the 9 concrete mixes other than controlled concrete steel fibers were reinforced randomly and at one third position from top of the pavement surface. ACI method was used for mix design to prepare concrete having cube compressive strength greater than 60N/mm2.

4. Result and discussion

4.1. Compressive strength test:

Cube specimens of size 150mm X150mm X 150 mm were casted and compressive strength test was taken on compressive testing machine. The compressive testing machine has capacity of 3000 KN. The results of compressive strength test were summarized in table 7 as mean value of three tests.

Table 1. Compressive strength test

% of ST	0.5			1.0			1.5			Position
% of SF	5	10	15	5	10	15	5	10	15	
Comp. Strength (N/mm^2)	72.52	73.12	73.67	74.43	75.55	76.10	74.18	75.33	76.46	Random
Comp. Strength (N/mm^2)	69.77	62.22	73.33	72.22	73.11	73.70	72.88	74.20	74.66	One third

ST: Steel fiber SF silica fume

4.2 Flexural strength test:

The specimens of size 150mmX150mmX700mm were tested on universal testing machine of capacity 1000KN. The deflection of beam at centre of span was directly measured by using digitized UTM of speed at no load is 0 to 150 mm per minute. The flexural strength of fiber reinforced concrete was found to be increasing as the percentage of fiber increases in both the position of steel fiber.

Table 2. Flexural strength test

% of ST	0.5			1.0			1.5			Position
% of SF	5	10	15	5	10	15	5	10	15	
Flexural Strength (N/mm^2)	6.93	6.07	6.72	6.13	6.15	7.28	8.53	8.62	8.92	Random
Flexural Strength (N/mm^2)	5.87	5.48	6.06	6.15	6.63	7.03	7.66	7.57	8.75	One third

4.3. Multiple Regression Technique to predict compressive strength with randomly reinforced steel fiber
The strength of correlation between the predictor variables (% of steel fiber and % of silica fume) and criterion variable (compressive strength) was calculated as shown below in figure 1 and figure 2 the scatter plots.

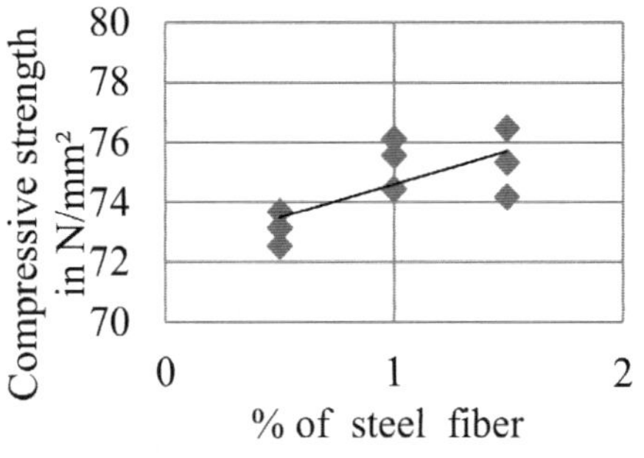

Figure 1. Strength of correlation between percent of randomly reinforced steel fiber and compressive strength

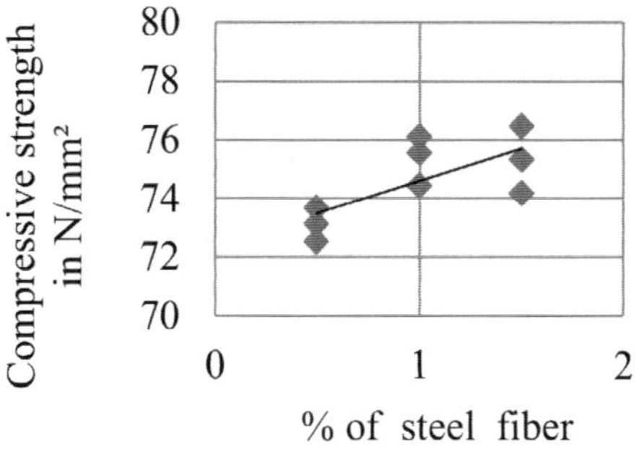

Figure 2. Strength of correlation between percent of silica fume and compressive strength

The strength of correlation of percentage steel fiber and percentage silica fume with compressive strength is 0.71 and 0.54 respectively. It shows a strong correlation between the two types of variables.
The low value of Significance-F parameter (0.008) in ANOVA table shows that the regression result is not out of randomness. An adjusted R-square value of 0.73 in the regression statistics table shows that 73% of the change in the compressive strength is explained by the two predictor variables percentage steel fiber percentage silica fume. The p-values of both the independent variables, which are 0.01 and 0.03respectively, are less than 0.05 (i.e. 5% significance level).

Table 3. Regression statics for Compressive strength test results with silica fume and randomly reinforced steel fiber

Multiple R	0.89
R square	0.80
Adjusted R square	0.73
Standard error	0.71

Table 4. Significance F for Compressive strength test results with silica fume and randomly reinforced steel fiber

	Df	SS	MS	F	Significance F
Regression	2	11.7276	5.8638	11.711118	0.008480
Residual	6	3.004222	0.5007037		
Total	8	14.7318222			

Table5.P-value for Compressive strength test results with silica fume and randomly reinforced steel fiber

	Coefficients	Standard Error	t- stat	P- value
Intercept	70.68	0.85	83.11	0.00
% of steel	2.22	0.58	3.84	0.01
% of silica	0.17	0.06	2.94	0.03

The regression equation based on the regression model is as follows:

Compressive strength = 70.68 + (2.22 x % steel fiber) + (0.17 x % silica fume) + e. (1)

Finally, the graph below shows both the actual output and the output calculated from the fitted regression line. Both the curves show similar trends. Thus, the model demonstrated by the equation above explains the relationship between compressive strength and the two-independent variable i.e. % of steel fiber and % of silica-fume.

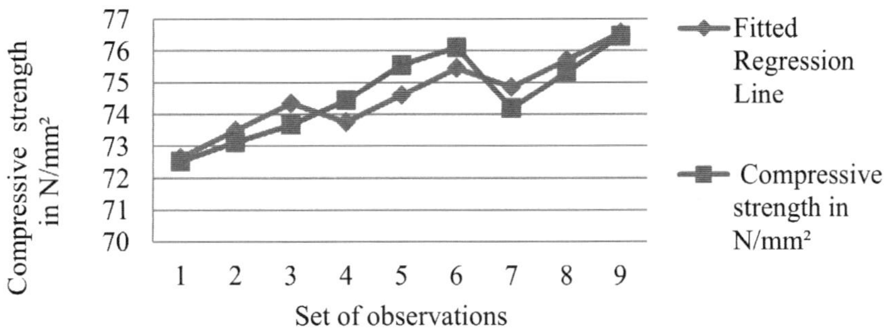

Figure 3. Fitted line equation curve for compressive strength test results with silica fume and randomly reinforced steel fiber.

4.4 Multiple Regression Technique to predict compressive strength with one third reinforced steel fiber
Scatterplot is drawn between the independent and dependent variables as shown below. The logarithmic relationship was determined between % of fiber and average compressive strength.

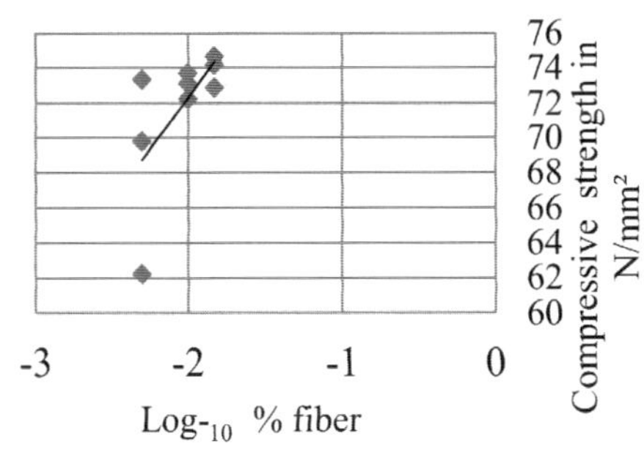

Figure 4. Strength of correlation between Log 10 percent of one third reinforced steel fiber and compressive strength

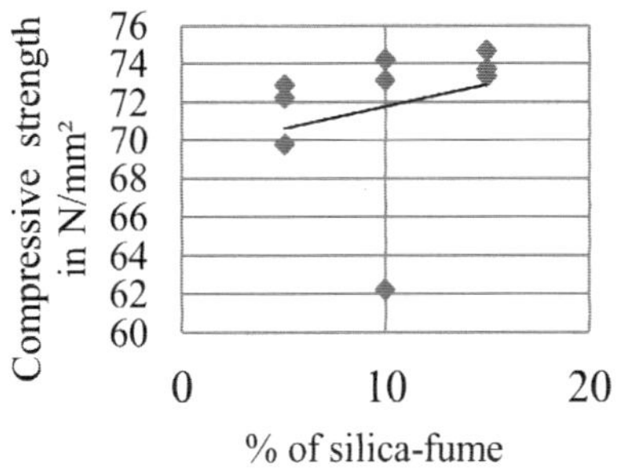

Figure 5. Strength of correlation between percent of silica fume and compressive strength when steel fiber reinforced at one third depth

The two scatter-plots demonstrate that compressive strength improves with logarithmic increase in % of fiber and linear increase in % of silica-fume. The strength of correlation of Log10 percentage steel fiber and percentage silica fume with compressive strength is 0.64 and 0.25 respectively. It shows a strong correlation between the first independent variable however a weak correlation with the second independent variable with average compressive strength. An adjusted R-square value of 31% in the regression statistics table shows that only 31% of the deviation in the compressive strength is explained by the two independent variables % of fiber and % of silica fume. The p-value of log of % fiber is close to 0.05 while the p-value of % silica-fume is higher than 0.05 (i.e. 5% significance level). Also, the t-stat value for log of % fiber is more than 2.

Table 6. Regression statics for compressive strength test results with silica fume and one third reinforced steel fiber

Multiple R	0.69
R square	0.48
Adjusted R square	0.31
Standard error	3.71

Table 7. Significance F for Compressive strength test results with silica fume and one third reinforced steelfiber

	df	SS	MS	F	Significance
Regression	2	56.972000	28.486000	2.762352	0.141112
Residual	6	61.87335	10.31222		
Total	8	118.845355			

Table 8. P- value for compressive strength test results with silica fume and one third reinforced steelfiber

	Coefficients	Standard Error	t- stat	P- value
Intercept	93.75090	11.44945	8.1882	0.000178
Log 10 % of fiber	11.8710	5.43368	2.1847	0.071594
% of silica fume	0.22733	0.262198	0.8670	0.419245

The regression equation based on the regression model is as follows:

Compressive-strength = 93.75 + [11.87 x log (fiber)] + (0.227 x % of silica-fume) + e(2)

Finally, the graph below shows both the actual output and the output calculated from the fitted regression line. Both the curves show similar trends. Thus, the model demonstrated by the equation above explains the relationship between compressive strength and the two-independent variable log 10 of % of steel fiber and % of silica-fume

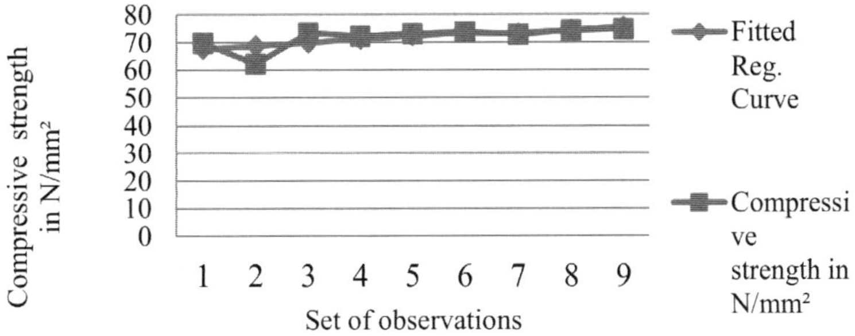

Figure 6. Fitted line equation curve for compressive strength test results with silica fume and one third reinforced steel fiber

4.5 Multiple Regression Technique to predict flexural strength when steel fiber randomly reinforced

Scatterplot is drawn between the independent and dependent variables as shown below and strength of correlation between the independent and dependent variables was calculated.

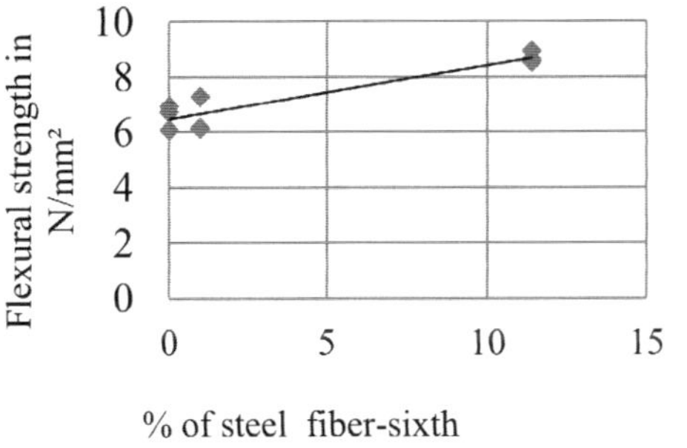

Figure 7. Strength of correlation between Percent of randomly reinforced steel fiber- sixth and flexural strength

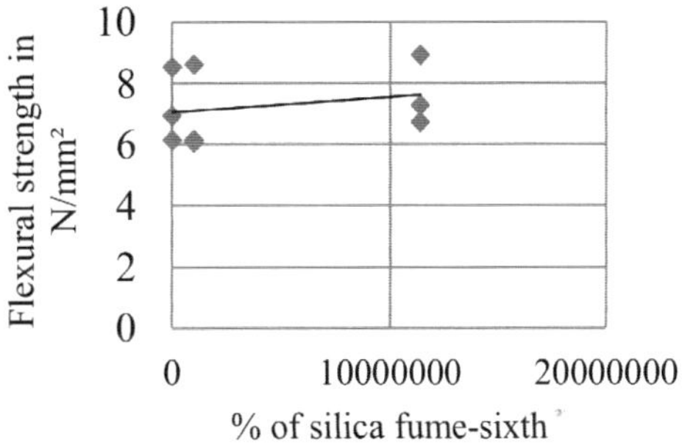

Figure 8. Strength of correlation between percent of silica fume-sixth and flexural strength

The strength of correlation of percentage steel fiber and percentage silica fume with flexural strength is 0.93 and 0.24 respectively. It shows a strong correlation between the % of steel fiber variable and flexural strength and a medium strong correlation between % silica-fume and flexural strength. The low value of Significance-F parameter in ANOVA table shows that the regression result is not out of randomness. An adjusted R-square value of 89% in the regression statistics table shows that 89% of the variation in the average flexural strength is explained by the sixth order of two independent variables % of fiber and % of silica-fume. This is a good indicator of fitness of model. The p-values of the first independent variable is less than 0.05 (i.e. 5% significance level) and the second one is close to it.

IOP Conf. Series: Materials Science and Engineering **814** (2020) 012043 doi:10.1088/1757-899X/814/1/012043

Table 9.Regression statics for flexural strength test results with silica fume and randomly reinforced steel fiber.

Multiple R	0.96
R square	0.92
Adjusted R square	0.89
Standard error	0.37

Table 10.Significance F for flexural strength test results with silica fume and randomly reinforced steel fiber

	df	SS	MS	F	Significance F
Regression	2	9.70523	4.852617	34.70890	0.00050
Residual	6	0.83885	0.13980		
Total	8	10.544088			

Table11.P-value for flexural strength test results with silica fume and randomly reinforced steel fiber

	Coefficients	Standard Error	t- stat	P- value
Intercept	6.2447	0.19	33.10	0.00
% of fiber-sixth	0.19541	0.02	8.07	0.0002
% of silica fume-sixth	0.00000005	0.00	2.08	0.0828

The regression equation based on the regression model is as follows:

$$\text{Flexural-strength} = 6.244 + (0.195 \text{ x } \% \text{ of fiber}^6) + (5\text{x } 10^{-8}\text{x } \% \text{ of silica-fume}^6) + e \quad (3)$$

Finally, the graph below shows both the actual output and the output calculated from the fitted regression line. Both the curves show similar trends. The model demonstrated by the equation above explains the relationship between average flexural strength and the two-independent variables i.e. % of fiber and % of silica-fume.

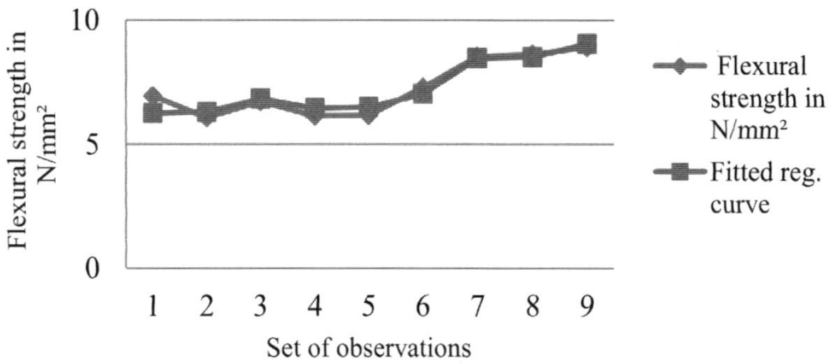

Figure 9. Fitted line equation curve for flexural strength test results with silica fume and randomly reinforced steel fiber

4.6 Multiple Regression Technique to predict flexural strength when steel fiber reinforced at one third.
Scatterplot is drawn between the independent and dependent variables as shown below.

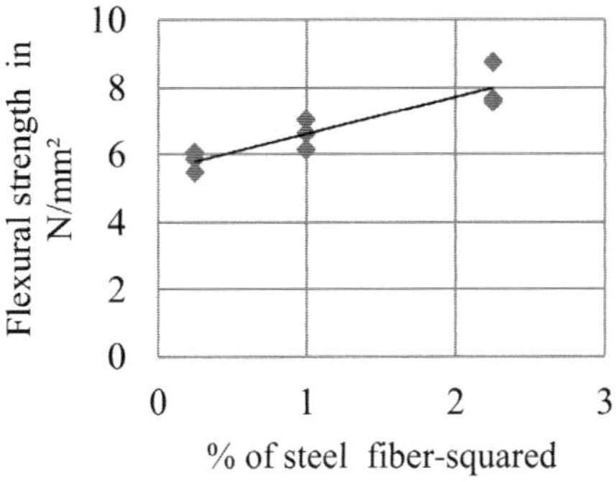

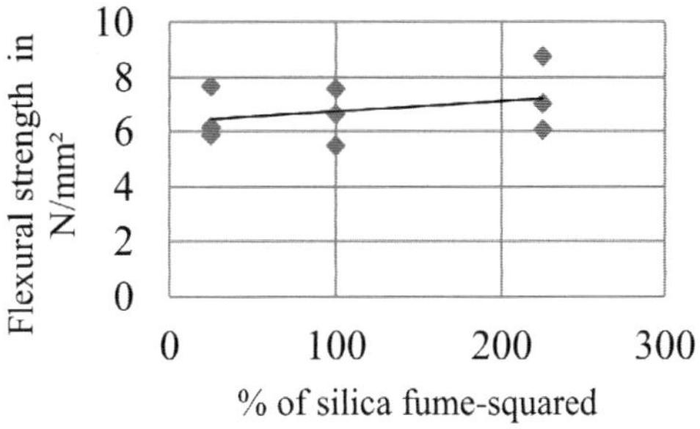

Figure 10. Strength of correlation between percent of one third reinforced steel fiber- squared and flexural strength

Figure 11. Strength of correlation between percent of silica fume squared and flexural strength when steel fiber reinforced at one third depth

The strength of correlation of percentage steel fiber when reinforced at one third depth and percentage silica fume with flexural strength is 0.92 and 0.32 respectively.The low value of Significance-F parameter in ANOVA table shows that the regression result is not out of randomness. An adjusted R-square value of 0.92 or 92% in the regression statistics table shows that 92% of the variation in the flexural strength is explained by the second order of two independent variables % of fiber and % of silica-fume. This is a good indicator of fitness of model, and hence, it can still be surmised that the two variables explain the variation in the model. The p-values of both the independent variables are less than 0.05 (i.e. 5% significance level).

Table 12. Regression statics for flexural strength test results with silica fume and one third reinforced steelfiber.

Multiple R	0.97
R square	0.94
Adjusted R square	0.92
Standard error	0.30

Table 13. Significance F for flexural strength test results with silica fume and one third reinforced steel fiber

	df	SS	MS	F	Significance F
Regression	2	8.261290	4.13064	46.5067	0.00022
Residual	6	0.53290	0.08881		
Total	8	8.7942			

Table 14.P- Value for flexural strength test results with silica fume and one third reinforced steel fiber

	Coefficients	Standard Error	t- stat	P- value
Intercept	5.0748	0.2221	22.8455	0.00
% of fiber-sixth	1.0967	0.1204	9.1076	0.0001
% of silica fume-sixth	0.0038	0.0012	3.1726	0.0193

The regression equation based on the regression model is as follows:

Flexural-strength = 5.075 + (1.0967 x % fiber^2) + (0.0038 x % silica-fume^2) + e(4)

The graph below shows both the actual output and the output calculated from the fitted regression line. Both the curves show similar trends.

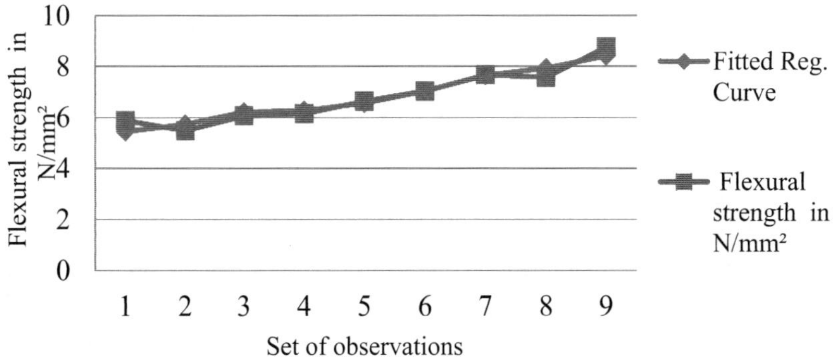

Figure12. Fitted line equation curve for flexural strength test results with silica fume and one third reinforced steel fiber

5. Conclusions

- Coefficient of correlation values for compressive strength and flexural strength test is 0.80, 0.48, 0 .92 and 0.94 it shows regression model best suited for experimental analysis.
- The experimental results were compared with the outcomes obtained from regression technique and found to be tolerable.
- Therefore regression technique can be applied for computation of strength properties of steel fiber reinforced concrete.

References

[1] M.F.M. Zaina M.F.M. Zain, H.B. Mahmud, Ade Ilham, 2002 *Cement and Conc. Res.* **32** 1251– 58
[2] Mahmoud Sayed-Ahmed 2012 *Conc. Reserch Letters* **3** part 2.
[3] Julien Michels, D.Waldmann, Stefan Maas, A. Zurbesa, 2012 *Const. and Build. Mater* **26** 145 –55.
[4] Lidia Rizzuti and Francesco Bencardino *Contemporary Engineering Sciences***7**379 – 90.
[5] A O Baarimah and S M Syed Mohsin, 2017 *IOP Conf. Ser.: Mater. Sci. Eng.***271** 012099.
[6] Abdel latif Selmi 2014 *American Journal of Applied Sciences, Science Publication***11** 145-51
[7] M. A. Islam and K Begam, 2011 *Journal of Scientific Research***3** 225-38
[8] Kiambigi, Maina, Gwaya, A.O, Koteng, D.O 2019*Int. J. Soft Computing and Engineering* **8**
[9] Palika Chopra, Dr. R. K. Sharma, Dr. Maneek Kumar 2014 *International Journal of Latest Trends in Engineering and Technology* **3**
[10] Serkan Suba2009 *Scientific Research and Essay* **4** 289-97

IOP Conf. Series: Materials Science and Engineering **814** (2020) 012044 doi:10.1088/1757-899X/814/1/012044

Thermal Management Materials for Advanced Heat Sinks used in Modern Microelectronics

S V Jadhav[1], P M Pawar[2], S S Wangikar[3], N N Bhostekar[4] and S T Pawar[5]

[1,2,3] SVERI's College of Engineering, Pandharpur, India
[4]Rajiv Gandhi Institute of Technology, Mumbai, India
[5]K. B. P. College, Pandharpur, India

E-mail: svjadhav@coe.sveri.ac.in

Abstract. Heat sinks are used for dispensing the heat from a system in many modern microelectronic components. It helps in systematically absorbing the heat from the system and dissipating it to the atmosphere, thereby ensuring system performance, extended life and reliability. The material used for the heat transfer device is one of the utmost significant parameter, which decides the size of the heat transfer device and its heat transfer efficiency. Lower values of density, overall cost, and coefficient of thermal expansion along with higher thermal conductivity, are considered as the essential properties required for any heat sink material. Though, the Copper and Aluminium materials are generally observed to be the best materials for a heat sink applications, some of their alloys are also found to be suitable. The manufacturing difficulty, the higher cost involved, comparatively poor thermo-physical properties and insufficient investigation are some of the factors that have limited the use of advanced composites in heat sink applications. A comparative numerical analysis for different thermal management materials like Copper, Aluminium, Al2O3, AlN, and Si3N4, is presented in this paper. COMSOL MULTIPHYSICS software is used for the present analysis to check the suitability of these materials for construction of heat sink. Moreover, the study of the cost-effectiveness of the material selected is important before the actual application of that material for heat sinks.

1. Introduction

Extensive research has been initiated in the area of microchannel heat transfer due to the requirement of the high-performance cooling technologies in modern microelectronics. Recently, the microchannels have been recommended by many researchers as an effective technique for the modern micro cooling systems. The larger heat transfer area per unit volume of microchannels and corresponding greater heat removal capacity makes it suitable to be used in heat sink applications. Therefore, it is essential to understand the fluid flow and heat transfer performance characteristics of the microchannels for the proper design of any microchannel cooling systems. The heat exchanger transfers the heat from a higher temperature surface of the electronic component to the lower temperature surrounding medium, preferably air. A heat transfer device (heat sink) is connected to the heat-generating equipment for enhancing the heat dissipation from the hot surface to the coolant medium. The heat sinks can be directly mounted on the enclosure or in the base of microelectronic devices to provide the required heat transfer area. In microelectronics, the surface area available for heat transfer is very less, and the amount of heat needed to be dissipated is very high. Therefore, the liquid-cooled microchannel heat sinks were introduced to cope up the requirement of higher surface area and higher heat transfer coefficient simultaneously.

2. Literature review

Tuckerman et al. [1] proposed the heat sinks, with microchannels, for the micro-electronic cooling applications. Theoretical and experimental study related to heat transfer (single-phase) in the micro-channels were done in 2015 by Hetsroni et al. [2] to understand the influence of the channel geometry on microchannel performance. Wang et al. [3] in 2008 analysed the performance of microchannel heat sink integrated with a high efficient copper heat spreaders. Theoretical and experimental investigation of the water-cooled straight microchannel heat sink was carried out by Gawali et al. [4] in 2014 to understand the effect of different parameters like pressure drop, heat transfer coefficient, thermal resistance, etc. Pietrak et al. [5] in 2015 reviewed the empirical, analytical and numerical expressions for predicting the thermal conductivity of composite materials. After investigating the copper microchannels for thermal applications, Jarosinski et al. [6] in 2017 examined the possibility of enhancing the thermal conductivity of epoxy resin systems by using composites filled with graphene nano-particles for electrical insulation application. Numerical analysis of microchannels with pin fins were done by Jadhav et al. [7] in 2018 and reported that pin fin size, shape, length and coolant flow velocities should be considered for the proper selection of pin fins for the enhancement of heat dissipation in microchannels. The theoretical, computational and the experimental investigation was done for the enhancement of thermal conductivity of marble powder-filled epoxy composites by Ray et al. [8] in 2018. Numerical analysis of the microchannel heat sinks was done by Jadhav et al. [9] in 2019 to analyse the influence of the various pin-fin configurations (layouts) on the microchannel performance and concluded that the arrangement of the pin fins in the microchannel has a considerable impact on the heat sink performance. Various studies have reported about the use of microchannel for mixing of multi-fluids and related mixing performance [10-13]. Manufacturing of a microchannel heat sink is also a crucial task. The fabrication aspect of microchannels and suitable processes have been reported by Wangikar et al. [14] in 2017.

For deciding the thermal and geometrical parameters related to a heat sink, it is essential to examine essential different parameters that may affect the performance of a heat sink. The properties of the material used for the manufacturing of the heat sink also decides the performance of a cooling system. While designing a heat sink, the different parameters need to be taken into consideration. Some of them are coolant inflow velocity, pressure drop induced in the microchannel, the size and shape of the microchannel, amount of cooling required in the application, maximum allowable temperature, atmospheric conditions, and the cost involved in the heat sink manufacturing. The thermal performance of microchannel is generally case-specific, resulting in many inconsistencies in the microchannel research and therefore, requires further research in this area.

3. Description of the model

In this paper, numerical analysis of heat sink is carried out using commercial software, COMSOL MULTIPHYSICS. The work is performed for comparing appropriateness of the various materials which can be used for the heat sink construction. Instead of analysing the complete heat sink, a representative elemental volume can be analysed to save the computational time and cost. However, in such a case, selecting a proper elemental size and applying the correct boundary conditions become very important for correctly predicting the behaviour of the complete process. In this work, a heat sink having several microchannels and with water used as a coolant is considered for the study. Since all the channels in the heat sink are identical in shape and size, a three-dimensional microchannel with water in it is considered as a representative elemental volume here. To the bottom surface of the channel, heat flux is applied. The schematic of the microchannel model employed in the analysis is depicted in Figure 1. After passing through the channel base and the vertical walls, the heat is transferred into the fluid in the channel. Therefore, it can be considered as a conjugate heat transfer problem. Considering this, the conjugate heat transfer module of COMSOL Multiphysics software is employed in this study. The schematic of the microchannel model and the meshed model used in this study are presented in Figure 2.

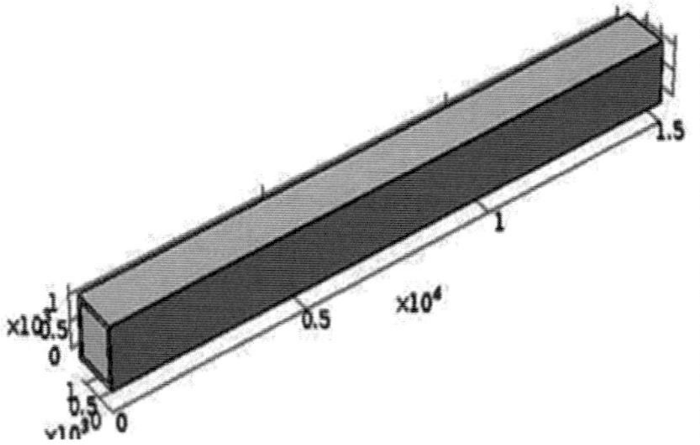

Figure 1. The base model employed in the analysis.

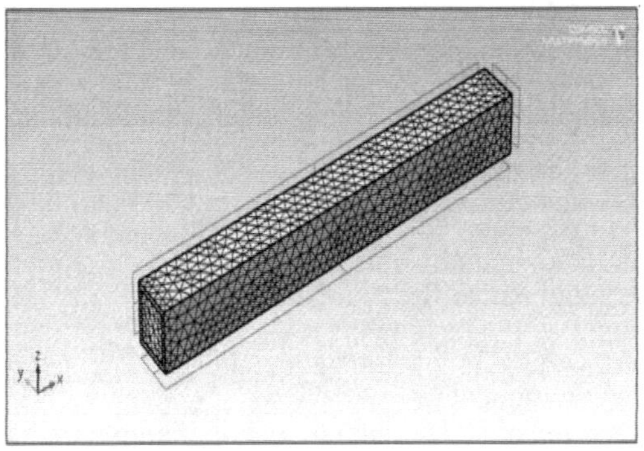

Figure 2. Meshed model.

For simplifying the problem, some assumptions are considered. The flow considered is incompressible and steady. The fluid flow considered is laminar with constant velocity at the entrance of the channel. The thermo-physical characteristics of both the solid and fluid-solid interface are assumed to be independent of the temperature. Heat is given (applied) to channel substrate, and there is no radiation heat transfer.

The below mentioned governing equations are employed for describing the heat transfer and flow of the fluid.

Continuity equation

$$\nabla U = 0 \quad (1)$$

Conservation of momentum

$$\rho(U \cdot \nabla U) + \Delta P - \mu \nabla^2 T = 0 \quad (2)$$

Conservation of energy

$$U . \nabla T = 0 \quad \text{(for solids)} \quad (3)$$

$$\rho Cp(U . \nabla T) - k_f(\nabla^2 T) = 0 \quad \text{(for fluid)} \quad (4)$$

The coolant, in the form of a laminar flow, with uniform velocity enters the microchannel at its front end with a temperature of 293.15K. At the outlet boundary of the microchannel, atmospheric pressure is allocated. Table 1 presents the boundary conditions for control volume considered in the analysis. The governing equations (mentioned above) are solved after applying the boundary conditions using the software to determine the average velocity of the coolant, the pressure drop achieved in microchannel and different temperatures in the fluid domain and heat sink. Further, the values of the average velocity, the pressure drop and the different temperatures are used for calculating different parameters like Reynolds number, friction factor and the heat transfer coefficient for the microchannel. The value of heat transfer coefficient is calculated using inflowing heat flux from the simulation results. The Fanning friction factor is derived from Kakac et al. (1987), where Po is Poiseuille number.

IOP Conf. Series: Materials Science and Engineering **814** (2020) 012044 doi:10.1088/1757-899X/814/1/012044

Table 1. Boundary Conditions employed for the model.

Boundary	Fluid Boundary condition	Thermal Boundary Condition
Front Fluid inlet	Inlet	Adiabatic
Front Fluid outlet	Outlet	Adiabatic
Channel Front solid	Wall	Adiabatic
Channel Back solid	Wall	Adiabatic
Channel Left surface	Symmetry	Symmetry
Channel Bottom surface	Wall	Uniform heat flux
Channel Right surface	Symmetry	Symmetry
Channel Top surface	Wall	Adiabatic

4. Analysis of the baseline model

Table 2. Parameters set microchannel used for analysis.

Parameter	Baseline value
Channel wall thickness (W)	200μm
Channel length (L)	15000μm
Channel fluid flow area	1000μm X 1000μm
Heat flux	$20W/cm^2K$
Thermal conductivity of water	0.608W/m K
Density of water	$1000kg/m^2$

The parameters employed for the basic model analysis are presented in table 2. Due to its higher thermal conductivity value, the copper material is chosen for the basic model. Initially, the basic model is checked for grid independence. As depicted in Figure 3, the numerical results for the coolant temperature at the outlet (Figure 3a) and the pressure drop (Figure 3b) from the microchannel are considered for the grip independency. It can be observed from the simulation results that after the third configuration, an increase in the number of elements has an insignificant influence on the results. Therefore, the third configuration can be considered for further analysis.

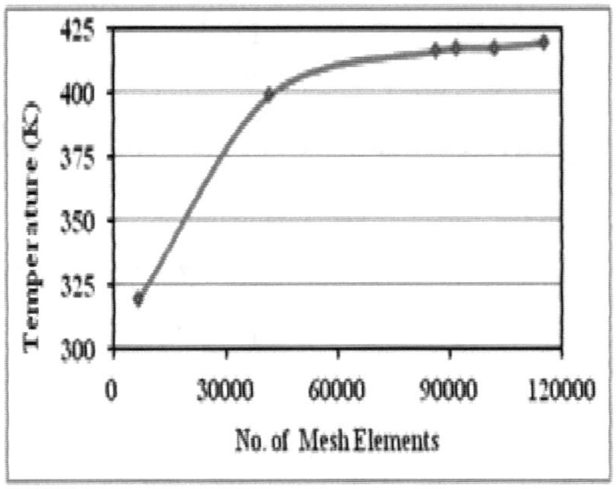

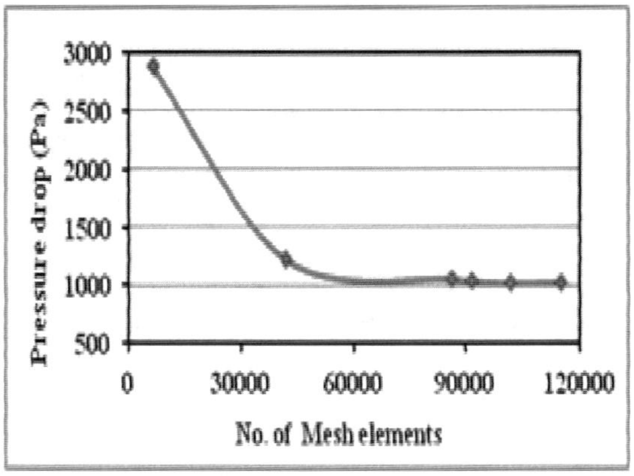

Figure 3. Mesh Independent Test (a) For outlet temperature (b) For pressure drop.

For the basic model validation, the simulation results are compared with the results from Gawali et al., as shown in Figure 4a and 4b. Figure 4a compares the friction factor and Figure 4b compares the heat transfer coefficient. The present work is found to be in better agreement with the results of Gawali et al. [4].

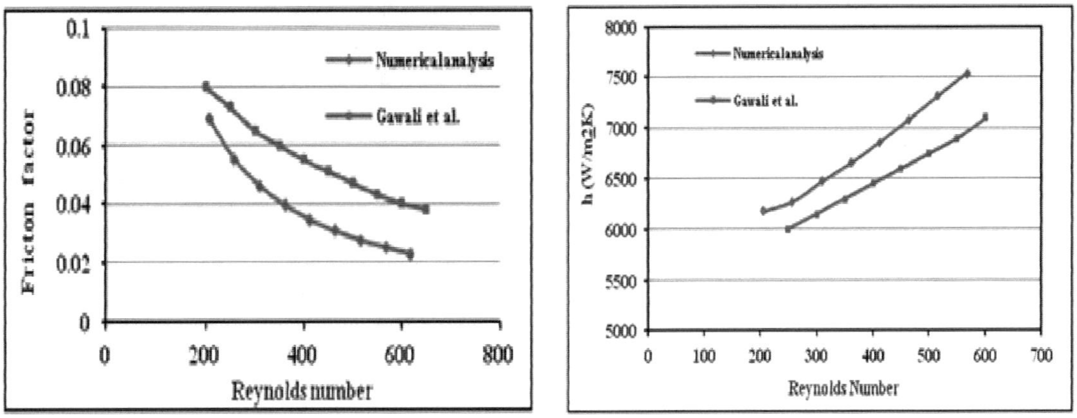

Figure 4. Validation of the basic model for **(a)** Friction factor and **(b)** Heat transfer coefficient.

5. Results and Discussion

This work is performed to analyse the appropriateness of various thermal management materials for the heat sink applications. The various thermal management materials like Copper, Aluminium, Al_2O_3, AlN, and Si_3N_4, are the materials considered for this study. The coolant flow through the microchannel is considered to be fully developed laminar flow. The values of inlet velocity range from 0.2m/s to 1m/s. The different performance parameters used for understanding the effect of different heat sink materials are pressure drop, heat transfer coefficient and friction factor.

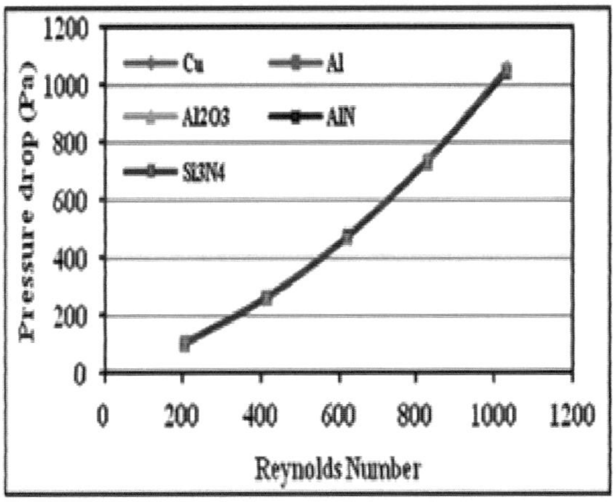

Figure 5. Pressure drop in different microchannels.

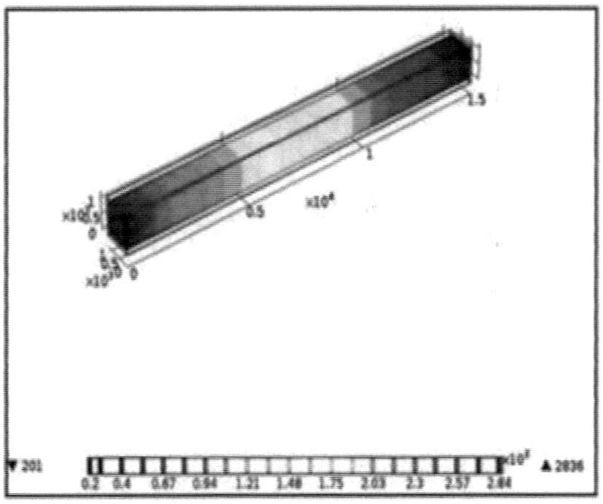

Figure 6. Pressure contours in a microchannel.

The pressure drop in the microchannel with different Reynolds number values is portrayed in Figure 5. As the Reynolds number increases, pressure drop increases. Since the flow cross-section area is the same for all microchannels with different materials, the pressure drop values observed also remain similar. The pressure contours for the microchannel, is as shown in figure 6. The Fanning friction factor is a function of loss of pressure because of the wall friction. Since, all microchannels have the same amount of surface contact area with fluid, wall friction in all microchannels remains the same. Thus, the nature of the friction factor curve remains similar for all the microchannels with different materials, as shown in Figure 7. Higher Reynolds number corresponding to higher coolant flow velocities, increases the thermal boundary layer disturbance in the microchannel. Increase in the flow disturbance results in a higher value of heat being carried away from the microchannel walls. Thus, the heat transfer coefficient value generally increases with increase in Reynolds number.

Figure 8 depicts the heat transfer coefficient curves obtained for microchannels with different heat sink materials. The heat transfer coefficient performance of copper microchannel heat sink observed is best among the different materials considered in this study. The higher value of thermal conductivity of the copper material, can be regarded as one of the most important reason for the higher performance of the copper microchannel. The Velocity profile in the microchannel is shown in Figure 9, and the surface temperature of the fluid in the microchannel is portrayed in Figure 10.

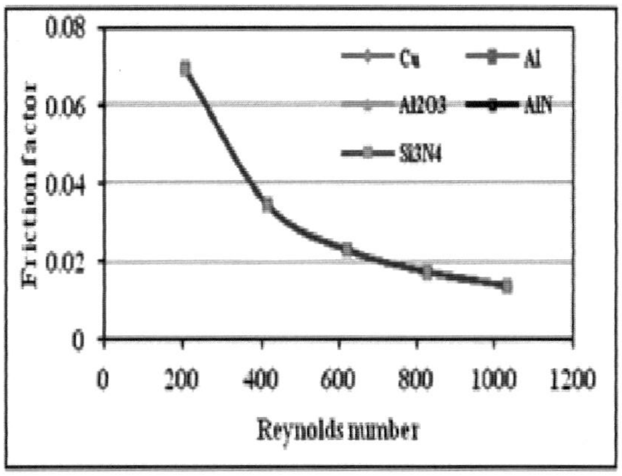

Figure 7. Friction factor for different microchannels.

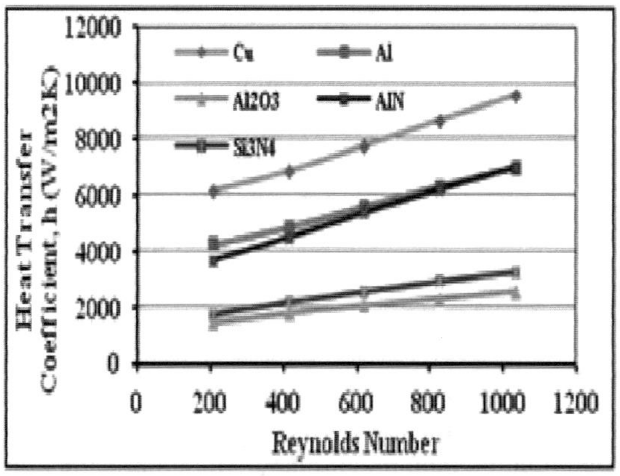

Figure 8. Heat transfer coefficient for different microchannels.

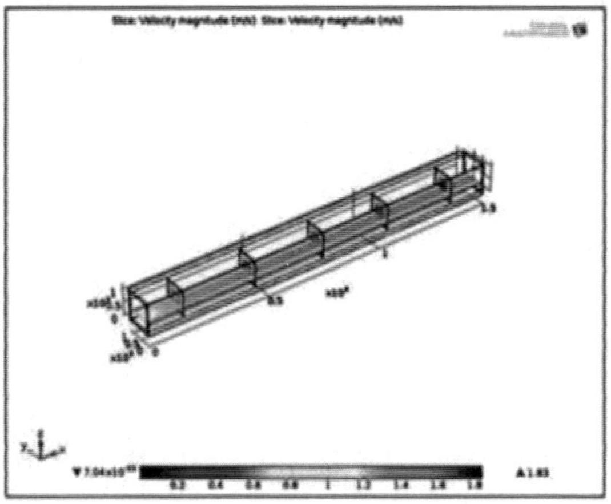

Figure 9. The velocity profile in the microchannel.

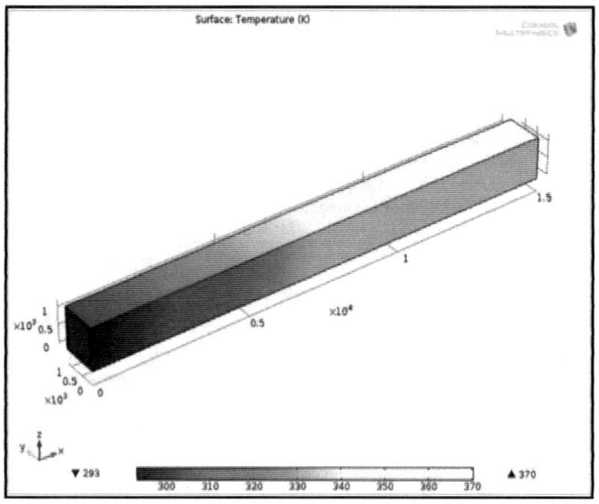

Figure 10. The surface temperature of the fluid in the microchannel.

6. Conclusion

Three-dimensional square microchannels of different thermal management materials are compared to observe the effect of change in the material on the microchannel heat sink performance. Firstly, a basic microchannel model of copper is validated using the results of Gawali et. al. The basic model is further compared with microchannels of other materials like Aluminium, Al_2O_3, AlN, and Si_3N_4. Among the different material chosen for the study, the copper microchannel is observed to be having the best performance. Copper microchannel with a comparatively higher value of thermal conductivity can be considered as the reason for its best performance, when compared to microchannels made of other materials.

IOP Conf. Series: Materials Science and Engineering **814** (2020) 012044 doi:10.1088/1757-899X/814/1/012044

References

[1] Tuckerman D B and Pease R F W 1981 *IEEE Electron Device Lett.* **2 (5)** 126 -129
[2] Hetsroni G Mosyak A Pogrebnyak E and Yarin L 2005 *Int. J. Heat Mass Tran.* **48** 5580–5601
[3] Wang y and Ding G 2008 *J. Micromech. Microeng.* **18(3)** 1-8
[4] Gawali B S Swami V B Thakre S D 2014 *Int. j. innov. res. sci. eng. technol.* **3(8)** 15631-15640
[5] Pietrak K and Wisniewski T S 2015 *J. Power Technol.* **95 (1)** 14–24
[6] Jarosinski L Rybak A askai K Kmita G Porebska R and Kapusta C 2017 *Mater. Sci-Poland.* **35(2)** 382-389
[7] Jadhav S V Pawar P M Ronge B P 2018 *Chem. Prod. Process Model.* 20180016 **1-15** DOI: 10.1515/cppm-2018-0016
[8] Ray S Rout A K Sahoo A K 2018 J. Chem **11(1)** 80-87
[9] Jadhav S V and Pawar P M *Int. J. Numer. Model. EL.* **1-19**
[10] Wangikar S S Patowari P K. and Misra R D 2018 *Microsyst. Technol.* **24(8)** 3307-3320
[11] Das S S Tilekar S D Wangikar S S and Patowari P K 2017 *Microsyst. Technol.* **23(12)** 5977-5988
[12] Gidde R R Pawar, P M Ronge, B P Shinde, A B Misal N D and Wangikar S S 2019 *Microsyst. Technol.* **25(3)** 1017-1030
[13] Jadhav S V Pawar P M Shinde A B and Wangikar S S 2020 *Int. Conf. on Advanced Technologies for Societal Applications Springer* 295-304
[14] Wangikar S S Patowari P K. and Misra R D 2017 *Mater. Manuf. Process* **32(15)** 1747-1755

IOP Conf. Series: Materials Science and Engineering **814** (2020) 012045 doi:10.1088/1757-899X/814/1/012045

Effect on bubble dynamics during pool boiling heat transfer in a novel aqueous binary mixture of surfactants

Prashant B. Pawar[1], Ganesh N. Rakate[2], Sujit V. Yadav[3], Vivek S. Gondkar[4], Suresh L. Chittewar[5]and Prashant S. Daingade[6]

Department of Mechanical Engineering, Annasaheb Dange College of Engineering and Technology Ashta, Maharashtra, 416301, India

E-mail: pbpawar.mech@gmail.com

Abstract Temperature near the heating region changes with the water boiling curve when the heat flux applied. In this study, experiments were carried out by maintaining a cylindrical subcutaneous hypodermic needle as a single active nucleation site. By controlling heat supplied to the site, a single bubble was formed in the saturated water. The boiling phenomenon was recorded. The surfactants used were sodium dodecyl sulfate (SDS) and centrimonium bromide (CTAB). Aqueous binary mixture SDS-CTAB was prepared on a volume percent basis. The effect was measured by studying a single bubble. The vapor bubble resulting from the nucleation site was observed. Bubble parameters, without and with a surfactant, were plotted in two heat fluxes concerning the excess temperature. It is found that, compared with saturated water, the average measurements of bubble detachment diameters and terminal velocities decreased, but the bubble departure frequency increased. On the other side of analysis, with increasing excess needle temperature of site, at 601 kW/m^2, the detachment diameter decreased, departure frequency increased, rising velocity was increased and at 950 kW/m^2, the detachment diameter increased, departure frequency increased, rising velocity decreased.

1. Introduction

Nuclear boiling heat transfer transfers a huge amount of heat, utilized in many applications such as rocket engines, boilers, and nuclear reactors. Single bubble formation is very important as it generates the slag and column for boiling enhancement. Therefore, to understand nuclear boiling, it is crucial to understand the dynamics of an isolated bubble.

Estimation of nucleate boiling parameters can be interpreted by the study of the single bubble dynamics by preparing artificial nucleation sites of different shapes. Shoji & Takagi [1] carried boiling experiments and produced an artificial cavity of three different shapes. Cylindrical cavities show continuous foaming from low heated wall temperatures. Cylindrical nucleation sites showed anticipated behaviour compared with concave and conical cavities. Qiu [2] formed cavities in polished wafers of silicon and found large forces of the surface tension and buoyancy. Lee [3] investigated uniform heat flux boundary conditions applied for forced convection boiling. Departure frequency increased, divergence size of the bubbles was nonaligned with the rising temperature of the wall. Siedel [4] investigated a single nucleation site, detachment diameter of bubble found nonaligned with the wall temperature. The frequency detachment of the bubble was proportional to the wall temperature. Robinson and Judd [5] studied intercommunication of vapor-liquid for a bubble, suggested that initial bubble development is managed by the inside bubble surface tension.

1.1. Review on boiling including surfactant
Acceleration of studies on boiling with a surfactant solution began several years ago. Yang [6] suggested twelve measures in boiling acceleration with three surfactants. Gajghate [7] examined that the ultimate heat transfer level in water-based ammonium chloride at 2600 ppm concentration. King [8] investigated the reflex of surfactants on the development and detachment of the bubble, added surfactant lowered the detachment diameter of the bubble. Najim [9] calculated the dynamics of the bubble linked with changes in the nucleation sites diameter [10]. Also observed that the increase in the frequency of detachment at higher heat flux is greater than that in the lower heat flux [13]. Pawar [11], [12] and [25] conducted unsteady boiling experiments and used ammonium chloride as a surfactant to investigate the effect of the excess needle temperature, suggested that the detachment diameter was nonaligned with increasing excess needle temperature and departure frequency found increased in an aqueous surfactant solution. Acharya [17] studied a single bubble as an interest of the region. Hoang [15] investigated a model for the evaluation of the departure bubble diameter. Researchers [20] and [21] studied on natural heating surfaces, whereas [4,11, 17,18] studied artificial heating surfaces. When compared to natural sites with the artificial, different behaviour for the detachment diameter of the bubble was observed. Their contradictions are tabulated in Table 1.

Table 1. Behaviour of departure frequency, detachment diameter (with increasing excess needle temperature)

Type of nucleation site	Author	Detachment diameter			Departure frequency
Natural	[4] Siedel				Increasing
	[18] Hutter	Increasing			
	[19] Gorenflo		Decreasing		
Artificial	[13] A. Najim				Increasing
	[20] Kutateladze		Decreasing		Increasing
	[21] Cole			Independent	
	[25] Pawar	Increasing at 601 kW/m^2	Decreasing at 950 kW/m^2		Increasing

Table 1. shows that a minimum of two values of heat flux should be applied to heating surfaces during pool boiling. The experiment would be carried with the increase in the wall temperature in the heat flux. The bubble detachment diameter, the departure frequency, and the rising velocity were plotted. Bubble criterion measured in the solution with surfactant were correlated with those with the water. The effect of the addition of surfactants in water and the effect of heat flux are discussed.

2. Experimental Measurement

2.1. Experimental set-up
As shown in figure 1, a glass beaker of borosilicate material filled with 1800 ml of saturated water followed by an aqueous binary mixture of surfactant solution was used. A needle of DISPO VAN®, which acts as a bubble generation site, acts as the heat input surface. The needle of inner diameter 514 μm and length of 25 mm was placed perpendicular to the ground at the bottom of the middle of glass

containers. The needle top was cut precisely to allow flow through the needle and the bottom was closed to prevent flow it.

2.2. The surface tension of aqueous binary mixture of surfactant solution

The surface tension of aqueous binary systems was measured minimum five times and final average values were recorded. The percentage of the amount of primary SDS surfactant and the percentage of the amount of corresponding secondary CTAB surfactant in the same aqueous binary system shows the variation in surface tension. The surface tension of anionic-catanionic binary mixture of 25: 75 vol % SDS - CTAB was found least, those of 50: 50 vol % and 75: 25 vol % combinations of volume fractions were higher. Measurement shows per results of Ravikumar [26].

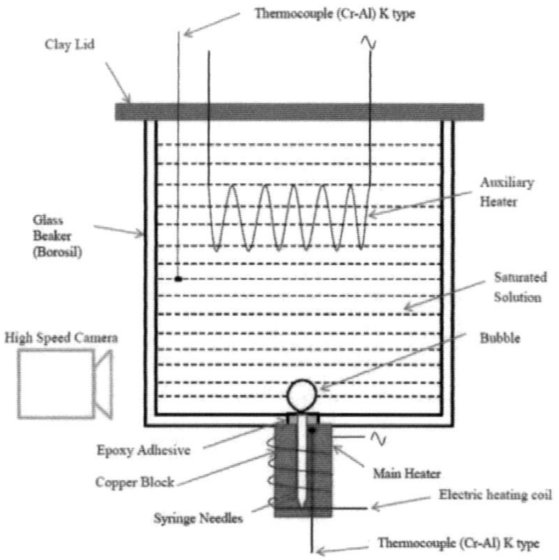

Figure 1. Schematic diagram of the experimental set-up

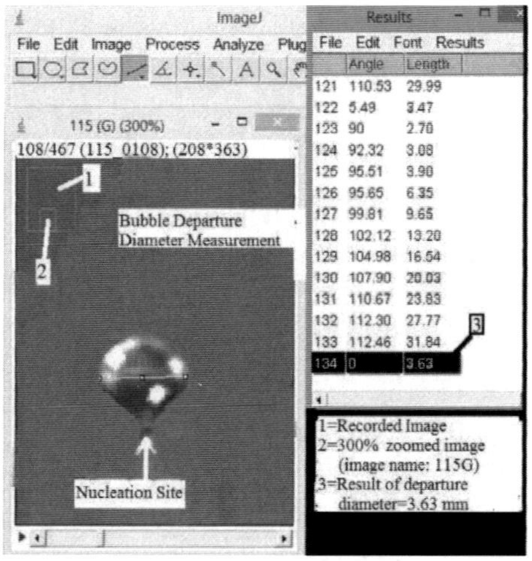

Figure 2. Procedure for Detachment diameter measurement using a high-speed camera

2.3. Experimental procedure

In the present study, a wide range of pool boiling experiments was carried out. A binary mixture of anionic - catanionic SDS-CTAB of 25: 75 vol % is used as a surfactant additive [26]. From the experimental measurement obtained, which are plotted for the bubble parameters versus excess needle temperature. The bubble generation has seen and accessed by a PCO AG 200 camera at 100 fps.

2.4. Measurement of bubble parameters and Uncertainty of the measurements

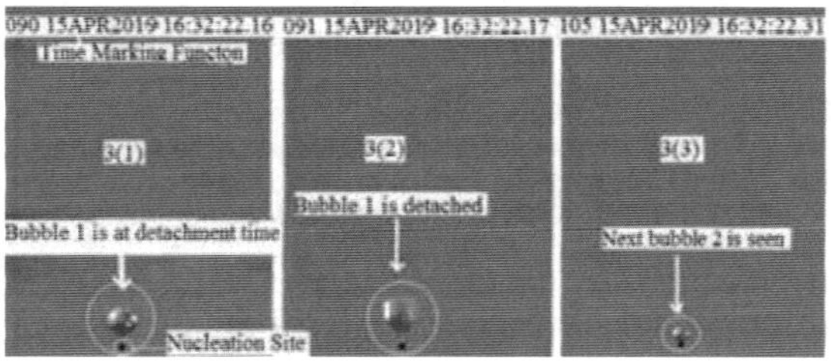

Figure 3. Departure frequency measurement

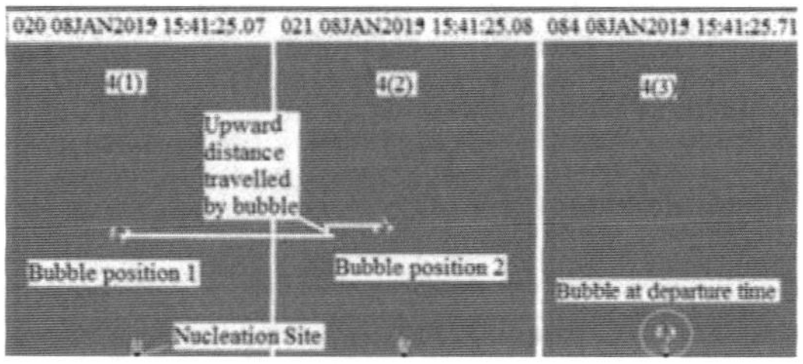

Figure 4. Rising velocity measurement

Bubble parameters were evaluated using ImageJ software. As shown in figure 2, a number of pixels on the bubble image counted and dimensions were measured. The detachment diameter of the bubble was measured at the center of gravity, which located about 75% height from the bottom point [15]. The bubble shape found balloon-shaped before departure.

Voltmeters, ammeters, camera, temperature indicators with least counts of 1 V, 1/100 amps, 0.01 sec, 1/100 °C. respectively. The maximum uncertainty of the heat flux was less than 0.02. The detachment diameter uncertainty measurement is ± 125 μm. The uncertainty in departure frequency is ±1/100 Hz. Boiling phenomena is very important for repeatability, to ensure it, all measurements were repeated at weekly intervals with these solutions. The data found within the aforementioned measurement tolerance and in good agreement.

The time marking function displayed at the top of the screen to determine the departure frequency, as shown in figure 3. The height traveled by the bubble is the average of the distances traveled until the next bubble seen, as shown in figure 4.

3. Results and Discussions

3.1. Detachment diameter (D_b)

The detachment diameter of the bubbles varies as tabulated in Table 2. The addition of a binary mixture of a surfactant solution reduces the downward force acting on the bubbles. Bubbles separate in smaller volumes than in saturated water, increasing the area of contact between the heated bubbles and saturated water, finally increasing the heat transfer rate from the needle to the water. The detachment diameter from the needle is reduced by 25% to 33% with the addition of surfactants and is shown in figure 5.

Table 2. Experimental Results

	601 kW/m^2		950 kW/m^2	
	Water	Binary mixture	Water	Binary mixture
D_b, mm	3.42±0.125	2.53±0.125	3.41±0.125	2.34±0.125
f, Hz	2.14±0.01	3.58±0.01	2.03±0.01	4.49±0.01
V_b, m/s	0.51±0.125	0.43±0.125	0.51±0.125	0.44±0.125

As the excess needle temperature increased, the detachment diameter decreased to 601 kW/m^2. The detachment diameter decreases due to excessive wall overheating [19], [20]. The detachment diameter increased at 950 kW/m^2 [18], [21] due to the capillary effect, which has been found to dominate the surfactant effect at higher heat fluxes [16]. It has been found a nearly constant behaviour of the detachment diameter at these conditions. The microlayer between the bubble and the heating surface is escaped not because of its vaporization but because of the capillary effect.

3.2. Bubble departure frequency (f)

The bubble departure frequency from the wall varies as tabulated in Table 2. This explains that the addition of a binary mixture of surfactant solutions increases the departure frequency considerably. The bubble departure frequency increases by 139% to 142% for heat fluxes of 601 kW/m^2 and 950 kW/m^2, when compared to saturated water.

The addition of the surfactant lowers the surface tension. Thus, the surface tension in the bubble also lowered. As the heat flux is uniform, vapor nucleation at this heat flux is uniform for both water and the surfactant solution. This lowered the downward surface tension force and the detachment diameter and increases the departure frequency.

Figure 6. shows that departure frequency is independent at higher heat flux. Siedel [4] previously concluded relation that parameter (D_b*f) is proportional to excel needle temperature. Here, it was seen improving from 15^0C to 22^0C, D_b found lowered at lower heat flux with excess needle temperature, positive value of the gradient of departure frequency is found higher as compared to that at lower heat flux. Also, D_b found improved at higher heat flux, gradient of departure frequency at low heat flux is found less than higher heat flux. Finally, measurements of departure frequency were found by earlier observations by [14, 13,25]..

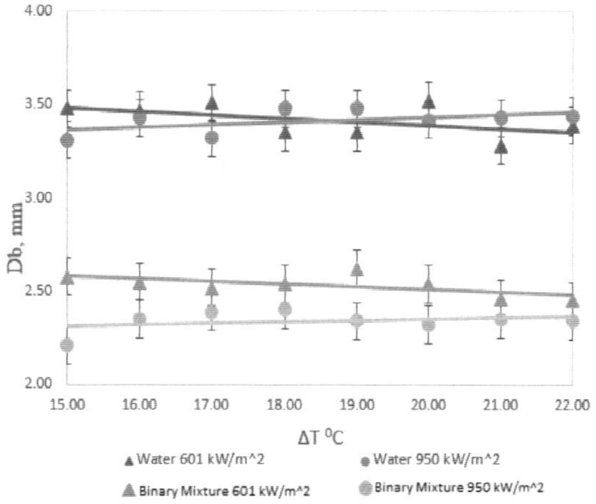

Figure 5. Detachment diameter vs excess needle temperature

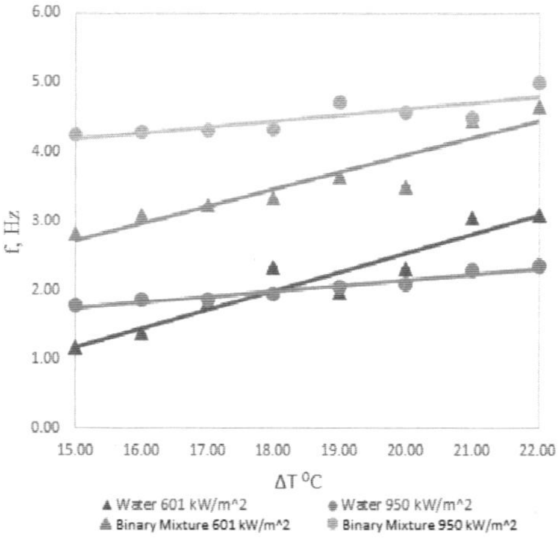

Figure 6. Departure frequency vs excess needle temperature

3.3. Rising velocity of bubbles (V_b)

The rising velocity of air bubbles changes is tabulated in table 2. This interprets that the binary mixture of surfactants solution reduces the rising velocity of the bubbles, which reduces 15% to 12% for heat fluxes of 601 kW/m^2 and 950 kW/m^2 ss shown in figure 7. The addition of a binary mixture of surfactant solution reduces the detachment diameter, and the detached bubbles become less buoyant and take extra time to reach the liquid surface [7].

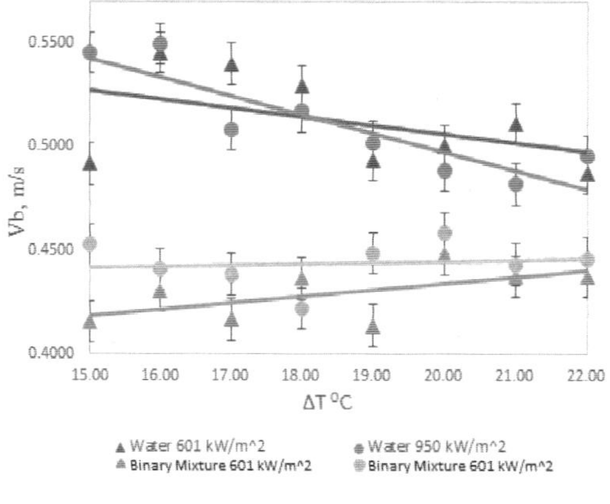

Figure 7. Rising velocity vs excess needle temperature

It can be seen that the average rising velocity is nearly constant for water and surfactant solutions. D_b is found to be nearly constant [23, 24]. Therefore Mchale relationship between V_b and D_b, $V_b \alpha D_b^{1/2}$ is satisfied [14].

IOP Conf. Series: Materials Science and Engineering **814** (2020) 012045 doi:10.1088/1757-899X/814/1/012045

4. Conclusion

The addition of the surfactant lowered the detachment diameter by up to 33%, the departure frequency by up to 142% and the rising velocity by up to 12%.

Different behaviour was observed at the two heat fluxes, revealed some contradiction with the previous results. At 601 kW/m^2, the detachment diameter decreased (slope= -0.01), the departure frequency improved (slope = 0.27), and the rising velocity was increases (slope= 0.003) with addition of surfactant. At 950 kW/m^2, the detachment diameter increased (slope= 0.007), the departure frequency improved (slope = 0.08), and the rising velocity was nearly uniform (slope= 0.0006).

ACKNOWLEDGMENT

This experiment was conducted at Annasaheb Dange College of Engineering and Technology, Ashta, 415301 and Government College of Engineering, Karad, 415124, India.

REFERENCES

[1] D. M. Qiu and V. K. Dhir 2002 Single-Bubble Dynamics during Pool Boiling under Low Gravity Condition *J. of Thermo physics and Heat Transf.* **16** 336-345.

[2] M. Lee, L. S. Cheung, Y. Lee and Y. Zoha 2005 Height Effect on Nucleation-Site Activity and Size-Dependent Bubble Dynamics in Micro-Channel Convective Boiling J. *of Micromechanics and Micro-eng.* **15** 2121-2129.

[3] S. Siedel, S. Cioulachtjian and J. Bonjour 2008 Experimental Analysis Growth, Departure and Interactions during Pool Boiling on Artificial Nucleation Sites *Experimental Thermal and Fluid Sci.* **32** 1504-1511.

[4] A.J. Robinson and R.L. Judd 2004 The dynamics of spherical bubble growth *Int. J. Heat Mass Transf.* **47** 5101– 5113.

[5] Y. M. Yang and J. R. Maa 2001 On the criteria of nucleate pool boiling enhancement by surfactant addition to water *Trans. of IChemE.* **79** 409-416.

[6] S.S. Gajghate, A.R. Acharya and A.T. Pise 2014 Experimental study of aqueous ammonium chloride in pool boiling heat transfer *Expt. Heat Trans.* **27** 113-123.

[7] L. King and S. Sadhal 2014 Effect of surfactant on the growth and departure s from solid surfaces *J. Heat Mass Transf.* **50** 373–382.

[8] A. Najim, A.R. Acharya, A.T. Pise and S.S. Gajghate 2014 Experimental study dynamics in pool boiling heat transfer using saturated water and surfactant solution *Proc. of IEEE Int. Conf. on Advances in Eng and Tech* 1-6.

[9] A. Najim and A. Acharya 2014 Experimental study of effect of nucleation site size on bubble dynamics during nucleate pool boiling heat transfer *Appl. Mech. Mater.* 592–595.

[10] P. B. Pawar, A.R. Acharya and A.T. Pise 2015 Experimental study of single bubble dynamics during nucleate pool boiling using ammonium chloride for heat transfer enhancement *J. of Mat. Sci. and Mech. Eng.* **2** 298-302.

[11] P. B. Pawar, A. R. Acharya, S. S. Gajghate and A. T. Pise 2016 Experimental study of single bubble dynamics during nucleate pool boiling using saturated water for heat transfer enhancement *Int. J. on theoretical and applied research in mech. Eng.* **2** 21-23.

[12] A. Najim and Saiprasad Pise 2016 Boiling heat transfer enhancement with surfactant on the tip of a submerged hypodermic needle as nucleation site *Applied Thermal Eng.* **103** 989–995.

[13] J. Mchale and S. V. Garimella 2010 Bubble Nucleation Characteristics in Pool Boiling of a Wetting Liquid on Smooth and Rough Surfaces *Int. J. of Multiphase Flow* **36** 249-260.

[14] N. H. Hoang, V. T. Nguyen, I. C. Chu, D. J. Euh and C.-H. Song 2013 Assessment of bubble departure diameter models for nucleate boiling heat transfer *Trans. of the Korean nuclear society autumn meeting Gyeongju, Korea.*

[15] T. Inoue, Y. Teruya and M. monde 2004 Enhancement of pool boiling heat transferrin water and ethanol water mixture with surface active agent *Int. J. of Heat and Mass Trans* **47** 5555–5563.

IOP Conf. Series: Materials Science and Engineering **814** (2020) 012045 doi:10.1088/1757-899X/814/1/012045

[16] A. R. Acharya, and A. T. Pise 2016 A review on augmentation of heat transfer in boiling using surfactant additives *heat and Mass Transf.* 1-21.
[17] C. Hutter 2013 Vertical coalescence during nucleate boiling from a single artificial cavity *Experimental Therm. and Fluid Sci.* **51** 94–102.
[18] D. Gorenflo, V. Knabe and V. Bieling 1986 Bubble density on surfaces with nucleate boiling-its influence on heat transfer and burnout heat flux at elevated saturation pressure *8th International Heat Transfer Conference, San Francisco, USA* **4** 1995–2000.
[19] S.S. Kutateladze and I.I. Gogonin 1980 Growth rate and detachment diameter of a vapour bubble in free convection boiling of a saturated liquid *high temp.* **17** 667–671.
[20] R. Cole and W. Rohsenow 1969 Correlations of bubble departure diameters for boiling of saturated liquids *Chem. Eng. Prog.* **65** 211-213.
[21] G. Hetsroni 2006 Bubble growth in saturated pool boiling in water and surfactant solution *Int. J. of Multiphase Flow* ***32*** 159–182.
[22] H. J. Ivey 1967 Relationships between bubble frequency, departure diameter and rise velocity in nucleate boiling *Int. J. Hear Mass Transf.* **10** 1023-1040.
[23] R. Darby 1964 The dynamics of vapour bubbles in nucleate boiling *Chemical Eng. Sci.* **19** 39-49.
[24] V. K. Dhir 1998 Boiling heat transfer *Ann review of fluid mechanics* **30** 365-401.
[25] P. B. Pawar, S. S. Ahankari and A. R. Acharya 2017 Effect of ammonium chloride as a surfactant additive on bubble dynamics during nucleate boiling of water *Proc. Int. Conf. for Convergence in Tech. 978-1-5090-4307-1/17.*
[26] S. V. Ravikumar, J. M. Jha, I. Sarkar, S. S. Mohapatra, S. K. Pal and S. Chakraborty 2013 Achievement of ultrafast cooling rate in a hot steel plate by air-atomized spray with different surfactant additives *Exp. Therm. Fluid Sci.* **50** 79-89.

IOP Conf. Series: Materials Science and Engineering **814** (2020) 012046 doi:10.1088/1757-899X/814/1/012046

Experimental study of partial replacement of cement by waste marble powder in a concrete prepared with artificial sand

I Korabu[1] and C Pise[2]

[1]Postgraduate student, SKN Sinhgad College of Engineering, Pandharpur, India
[2]Professor, Department of Civil Engineering, SKN Sinhgad College of Engineering, Pandharpur, India

E-mail:imamkorabu0@gmail.com

Abstract. Marble is the most common building material having a huge amount of waste. This waste causes dust pollution also it makes soil infertile etc. Day by day marble waste goes on increasing. Another problem is regarding no availability of natural sand as resources of natural sand exhausting very rapidly also it contains a lot of deleterious materials to overcome this issue we can make partial replacement of cement by waste marble powder and natural sand by artificial sand. The concrete is made with heterogeneous material therefore compressive strength of concrete depends upon various materials with various trials.so the process is simplified in a mathematical technique called the regression method, which gives predicted compressive strength, if we get regression equation, it reduces the trials of concrete mix.

1. Introduction

Manpreet Singh [1-2, 6-7, 12] gives result up to 10-15% use of marble powder increases strength of concrete. Valeria Corinaldesi et. al [3] shows 10% substitution of sand by marble powder with plasticizer gives maximum strength with workability. A. S. E. Belaidi [4] study shows up to 40 % (natural pozzolona + marble powder) improve workability with suitable strength. Rajendra Kumar Khayaliya [5] study concluded marble waste could be harmless used in aggressive and non-aggressive environment by replacing river sand by 25% and 50%. Ali khodabakhshian [8, 10] study concluded 20% waste marble powder and 10% silica fumes partially replacement of cement decreases detrimental effect of cement industry on environment. Deepankar Kumar Ashish [10] noticed optimal replacement with 10% cement and 10% sand by 20% marble powder. V. Nezerka [11] indicates that elimination of the interfacial transition zone (ITZ) and minimization of porosity would increase the strength of cement-based materials carrying waste marble powder.

Present study focuses on experimental investigation of compressive strength and workability of concrete with substitute of cement by waste marble powder and sand by artificial sand, from this study we will come to know effect of waste marble powder with artificial sand on compressive strength and workability of concrete.

2. Why Waste M. P as alternative for cement

There is basic reason to select waste M.P as alternative for cement because both cement and waste marble powder having same chemical composition.

IOP Conf. Series: Materials Science and Engineering **814** (2020) 012046 doi:10.1088/1757-899X/814/1/012046

Table 1. Chemical properties of both cement and waste M.P

Chemical constituents	Cement	Waste M.P
Alumina(Al_2O_3)	4.6%	0.7%
Iron oxide(Fe_2O_3)	3%	0.33%
Calcium oxide(CaO)	61.40%	51.49%
magnesium oxides(MgO)	2.3%	0.36%
Sodium oxide(Na_2O)	0.19%	0.19%
Potassium oxide(K_2O)	0.83%	0.25%
Sulphur trioxide(SO_3)	3.8%	0.1%
Loss on ignition	2.7%	44.6%

Table 2. State wise marble production in tonnes

	Year		
State	2015-16	2016-17	2017-18
Rajasthan	15676902	13523759	13199265
Gujarat	-	983401	829711
Madhya Pradesh	104301	-	-
Andhra Pradesh	-	357	-

3. Methodology

3.1 Materials
O.P.C confirming to I.S.269-1989.the specific gravity of O.P.C was 3.15, artificial sand confirming to I.S.383-1970 is used as fine aggregate. Waste marble powder make available from Pandharpur located in Maharashtra. The waste was generated from cutting the marble into pieces; obtained waste was dried and then used. Coarse aggregate having size 20mm is used.

3.2 Mix proportion
The research was carried out on concrete mixes in which some amount of cement was substitute with waste marble powder and natural sand by artificial sand. The substitute of waste marble powder was 0%, 5%, 10%, 15%, 20%, 25% and 30% by weight of cement. The concrete mix was designed as per I.S.10262-2009 with target 28 compressive strength of 20 MPa. The W/C ratio was kept as 0.5; further subsequent trials are prepared with substituting cement with M.P.

3.3 Testing details
The tests was conducted for examining the properties, as workability was measured through slump test and actual strength was measured at different ages after casting.

4. Result and discussion

4.1 Workability
The different values of slump are presented in figure1. It was concluded that workability decreases as amount of M.P increases. The workability also decreases due to artificial sand.

The workability was decrease due to two major facts:

1. Artificial sand having rough surface texture that requires maximum amount of water
2. Waste marble powder containing large amount of fine particles requires more water to wet.
   Due to these combined effects, more water is required for workability.

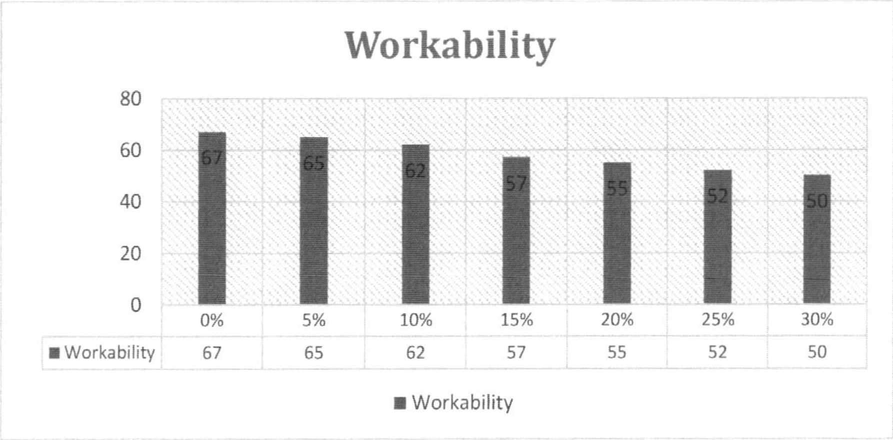

Figure 1. Workability with increase in waste marble powder

4.2 Compressive Strength

The major strength of concrete at different phases are determined and are shown in table no 3. from fig compressive strength goes on increasing with up to 10%, thereafter compressive strength goes on decreasing as amount of waste marble powder increases. The 28 days compressive strength of mixes are 28.29,31.22,32.35,27.32,26.97,25.13and21.24N/mm^2 respectively for 0%,5%,10%,15%,20%,25% and 30% from the observation the optimum compressive strength of concrete at 10% substitute of waste marble powder against cement.

The compressive strength of concrete added on due to fine marble powder, which increases binding property of concrete. The artificial sand distinctly make impact on compressive strength.

Table 3. Compressive strength at varying % of M.P

Sr. No	% of Marble Powder	W/C Ratio	Water lit/m3	Cement kg/m^3	Fine aggregates /kg/m3	Coarse aggregates kg/m3	Marble Dust Powder kg/m3	Compressive strength in (N/mm^2)	
								Expt./ Obs	Predicted
1	0%	0.5	191.58	368.42	690.93	1147.1	0	28.29	31.83
2	5%	0.5	191.58	350	690.93	1147.1	18.42	31.22	30.34
3	10%	0.5	191.58	331.58	690.93	1147.1	36.84	32.35	28.85
4	15%	0.5	191.58	313.16	690.93	1147.1	55.26	27.32	27.36
5	20%	0.5	191.58	294.74	690.93	1147.1	73.68	26.97	25.87
6	25%	0.5	191.58	276.32	690.93	1147.1	92.1	25.13	24.39
7	30%	0.5	191.58	257.89	690.93	1147.1	110.53	21.34	22.9

4.3 Regression method

Now a day's construction work is growing rapidly as concrete is made up of different materials. The important parameter in construction most probably is compressive strength. Compressive strength depends upon different parameters like cement, W/C ratio, F.A, C.A, surface structure etc. In this fast growing industry, there is lot of time waste for trial mix of concrete and checking of compressive

IOP Conf. Series: Materials Science and Engineering **814** (2020) 012046 doi:10.1088/1757-899X/814/1/012046

strength at different ages.so to overcome this some mathematical model is established to check predicated compressive strength called as regression techniques.

Multi- linear regression resolves the interrelationship between different independent variables and dependent variable with providing a linear equation with actual data.

The correlation between dependent and independent variable is expressed in equation.

$P = a_0 + a_1 w© + a_2 MP$.... ...multi-linear regression

Where, P= Predicted strength of concrete

©= amount of cement

M.P = Waste Marble Powder

From multilinear regression, methods following (Table 4) coefficients are determined to calculate predicated compressive strength

Table 4. Coefficients of compressive strength

Specifications	coefficients	values
Constant	a0	3.095
Cement	a1	0.078
Waste Marble Powder	a2	-0.0028

5. Conclusion

With the available research, we conclude:

- The strength of concrete with waste marble powder with artificial sand gives higher strength in comparison to natural sand.
- Workability do affect due to artificial sand.
- 20% waste marble powder gives optimum strength.
- Regression technique gives standard equation to infer the strength of concrete with waste marble powder.
- It is observed the infer values of compressive strength achieved by suggested regression equation are in good compatibility with the actual results.

References

[1] Manpreet Singh, Kailash Chaudhary, b, Anshuman Srivastava, Kuldip Singh Sangwan, Dipendu Bhunia 2017 *J. of Building Engineering* **13** 87.

[2] Manpreet Singh, AnshumanSrivastava, DipenduBhunia 2017 *J. of Building Engg.* **134** 471.

[3] Valeria Corinaldesi, GiacomoMoriconi, Tarun R. Naik. 2010 *J. of Building Engg.* **24** 113.

[4] A.S.E. Belaidi L. Azzouz, E. Kadri , S. Kenai. 2012 *J. of Building Engineering* **31** 251–257

[5] Rajendra Kumar Khyaliya, K.I. Syed Ahmed Kabeer, Ashok Kumar Vyas. 2017 *Journal of Building Engineering* **25** 598.

[6] Omar M.Omar, Ghada D. AbdElhameed, Mohamed A.Sherif, Hassan A.Mohamadien *HBRC Journal* **8** 193.

[7] Gulden CaginUlubeyli, RecepArtir 2015 *Procedia - Social and Behavioral Sciences* **195** 2181.

[8] Ali Khodabakhshian, Mansour Ghalehnovi, Jorge de Brito ,ElyasAsadiShamsabadi 2018 *Journal of Cleaner Production* **170** 42.

[9] Deepankar Kumar Ashish 2018 *Journal of Building Engineering.* **15** 236.

[10] Ali Khodabakhshian, Jorge de Brito, Mansour Ghalehnovi, Elyas Asadi Shamsabadi 2018 *Construction and Building Materials* **169** 237.

[11] V. Ne V. Hrbek, Z. Prosek, M. Somr, P. Tesarek, J. Fladr 2018 *J.Cleaner Production* **195** 1081.

[12] Hasan Şahan Arel. 2018 *Journal of Cleaner Production* **131** 179.

AUTHOR INDEX